# Metals and Mines

Studies in Archaeometallurgy

# Metals and Mines

Studies in Archaeometallurgy

Selected papers from the conference *Metallurgy: A Touchstone for Cross-cultural Interaction* held at the British Museum 28–30 April 2005 to celebrate the career of Paul Craddock during his 40 years at the British Museum

*Edited by Susan La Niece, Duncan Hook and Paul Craddock*

Archetype
Publications

in association with
THE BRITISH MUSEUM

First published 2007 by Archetype Publications Ltd. in association with the British Museum

Archetype Publications Ltd.
6 Fitzroy Square
London W1T 5HJ
www.archetype.co.uk

Tel: 44(207) 380 0800
Fax: 44(207) 380 0500

ISBN: 978-1-904982-19-7

**British Library Cataloguing in Publication Data**
A catalogue record for this book is available from the British Library.

The views and practices expressed by individual authors are not necessarily those of the editors or the publisher.

Typeset by Kate Williams, Swansea
Printed and bound in Italy by Printer Trento srl.

# Contents

# Preface and acknowledgements

This book arose from the conference *Metallurgy: A Touchstone for Cross-cultural Interaction*, which took place at the British Museum on 28–30 April 2005 to celebrate the enormous contribution to the study and understanding of metallurgy made by Paul Craddock during his 40 years at the Museum. The conference had an extremely wide remit, as befits Paul's interests and international reputation, and brought together scholars from four continents and from all the many disciplines which contribute to the study of archaeometallurgy. Only some of the papers presented on that occasion appear here because of the practical necessity to present a single volume. That sadly means we have not been able to include many excellent papers which were given at the conference.

There are many colleagues and friends we wish to thank: Ian Freestone for initial encouragement in planning the conference, Nigel Meeks for all his hard work as a co-organiser of the conference, and all the other colleagues and ex-colleagues in the Department of Scientific Research without whom the organisation would have been impossible. Input and assistance from the Advisory Committee of Alessandra Giumlia-Mair, Peter Northover, Thilo Rehren and Michael Wayman was much appreciated. In addition, we owe a debt to many others, especially the referees, for their advice on the contents of the volume and also for practical help in its production. So many have contributed but special thanks must go to Antony Simpson for his imaging expertise and to Eleanor Beyer.

We would like to thank the Applied Mineralogy Group of the Mineralogical Society for supporting the conference by the provision of a grant for student bursaries.

Unusually for a recipient of a dedicated volume, Paul chose to be a joint editor: retirement does not seem to have affected him unduly.

Susan La Niece and Duncan Hook
The British Museum, September 2007

# Introduction: achievements and challenges

*Paul Craddock*

It is almost 30 years since the first British Museum volume dedicated to the scientific study of early extractive metallurgy appeared and this will be the fifth. It is perhaps appropriate to look back at what has been achieved in that time and the changes in approach to the subject. The 1979 *Archaeometry* conference was co-hosted by the British Museum and published as a series of British Museum Occasional Papers, including the volume on extractive metallurgy (Craddock 1980). As was noted in the Foreword, there was a real potential for learning more of the technology of early artefact production by scientific examination. However, especially on synthetic materials such as metals, there is only so much that the artefact could tell you. If you wanted the whole story then it was necessary to investigate the production sites. With ephemeral operations such as primitive metal smelting, the surviving evidence was likely to be fairly uninspiring but scientific studies could perhaps go some way to elucidating the process parameters. There had already been some works on early metallurgy combining archaeology and science, notably Tylecote (1963), and some scientific collaboration, especially at the early copper mines and smelters at Timna in Israel (Rothenberg 1972). The British Museum's 1980 Occasional Paper was soon followed by other monographs devoted to the subject, notably Bachmann's 1982 study of early slags. In 1982 the Historical Metallurgy Society and the British Museum held a conference on early smelting which resulted in another Occasional Paper (Craddock and Hughes 1985). A more specific interest in zinc and brass led to another meeting of the Historical Metallurgy Society and the British Museum, this time held in Bristol, the home of the British zinc industry, which in turn generated a further Occasional Paper (Craddock 1998). The growing realisation of the survival of prehistoric copper mines in Britain led to a meeting organised jointly by the Early Mines Research Group and the British Museum, held in London and North Wales, which produced a more general volume on early mining and smelting (Craddock and Lang 2003).

Over the last 30 years significant progress has been made in understanding the diversity of processes by which metals were smelted in the past. These include the smelting of non-ferrous metals especially in the Mediterranean and Middle East from sites such as Timna (Conrad and Rothenberg 1980; Rothenberg 1990) and Feinan (Hauptmann 2000, 2003; Weisgerber 2003) and the smelting of iron in early Europe (Pleiner 2000), where the excavation and scientific study of materials and production debris have documented the processes. Together with the more general elucidation of the early processes of metals smelted by slag-forming processes in bowl or shaft furnaces, there has been a succession of important individual sites that have greatly extended our knowledge of the range of early metallurgy. They are diverse in period, process and place and include the Copper Age copper mines and smelters at Ross Island in Ireland (O'Brien 2004); Early Bronze Age copper smelting at Chrysokamino on Crete (Betancourt 2006); and Early Bronze Age tin production at Kestel and Gültepe in Turkey (Yenner *et al.* 2003). From later periods has come the discovery and exploration of the extensive windblown iron furnaces in early medieval Sri Lanka (Juleff 1998); early Islamic crucible steel production at Merv in Turkmenistan (Feuerbach *et al.* 2003) and Achsiket in Uzbekistan (Rehren and Papachristou 2003); and medieval zinc distillation at Zawar in India (Craddock *et al.* 1998) and more recently in China both archaeologically (Liu Haiwang *et al.*, this volume, pp. 170–78) and as presently operated (Craddock and Zhou Weirong 2003). These are all sites whose significance or even existence was unsuspected 30 years ago.

Whole areas of new enquiry have emerged during this period, most noticeably the evidence for the earliest smelting processes. These did not regularly produce large quantities of durable debris and thus had not been previously recognised. This situation had resulted in a long gap of several thousand years between the first metal artefacts and the first evidence for extractive metallurgy. There was thus little knowledge of how the metal was smelted beyond the composition of the artefacts, where the very low iron content suggested processes that were only moderately reducing (Craddock and Meeks 1987). Clearly further progress could only be made by fieldwork to locate and excavate the sites, followed up by scientific examination and experimental replication, and this is now being achieved in the Mediterranean and Middle East (Betancourt 2006; Craddock 2001; Hauptmann 2003).

Through much of western Europe and particularly in the British Isles, it was generally believed that most evidence of prehistoric metal mining and smelting was irretrievably lost through later mining activities. Such work as had been published by pioneers such as the geologist John Jackson (1980) at Mount Gabriel in southwest Ireland was called into question by some archaeologists. There were even articles in mainstream, supposedly peer-reviewed, journals such as the *Proceedings of the Prehistoric Society*, where in all seriousness it was suggested that the ubiquitous stone mining hammers were to be dated to the mid-19th century AD (Briggs 1985). From this low point investigations, initially by amateur archaeologists, began to establish the reality of surviving Bronze Age mine systems in Britain (O'Brien 2004; Timberlake 2003a; Timberlake and Prag 2005) and across Europe as exemplified by the work of Rovira Llorens *et al.* (1997, 2003) and Delibes de Castro and Montero-Ruiz (1999) in Iberia. As a result of these and other works we know much of Bronze Age mining technology and are beginning to understand the smelting processes.

Similarly in the eastern Mediterranean and Middle East, the excavations of sites such as Feinan, Jordan (Hauptmann 2000, 2003; Weisgerber 2003), the Wadi Dara, in Egypt (Castel *et al.* 1998) and at Chrysokamino on Crete (Betancourt 2006) are revealing the early processes (Craddock 2001).

The latter sites have one thing in common: they are all wind-assisted, if not windblown smelting installations. The prevalence of windblown furnaces for smelting a variety of metals at sites across the world is one of the major discoveries of the past few decades, as further exemplified by the major iron-smelting furnaces on the west-facing slopes of the hills at Balangoda, in Sri Lanka (Juleff 1998) and should perhaps be more seriously considered for prehistoric smelters in temperate Europe.

There have also been advances in the methods of study of the early processes. The continued improvement in scientific apparatus in general is well exemplified by the scanning electron microscopy micrograph on the cover of this volume. The development of laser-ablation inductively coupled plasma mass spectrometry (Watling *et al.* 1999) with great sensitivity may make meaningful provenancing based on the trace element content of the metal feasible in some instances. Lead isotope analysis was already established 30 years ago, and although the work on British copper and lead sources showed that caution was needed in sampling and interpreting the results (Rohl and Needham 1998), the method continues to document the sources of metals, particularly in the Mediterranean world as exemplified by the work of the Gales in this volume (pp. 103–11). The results can sometimes be surprising, as evidenced by the apparent import of copper ingots from Cyprus to copper-rich Sardinia in the Bronze Age (Sumna *et al.* 2004) and the 'foreign' copper in Bronze Age Oman, including at the major mining site of Maysar (Weeks, this volume, pp. 89–96).

The experimental replication of ancient processes has also assumed greater importance (Crew 1991) with the major studies of Merkel (1990) for example re-establishing the processes practised at Timna based on the study of the surviving debris. Experimental studies assume even greater importance as the study of early metallurgy moves into areas where there is less surviving debris.

The recording of simpler, sometimes traditional technologies has always formed an important component of our understanding of early processes. Much of the Early Mines Research Group's experimental smelting practice is based on first-hand observations made in the 19th century, particularly in Iran and India (Timberlake 2003b and this volume, pp. 27–36). Latterly, archaeologists and ethnographers have had to rely more on the memory of traditional practices rather than on current practice (Juleff 1998; Schmidt 1997).

However, even in the late 20th century some primitive mining and smelting processes were still being practised and have been recorded as exemplified by the digging and smelting of tin in India (Babu 2003) and the distillation of zinc in China (Craddock and Zhou Weirong 2003). Very often the present primitive technologies do not represent the survival of a traditional technology but are instead a modern new response to changed economic conditions. For example, the current primitive activities among the collapsed ancient mine workings near Kuang Shan Zhen in Yunnan, China are of very recent origin carried out by local farm workers with little or no tradition or experience of mining (Craddock 1997). It is likely that many of the so-called traditional primitive mining and smelting technologies so pejoratively described by geologists and other colonial officials in 19th-century India and Iran were in fact similar low-tech new enterprises.

Even where a real traditional process is still being practised it is unlikely to be totally unchanged: for example, the primitive tribal tin smelters operating deep in the forests of Orissa in India now use hand-operated rotary fan bellows, replacing the bag bellows used previously. In China the sophisticated traditional zinc-distillation process developed continuously through the 20th century (Craddock and Zhou Weirong 2003). Although these changes are extremely important in themselves, they do illustrate the dangers of relying too uncritically on supposedly unchanged tradition as indicators of past practice.

The 2005 British Museum conference, *Metallurgy: A Touchstone for Cross-cultural Interaction*, generated such a huge response across the whole subject of archaeometallurgy that it was necessary to restrict contributions to a few representative themes in order to keep the volume manageable. These were all related to mining and extractive metallurgy, more specifically the production of zinc and brass, African metallurgy, aspects of early iron and steel and the inception and nature of the first smelting technologies.

The papers on brass include several documenting early occurrences of zinc in copper alloy metalwork (Thornton, pp. 123–35; Montero-Ruiz and Perea, pp. 136–9) and the pivotal moment in the 1st century BC when, for the first time, brass of a reasonably regular composition was adopted for a specific class of metalwork (Istenič and Šmit, pp. 140–47). Excavations in China have revealed the remains of zinc distillation dating from the 15th to the 17th century AD (Liu Haiwang *et al.*, pp. 170–78). The process was already at a developed industrial stage, suggesting the origins could be much earlier. This is rather at odds with the documentary evidence which suggests that zinc was unfamiliar before the end of the 16th century, at least for the production of coinage (Zhou Weirong, pp. 179–86).

Iron and steel are represented here by a variety of process and places from China, Africa and Europe. The apparent absence of crucible steel from Europe, while it was in regular use throughout central, southern and western Asia, had been commented on (Craddock 2003) and thus the discovery that certain Viking swords were forged from crucible steel

(Williams, pp. 233–41) is of some significance and raises the question of what other usage of crucible steel there may have been in Europe prior to the 18th century. The excavation and scientific study of the first Early Iron Age iron-smelting site to be recognised in the Middle East (Veldhuijzen and Rehren, pp. 189–201) is of considerable importance, not only for documenting the process but also for related problems such as the origins of iron smelting in sub-Saharan Africa. The research and experimentation on the solid-state direct iron reduction technologies in Europe are continuing to demonstrate the variety of processes (Paynter, pp. 202–10) and the value of experimentation to establish the true potentials of the processes (Crew and Charlton, pp. 219–25).

The studies on the inception of metallurgy and the evidence of the processes have been charted here for the Cyclades and Crete (Gale and Gale, pp. 103–11; Muhly, pp. 97–102), Thailand (Pigott and Ciarla, pp. 76–88), southern Iberia (Müller et al., pp. 15–26) and across Europe and the Middle East (Bougarit, pp. 3–14). The recognition of these early smelting places – where considerably less permanent debris was produced than in the later processes – has created problems in reconstructing the technology. Experimental replications have proved useful in rather differing ways for the mining and smelting operations (Timberlake, pp. 27–36). In the case of mining, where much surviving evidence was only belatedly discovered, experimental reconstructions have been essential to establish the effectiveness of techniques such as firesetting, and the best way to make and use the surviving tools (Timberlake 2003b). For the smelting the situation is very different. With little if any surviving material evidence, the challenge is to establish what methods, if any, are feasible to smelt metal with compositions commensurate to that of the early metal, in processes which leave little debris. Experiments have produced copper with only minor traces of iron from oxidised ores, leaving little or no permanent debris with the possible exception of the tuyeres. The evidence from the mines, however, suggests that the mixed copper-iron sulphides were exploited from the inception of metallurgy at least in Europe. Smelting experiments using chalcopyrite have so far failed to produce copper with only traces of iron and without producing quantities of slags (Craddock et al., pp. 37–45).

As the investigation of ancient processes tackles ever more ephemeral evidence, the role of experimental replication will become more vital as a means of demonstrating the feasibility of the postulated processes. Centres such as those that already exist in France, Holland and Scandinavia are likely to attain great importance in archaeometallurgy. As well as establishing many of the physical parameters, actually carrying out the processes confronts the practitioners with the broader issues raised by the sometimes enigmatic debris of early metallurgy as described by Rehren et al. (pp. 211–18) for early African slags in this volume. Such considerations beyond the physical and chemical properties of the materials are necessary if the study of archaeometallurgy is to be integrated into archaeology as a whole rather than just another specialist discipline whose practitioners speak only among themselves. We are now beginning to understand something of the process metallurgy; the next stage will be to ascertain what determined those technical choices and the dynamics of technical change in the societies in which they operated.

## References

Babu, T.M. 2003. Advent of the Bronze Age in the Indian subcontinent. In Craddock and Lang 2003, 174–80.

Bachmann, H.G. 1982. The Identification of Slags from Archaeological Sites. Institute of Archaeology Occasional Paper 6. London: Institute of Archaeology.

Betancourt, P.B. 2006. Chrysokamino I: The Metallurgy Workshop and its Territory. Athens: Hesperia Supplement 30.

Briggs, C.S. 1985. Copper mining at Mt. Gabriel, Co. Cork: Bronze Age bonanza or post famine fiasco? Proceedings of the Prehistoric Society 49: 317–35.

Castel, G., Köhler, E.C., Mathieu, B. and Pouit, G. 1998. Les mines du ouadi UM Balad. Bulletin de l'Institut Français d'Archéologie Orientale 98: 57–87.

Conrad, H.-G. and Rothenberg, B. (eds) 1980. Antikes Kupfer im Timna-Tal. Der Anschnitt Beiheft 1. Bochum: Deutsches Bergbau-Museum.

Craddock, P.T. (ed.) 1980. Scientific Studies in Early Mining and Extractive Metallurgy. British Museum Occasional Paper 20. London.

Craddock, P. 1997. Low technology zinc mining in the south-west of China. Mining History: Bulletin of the Peak District Mines Historical Society 13(3): 41–51.

Craddock, P.T. (ed.) 1998. 2000 Years of Zinc and Brass, 2nd edn. British Museum Occasional Paper 50. London.

Craddock, P.T. 2001. From hearth to furnace: evidences for the earliest metal smelting in the eastern Mediterranean. Paléorient 26(2): 151–65.

Craddock, P.T. 2003. Cast iron, fined iron, crucible steel: liquid iron in the ancient world. In Craddock and Lang 2003, 231–57.

Craddock, P.T. and Hughes, M.J. (eds) 1985 Furnaces and Smelting Technology in Antiquity. British Museum Occasional Paper 48. London.

Craddock, P.T. and Lang, J. (eds) 2003. Mining and Metal Production through the Ages. London: British Museum Press.

Craddock, P.T. and Meeks, N. 1987. Iron in ancient copper. Archaeometry 29(2): 187–204.

Craddock, P.T. and Zhou Weirong, 2003. Traditional zinc production in modern China: survival and evolution. In Craddock and Lang 2003, 267–92.

Craddock, P.T., Freestone, I.C., Gurjar, L.K., Middleton, A.P. and Willies, L. 1998. Zinc in India. In Craddock 1998, 27–72.

Crew, P. 1991. Experimental production of prehistoric bar iron. Journal of the Historical Metallurgy Society 25: 21–35.

Delibes de Castro, G. and Montero-Ruiz, I. 1999. Las Primeras Etapas Metalurgicas en la Peninsula Iberica II: Estudios Regionales. Madrid: Institutio Universitario Ortega y Gasset.

Feuerbach, A.M., Griffiths, A.M. and Merkel, J.F. 2003. Early Islamic crucible steel production at Merv. In Craddock and Lang 2003, 258–66.

Hauptmann, A. 2000. Zur frühen Metallurgie des Kupfers in Fenan, Jordanien. Der Anschnitt Beiheft 11. Bochum: Deutsches Bergbau-Museum.

Hauptmann, A. 2003. Developments in copper metallurgy during the fourth and third millennia BC at Feinan, Jordan. In Craddock and Lang 2003, 90–100.

Jackson, J.S. 1980. Bronze Age copper mining in counties Cork and Kerry, Ireland. In Craddock 1980, 9–30.

Juleff, G. 1998. Early Iron and Steel in Sri Lanka. Mainz: KAVA-Materialien 54.

Merkel, J.F. 1990. Experimental reconstruction of Bronze Age copper smelting based on archaeological evidence from Timna. In Rothenberg 1990, 78–122.

O'Brien, W. 2004. Ross Island: Mining Metal and Society in Early Ireland. Department of Archaeology, National University of Ireland, Galway.

Pleiner, R. 2000. *Iron in Archaeology: The European Bloomery Smelters*. Praha: Archeologický Ústav Avčr.

Rehren, Th. and Papachristou, O. 2003. Similar like white and black: a comparison of steel-making crucibles from Central Asia and the Indian subcontinent. In *Man and Mining*, T. Stöllner, G. Körlin, G. Steffens and J. Cierny (eds). *Der Anschnitt* Beiheft 16. Bochum: Deutsches Bergbau-Museum, 393–404.

Rohl, B. and Needham, S.B. 1998. *The Circulation of Metal in the British Bronze Age: The Application of Lead Isotope Analysis.* British Museum Occasional Paper 102. London.

Rothenberg, B. 1972. *Timna*. London: Thames and Hudson.

Rothenberg, B. (ed.) 1990. *The Ancient Metallurgy of Copper*. London: IAMS.

Rovira Llorens, S. and Gomez Ramos, P. 2003. *Las Primeras Etapas Metalugicas en la Pemninsula Iberica III*. Madrid: Estudios Metalograficos.

Rovira Llorens, S., Montero Ruiz, I. and Consuegra Rodriguez, S. 1997. *Las Primeras Etapas Metallurgicas en la Peninsula Iberica 1. Analise de Materiales*. Madrid: Instituto Universitario.

Schmidt, P.R. 1997. *Iron Technology in East Africa*. Bloomington, IN: Indiana University Press.

Sumna, U., Valera, R. and Lo Schiavo, F. (eds) 2004. *Archaeo Metallurgy in Sardinia*. Rome: CUEC Editrice.

Timberlake, S. 2003a. *Excavations on Copa Hill, Cwmystwyth (1986–1999)*. BAR 348. Oxford: BAR.

Timberlake, S. 2003b. Early mining research in Britain. In Craddock and Lang 2003, 21–42.

Timberlake, S. and Prag, A.J.N.W. 2005. *The Archaeology of Alderley Edge*. Oxford: John and Erica Hedges.

Tylecote, R.F. 1963 *Metallurgy in Archaeology*. London: Arnold.

Watling, R.J., Taylor, J., Shell, C.A., Chapman, R.J., Warner, R.B., Cahill, M. and Leake, R.C. 1999. The application of laser ablation inductively coupled plasma mass spectrometry (LA–ICP–MS) for establishing the provenance of gold ores and artefacts. In *Metals in Antiquity*, S.M.M. Young, M. Pollard, P. Budd and R. Ixer (eds). BAR International 792. Oxford: Archaeopress, 53–62.

Weisgerber, G. 2003. Spatial organisation of mining and smelting at Feinan, Jordan. In Craddock and Lang 2003, 76–89.

Yenner, K.A., Adriaens, A., Earl, B. and Özbal, H. 2003. Analyses of metalliferous residues, crucible fragments, experimental smelts, and ores from Kestel tin mine and the tin processing site of Göltepe, in Turkey. In Craddock and Lang 2003, 181–97.

# Mining and smelting

# Chalcolithic copper smelting

*David Bourgarit*

*ABSTRACT* Over the last decade, our understanding of the first copper-smelting processes has considerably evolved, thanks mostly to a dramatic increase in available archaeological and the related archaeometallurgical data. Copper-smelting activities from the Late Neolithic to the very first phases of the Early Bronze Age (EBA) have been discovered and investigated on some 20 archaeological sites located in the Old World, from the Iberian Peninsula to the Iranian plateau. By summing up most recent studies done in France by the author and by reviewing the published literature concerning the other areas, the present paper reports and discusses the prominent technical features of what may be called, for the sake of convenience, the 'chalcolithic' copper-smelting processes. The main finding of this survey is the lack of technological consistency encountered at the beginning of copper extractive metallurgy, including quite technically advanced processes; this supports the considerable variety in copper-production modes reported by others.

*Keywords:* copper, smelting, chalcolithic, Bronze Age, slag, archaeometallurgy, process reconstruction.

## Introduction

During the last ten years, three sites revealing the earliest copper-smelting activities so far recorded in France have been excavated and investigated using integrated approaches including archaeology, analytical chemistry and experimental reconstructions. These are the settlement of Al Claus, Tarn et Garonne (Carozza *et al.* 1997), the pits of Roque-Fenestre within the Cabrières mining district, Herault (Bourgarit and Mille 1997; Espérou 1993), and the 'metallurgists' village' of La Capitelle du Broum, also at Cabrières (Ambert *et al.* 2005; Bourgarit and Mille 2005; Bourgarit *et al.* 2003). All three sites are dated to the Late Neolithic, 3rd millennium BC. Numerous other sites providing evidence of copper-smelting activities in 'primitive' technological stages, covering major parts of the ancient world, have been investigated or revisited using similar interdisciplinary integrated approaches. The present paper aims at reviewing this new set of data by focusing mainly on technological aspects. In particular, the concept of a 'chalcolithic copper-smelting model' will be further tested: indeed, although quite a variety of processes has been observed, some recurrent features have emerged when considering the first smelting processes (Craddock 1995, 1999). Hence, processes are systematically described as non-slagging, carried out under relatively low temperatures and poorly reducing conditions. In the current paper, these technical features are looked at step-by-step, starting from the concept of non-slagging and moving on to its main process corollaries such as fluxing, working temperatures and working atmospheres.

Only technological aspects are examined. Given that their study relies mainly on so-called slag investigations, only sites providing both slags and analytical slag investigations have been reviewed (in the following, the term 'slag' is used somewhat incorrectly to designate smelting debris). The geo-graphical region surveyed extends from western Europe to the Middle East, covering periods from the Late Neolithic to the very beginning of the Early Bronze Age (EBA).[1] For the sake of convenience, 'chalcolithic' is used here as a generic term to describe this time period. Finally, 20 archaeological sites have been chosen (Table 1 and Fig. 1), exhibiting quite different socio-economic *statii*, from small Neolithic-like polyfunctional settlements, more specialised 'metallurgist' communities, to metal workshops in proto-urban contexts (Table 1).

## Slagging?

Depending on the site, the smelting debris appear either as separate slag fragments, as thin slaggy layers covering the inner side of overheated ceramic potsherds, or as layers of varying thickness on what are interpreted as furnace walls; all these forms may be encountered on the same site (Table 1). In the following paragraph, all three forms will be considered as similar and called slags. In almost all chalcolithic sites under review, slag size rarely exceeds several centimetres. Moreover, only minute quantities of such smelting debris are usually recovered (Table 1), one of the extreme cases being Ross Island, Ireland, where not a single ounce of slag could be found (O'Brien 2004). These derisory amounts beg the question of how representative are the recovered slags, a question that is crucial since the reconstruction of the smelting process relies almost exclusively on these debris.

According to the most widely accepted theory, the scarcity of slags in chalcolithic times is due to the fact that high-grade ores were used for smelting, thus producing very little debris. Yet 'very little' is still much more than is usually found in the

**Table 1** Main data concerning the chalcolithic copper-smelting sites reviewed. Only those associated with copper smelting are reported. Note that all the slags found are so-called 'furnace slag' (heterogenous, partly fused with unfused minerals remnants).

| Site | Region/district | Site status (1) | Phase | Date (BC) | Smelted ore (2) | Slag type (3) | Recovered state of slag (4) | Total weight (number) of slags recovered (5) | Slag size or layer thickness (cm) | Slag weight (g) | Number of slags investigated | Slag mineralogy (6) | Major elements in the slags | CuO in slag (wt%) | Smelting reactor | Tuyere/blowpipe nozzle | Reference |
|---|---|---|---|---|---|---|---|---|---|---|---|---|---|---|---|---|---|
| Tepe Sialk (III/6-7) | Iran | U | | 3700–3500 | M? | S | F | ×00 g | nut | ? | 16 | PMS | AlSiCaFe | 6–30 | ? | – | Schreiner et al. 2003 |
| Murgul mining district (5 sites) | Eastern Anatolia | W | Late Chalcolithic/EBA-IA | 4th m. | M? | S | C | 10–50 t | 15–20 | 1500–3000 | 7 | FS | SiFe | 1–3 | ? | – | Lutz 1990; Lutz et al. 1994; Hauptmann et al. 1993 |
| Norsun Tepe (level?) | Eastern Anatolia | U | Late Chalcolithic | 4th m. | M? | L+S | F | 1.5 kg (50) | nut to 8 | ? | 8 | MCD | SiFe | 12–33 | crucible | – | Hauptmann et al. 1993 |
| Arslan Tepe (VII-VIA) | Eastern Anatolia | U | Late Chalcolithic/EBA-IA | 3700–3000 | O | S | F | ? | 1–8 | 5–150 | 3+1 | PMFeSp | AlSiCaFePb | 0.8 | crucible? | ? | Palmieri et al. 1999; Hess 1998 |
| Nevali Cori | Eastern Anatolia | U | EBA-1 | 3000 | M? | S | E | 10 kg | 10 | 150–250 | | FPMS | AlSiCaFe | 1–3 | crucible | ? | Hauptmann et al. 1993 |
| Abu Matar | Beersheva valley | U | Late Chalcolithic | 2nd half 5th m. | M | L+S | F | 2 kg | 1–4 | 1 | 24+23 | FMCDS | AlSiCaFe | 1–10 | furnace | T? | Shugar 2000, 2003 |
| Shiqmim | Beersheva valley | U | Late Chalcolithic | 2nd half 5th m. | M? | S+L | F | (28) | 1–3 | ? | 8 | FMCDS | SiCaFe | 13–40 | furnace | – | Shalev and Northover 1987; Golden et al. 2001 |
| Timna sites 39 a + b | Western Arabah | W | Late Chalcolithic / EBA ? | 2nd half 4th m? | O | S | F | ×0 kg | 1–5 | ? | 21+2+1 | FM | SiFe | 16 | furnace | – | Lupu 1970; Bachmann 1978, 1980; Rothenberg 1990; Merkel and Rothenberg 1999 |
| Feinan, Wadi Fidan 4 | Eastern Arabah | PS | EBA-1 | 2nd half 4th m. | M? | S | F | (1 hand) | nut | ×0 | 33 | PMCDS | SiFe+MgSiCa | 13–60 | crucible | – | Hauptmann et al. 1996; Hauptmann 2000 |
| Dolnoslav | Bulgaria | PS | Late Eneolithic | 2nd half 5th m. | M | L | F | (1) | 1 | – | 1 | MFCOxS | AlSiCaFe | ? | crucible + furnace? | – | Ryndina et al. 1999 |
| Brixlegg-Mariahilfbergl | Northern Tirol | PS | Late Neolithic | 2nd half 5th m. | M | S | F | 250 g | 1–6 | ? | 6 | PCS | MgSiCaFe | ? | ? | B | Bartelheim et al. 2002; Huijsman et al. 2004 |
| Milland | Trentino/ Alto Adige | ? | Late Eneolithic | 1st half 3rd m. | S | S | C+F | ×0 kg | 5–15 | ? | 20 | FMS | ? | few | ? | B | Dal Ri et al. 2005; Artioli et al. 2005; Colpani et al. in prep. |
| Riparo Gaban | Trentino | PS? | Late Eneolithic | 3rd m. | S | S | C+F | (few) | 4–20 | ? | 12+ | FOIMS | MgSiCaFeZn | 2–3 | ? | – | Perini 1992; D'Amico et al. 1997; Anguilano et al. 2002; Colpani et al. in prep. |
| La Capitelle du Broum | Cabrieres, Herault | SS | Late Neolithic | begin 3rd m. | M | S | F | 644 g (514) | 1–5 | 1 | 77 | P(OI)MCS | AlMgSiCaFe | 2–30 | furnace | -- | Ambert et al. 2002, 2005; Bourgarit and Mille 2005; Bourgarit et al. 2003 |
| Roque-Fenestre | Cabrieres, Herault | W | Late Neolithic | 3rd m. | M | S | F | 50 g | 1–5 | 1 | 8 | PMCS | AlMgSiCaFe | 6–30 | ? | – | Esperou 1993; Bourgarit and Mille 1997 |
| Al Claus | Tarn et Garonne | PS | Late Neolithic | 3rd m. | S | L | F | 30 g | 2 | 1 | 5 | FMS | SiFe | 1–23 | vase | T? | Carozza et al. 1997 |
| Los Millares | Almeria | PS | Chalcolithic | 3rd m. | O | L | F | (hundreds) | 0.1–0.3 | ? | 25 | MCDSSp | SiCaFe | ? | vase | – | Hook et al. 1991; Keesmann and Onorato 1999 |
| Almizaraque | Almeria | PS | Chalcolithic | 2nd half 3rd m. | O | L+S | F | 10 kg | 1–2 | ? | 10 | PMCDSp | MgSiCaFe | 20–50 | vase | – | Schubart 1988–90; Delibes et al. 1996; Rovira Llorens 2002; Müller 2002; Müller et al. 2004a, 2004b |
| Cabezo Juré | Huelva | SS | Chalcolithic | 3rd m. | O | L+S | C+F | 10 kg | ? | 10–1000 | ? | PP(F)MCDS | AlSiCaFe | 0.1–10 | furnace | T | Saez et al. 2003; Nocete et al. 2005; Nocete 2006 |
| La Ceñuela | Murcia | PS | Chalcolithic | 3rd m. | S | L | F | ? | ? | ? | 10 | SF? | AlSiCaFe | 5–85 | vase | – | Rovira Llorens 2002 |

(1) Site status: **P**olyfunction settlement, **S**pecialised settlement, **U**rban workshop, **W**orkshop; (2) Smelted ore: **S**ulphide / **O**xide / **M**ixed sulphide and oxide; (3) Slag type: slagged **L**ayer on ceramic material (crucible, vase, furnace); (4) Slag state: **F**ragmented / **C**omplete; (5) Total weight (number) of slags recovered; '×0 g' = tenths of grams, '×00 g' = hundreds of grams; (6) Slag mineralogy: **P**yroxene / **F**ayalite / other **O**livine / **P**lagioclase / **A**kermanite / **F**eldspar / **M**agnetite / **C**uprite / **D**elafossite / copper **S**ulphide inclusions / **Sp**eiss

4

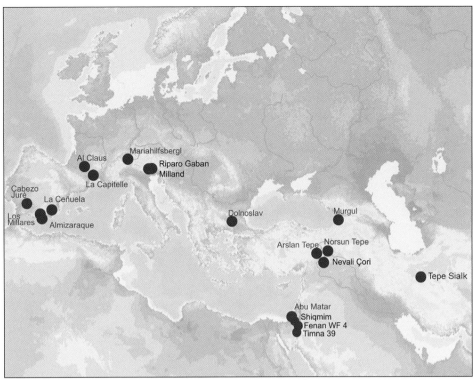

**Figure 1** Map showing the chalcolithic copper-smelting sites reviewed.

field: a rough calculation shows, for example, that the 500 slag fragments recovered at La Capitelle represent at the most two smelting operations![2] Alternative explanations for the scarcity of slags may then be twofold. The first hypothesis is that the process did not produce any slag at all. The recovered slags would then originate from low-grade or not well-sorted ore being *accidentally* smelted, with such accidents being part of a 'trial-and-error' technological stage. This hypothesis may be discarded, since 'ore accidents' are likely to have been rare. Ore beneficiation indeed seems to have been an important step in the chalcolithic *chaîne opératoire* of copper extractive metallurgy, as shown notably by the four pits at Roque-Fenestre, mainly devoted to this activity. Note also the report by Barbara Ottaway of copper sulphides being intentionally sorted out from oxide ores in the Rudna Glava mines (Jovanovic and Ottaway 1976).

The second hypothesis is that the slags revealed by the excavations do belong to a routine process, but a major share of them is missing from the archaeological record, for several reasons. The distortion of the archaeological record may play a significant role.[3] A primary transformation or displacement of the data may have occurred because, for example, of the further processing of the slags away from the smelting site, such as at the Roque-Fenestre pits: the presence of slag remnants (still in derisory amounts) and sand consisting of finely crushed slags suggest that slags were transported around this specific area for crushing and sorting (Bourgarit and Mille 1997).[4] This second assumption which supports the idea of a real slagging process seems much more probable. Furthermore, it may be asserted that the archaeological slags do represent the smelting process correctly in as far as the studied population meets statistical requirements regarding heterogeneity, which seems to

be the case for most of the sites reviewed. Small-scale metal production surely accounts for the systematic absence of giant slag heaps, such as those encountered in later periods. Lower production of waste due to higher grade ore being smelted probably also plays a role. It should be noted that the smelting debris may even help in differentiating chalcolithic metal production modes. Thus, at sites where metallurgy still belongs to the domestic setting, such as at Al Claus or Los Millares, the debris is very scarce, whereas in more specialised contexts the slag quantities are much more significant (Murgul, Cabezo Juré: see Table 1).

That said, the main chalcolithic processes clearly demonstrate immature slagging. Almost all chalcolithic slags are highly viscous, partially fused material bearing numerous remains of unfused minerals. These are the so-called 'furnace slags' (Bachmann 1980), where numerous and/or large millimetric metallic copper prills are still entrapped and account, at least partly, for the high copper content systematically measured (Fig. 2). These features clearly show that slagging was not well mastered, requiring the subsequent crushing of the smelting product in order to recover the remaining valuable material, as borne out by the high fragmentation of the slags (the original size of the smelting product corresponds to the size of the reactor, that is in most cases more than 10 cm in diameter). The working conditions responsible for this immature slagging are discussed in the following paragraphs: these conditions represent the basis of the process reconstruction. Yet, some recent studies clearly question the universality of the chalcolithic slag's 'nut-size': at Murgul, Cabezo Juré and Milland, much larger slags have been recovered (Table 1), some probably barely crushed, if at all, thus providing evidence already in chalcolithic times of varying degrees in the mastery of slagging.

## Fluxing?

To carry out proper slagging, the chemical composition of the system may be crucial: it can be adjusted by fluxing. But whether deliberate fluxing was used in chalcolithic copper-smelting processes is still highly debated.

Arguments encountered in the literature supporting deliberate fluxing are of two types (if one discards the intriguing high lead contents, more than 20 wt% PbO, in the slags from Arslan Tepe level VII; see Hess 1998). First, in some rare cases such as at Cabezo Juré, the slag compositions are claimed not to fit the composition of the processed ores (Saez *et al.* 2003).

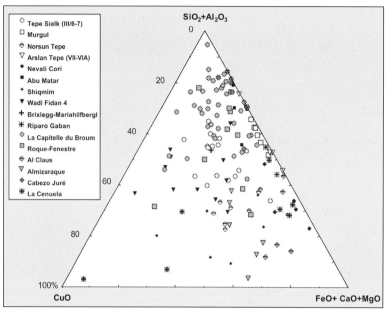

**Figure 2** Bulk chemical compositions of the slags from most of the chalcolithic sites reviewed (141 analyses) plotted in a $CuO$–$(SiO_2+Al_2O_3)$–$(FeO+CaO+MgO)$ diagram (compositions in wt%). All major elements are plotted, thus clearly showing the high copper contents in the slags (expressed in $CuO$ wt%).

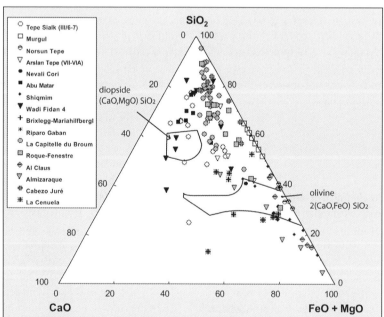

**Figure 3** Bulk chemical compositions of the slags from most chalcolithic sites reviewed (141 analyses) plotted in the $CaO$–$SiO_2$–$(FeO+MgO)$ diagram (compositions in wt%). The liquidus fields of the two most usual neo-formed silicates encountered in the slags reviewed are reported: the diopside field, representing clinopyroxenes, as seen in a $CaO$–$SiO_2$–$MgO$ diagram (after Muan and Osborn 1965), and the olivine field as seen in a $CaO$–$SiO_2$–$FeO$ diagram (after Muan and Osborn 1965). Two things need to be stressed. First, a large discrepancy of slag compositions can be clearly seen at most of the sites. Secondly, most of the compositions lie far from the two ideal eutectic fields. High silica contents are mostly responsible for this, leaving aside the fact that the effective quantity of divalent iron is in most cases much lower than quoted, as shown notably by the high magnetite contents in the slags.

One argument against this assertion is that contamination during the pyrometallurgical process by fuel, building or soil materials may, accidentally, have further introduced elements such as aluminium (Al), magnesium (Mg) and calcium (Ca). The partially liquefied state of the chalcolithic melts surely does not favour such contamination (Hauptmann 2000). Also, very high amounts of 'flux' as calculated by Saez (up to 50% in the Cabezo Juré slags) could rule out such a contamination hypothesis. Yet, field experiments reproducing chalcolithic smelting of silica ores bearing tetrahedrite (Cu-As-Ag-S) and malachite have produced high calcium and magnesium contents in the partially fused slags (Bourgarit and Mille 1997): the experimental dolomite furnace walls were heavily corroded, whereas the Ca+Mg contents reached 20 wt% of the total slag weight (with the atomic Ca:Mg ratio respecting the dolomite stoichiometry). Subsequent arguments sometimes encountered supporting deliberate fluxing include the presence of fragments of non-copper ores such as iron oxides in the archaeological assemblage (see for example Shugar 2000). Such evidence remains debatable since it depends on the well-known question of how representative are the surviving ore fragments found at smelting sites, and their exact nature: raw material or debris.

Yet, both major discrepancies of bulk compositions of chalcolithic slags within each site, and compositions located far from the eutectic fields (Fig. 3) are the most convincing arguments so far in the fluxing debate, leaving aside the large amounts of copper left in the slags (Fig. 2). They contradict the idea of deliberate fluxing. At the least, they clearly reflect a poor mastery of the initial feed composition, even at sites showing the most 'mature' slags such as in the Feinan district where non-ideal slag composition, interpreted as witnessing a non-fluxing process, is found up to Roman times (Hauptman 2000). Moreover, in most situations, slag compositions do account for the local iron and silica-bearing ores (Fig. 3), which may be considered as 'self-fluxing' in the sense that the physical properties of the slags thus obtained met the chalcolithic productivity requirements.

## Blowing?

Petrographical and physico-chemical investigations of chalcolithic slags usually point towards working temperatures in the range 1100–1200 °C. These are quite low temperatures with regard to the melting points of the non-ideal compositions of the slag-forming minerals as seen above, thereby explaining the partly liquefied state of the early slags and also the probable high viscosity of the fused phases. Yet, these are still high enough temperatures to require some minimum technological skill in heat production (Rehder 1999).

Heat production is governed both by fuel and air supply. The overall nature of the fuel is actually never discussed for chalcolithic smelting, but is assumed to be charcoal. In the rare studies where attention is devoted to the air-supply system, an almost universal model based on blowpipes is emerging (Weisgerber 2004). Yet, according to Roden (1988), the only clear archaeological evidence of ancient copper smelting using blowpipes is to be found in the pre-Colombian site of Cerro de

los Cementerios, Batan Grande, Peru, where huge quantities of clay tips have been found. At the other sites, such evidence remains scarce (Table 1). The scarcity of direct archaeological evidence for clay tips may not be that disturbing if a 'heating from above' process is considered, in which corrosion of the tuyere by the melt and thus consumption of tuyeres can be avoided.[5] Experimental investigations have demonstrated the technical feasibility of such an air-supply system (Fasnacht 1999; Hauptmann 2000; Merkel and Shimada 1988; Zwicker et al. 1985). Moreover, theoretical considerations (Rehder 1994) tend to discount more efficient systems using bellows.

That said, alternatives may have existed. First, the 'blow-pipes versus bellows' issue may not be that easily addressed, in particular since upper temperature limits are quite hard to infer within the complexity of chalcolithic chemical systems, leaving aside all kinetic effects.[6] While to my knowledge, the only chalcolithic tuyere so far recovered comes from Cabezo Juré, if one discards the contentious findings of Al Claus and Abu Matar, the lack of archaeological clay tuyeres may be explained in the same way as the lack of blowpipe nozzles. An alternative air-supply system may have been in use at La Capitelle as well as at Cabezo Juré, where the location of the hearths in relation to the prevailing winds suggests the reactors may have been powered by wind, at least partially. Some preliminary smelting experiments performed at La Capitelle du Broum with wind as a sole air supplier in a simple bonfire (Happ 2005) did at least show the feasibility of the use of the wind. These experiments confirm previous observations during experimental copper-sulphide roasting in open hearths, where some metallic copper was formed (Doonan 1994) together with a fused phase very similar to the one encountered in the slags (Bourgarit and Mille 1997).

Unfortunately, the current lack of relevant archaeological data prevents further discussion of chalcolithic air-supply systems and particularly of a possible correlation between the system used and the degree of slagging mastery achieved. For the same reason, the role of the reactor design is difficult to investigate (Table 1). Nevertheless, on sites where 'immature' slags were produced, it seems the reactor design may not have been that crucial a parameter. The fact that a large variety of reactor morphologies leads to similar 'immature' slags may at least support this hypothesis. Indeed, open forms (such as small crucibles or large ceramic vessels), as well as more closed reactors (such as bowl furnaces) are inferred for smelting (Fig. 4). Reactor heights may not play any significant role, given that a natural draught is prevented in most cases by the closed profile of the reactor (although true profiles of the furnaces are never ascertained). Does this mean temperature and, notably, thermal insulation were not that crucial? Since upper temperatures are not known, as previously discussed, this question cannot be answered. Yet, technological aspects may not fully account for the variety of reactor designs encountered at chalcolithic smelting sites, and particularly for the dichotomy between ceramic crucibles and vase furnaces on the one hand, and furnaces on the other, while both types are today quite equally distributed throughout the archaeological record (Table 1). Socio-economic aspects should then be evoked, keeping in mind that a large number of sites still provide no clear indication, either of the nature of the reactor or of the socio-economic nature of the smelting activity. At least it may be noticed that in domestic settings such as Al Claus, La Ceñuela and Almizaraque – although the status of Almizaraque is still debated (see references cited in Table 1) – smelting reactors are invariably non-refractory ceramic common wares, whereas at more specialised metallurgical sites such as La Capitelle, Abu Matar and Cabezo Juré, smelting is carried out in furnaces, i.e. with proper metallurgical equipment.

## Oxidising?

Redox conditions also greatly influence slag formation, at least in slags containing much iron like most of the ones reviewed here (Fig. 3). Hence, at almost all sites high magnetite contents have been observed (Table 1), thus partly explaining the high viscosities of the slags (Davenport *et al.* 2002). The corresponding slightly reducing conditions are one of the most typical features of the 'chalcolithic copper-smelting model'. Nevertheless, this universal picture needs to be qualified. Redox conditions as seen from the compositions of the chalcolithic slags show large discrepancies. Two main composition groups can be discerned: an 'oxidised' composition bearing delafossite (CuII+FeIII), and a 'reduced' composition without any CuII and less FeIII, where fayalite (FeII) may appear as the main crystallised silicate phase (Table 1). As a consequence, the generally accepted high oxygen partial pressures of $10^{-5}$–$10^{-6}$ atm may be lowered in extreme cases by three orders of magnitude (Fig. 5).[7] This in turn explains the specific low iron

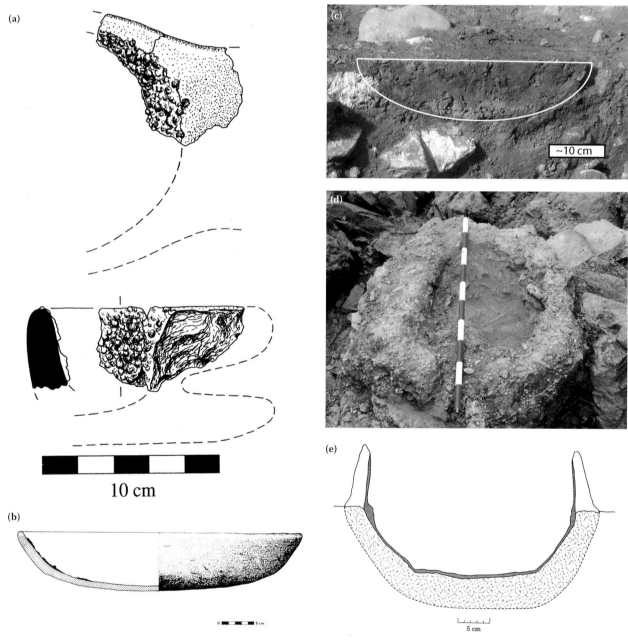

(a)

(b)

(c)

(d)

(e)

10 cm

5 cm

5 cm

**Figure 4** Examples of chalcolithic smelting reactors, most of them (except Cabezo Juré) producing the typical 'immature' slag: (a) a crucible from Wadi Fidan 4 (after Hauptmann 2000); (b) a vase furnace from Almizaraque (after a drawing by Arturo Ruiz Taboada in Montero Ruiz 1993), scale 5 cm; (c) a cross-section of a furnace from La Capitelle, where the burnt soil area is indicated by the white outline (photo: Benoît Mille, C2RMF); (d) a furnace from Cabezo Juré (after Nocete *et al.* 2005), scale 1 m; (e) a reconstruction of a furnace from Shiqmim (after Golden *et al.* 2001).

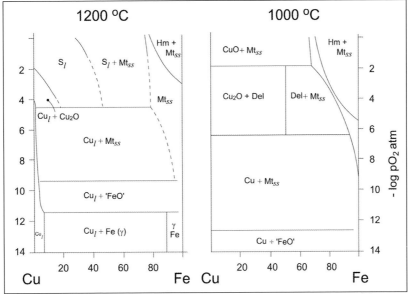

**Figure 5** Equilibrium Cu-Fe diagram at 1000 °C and 1200 °C as function of the $pO_2$ (after Hauptmann 2000). Note in particular the high $pO_2$ delimiting the stability domain of delafossite at 1000 °C, and the low $pO_2$ associated with fayalite (appearing here as FeO) at 1200 °C.

contents of chalcolithic copper (Craddock and Meeks 1987), at least in the slags: at La Capitelle and Al Claus, copper prills within the slags frequently bear up to 3 wt% iron, while high iron contents in the copper prills are observed even in delafossite-containing slags (Müller *et al.* 2004a, 2004b), accounting for local redox variations and, more generally, non-equilibrium conditions.

Working conditions such as temperature, height of the charcoal layer, oxygen penetration in the system and cooling rates are not precisely known and may, if combined, gener-

ate some of the discrepancies observed. Yet, one parameter seems to correlate quite well with the two composition groups, namely the sulphidic versus oxidic nature of the ore (Fig. 6). The influence of the oxygen to sulphur ratio in the ore (O:S) has been noticed in the field of modern metallurgy, notably by those exploring the direct smelting process (Odle *et al.* 1983). Laboratory and field experiments investigated this O:S effect further (Bourgarit *et al.* 2002; Burger 2005; Forel 2002; Wirth 2003) by demonstrating the equivalent role of gaseous oxygen ($pO_2$) and the solid oxygen brought by oxide ores. As

**Figure 6** Map showing the degree of correlation, within the chalcolithic smelting sites reviewed, between the redox composition of slags (expressed by the presence/absence of delafossite $CuFeO_2$, that is of CuII and FeIII) and the type of copper ore inferred to have been smelted (copper sulphide or oxide).

a result, the variation in the composition of the smelted ore alone may explain the variation of the oxidation states in the chalcolithic slags. Hence, the working conditions such as the external redox potential imposed by the charcoal cover may not have played any role, and may thus have remained more or less similar on every site. The low reducing conditions may ultimately still pertain to the chalcolithic model, fitting the generally inferred low height of charcoal layer.[8]

Such constancy in the redox working conditions appears paradoxical when compared to the variety of smelted ores, including sulphides (Table 2). Some time ago the idea entered the chalcolithic model that copper sulphides, or at least copper ore containing copper sulphides, were being smelted at the very beginning of metallurgy (Craddock 1995),[9] yet the implied single chalcolithic pyrometallurgical *chaîne opératoire* capable of dealing with sulphides has been little investigated (but see Timberlake, this volume, pp. 33–5). While Rostoker *et al.* (1989) first tested the feasibility of a one-step smelting process for mixed sulphide and oxide copper ores by experiments in a gas furnace, they overlooked and/or omitted two parameters which drastically restrict the operating conditions of such a simple model: the O:S ratio in the feed and the atmosphere in a chalcolithic smelting reactor. In investigating both parameters, recent laboratory and field experiments (see above and Table 2) have shown that the sulphur content in the initial charge rapidly prevents reduction to metallic copper, thus requiring additional treatment of the smelting product, namely the so-called matte (copper-rich sulphide). Moreover, at La Capitelle du Broum, an important part of the metallurgical debris occurred as separate droplets of matte or so-called white metal. Whatever the exact origin of this matte, partially reacted ore or smelted ore, it represents a metallurgical waste which seems to have been sorted and isolated. Ethnographic evidence suggests a multi-step process whereby the white metal is reprocessed several times under the same conditions (Bourgarit *et al.* 2003).

However, such a process, though multi-stepped, is still within the bounds of the 'chalcolithic simplicity', whereas some chalcolithic smelting sites already exhibit a complex *chaîne opératoire*. Hence, at Abu Matar, at Cabezo Juré and probably also at Almizaraque (Müller *et al.* 2004a), two distinctive pyrometallurgical operations working under different redox conditions are proposed. This is evidenced by two distinctive compositions of metallurgical debris: on the one hand, copper ore (mixed or pure oxide) was smelted in a furnace (in one step?), and on the other, the metallic copper thus produced was further processed in a ceramic crucible, possibly for refining. Inevitably, these complex *chaîne opératoire* raise questions which used to be restricted to later time periods, notably the question of the exact nature of the final product leaving the 'smelting' site, a particularly relevant issue for Bronze Age sites dealing with sulphides (Doonan 1999; Mette 2003). The case of Arslan Tepe for the chalcolithic and EBA-1A levels may be recalled, in particular, where the separate metallic Cu-As and Ni-As fragments which were recovered (the so-called *speiss*) raised the possibility of their use as 'master alloys' (Hess 1998).

## Conclusions

Despite the scarcity of archaeological material such as slags and the immature, non-equilibrium physico-chemical state of the slags that are found, researchers and scholars have demonstrated a growing capacity over the last decade to reconstruct the first copper-smelting processes by following integrated approaches. As a result, the 'chalcolithic copper-smelting model' has considerably gained in prominence, while losing some of its 'primitive' features. First, the 'chalcolithic' smelting process is a true slagging process with a growing amount of evidence pointing towards already advanced slagging control, although a majority of sites still exhibit the traditional low amounts of immature, crushed slags. Second, evidence of a growing variety of working conditions and tools is emerging, be it the nature of the smelted copper ore including sulphides, the smelting reactor design or the air-supply system. Only two features seem to remain universal, namely the absence of deliberate fluxing and the low reducing conditions, although for the latter the effective redox conditions in the reacting system show large discrepancies, depending on the sulphide to oxide ratio in the ore.

The perceived break with the following EBA metallurgical phase remains important, but may not be that abrupt. From

**Table 2** Iron and copper crystalline phases encountered in laboratory experimental copper-smelting slags, according to the oxygen partial pressure imposed in the tubular electrical furnace (vertical axis) and the composition of the initial ore expressed as a molar ratio oxygen/sulphur (horizontal axis) (after Burger 2005). It is noticeable that thermodynamic equilibrium is rarely achieved, which is quite usual in such liquid/gas reactions.[10]

| $pO_2$ (-log) | O/S=0 | 1 | 2 | 2.5 | 3 | 4 | ∞ |
|---|---|---|---|---|---|---|---|
| 3 | Cu–Fe–S | $Fe_2SiO_4$ | $Fe_2SiO_4$ + $Fe_3O_4$ | $Fe_3O_4$ | $CuFeO_2$ + $Fe_3O_4$ | $CuFeO_2$ + $Cu_2O$ | $Cu_2O$ |
| 4 | | | $Fe_2SiO_4$ + $Fe_3O_4$ | | $CuFeO_2$ + $Fe_3O_4$ | $CuFeO_2$ + $Cu_2O$ | |
| 7 | | | $Fe_2SiO_4$ + $Fe_3O_4$ | $Fe_3O_4$ | $CuFeO_2$ + $Fe_3O_4$ | $CuFeO_2$ + $Cu_2O$ | |
| 10 | | | | $Fe_2SiO_4$ + $Fe_3O_4$ | | | |

the technological point of view alone, the various degrees of slag mastery witnessed in chalcolithic periods may already herald the EBA perfect low viscous slags, such as the thin homogenous and dense fayalitic slags recovered at St Véran, Hautes Alpes (Burger *et al.* in preparation; Ploquin *et al.* 1997; Rostan *et al.* 2006). These are already identical to the much later *plattenschlacke* from the Alpine Late Bronze Age. On the socio-economic level, the chalcolithic production cannot by its nature compete with EBA mass production, as seen in the Feinan and Timna mining districts (Hauptmann 2000; Rothenberg 1990), in the Oman region (Hauptmann *et al.* 1988), in the Aegean world at the Cycladic Kythnos Island and at the Cretan Chrysokamino smelting site (Bassiakos and Catapotis 2006; Bassiakos and Philaniotou-Hadjianastasiou in press), or at St Véran, where copper production is estimated to have been running at some seven tonnes a year (Rostan *et al.* 2002, 2006). Yet, copper production modes belonging to extra communitarian systems have already been shown to exist in chalcolithic times, which may be underlined by several specific archaeometallurgical features, such as the quantities of metallurgical debris and/or the type of smelting reactor. Hence, larger quantities of slags are recovered from the more specialised metallurgical sites, Murgul being an extreme case. Also, at La Capitelle, Cabezo Juré and Abu Matar, real furnaces were already in use, whereas in domestic settings, such as at Al Claus and at some Millarian sites, smelting reactors were common ceramic wares. Similar trends are still observed for the EBA. Thus, at mass production sites, furnaces were generally used, whereas smelting was still performed in ceramic vessels where more domestic practices survived, as seen at Shahr-i-Sokhta, Iran (Hauptmann *et al.* 2003) and, subject to confirmation, at Daskaleio-Kavos on the Cycladic Keros Island (Georgakopoulou 2004, 2005) and at Buchberg, Northern Tirol (Martinek and Sydow 2004).

## Acknowledgements

First, the author would like to thank Benoit Mille with whom he has been collaborating on most parts of the 'chalcolithic copper-smelting mystery' for almost ten years. He is also very grateful to Laurent Carozza and Albane Burens both for their trust in providing the Al Claus material for study, and for their invaluable, active and enthusiastic participation in all field experiments, whatever the weather conditions. Many scholars and researchers have participated in compiling the present review by providing much published and unpublished data. For all their help, the author sincerely thanks Gilberto Artioli, Yannis Bassiakos, Paolo Bellintani, Fabio Colpani, Myrto Georgakopoulou, Andreas Hauptmann, Thomas Levy, Ignacio Montero, Roland Müller, Robert Odle, Barbara Ottaway, Reinaldo Saez, Sariel Shalev, Aaron Shugar and Xander Veldhuijzen. Last but not least, Evanthia Ioannidou is warmly acknowledged.

## Notes

1. The upper chronological limit of the Early Bronze Age (EBA) is fixed here by the shift to large-scale copper production, which might correspond to Strahm's *Metallikum* in Europe (Strahm 1996) or to Yalçin's phase V in the Anatolian metallurgy (Yalçin 2004). On most of the smelting sites (see Conclusions), this corresponds to the appraisal of high quantities of visible slag (i.e.

slag heaps), as for example at the Feinan district at the EBA2 (Hauptmann 2000).
2. A single piece of slag cannot be produced alone – it is part of at least one complete layer in the reactor. With an average size of one piece of slag of around 1cm$^3$, and an internal reactor diameter of some 20 cm, one layer of slag should represent at least 314 cm$^3$: the total amount of slag produced in one run should be more than 300 cm$^3$.
3. Rovira Llorens and Ambert (2002) report that at Almizaraque, early excavations failed to find any slag. Also, natural processes such as erosion may eliminate part of the slags, as suggested notably for Cabezo Juré (Saez, pers. comm.).
4. Note that the 'slag sand' almost escaped the metallurgical debris record, while primarily interpreted as altered sand being part of a furnace wall (Espérou 1993).
5. We never changed the tuyere after more than 20 field experiments on chalcopyrite vase smelting using a one-bag bellows (Bourgarit *et al.* 2002). In contrast, thousands of slag-corroded tuyeres were found at the Iron Age iron-smelting site of Tell Hammeh, Jordan (Veldhuijzen, this volume, pp. 189–201) where blowing was performed from the bottom (moreover, tuyeres are supposed to be intentionally melted to help fluxing the smelt (Veldhuijzen 2005)).
6. Laboratory experiments (Wirth 2003) confirmed that at 1250 °C – i.e. at higher temperatures than usually inferred – a small distortion to the fayalitic stoichiometric composition leads to highly viscous slags, under proper reducing conditions. Moreover, repeated field experiments in ceramic vases using a one-bag bellows (Bourgarit *et al.* 2002) revealed that 1350 °C was largely reached in the melt, while still 'immature' partially fused slags were obtained despite both a stoichiometric Fe:Si ratio in the feed, and sufficient reducing conditions. Similar products were obtained in a bowl furnace operated with a one-bag bellows (Bourgarit and Mille 1997).
7. This drop towards the low pO$_2$ echoes the rise towards the high pO$_2$ demonstrated several times by Hauptmann (2000), when he proved that delafossite could also be encountered in smelting slags, not just in melting slags.
8. Returning to the slagging issue, the presence or not of sulphide in the ore does not seem to influence either the viscosity or the overall 'furnace slag macrostructure' of the chalcolithic slags reviewed, though sulphides may play a direct role in the separation of copper or matte from the so-called slag through the liquation process, as proposed by Hauptmann (2003).
9. It is usually attributed to Rostoker's experiments (Rostoker *et al.* 1989). Yet according to Hauptmann (Hauptmann *et al.* 2003), Heskel (1982) had earlier proposed the concept of mixed sulphide and oxide ore for early copper smelting.
10. R.R. Odle, pers. comm.

## References

Ambert, P. and Vacquer, J. (eds) 2005. *La première métallurgie en France et dans les pays limitrophes.* Carcassonne: Société Préhistorique Française.

Ambert, P., Coularou, C., Cert, C. *et al.* 2002. Le plus vieil établissement de métallurgistes de France (IIIe millénaire av. J.C): Péret (Hérault). Compte-rendu Académie des Sci, Palevol 1, 67–74.

Ambert, P., Bouquet, L., Guendon, J. L. and Mischka, D. 2005. La Capitelle du Broum (district minier de Cabrières-Péret, Hérault): établissement industriel de l'aurore de la métallurgie française (3100–2400 BC). In Ambert and Vacquer 2005, 83–93.

Anguilano, L., Angelini, I., Artioli, G., Moroni, M., Baumgarten, B. and Oberrauch, H. 2002. Smelting slags from copper and Bronze Age archaeological sites in Trentino and Alto Adige. In *Atti del II Congresso Nazionale di Archeometria, Bologna, 29 gennaio–1 febraio 2002*, C. D'Amico (ed.). Bologna: Patron, 627–38.

Artioli, G., Colpani, F., Angelini, I. and Anguilano, L. 2005. Metallurgical activity at Millan, Bressanone valley, Italy: mineralogical analysis of the smelting slags. Paper presented at the conference *Il sito fusorio della tarda età del Rame di Millan presso Bressanone dell'area alpina*, Bolzano, 15 June 2005.

Bachmann, H.G. 1978. The phase composition of slags from Timna site 39. In *Chalcolithic Copper Smelting*, B. Rothenberg and R. Tylecote (eds). Institute of Archaeo-Metallurgical Studies Monograph 1. London: Institute of Archaeo-Metallurgical Studies, 21–3.

Bachmann, H.G. 1980. Early copper smelting techniques in Sinai and in the Negev as deduced from slag investigations. In *Scientific Studies in Early Mining and Extractive Metallurgy*, P.T. Craddock (ed.). British Museum Occasional Paper 20. London: British Museum Press, 103–34.

Bartelheim, M., Eckstein, K., Huijsmans, M., Krause, R. and Pernicka, E. 2002. Kupferzeitliche Metallgewinnung in Brixlegg, Österreich. In *Die Anfänge der Metallurgie in der Alten Welt/The Beginnings of Metallurgy in the Old World*, M. Bartelheim, E. Pernicka and R. Krause (eds). Rahden: Verlag Marie Leidorff, 33–82.

Bartelheim, M., Eckstein, K., Huijsmans, M., Krause, R. and Pernicka, E. 2003. Chalcolithic metal extraction in Brixlegg, Austria. In *Archaeometallurgy in Europe, Milan, 24–26 September 2003*, vol. 1. Milan: Associazione Italiana di Metallurgia, 441–7.

Bassiakos, Y. and Catapotis, M. 2006. Reconstruction of the copper smelting process at Chrysokamino based on the analysis of ore and slag samples. In *The Chrysokamino Metallurgy Workshop and its Territory, Hesperia Supplement 36*, P.P. Betancourt (ed.). Princeton, NJ: American School of Classical Studies at Athens, 329–53.

Bassiakos, Y. and Philaniotou-Hadjianastasiou, O. in press. Early copper production and metallurgy on Kythnos. Archaeological evidence – material and analytical reconstruction of metallurgical processes. In *Metallurgy in the Early Bronze Age Aegean*, P.M. Day and R.C.P. Doonan (eds). Oxford: Oxbow Press.

Bourgarit, D. and Mille, B. 1997. La métallurgie chalcolithique de Cabrières: confrontation des données expérimentales et archéologiques en laboratoire. *Archéologie en Languedoc* 21: 51–63.

Bourgarit, D. and Mille, B. 2005. Nouvelles données sur l'atelier métallurgique de la Capitelle du Broum (district minier de Cabrières, Hérault, France): la transformation de minerais de cuivre à base de sulfures se précise! In Ambert and Vacquer 2005, 97–108.

Bourgarit, D., Mille, B., Burens, A. and Carozza, L. 2002. Smelting of chalcopyrite during chalcolithic times: some have done it in ceramic pots as vase-furnaces. In *33rd International Symposium on Archaeometry, Amsterdam, 22–26 April 2002*, H. Kars and E. Burke (eds). Amsterdam: Vrije Universiteit, 297–302.

Bourgarit, D., Mille, B., Prange, M., Ambert, P. and Hauptmann, A. 2003. Chalcolithic fahlore smelting at Cabrières: reconstruction of smelting processes by archaeometallurgical finds. In *Archaeometallurgy in Europe, Milan, 24–26 September 2003*, vol. 1. Milan: Associazione Italiana di Metallurgia, 431–40.

Burger, E. 2005. *La première métallurgie extractive du cuivre en France: caractérisation analytique de produits de réduction expérimentaux*. Master Dissertation. Physico-chimie analytique, Université de Paris VI.

Burger, E., Bourgarit, D., Rostan, P. and Carozza, L. in preparation. The mystery of Plattenschlacke in protohistoric copper smelting: precocious evidence at the EBA site of Saint-Véran, French Alps, proposal for the international conference *Archaeometallurgy in Europe*, Aquileia and Grado, Italy, 17–21 June 2007.

Carozza, L., Bourgarit, D. and Mille, B. 1997. L'habitat et l'atelier de métallurgiste chalcolithique d'Al Claus: analyse et interprétation des témoins d'activité métallurgique. *Archéologie en Languedoc* 21: 147–60.

Colpani, F., Burger, E., Artioli, G., Bourgarit, D., Angelini, I. and Anguilano, L. in preparation. The earliest copper smelting slags in Italy: petrographic and chemical investigations towards a reconstruction of the metallurgical processes, proposal for the international conference *Archaeometallurgy in Europe*, Aquileia and Grado, Italy, 17–21 June 2007.

Craddock, P.T. 1995. *Early Metal Mining and Production*. Edinburgh: Edinburgh University Press.

Craddock, P.T. 1999. Paradigms of metallurgical innovation in prehistoric Europe. In Hauptmann *et al.* 1999, 175–92.

Craddock, P.T. and Meeks, N.D. 1987. Iron in ancient copper. *Archaeometry* 29(2): 187–204.

D'Amico, C., Gasparotto, G. and Pedrotti, A. 1997. Scorie eneolitiche di Gaban e Acquaviva (Trento). Caratteri, provenienza ed estrazione del metallo. In *Le scienze della Terra e l'Archeometria, Napoli, 20–21 febbraio 1997*, Istituto Universitario Suor Orsola Benincasa, 31–8.

Dal Ri, L., Rizzi, G. and Tecchiati, U. 2005. Lo scavo di una struttura della tarda età del Rame connessa a processi estrattivi e di riduzione del minerale a Milland presso Bressanone. Paper presented at the *Il sito fusorio della tarda età di Rame di Milland presso Bressanone dell'area alpina*, Bolzano, 15 June 2005.

Davenport, W.G., King, M., Schlesinger, M. and Biswas, A.K. 2002. *Extractive Metallurgy of Copper*, 4th edn. Oxford: Elsevier.

Delibes, G., Diaz-Andreu, M., Fernandez-Posse, M.D., Martin, C., Montero, I., Munoz, I.K. and Ruiz, A. 1996. Poblamiento y desarrollo cultural en la Cuenca de Vera durante la prehistoria reciente. *Complutum Extra* 6(1): 153–70.

Doonan, R.C.P. 1994. Sweat, fire and brimstone: pre-treatment of copper ore and the effects on smelting techniques. *Historical Metallurgy* 28(2): 84–97.

Doonan, R.C.P. 1999. Copper production in the eastern Alps during the Bronze Age: technological change and the unintended consequences of social reorganization. In *Metals in Antiquity*, S.M.M. Young, A.M. Pollard, P. Budd and R.A. Ixer (eds). BAR International 792. Oxford: Archaeopress, 72–7.

Espérou, J.L. 1993. La structure métallurgique de Roque-Fenestre (Cabrières, Hérault). *Archéologie en Languedoc* 17: 32–46.

Fasnacht, W. 1999. Experimentelle Rekonstruktion des Gebrauchs von frühbronzezeitlichen Blasdüsen aus der Schweiz: Kupferverhüttung und Bronzeguss. In Hauptmann *et al.* 1999, 291–4.

Forel, B. 2002. *Etude expérimentale de la réduction de minerais de cuivre sulfurés et oxydés*. Masters dissertation, Université de Paris XI.

Georgakopoulou, M. 2004. Examination of copper slags from the Early Bronze Age site of Daskaleio-Kavos on the island of Keros (Cyclades, Greece). *Institute of Archaeometallurgical Studies Newsletter* 24: 3–12.

Georgakopoulou, M. 2005. *Technology and Organisation of Early Cycladic Metallurgy: Copper on Seriphos and Keros, Greece*. PhD dissertation, Institute of Archaeology, University College London.

Golden, J., Levy, T.E. and Hauptmann, A. 2001. Recent discoveries concerning chalcolithic metallurgy at Shiqmim, Israel. *Journal of Archaeological Science* 28(9): 951–63.

Happ, J. 2005. Comment j'ai enterré ma tuyère et mon soufflet à la Capitelle du Broum. *Cu +*: 29–31.

Hauptmann, A. 2000. Zur frühen Metallurgie des Kupfers in Fenan, Jordanien. *Der Anschnitt* Beiheft 11. Bochum: Deutsches Bergbau-Museum.

Hauptmann, A. 2003. Rationales of liquefaction and metal separation in earliest copper smelting: basics for reconstructing chalcolithic and Early Bronze Age smelting processes. In *Archaeometallurgy in Europe, Milan, 24–26 September 2003*, vol. 1. Milan: Associazione Italiana di Metallurgia, 459–68.

Hauptmann, A., Weisgerber, G. and Bachmann, H.G. 1988. Early copper metallurgy in Oman. In *The Beginning of the Use of Metals and Alloys. Papers from the Second International Conference on the Beginning of the Use of Metals and Alloys, Zhengzhou, China, 21–26 October 1986*, R. Maddin (ed.). Cambridge, MA: MIT Press, 34–51.

Hauptmann, A., Lutz, J., Pernicka, E. and Yalçin, Ü. 1993. Zur technologie der frühesten Kupferverhüttung im östlichen Mittelmeerrau. In *Between the Rivers and Over the Mountains – Archaeologica Anatolica et Mesopotamica Alba Palmieri Dedicata*. Rome: Universita La Sapienza, 541–72.

Hauptmann, A., Bachmann, H.G. and Maddin, R. 1996. Chalcolithic copper smelting: new evidence from excavations at Feinan/Jordan. In *Archaeometry'94*, S. Demirci, A.M. Özer and G.D. Summers (eds). Ankara: Tubitak, 3–10.

Hauptmann, A., Pernicka, E., Rehren, Th. and Yalçin, Ü. (eds) 1999. *The Beginnings of Metallurgy*. *Der Anschnitt* Beiheft 4. Bochum: Deutsches Bergbau-Museum.

Hauptmann, A., Rehren, Th. and Schmitt-Strecker, S. 2003. Early Bronze Age copper metallurgy at Shahr-i-Sokhta (Iran), reconsidered. In Stöllner *et al.* 2003, 197–213.

Heskel, D. 1982. *The Development of Pyrotechnology in Iran during the Fourth and Third Millennium B.C.* Dissertation, Harvard University.

Hess, K. 1998. *Zur frühen Metallurgie am oberen Euphrat: Untersuchungen an archäometallurgischen Funden vom Arslantepe aus dem 4. und 3. Jahrtausend v. Chr.* PhD dissertation, Universität Frankfurt/Main.

Hook, D.R., Freestone, I.C., Meeks, N.D., Craddock, P.T. and Moreno, A. 1991. The early production of copper-alloys in south-east Spain. In *Archaeometry '90*, E. Pernicka and G.A. Wagner (eds). Basel: Birkhäuser Verlag, 65–76.

Huijsmans, M., Krauss, R. and Stibich, R. 2004. Prähistorischer Fahlerzbergbau in der Grauwackenzone. Neolitische und Bronzezeitliche Besiedlungsgeschicjte und Kupfermetallurgie im Raum Brixlegg (Nordtirol). In Weisgerber and Goldenberg 2004, 53–62.

Jovanovic, B. and Ottaway, B.S. 1976. Copper metallurgy and mining in the Vinca group. *Antiquity* 50: 104–13.

Keesmann, I. and Onorato, A.M. 1999. Naturwissenschaftliche Untersuchungen zue frühen Technologie von Kupfer und Kupfer-Arsen-Bronze. In Hauptmann *et al.* 1999, 317–32.

Lupu, A. 1970. Metallurgical aspects of chalcolithic copper working at Timna (Israel). *Bulletin of the Historical Metallurgy Group* 4: 21–3.

Lutz, J. 1990. *Geochemische und mineralogische Aspekte der frühen Kupferverhüttung in Murgul/Nordost-Türkei*. PhD dissertation, University of Heidelberg.

Lutz, J., Wagner, G. and Pernicka, E. 1994. Chalkolitische Kupferverhüttung in Murgul, Ostanatolien. In *Handwerk und Technologie im Alten Orient*, R.B. Wartke (ed.). Mainz: Zabern, 60–66.

Martinek, K.P. and Sydow, W. 2004. Frühbronzezeitliche Kupfermetallurgie im Unterinntal (Nordtirol). In Weisgerber and Goldenberg 2004, 199–211.

Merkel, J. and Rothenberg, B. 1999. The earliest steps to copper metallurgy in the western Arabah. In Hauptmann *et al.* 1999, 149–65.

Merkel, J.F. and Shimada, I. 1988. Arsenical copper smelting at Batan Grande, Peru. *Institute of Archaeometallurgical Studies Newsletter* 12: 4–7.

Mette, B. 2003. Beitrag zur spätbronzezeitlichen Kupfermetallurgie im Trentino (Südalpen) im Vergleich mit anderen prehistorischen Kupferschlacken aus dem Alpenraum. *Metalla* 10(1/2): 1–122.

Montero Ruiz, I. 1993. Bronze Age metallurgy in southeast Spain. *Antiquity* 67: 46–57.

Muan, A. and Osborn, E.F. 1965. *Phase Equilibria among Oxides in Steelmaking*. Reading, MA: Addison-Wesley.

Müller, R. 2002. *Chalcolithic Metallurgy in South-East Spain: A Study of Archaeometallurgical Remains from Almizaraque*. MSc dissertation, Institute of Archaeology, University College London.

Müller, R., Rehren, T. and Rovira Llorens, S. 2004a. Almizaraque and the early copper metallurgy of southeast Spain: new data. *Madrider Mitteilungen* 45: 33–56.

Müller, R., Rovira Llorens, S. and Rehren, Th. 2004b. The question of early copper production at Almizaraque southeast Spain. In *Proceedings of the 34th International Symposium on Archaeometry, 3–7 May 2004, Zaragoza, Spain*, J. Pérez-Arantegui (ed). Zaragoza: Institución 'Fernando el Católico', 209–16.

Nocete, F. 2006. The first specialised copper industry in the Iberian Peninsula: Cabezo Jur 0233 (2900–2200 BC). *Antiquity* 80(309): 646–57.

Nocete, F., Escalera, P., Linares, J.A., Lizcano, R., Orihuela, A., Otero, R., Romero, J.C. and Saez, R. 1998. *Estudio del material arqueologico del yacimiento de Cabezo Juré (Alosno, Huelva). Primera Campana* 1994, 67–78.

Nocete, F., Alex, E., Nieto, J.-M., Saez, R. and Bayona, M.R. 2005. An archaeological approach to regional environmental pollution in the south-western Iberian Peninsula related to third millennium BC mining and metallurgy. *Journal of Archaeological Science* 32: 1566–76.

O'Brien, W. 2004. *Ross Island. Mining, Metal and Society in Early Ireland*. Galway: National University of Ireland.

Odle, R.R., Morris, A.E. and McClincy, R.J. 1983. Investigation of direct smelting of copper concentrates. In *Advances in Sulfide Smelting*, vol. 1, R.A. Warrendale (ed.). Minerals, Metals and Material Society of the American Institute of Mining, Metallurgy and Petroleum Engineers (AIME), 57–72.

Palmieri, A.M., Frangipane, M., Hauptmann, A. and Hess, K. 1999. Early metallurgy at Arslantepe during the Late Chalcolithic and the Early Bronze Age IA-IB periods. In Hauptmann *et al.* 1999, 141–8.

Perini, R. 1992. Evidence of metallurgical activity in Trentino from chalcolithic times to the end of the Bronze Age. In *Archeometallurga Ricerche e Prospettive*, E. Antonacci Sanpaolo (ed.). Bologna: Assoziazione Italia di Metallurgia, 53–80.

Ploquin, A., Happ, J., Barge, H. and Bourhis, J.-R. 1997. Scories archéologiques et reconstitution expérimentale de réduction de sulfure de cuivre (minerai de St Véran, Hautes Alpes): prémices d'une approche pétrographique. *Archéologie en Languedoc* 21: 111–20.

Rehder, J.E. 1994. Blowpipes versus bellows in ancient metallurgy. *Journal of Field Archaeology* 21(3): 345–50.

Rehder, J.E. 1999. High temperature technology in antiquity: a sourcebook on the design and operation of ancient furnaces. In Hauptmann *et al.* 1999, 305–15.

Roden, C. 1988. Blasrohrdüsen. Ein archĠologischer Exkurs zur Pyrotechnologie des Chalkolithikmus und der Bronzezeit. *Der Anschnitt* 40(3): 62–82.

Rostan, P., Rossi, M. and Gattiglia, A. 2002. Approche économique et industrielle du complexe minier et métallurgique de Saint-Véran (Hautes-Alpes) dans le contexte de l'Age du Bronze des Alpes du Sud. In *Les Alpes dans L'Antiquité, Tendes, France*, Société Valdôtaine de Préhistoire et d'Archéologie, 77–96.

Rostan, P., Bourgarit, D., Burger, E., Carozza, L. and Artioli, G. 2006. The beginning of copper mass production in the southern part of western Alps: the St Véran mining area reconsidered (Hautes Alpes, France). In *Proceedings of the International Conference Beginning of the Use of Metals and Alloys (BUMA VI)*, Beijing, 16–19 September 2006.

Rostoker, W., Pigott, V.C. and Dvorak, J.R. 1989. Direct reduction to copper metal by oxide-sulfide mineral interaction. *Archeomaterials* 3: 69–87.

Rothenberg, B. (ed.) 1990. *The Ancient Metallurgy of Copper*. London: Institute of Archaeology.

Rovira Llorens, S. 2002. Early slags and smelting by-products of copper metallurgy in Spain. In *Die Anfänge der Metallurgie in der Alten Welt / The Beginnings of Metallurgy in the Old World*, M. Bartelheim, E. Pernicka and R. Krause (eds). Rahden, Westfalen: Marie Leidorf Gmbh, 83–98.

Rovira Llorens, S. and Ambert, P. 2002. Les céramiques à réduire le minerai de cuivre: une technique métallurgique utilisée en

Ibérie, son extension en France méridionale. *Bulletin de la Société Préhistorique Française* 99(1): 105–26.

Ryndina, N., Indenbaum, G. and Kolosova, V. 1999. Copper production from polymetallic sulphide ores in the north-eastern Balkan Eneolithic culture. *Journal of Archaeological Science* 26: 1059–68.

Saez, R., Nocete, F., Nieto, J.M., Capitan, M.A. and Rovira Llorens, S. 2003. The extractive metallurgy of copper from Cabezo Juré, Huelva, Spain: chemical and mineralogical study of slags dated to the third millennium B.C. *Canadian Mineralogist* 41: 627–38.

Schreiner, M., Heimann, R.B. and Pernicka, E. 2003. Mineralogical and geochemical investigations into prehistoric smelting slags from Tepe Sialk/Central Iran. In *Archaeometallurgy in Europe, Milan, 24–26 September 2003*, vol. 1. Milan: Associazione Italiana di Metallurgia, 487–96.

Schubart, H. 1988–90. Almizaraque und Zambujal als kupferzeitliche Hafenplätze. *Maburger Beiträge zur Archäologie* 15:17–35.

Shalev, S. and Northover, P. 1987. Chalcolithic metal and metal working from Shiqmin. In *Shiqmim I*, T.E. Levy (ed.). London: British Museum, 357–71.

Shugar, A.N. 2000. *Archaeometallurgical Investigation of the Chalcolithic Site of Abu Matar, Israel*. PhD dissertation, Institute of Archaeology, University College London.

Shugar, A.N. 2003. Reconstructing the chalcolithic metallurgical process at Abu Matar. In *Archaeometallurgy in Europe, Milan, 24–26 Sept 2003*, vol. 1. Milan: Associazione Italiana di Metallurgia, 449–58.

Stöllner, T., Körlin, G., Steffens, G. and Cierny, J. (eds) 2003. *Man and Mining. Der Anschnitt* Beiheft 16. Bochum: Deutsches Bergbau-Museum.

Strahm, C. 1996. Le concept bronze ancient. In *Cultures et sociétés du bronze ancien en Europe*, C. Mordant and O. Gaiffe (eds). Paris: Comité des Travaux Historiques et Scientifiques, 667–74.

Veldhuijzen, H. A. 2005. Technical ceramics in early iron smelting: the role of ceramics in the early first millennium BC iron production at Tell Hammeh (az-Zarqa), Jordan. In *Understanding People through their Pottery: Proceedings of the 7th European Meeting on Ancient Ceramics (Emac '03)*, I. Prudêncio, I. Dias and J.C. Waerenborgh (eds). Lisbon: Instituto Português de Arqueologia (IPA), 295–302.

Weisgerber, G. 2004. Schmelzanlagen früher Kupfergewinnung – ein Blick über den Alpen. In Weisgerber and Goldenberg 2004, 15–36.

Weisgerber, G. and Goldenberg, G. 2004. *Alpenkupfer – Rame delle Alpi*. Bochum: Deutsches Bergbau-Museum.

Wirth, E. 2003. *Etude expérimentale de la réduction de minerais de cuivre*. Engineer dissertation, Ecole Nationale Supérieure des Ingénieurs en Arts Chimiques et Technologiques (ENSIACET).

Yalçin, Ü. 2004. Metallurgie in Anatolien. In Stöllner *et al.* 2003, 527–37.

Zwicker, U., Greiner, H., Hofman, K.H. and Reithinger, M. 1985. Smelting, refining and alloying of copper and copper alloys in crucible furnaces during prehistoric up to Roman times. In *Furnaces and Smelting Technology in Antiquity*, P.T. Craddock and M.J. Hughes (eds). British Museum Occasional Paper 48. London: British Museum Press, 103–16.

## Author's address

David Bourgarit, Centre de Recherche et de Restauration des Musées de France (C2RMF), Palais du Louvre, 14 quai François Mitterrand, 75001 Paris, France (david.bourgarit@culture.gouv.fr)

# Zambujal and the beginnings of metallurgy in southern Portugal

*Roland Müller, Gert Goldenberg, Martin Bartelheim, Michael Kunst and Ernst Pernicka*

*ABSTRACT*   This paper presents the first results of a larger research project investigating the innovation of copper metallurgy at the Chalcolithic fortified settlement of Zambujal, the neighbouring settlements of Penedo and Fórnea, and in southern Portugal in general. Copper casting and working took place at all three sites. Copper smelting is more difficult to identify; the first traces of this metallurgical process may be found at Zambujal House V. The correlation between certain artefact types and the arsenic content indicates a conscious selection of the metal produced. Lead isotope analyses point towards the so-called Ossa Morena Zone as the most likely source of the copper where surveys have revealed a series of mines which were most probably exploited in prehistory. Hence, at this stage of research it seems that the copper from Zambujal was imported over a distance of at least 100 km, either as metal or as ore mineral.

*Keywords:* Copper Age, innovation, metallurgy, slag, crucible, melting, smelting, lead isotope analyses, mining.

## Introduction

In southern Portugal, the first objects of gold and copper appeared at the end of the 4th and the beginning of the 3rd millennium BC, a time of cultural change, when regional elites emerged whose rise to power is testified by fortified settlement centres with sophisticated wall and ditch systems and gigantic dolmen and *tholoi* tombs marking the landscape. Thus far, however, the circle of Chalcolithic copper ore procurement, metal extraction and processing as well as metal use and discard has not been clearly defined, limiting our perception of the sociocultural impact of the metal innovation in this cultural context (cf. Estacio da Veiga 1889; Soares *et al.* 1994).

In this paper we present a first resumé of a larger research project that investigates the archaeometallurgical remains of the central fortified settlement of Zambujal, including the neighbouring sites of Fórnea and Penedo, and relate them to potential prehistoric ore sources and mining sites in southern Portugal (Fig. 1). The research is based on field surveys (sampling of southern Portuguese ore deposits and tracing of prehistoric mining sites) and results of a new series of mineralogical, chemical and lead isotope analyses of metal objects, crucibles, slags and ores using energy-dispersive X-ray fluorescence analyses (EDX–XRF), neutron activation analyses (NAA), optical microscopy, scanning electron microscopy (SEM) and multi-collector inductively coupled plasma mass spectrometry (MC–ICPMS).

## The fortified settlement of Zambujal and the problem of early metallurgy

Even today, the majestic stone walls of Zambujal, standing up to 4 m high, dominate the valley of the Ribeira de Pedrulhos, a tributary to the Rio Sizandro near the Atlantic, about 50 km north of Lisbon (Figs 1 and 2a,b). The Chalcolithic settlement with its complex wall system presumably used to be a harbour site, located immediately adjacent to a bay which today is silted up with 18 m of sediment (Kunst and Trindade 1990). The settlement has a history of five construction phases, dating between the first half of the 3rd to the dawn of the 2nd millennium BC. Although the site was substantially modified during the occupation, its function as a fortification with four lines of defence seems to have been maintained throughout its existence (Kunst 2003 and in press; Sangmeister and Schubart 1981). Due to the good preservation of the wall system, it was possible to reconstruct certain concepts of defence, such as the embrasures of the central enclosure of the innermost defence wall (the so-called Zwinger) covering entrances of the second wall, both dating to Zambujal phase two (Fig 1); these concepts of defence were modified systematically in the course of the consecutive construction phases (Sangmeister and Schubart 1972, 1981). On the other hand, the extent of the site is still not known: excavations in the 1990s for instance revealed a possible occupation at the bottom of the fortified brow (Kunst and Uerpmann 1996).

The correlation of the construction phases with the regional ceramic sequence is still heavily disputed, especially when compared to the sequence of the other Estremadurian-type site Leceia (Cardoso and Guerra 1997–98). Yet, it is certainly possible to distinguish an early from a late Copper

**Figure 1** Aerial view of Zambujal from the northeast. Northeast of tower A, a working and living area, revealing fireplaces and some metallurgical debris, is currently being excavated. House V, which revealed the clay-ring structure with most of the metallurgical finds (Zambujal construction phase three), is located in front of and partly below tower B (built in construction phase four), hence between lines I and II of defence. The phase two defence concept of the Zwinger embrasures covering entrances of line two had been abandoned by then. Excavations in 2004 exposed a well-preserved prolongation of the innermost wall below the 17th-century AD farmstead.

Age, represented by the so-called cylindrical vessels (*copos canelados*) and the Bell Beaker horizon, respectively (Kunst 1995a; Lillios 1997; Uerpmann and Uerpmann 2003). In fact, the Estremadura is one of the regions with the oldest Bell Beaker dates in well-stratified settlement and burial contexts (Fig. 2a,b) (Müller and van Willigen 2001).

The size of Zambujal, its complex wall system, the accumulation of fine pottery and rare items of long-distance trade (a cowrie shell and ivory objects) on the one hand, and a dominant presence of copper objects and working remains as well as remains of amphibolite tool production on the other, seem to indicate a function as a central place, which probably controlled the estuary region of the Rio Sizandro (Kunst 1995b). Three smaller, though presumably also fortified settlements, were found a few kilometres further up the Rio Sizandro: Castro da Boiaca, Castro da Fórnea and Castro do Penedo (Gonçalves 1985; Kunst and Trindade 1990; Spindler 1969; Spindler and Gallay 1973). Excavations at the latter two sites revealed remains of copper metallurgy (Kunst and Trindade 1990). It becomes obvious that a better understanding of the innovation of copper metallurgy would add fundamentally to the general understanding of Chalcolithic Zambujal, the estuary of the Rio Sizandro and Copper Age dynamics in southern Portugal.

## The remains of early copper production

Since its discovery by Leonel Trindade in 1932, intermittent excavation at Zambujal has produced 900 copper objects weighing 3–4 kg. The most numerous items are metal droplets and unidentifiable, small, broken or semi-molten tool fragments, while only about 80 complete copper artefacts exist (the exact number depending on how one defines an entirely finished or still functional artefact). The great majority of these 80 objects are awls and small chisels, but there are also four copper axes, four saw blades, five arrow heads, three knives, a globular-headed pin and a dagger (Fig. 3) (Sangmeister 1995). One of the smaller axes – it may also have been used as a kind of chisel – has a bone handle. Copper droplets and working remains are scattered all over the site and are often associated with fireplace structures. The so-called 'clay-ring' (*Lehmring*) structure with associated fireplaces stands out in this context, since it revealed by far the biggest concentration of metal droplets (Figs 1 and 4) (Sangmeister 1995; Sangmeister and Schubart 1981). At the nearby settlements of Fórnea and Penedo, the few excavations revealed the very same kind of copper objects. While only few hundred grams of copper were found at Fórnea, almost 2 kg was recovered at Penedo.

The 364 optical emission spectroscopy analyses of copper artefacts from Zambujal conducted as part of the trans-European SAM programme (Studien zu den Anfängen der Metallurgie: Junghans *et al.* 1960, 1968, 1974) show that most of the artefacts are composed of arsenical copper containing very low concentrations of trace elements (Sangmeister 1995).

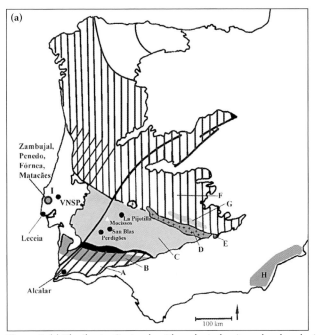

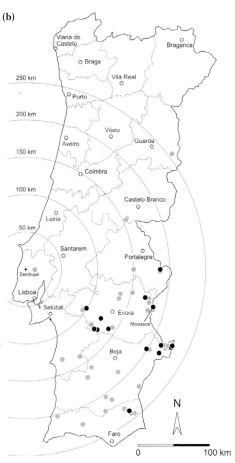

**Figure 2** (a) The Iberian Peninsula with geological units and archaeological sites mentioned in the text. A: Southportuguese Zone; B: Southwest Iberian Pyrite Belt which is part of the SPZ; C: Ossa Morena Zone; D: Los Pedroches Batholith; E: Linares – La Carolina district; F: Central Iberian Zone; G: Alcudia valley (D, E and G are summarised as 'south-central Spain'); H: ore deposits of southeast Spain (Betic Cordilleras); I: Postpaleozoic sediments of the Estremadura. The grey areas represent geological units or regions that contain copper occurrences of which lead isotope data exist (the data are discussed in the main text). There are small copper occurrences in the southwestern part of the CIZ which have not been studied thus far (*hatched zone*). ● – some of the largest Chalcolithic fortifications of southwest Iberia. △ – the mine of Mocissos which probably was already exploited in the Chalcolithic and Bronze Age (base map after Quesada and Munhá 1990). (b) Field surveys in the districts of Faro, Beja, Setúbal, Évora, Portalegre, Castelo Branco and Guarda covered the most important copper mining zones in Portugal. The ore samples (about 90) taken from the mines (●) can be considered as quite representative, at least for the southern part of the country; they will be analysed in the near future. Mining tools of stone were found at 14 mines (●) and indicate prehistoric phases of exploitation.

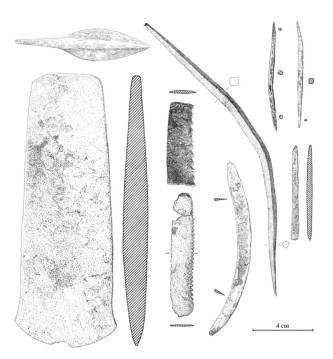

**Figure 3** Copper artefacts from Zambujal: awl, elongated awls, saw blades, knife blade, Palmela point and axe (after Sangmeister 1995).

The overall distribution of the arsenic content of the copper artefacts is very similar to the one represented by the 916 SAM analyses of Chalcolithic and Early to Middle Bronze Age copper artefacts from entire southern Portugal (region 01a in SAM 2: Junghans *et al.* 1968; or region 15 in Krause 2003), both approximating well to a lognormal distribution.[1] According to the SAM analyses, most of the artefacts analysed contain between 0.5 and 2% arsenic (Junghans *et al.* 1960, 1968, 1974) (Fig. 5). The lognormal distribution resembles the natural distribution of minor and trace elements in minerals, which indicates that the arsenic derived from smelted ores and could not be controlled directly (cf. Pernicka 1990: 90ff; Rovira Llorens 2005). Nevertheless, it is possible to show a correlation between arsenic content and certain artefact types (Fig. 5).[2] Axes and regular awls peak at low arsenical copper, while sheet metal fragments, saws, Palmela points and tanged daggers are much more frequently made of higher arsenical copper. A key break in the distribution seems to emerge at the border of 1% arsenic, a value often associated with a conscious production and use of arsenical copper (Lechtman 1996; Ottaway 1982: 131; cf. Soares 2005). The small group of long awls is nearly exclusively made of copper containing more than 2% arsenic. A long awl from Leceia analysed using neutron activation analyses fits the same criteria (Müller and Cardoso in press). It seems that a conscious selection of certain types of copper

17

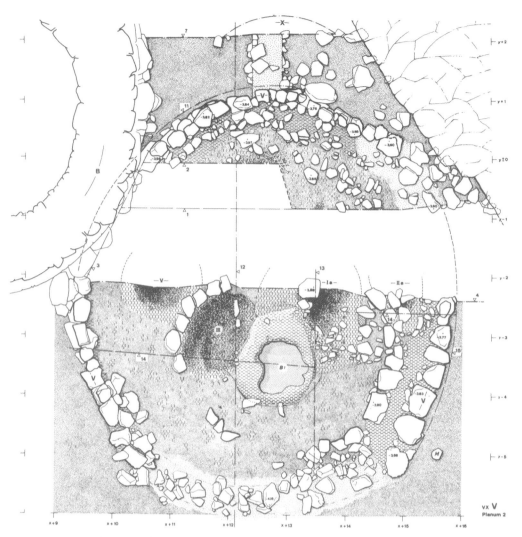

**Figure 4** House V with clay-ring (*Lehmring*) and associated fireplaces (dark areas). The clay-ring was filled with successive layers of ashy earth and sand. Hundreds of copper prills were scattered around the structure. A few pieces of slags and crucible fragments were found inside the clay-ring and in the associated fireplaces. Ores were not found. The clay-ring itself showed no signs of high temperature alteration: it is unlikely to have been part of a furnace. (Drawing from Sangmeister and Schubart 1981: pl. 98/99.)

took place after the metal had been smelted. Indeed, arsenical copper is harder than pure copper, but can nevertheless be shaped well by cold working. According to Lechtmann (1996: 501) all alloys from 0.5 to 7 wt% of arsenic maintain an almost constant level of ductility as the solid solubility limit is approached. They are also tougher at extreme deformations than pure copper: thin metal sheets and elongated objects would not brittle fracture as easily. Perhaps a trial-and-error production was practised in that copper which had the *right* colour and which could be easily deformed was reserved for certain artefacts (cf. Lechtmann 1996).

The question of correlations of certain metal types with certain artefact types of early Iberian metalwork has been discussed in previous papers (e.g. Harrison and Craddock 1981; Hook *et al.* 1991; Montero 1993). Indeed, the issue has often been one of the focal points in evaluating the complexity of early metallurgy and its impact on society. Lately, Rovira Llorens (2005) discussed the distribution of the overall arsenic content of a large number of Spanish Chalcolithic and Early Bronze Age copper artefacts analysed under the remit of

the Proyecto de Arqueometalurgia.[3] He explained the higher arsenic contents of daggers and halberds as the outcome of the use of recycled metal being confined to certain artefact types (less frequent re-melting of the copper would have limited the losses of the volatile element arsenic).

The iron content of the copper artefacts is another important technological parameter, particularly in regard to the processes of slag formation, which will be discussed below (Craddock and Meeks 1987). All the 916 SAM analyses show that the Chalcolithic and Early to Middle Bronze Age copper and arsenical copper artefacts contain very little iron, less than 0.01% for 98% of the cases and up to 0.2% for the remaining artefacts. The chemical analyses include the examination of 236 metal droplets and working debris. The low iron content of the early copper was also found by our own XRF and neutron activation analyses of Zambujal artefacts (unpublished data) as well as the neutron activation analyses published by Cardoso and Guerra (1997–98) of artefacts from Leceia.[4]

Slag fragments were identified in the Zambujal assemblage, but not from Penedo and Fórnea.[5] In an earlier small-scale

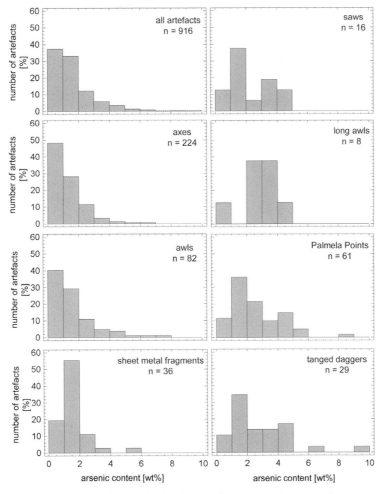

**Figure 5** Histograms showing the distribution of arsenic in various types of copper artefacts from Chalcolithic, Early and Middle Bronze Age of southern Portugal (region 01a according to SAM 2: Junghans *et al.* 1968; or region 15 according to Krause 2003). The data set was first studied by Sangmeister (1995). In this paper the artefacts made of tin bronze (around 100) were excluded.

study, Sperl (1981) interpreted them as remnants of a copper casting or refining metallurgy, mainly because major slag heaps, indicative of smelting sites of later time periods, are missing within or next to the settlement. Indeed, although many more pieces could be identified within the assemblage in the course of this analytical programme, all in all they do not account for more than about 30 g. It is however remarkable, not only that all the slags were from one house structure, but also that many of them were even associated with one fireplace complex, namely the clay-ring structure of House V mentioned earlier (Figs 1 and 4).

The slags are all black, have a glassy surface, are porous, occasionally show green spots of copper corrosion and are magnetic. They are all quite small (similar to the size of a small nut), except for sample FM-Z192, which is significantly bigger (Fig. 6). In some cases, small copper droplets sit on top of the slag pieces; in others spherical relic holes indicate their former presence. Chemical analyses of two and mineralogical examinations of four slag samples show that they are all very similar to each other: they contain iron oxides (mainly magnetite), relic quartz, copper prills and glass. Occasionally delafossite occurs in marginal zones and close to holes. Segregated arsenic-rich phases containing up to 20 wt% arsenic often

occur in the copper inclusions (Fig. 7). Hence, the crucible atmosphere seems to have been moderately oxidising, but not enough to prevent the formation of arsenical copper (cf. Hauptmann *et al.* 2003; Lechtman and Klein 1999; Müller *et al.* 2004). The quartz relics found in the slags are polycrystalline and repeatedly show sharp edges, so a sedimentary origin can be excluded. It appears that they do not represent remnants of sand added during a copper-refining process in order to support a slag formation (Fig. 8). Magnetite occurs as idiomorphic spinel or in agglomerates of irregularly shaped crystals; the latter are typical of early copper-smelting slags that developed from iron hydroxide-rich gangue materials (Hauptmann *et al.* 2003).

The chemical analyses of the two slag pieces FM-Z192 and FM-Z62, using SEM area scans underline the dominance of the iron oxides, followed by significant amounts of silica and copper and arsenic oxides (summary of area scans: 3–5% $Al_2O_3$; 35–50% $SiO_2$; 1–2% $K_2O$; 1–2% $CaO$; 30–40% $FeO$; 8–13% $CuO$; 3–10% $As_2O_3$). It is difficult to imagine that these iron oxide-rich slags developed during a refining process of the pure or very clean arsenical copper described above, especially considering the low iron content of the copper artefacts. Another argument against interpreting these as refining

19

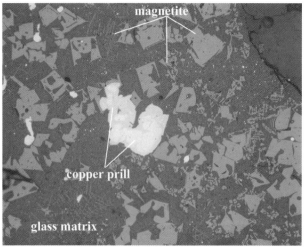

**Figure 6** Slag finds from Zambujal.

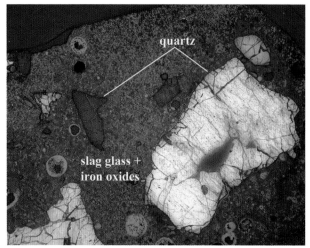

**Figure 7** Slag sample FM-Z192: photo under plane polarised light (width = 370 μm).

**Figure 8** Slag sample FM-Z151: photo under crossed polarised light (width = 4400 μm).

slags is the high silica to alumina ratio: assuming that no silica flux would have been added for refining, any silica present must have been derived from the crucible and should thus be accompanied by appropriate amounts of alumina. This is not the case. Gangue material of smelted ores seems to have formed the slags for the most part, next to minor contributions of fuel ash and the clay lining of the reactor vessel.

There are two major types of technical ceramic associated with metallurgy at Zambujal. The first is thin-walled and has heavily vitrified rims showing traces of copper (Fig. 9a). Based on the few fragments of this type of material found at Zambujal, Sangmeister (1995) reconstructed a plate-like crucible with a diameter of up to 30 cm. Very similar vitrified ceramic fragments found at other sites such as Perdigões (Lago *et al.* 1998), Três Moinhos, Castelo Velho de Safara, Porto Mourão and São Brás (Soares *et al.* 1994), however, show that these crucibles had various, often subrectangular shapes. The second type comprises thick-walled vessels often with much smaller capacities and varying shapes (some of them even have small feet) (Fig. 9b). Although copper droplets can clearly be seen in several spots, this type of ceramic bears only minor signs of secondary firing. In Chalcolithic settlements of southern Spain, it was demonstrated that heavily vitrified, plate-like vessels were used primarily for smelting, while the thick-walled ones presumably were used for re-melting the metal (Müller *et al.* 2006; Rovira Llorens 2005). In the case of the Zambujal assemblage, however, it is difficult to directly associate the heavily vitrified crucibles with smelting practices since there are no clear slag remains adhering to them. Even in thin section, only intensely altered ceramic with adhering traces of secondary copper phases are seen. On the other hand, it is again remarkable that the few fragments of this type of crucible that do exist in the Zambujal assemblage (six pieces) also all derive from the House V structure.[6] The thick-walled vessels show great variety, from extremely small spoon-like types to larger ones that could easily have been used as moulds for casting a metal blank from which, for example, a flat axe could have been forged and ground. These possible re-melting crucibles or moulds are much more frequent (60 pieces) and were found in several places on the site, often associated with fireplaces and copper droplets. One piece of a thick-walled crucible fragment was also found at the site of Penedo.

No ore minerals were identified within the Zambujal assemblage during this study but two pieces of calcareous sandstone bearing finely dispersed traces of malachite were found in the nearby settlement of Fórnea. Its mineralogy fits well with the local Estremadurian geology, which is dominated by carbonate-rich Jurassic and Cretaceous sedimentary rocks (limestone, calcareous sandstone). Indeed there is a very minor occurrence of secondary copper minerals at Matacães, immediately adjacent to the site of Fórnea (Fig. 2). Both the samples found at Fórnea, however, as well as the potential ores from Matacães are extremely low grade, hence quite unlikely to have been smelted in prehistory. Surveys around Matacães revealed neither primary copper ores nor any traces of mining (Goldenberg and Maaß forthcoming). With reference to geological maps, the archaeological literature mentions other Estremadurian copper occurrences, e.g. at Asfamil (Rio de Mouro) or close to Óbidos (Cardoso and Guerra 1997–98; Sousa *et al.* 2004) but no rich copper ores have been documented at any of them – in fact, careful prospection by Goldenberg and Maaß (forthcoming) in the area around Óbidos did not reveal any traces of a copper occurrence at all. At the fortified settlement of Vila Nova de São Pedro (VNSP; Azambuja), about 40 km west of Zambujal, 13.5 kg copper ore was found according to the excavators (Jalhay and Paço 1945); however, only some pieces of limonite showing traces of malachite could be identified at the Museu do Carmo (Lisbon), which hosts most of the VNSP finds.[7] Although again low grade, this type of ore would correspond much better to the slags discussed above in respect of its slag-forming minerals.

One can conclude that at Zambujal the innovation of copper seems to have had a decisive impact on Chalcolithic society. From the very beginning of the occupation, copper objects as well as remains of copper working and melting are scattered all over the site. A very similar picture emerges for the smaller neighbouring sites of Penedo and Fórnea, although we lack the clear stratigraphic contexts. New types of artefacts were created making use of the new material's specific properties. High arsenical copper seems to have been reserved for elongated awls, a range of sheet metal artefacts such as thin saws and for Palmela points and tanged daggers. On the basis of metallographic analyses of 64 copper objects from Zambujal, Wang and Ottaway (forthcoming) could demonstrate that the awls were left in an annealed soft state in most cases, while most of the chisels and the only saw examined were cast, cold-worked, annealed and finally cold-worked again. Very similar metallographic observations were made on equivalent artefacts from Chalcolithic and Early Bronze Age Spain (Hook *et al.* 1991; Rovira Llorens and Gómez Ramos 2003). It appears that the new material entered several spheres of sociocultural life: the regular small awls may have been used for leather working; the thin saws and knives for elaborate fine work of organic material, such as wood, bone or ivory; and the long awls, daggers and Palmela points as symbols of status and communication. The massive copper axes appear heavy and clumsy, especially when compared to the sophisti-

cated amphibolite axes (cf. Lillios 1997). The metallographic analysis of one axe from Zambujal reveals that it was left in an annealed, soft state, in agreement with results of metallographic analyses of axe blades from VNSP and Porto Mourão (Soares *et al.* 1996; Wang and Ottaway forthcoming). Hence, from their properties, the axes would have been much more suitable for use as ingots or objects of display, rather than tools (Cardoso and Guerra 1997–98; Soares *et al.* 1996).

So far there is not a single Chalcolithic site in southern Portugal, where the entire *chaîne opératoire* of primary copper production has been studied and could be reconstructed in detail.[8] Where and how was the copper smelted and where did the raw materials come from? Was there a need for craft specialisation or how much would a Chalcolithic society have invested in the procurement of the innovative material? Although we lack clear evidence for ores at Zambujal, the presence of slags in the assemblage does not seem to be explicable by a refining metallurgy, as argued above. Their clustered occurrence in House V, which at the same time produced most of the copper droplets, and which is the only structure revealing a different, heavily vitrified type of crucible, suggests that they are remnants of a different kind of metallurgical process. Since we lack a clear association to the potential ores used, we can only speculate about this other kind of metallurgy at this stage of research. It could have been primary smelting of arsenic-bearing secondary copper ores, co-smelting of secondary copper ores with arsenic-bearing sulphides or oxides, as well as an alloying process of pure copper with arsenic-rich minerals (cf. Craddock and Meeks 1987; Harrison and Craddock 1981; Hauptmann *et al.* 2003; Lechtmann and Klein 1999; Müller *et al.* 2004). The small amount of slag, the purity of the arsenical copper produced and the absence of sulphur, either as trace in the metal objects or as part of matte in the slag, are arguments for the use of rich secondary ore minerals. Direct smelting of arsenical copper from arsenic-bearing secondary copper ores could be demonstrated in several Chalcolithic settlement contexts of southern Spain, such as Almizaraque, El

(a)　(b)

**Figure 9**  (a) Thin-walled, heavily vitrified crucible from Zambujal; (b) thick-walled crucible from Zambujal.

Malagon and Los Millares (Müller *et al.* 2004; Rovira Llorens 2005). Nevertheless, one cannot exclude the possibility that the slags and the heavily vitrified crucible fragments entered the Zambujal strata only accidentally, as part of a larger mass of smelted copper.

## The provenance of the copper and first results of the survey for prehistoric mining sites in southern Portugal

Abundant and rich copper ore sources exist in many regions of the Iberian Peninsula. The metallogenetic map, Carta Mineira de Portugal (1 : 500.000) shows just over 80 copper ore deposits. Major occurrences closest to Zambujal are mineralisations of the so-called Ossa Morena Zone (OMZ), north Portuguese mineralisations of the Central Iberian Zone (CIZ), and mineralisations of the Southportuguese Zone (SPZ), which include the famous ore bodies of the Southwest Iberian Pyrite Belt (IPB) (Fig. 2a). Hence, in our field survey we focused on these zones and managed to sample ore deposits located in the districts of Faro, Beja, Setúbal, Évora, Portalegre, Castelo Branco and Guarda (Fig. 2b).[9] Samples from the very small occurrence at Matacães were also collected in order to test the possibility of a local prehistoric exploitation of this potential ore source (although doubts of a prehistoric exploitation of this mineralisation have already been expressed above). All in all, the survey covered about 85% of the area of potential copper occurrences according to the metallogenetic map of Portugal. Of the 65 copper mines visited, all abandoned today, 50 produced copper ore samples suitable for mineralogical and geochemical analyses. In addition, some samples from IPB deposits were given by the Departamento de Prospecção de Minérios Metálicos (INETI).[10] Depending on the situation in the field (most importantly the exposure of the mining dumps), the samples consist of minerals from the oxidised zone, the supergene enrichment zone or the primary zone of the corresponding ore body.

Figure 10 shows the isotope ratios of lead in the copper artefacts from Zambujal, Penedo and Fórnea in comparison to lead isotope ratios of copper artefacts from the fortified site of La Pijotilla and of ores of the south Iberian deposits. At this stage of research, we are bound to refer to already published ore data and our own analyses of ores from Matacães, since the main part of the ore samples gathered during the survey in southern Portugal still has to be analysed (Fig. 2b). Due to the lack of ore data and because only very little is known about 3rd-millennium copper production practices, the conclusions drawn here need to be considered only as working hypotheses (on principles of lead isotope analyses see: Gale and Stos-Gale 2000; Pernicka 1990).

The artefact data of all three Estremadurian sites scatter quite strongly, but overall show a fairly continuous distribution (Fig. 10). There is no significant correlation between the lead isotope signatures of the artefacts and their chronological contexts referring to the stratigraphy of Zambujal.[11]

The comparison of the artefact data with the isotope fields of southern Iberian ore deposits shows that individual artefact data points can often be allocated to various fields, i.e. various possible ore sources. Looking at the overall distribution of the artefact data, however, we see that some deposits and even entire ore regions are much less likely than others to have been the main resource for the Estremadurian settlements in the 3rd millennium BC (Fig. 10). The narrow and well-defined isotope fields of the ore regions of the Iberian Pyrite Belt, south-central Spain (districts of Los Pedroches, Alcudia, Los Linares and La Carolina) and southeast Spain cannot account for the main distribution of the more or less continuous artefact data cloud. Also the isotope field of the small deposit at Matacães coincides with few artefacts, although it must be noted that this field consists of only three analysed samples: more analyses are needed (Fig. 10). Two isotope fields representing small copper occurrences related to late Variscan hydrothermal mineralisations of the Southportuguese Zone coincide well with a large proportion of the copper artefacts. Since these isotope fields are defined by analyses of individual ore samples of different deposits, however, the validity and significance of these fields also need to be re-evaluated by future analyses. The lead isotope signatures of ore deposits that are located in the Ossa Morena Zone and that are related to metaluminous Variscan magmatism (Isotope Group 5 according to Tornos and Chiaradia 2004) show the best correlation with the signatures of the Estremadurian copper artefacts (4). The hypothesis that OMZ deposits played a major role in the supply for Chalcolithic copper is supported by the fact that copper artefacts from the contemporary fortified site of La Pijotilla, which is located in the Ossa Morena Zone (near Badajoz), show the same lead isotope pattern (cf. Hunt Ortiz 2003) (Fig. 10). Future investigations will have to underline this preliminary conclusion by further analyses of ores from the complex Ossa Morena Zone, but also by investigating the northern Portuguese occurrences and producing more data on the above-mentioned late Variscan hydrothermal occurrences of the SPZ.

A complete picture of the trace element characteristics of the different potential ore sources is still lacking. Yet, we can already say that the ores from the Matacães occurrence analysed thus far (using semi-quantitative XRF analyses) do not contain arsenic at all, which is another indication that this occurrence was not – if at all – a major source for the copper manufactured and used in Copper Age Estremadura.

During our prospections for copper-ore occurrences in the OMZ, we found rich copper ores that would have been ideal for the prehistoric copper smelters and compared well to the slags discussed above: copper carbonates (malachite, azurite), phosphates (pseudomalachite), oxides (cuprite, delafossite), simple sulphides (digenite, covellite) and sulphates (brochantite); the ore minerals are mostly associated with quartz and iron hydroxides (goethite). The fahlore, arsenopyrite and scorodite minerals that we found at some of the OMZ occurrences may have been the source of the varying arsenic content in the Chalcolithic copper artefacts. This hypothesis remains unsubstantiated, however, and requires more investigation (see smelting discussion above).

Furthermore, 14 mines were discovered where we believe prehistoric exploitation was carried out, most probably during the Copper and/or Bronze Age (Fig. 2a,b). So far our belief of their prehistoric date is based only on the presence of stone tools, e.g. hammer stones of quartzite with and without traces

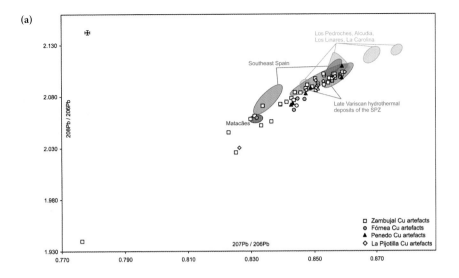

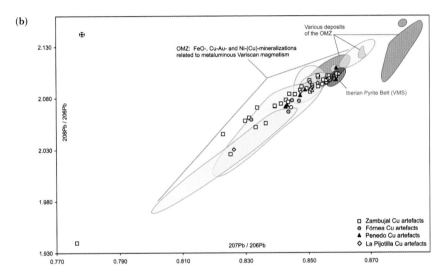

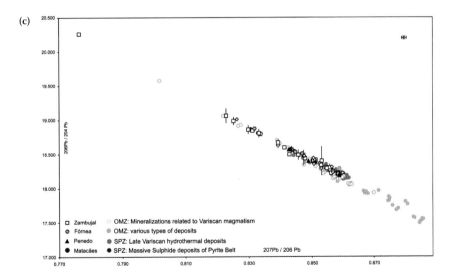

**Figure 10** Lead isotope data of copper artefacts from Zambujal, Penedo and Fórnea, as well as ores from Matacães in comparison to data of copper artefacts from La Pijotilla and of south Iberian ore deposits. The comparative data were taken from: Arribas and Tosdal 1994; Stos-Gale *et al.* 1995; Trincherini *et al.* 2001; Marcoux 1998; Hunt Ortiz 2003; Santos Zalduegui *et al.* 2004; Marcoux and Sáez 1994; Tornos and Chiaradia 2004. The isotope fields of the deposits found to be relevant in the discussion of the 207Pb/206Pb versus 208Pb/206Pb diagrams (a and b) are broken down to individual data points in the 206Pb/204Pb diagram (c) to allow the third dimension of the data set to be considered. The analytical error (2σ) rose above 0.1% regarding the 206Pb/204Pb ratios in a few cases (as the extended error bars indicate), due to the extremely low lead contents of some samples.

of shafting and anvil stones. We are planning to conduct small-scale excavations at some of the most promising sites to provide more solid evidence.

## Conclusions

From the very beginning of the occupation, copper casting and working occurred at several places within Zambujal, as well as at the neighbouring sites of Fórnea and Penedo. At House V of Zambujal, smelting slags and highly vitrified crucible fragments together with an accumulation of copper droplets indicate another type of metallurgical process. The lack of ores in this archaeological context does not allow reconstruction of the underlying metallurgical process at this stage of research. It seems that a deliberate choice of copper with varying arsenic contents was made after the metal had been smelted. Elongated awls, sheet metal objects, such as saws, as well as Palmela points and tanged daggers tend to have higher arsenic contents than small awls and axes. Lead isotope analyses indicate that at this stage of research the Ossa Morena Zone is much more likely to have been the geological source of the early copper than the Matacães occurrence near Zambujal, south Portuguese deposits (including Iberian Pyrite Belt ones) or the deposits in south-central and southeast Spain. This conclusion corresponds well with the results of the field surveys: in the Ossa Morena Zone particularly we encountered rich primary as well as secondary copper ores, occasionally associated with arsenic-bearing minerals. In this area we also discovered a number of mainly small-scale mines, which we believe comprise a prehistoric exploitation phase. Future excavations at some of the most promising sites should provide more information for dating. In any case, it seems that the geological source of the innovative material was located at least 100 km away from Zambujal.

It is not clear whether the copper ore or the copper metal was exchanged, since clear evidence for smelting has not been identified at any of the Chalcolithic sites of southern Portugal. Nevertheless, we do know that the metal was shaped to the final product at both larger and smaller settlements. A remarkable parallel to the procurement of copper is that of unworked amphibolite for stone tool production and use inside the Chalcolithic fortified settlements of the Estremadura. Based on geochemical analyses, Lillios (1997) showed that the Estremadurian amphibolite too probably derived from the Ossa Morena Zone. It is reasonable to suggest that in return for copper and amphibolite, alluvial gold of the lower Tejo and silex from Estremadurian occurrences were delivered to the interior (Cardoso 1999). On the other hand, the power of the Estremadurian fortified settlements may also have lain in the exchange of food or elaborated objects of organic materials, which may have been produced using the innovative copper tools.

## Acknowledgements

This research project ('Chalcolithic Metallurgy at Zambujal: from the ore deposit to the final product') is funded by Deutsche Forschungsgemeinschaft (German Research Council) and Studienstiftung des deutschen Volkes (German National Academic Foundation). We are very grateful to Isabel Luna, Rui Silva and Carlos Anuncianção of the Museu Municipal Leonel Trindade in Torres Vedras (Portugal) for their constant hospitality and kind support. The finds from Zambujal are stored and displayed at the museum. Roland Müller would like to thank Professor Richard Harrison for a profound and encouraging discussion of the first results of this project and for providing stimulating thoughts especially for interpreting the use of the early copper artefacts from Zambujal. Thanks also go to Jörg Adam and Bernd Höppner who gave technical advice for conducting lead isotope analyses. Gert Goldenberg would like to thank Fernando Real from the Instituto Português de Arqueologia, Lisbon, as well as João Matos from the Departamento de Prospecção de Minérios Metélicos, Instituto Nacional de Engenharia, Tecnologia e Inovação (INETI), Beja, for their kind support regarding the field survey. We thank Ben Roberts for his comments on the English text and the reviewers for their contribution.

## Notes

1. The analyses of the project Studien zu den Anfängen der Metallurgie (SAM; Junghans *et al.* 1960, 1968, 1974) still today form the most comprehensive set of quantitative chemical data that exists of prehistoric copper artefacts from southern Portugal. Since the culture-chronological transition from the Chalcolithic to the Early and Middle Bronze Age in southern Portugal is not well understood, the chemical data from artefacts dating to the latter two periods were included in the discussion. Indeed, the final construction phases of sites such as Zambujal and VNSP date to the Bronze Age. The exclusion of Late Bronze Age artefacts seemed less arbitrary as the upper chronological border of the data examination. Furthermore, the use of this statistical sample is supported by the fact that a major break in the technological development of metallurgy and the scale of metal production during the shift from the Chalcolithic to the Bronze Age does not seem to exist. Arsenical copper continued to be used alongside the first few tin-bronzes (Sangmeister 1995). Finally, since Sangmeister (1995) statistically evaluated early south Portuguese copper artefacts based on the very same data set, the chosen statistical sample has its relevance in terms of research history and comparability.
2. The high quality and reproducibility of the SAM data in general have been demonstrated in previous comparative studies (Ottaway 1982; Pernicka 1984, 1990). A large-scale, systematic evaluation of the analyses of the Iberian metalwork in particular is still lacking. However, because in this statistical examination we compare only SAM analyses with one another, we should avoid potential systematic errors.
3. The Proyecto de Arqueometalurgia is a long-term analytical programme directed by Prof. Germán Delibes des Castro (University of Valladolid). A large number of XRF analyses of prehistoric metal artefacts and archaeometallurgical remains were conducted and published (see e.g. S. Rovira Llorens, R. Montero Ruiz and S. Consuegra Rodriguez (1997) *Las primeras etapas metalúrgicas en la Peninsula Ibérica. I. Análisis de Materiales*, Imprenta Taravilla, Madrid).
4. Accounts on copper objects with high iron contents deriving from the Chalcolithic fortifications of Santa Justa and João Marques are a result of a misunderstanding: the chemical data derive from XRF analyses of unclean, corroded surfaces and hence do not reflect the actual composition of the artefact metal (cf. Gonçalves 1989; Hunt Ortiz 2003).

5. Slags are defined here as non-metallic metallurgical residues which consist mainly of silicates, glass and iron and copper oxides. They can develop, for example, during metal extraction from ores or during the purification of raw copper.

6. The two plate-like crucible fragments that were found in the lower-lying area GH of Zambujal are regarded as redeposited, originally coming from area V (Sangmeister 1995).

7. Monge Soares, pers. comm.

8. There are a few promising sites near the Guadiana basin where traces of extractive copper metallurgy seem to have been found, e.g. Perdigões (Reguengos de Monsaraz) and Fonte Ferrenha (Redondo) (Calado 2001; Lago *et al.* 1998). Future analyses will have to show how the materials relate to each other and what kind of metallurgical processes they represent.

9. Essential information for guiding the survey was gained from geological literature (e.g. Martins and Borralho 1998; Santos Oliveira *et al.* 2002) and through kind support from Dr João Matos of the Departamento de Prospecção de Minérios Metálicos, Instituto Nacional de Engenharia, Tecnologia e Inovação (INETI), Portugal.

10. See note 9 above.

11. Data analysis by Roland Müller, unpublished.

# References

Arribas, A. and Tosdal, R.M. 1994. Isotopic composition of Pb in ore deposits of the Betic Cordillera, Spain: origin and relationship to other European deposits. *Economic Geology* 89: 1074–93.

Calado, M. 2001. *Da Serra d'Ossa ao Guadiana: um estudo de pré-história regional*. Trabalhos de Préhistória 19. Lisbon: Instituto Português de Arqueologia.

Cardoso, J.L. 1999. Copper metallurgy and the importance of other raw materials in the context of Chalcolithic economic intensification in Portuguese Estremadura. *Journal of Iberian Archaeology* 1: 93–105.

Cardoso, J.L. and Guerra, M.F. 1997–98. Análises químicas não destrutivas do espólio metálico do povoado pré-histórico de Leceia, Oeiras e seu significado no quadro da intensificação económica calcolítica da Estremadura Portuguesa. *Estudos Arqueológicos de Oeiras* 7: 61–87.

Craddock, P.T. and Meeks, N.D. 1987. Iron in ancient copper. *Archaeometry* 29(2): 187–204.

Estacio da Veiga, S.P.M. 1889. *Antiguidades Monumentaes do Algarve. Tempos Prehistoricos*. Lisbon: Imprensa Nacional.

Gale, N.H. and Stos-Gale Z.A. 2000. Lead isotope analyses applied to provenance studies. In *Modern Analytical Methods in Art and Archaeology*, E. Ciliberto and G. Spoto (eds). Chemical Analyses Series, vol. 155. New York: John Wiley and Sons, 503–84.

Goldenberg, G. and Maaß, A. forthcoming. Prospektion nach kupferzeitlicher Rohstoffgewinnung (Kupfererz, Silex) in Estremadura im Jahr 2001. In *Kupferzeitliche Metallurgie in Zambujal, in Estremadura, Südportugal und Südwestspanien: Vom Fertigprodukt zur Lagerstätte. Arbeitstagung Alqueva-Staudamm, 27. bis 30. Oktober 2005*, M. Kunst (ed.), Madrider Beiträge. Mainz: Philipp von Zabern.

Gonçalves, J.L.M. 1985. O Castro da Fórnea (Matacães – Torres Vedras). In *Origens, Estruturas e Relações das Culturas Calcolíticas da Península Ibérica, Actas das I Jornadas Arqueológicas de Torres Vedras 3–5 Abril 1987*. Trabalhos de Arqueologia 7, M. Kunst (ed.). Lisbon: Instituto Português do Património Arquitectónico e Arqueológico, 123–40.

Gonçalves, V.S. 1989. *Megalitismo e metalurgia no alto Algarve oriental. Uma aproximação integrada*. Estudos e Memórias 2. Lisbon: Minerva do Comércio.

Harrison, R.J. and Craddock, P.T. 1981. A study of the Bronze Age metalwork from the Iberian Peninsula in the British Museum. *Ampurias* 43: 113–79.

Hauptmann, A., Rehren, Th. and Schmitt-Strecker, S. 2003. Early Bronze Age copper metallurgy at Shahr-i Sokhta (Iran) reconsidered. In *Man and Mining*, T. Stöllner, G. Körlin, G. Steffens and J. Cierny (eds). *Der Anschnitt* Beiheft 16. Bochum: Deutsches Bergbau-Museum, 197–213.

Hook, D., Freestone, I., Meeks, N.D., Craddock, P.T. and Moreno Onorato, A. 1991. Early production of copper-alloys in south-east Spain. In *Archaeometry '90*, E. Pernicka and G.A. Wagner (eds). Basel: Birkhäuser Verlag, 147–72.

Hunt Ortiz, M. 2003. *Prehistoric Mining and Metallurgy in South West Iberian Peninsula*. BAR International Series 1188. Oxford: Archaeopress.

Jalhay, E. and Paço, A.d. 1945. El castro de Vilanova de San Pedro. In *Actas y memorias de la Sociedad Española de Antropología Etnografía y Prehistoria. Tomo 20. Cuadernos 1–4*: 5–92.

Junghans, S., Sangmeister, E. and Schröder, M. 1960. *Metallanalysen kupferzeitlicher und frühbronzezeitlicher Bodenfunde aus Europa. Studien zu den Anfängen der Metallurgie (SAM)*. Berlin: Verlag Gebr. Mann.

Junghans, S., Sangmeister, E. and Schröder, M. 1968. *Kupfer und Bronze in der frühen Metallzeit Europas 1–3. Studien zu den Anfängen der Metallurgie (SAM)*. Berlin: Verlag Gebr. Mann.

Junghans, S., Sangmeister, E. and Schröder, M. 1974. *Kupfer und Bronze in der frühen Metallzeit Europa. Studien zu den Anfängen der Metallurgie 4*. Berlin: Verlag Gebr. Mann.

Krause, R. 2003. *Studien zur kupfer- und frühbronzezeitlichen Metallurgie zwischen Karpatenbecken und Ostsee*. Vorgeschichtliche Forschungen 24. Rahden: Verlag Marie Leidorf GmbH.

Kunst, M. 1995a. Zylindrische Gefäße, Kerbblattverzierung und Glockenbecher in Zambujal (Portugal). Ein Beitrag zur kupferzeitlichen Keramikchronologie. *Madrider Mitteilungen* 36:136–49, tables 13–14.

Kunst, M. 1995b. Central places and social complexity in the Iberian Copper Age. In *The Origins of Complex Societies in Late Prehistoric Iberia*, K. T. Lillios (ed.). Ann Arbor, MI: International Monographs in Prehistory, Archaeological Series 8, 32–43.

Kunst, M. 2003. Muralhas e derrubes. Observações sobre a fortificação calcolítica do Zambujal (Torres Vedras) e suas consequências para a interpretação estratigráfica. Um resumo. In *Recintos murados da pré-história recente. Técnicas construtivas e organização di espaço. Conservação, restauro e valorização patrimonial de arquitecturas pré-históricas. Mesa-redonda internacional. Realizada na Faculdade de Letras da Universidade do Porto nos dias 15 e 16 de Maio de 2003*, S. Oliveira Jorge (ed.). Porto: Coimbra, 169–75.

Kunst, M. in press. Zambujal (Torres Vedras, Lisboa). Relatório sobre as escavações de 2001. *Revista Portuguesa de Arqueologia*.

Kunst, M. and Trindade, L. 1990. Zur Besiedlungsgeschichte des Sizandrotals. Ergebnisse aus der Küstenforschung. *Madrider Mitteilungen* 31: 34–82.

Kunst, M. and Uerpmann, H.-P. 1996. Zambujal (Portugal). Vorbericht über die Grabungen 1994. *Madrider Mitteilungen* 37: 10–36.

Lago, M., Duarte, C., Valera, A., Alberaria, J., Almeida, F. and Carvalho, A.F. 1998. Povoado dos Perdigões (Reguengos de Monsaraz). Dados preliminares dos trabalhos arqueológicos realizados em 1997. *Revista Portuguesa de Arqueologia* 1(1): 45–152.

Lechtmann, H. 1996. Arsenic bronze: dirty copper or chosen alloy? A view from the Americas. *Journal of Field Archaeology* 23: 477–514.

Lechtmann, H. and Klein, S. 1999. The production of copper-arsenic alloys (arsenic bronze) by cosmelting: modern experiment, ancient practice. *Journal of Archaeological Science* 26: 497–526.

Lillios, K.T. 1997. Amphibolite tools of the Portuguese Copper Age (3000–2000 B.C.): a geoarchaeological approach to prehistoric economics and symbolism. *Geoarchaeology* 12(2): 137–63.

Marcoux, E. 1998. Lead isotope systematics of the giant massive sulphide deposits in the Iberian Pyrite Belt. *Mineralium Deposita* 33: 45–58.

Marcoux, E. and Sáez, R. 1994. Geoquímica isotópica del plomo de las mineralizaciones hidrotermales tardihercínicas de la faja pirítica Ibérica. *Boletín Sociedad Española Mineralogía* 171: 202–3.

Martins, L. and Borralho, V. 1998. *Mineral Potential of Portugal.* Lisbon: Instituto Geológico e Mineiro.

Montero, I. 1993. Bronze Age metallurgy in southeast Spain. *Antiquity* 67: 46–57.

Müller, J. and van Willigen, S. 2001. New radiocarbon evidence for European Bell Beakers and the consequences for the diffusion of the Bell Beaker phenomenon. In *Bell Beakers Today: Pottery, People, Culture, Symbols in Prehistoric Europe. Proceedings of the International Colloquium Riva del Garda (Trento, Italy) 11–16 May 1998*, F. Nicolis (ed.). Provincia Autonoma di Trento, Servizio Beni Culturali, Ufficio Beni Archeologici, 59–80.

Müller, R. and Cardoso, J.L. in press. The origin and use of copper at the Chalcolithic fortification of Leceia, Portugal. *Madrider Mitteilungen* 49.

Müller, R., Rehren, Th. and Rovira Llorens, S. 2004. Almizaraque and the early copper metallurgy of southeast Spain: new data. *Madrider Mitteilungen* 45: 33–56.

Müller, R., Rovira Llorens, S. and Rehren, Th. 2006. The question of early copper production at Almizaraque, SE Spain. In *Proceedings of the 34th International Symposium on Archaeometry, 3–7 May 2004, Zaragoza, Spain*, J. Pérez-Arantegui (ed). Zaragoza: Institución 'Fernando el Católico', 209–15.

Ottaway, B.S. 1982. *Earliest Copper Artifacts of the Northalpine Region: Their Analysis and Evaluation.* Schriften de Seminars für Urgeschichte der Universität Bern 7. Bern: Seminar für Urgeschichte.

Pernicka, E. 1984. Instrumentelle Multi-Elementanalyse archäologischer Kupfer- und Bronzeartefakte: Ein Methodenvergleich. *Jahrbuch des Römisch-Germanischen Zentralmuseums Mainz* 31: 517–31.

Pernicka, E. 1990. Gewinnung und Verbreitung der Metalle in prähistorischer Zeit. *Jahrbuch des Römisch-Germanischen Zentralmuseums* 37(1): 21–129.

Quesada, C. and Munhá, J. 1990. Metamorphism. In *Pre-Mesozoic Geology of Iberia*, R.D. Dallmeyer and E. Martínez García (eds). Heidelberg: Springer, 314–20.

Rovira Llorens, S., 2005. Tecnología Metalúrgica Campaniforme en la Península Ibérica. Coladas, moldeado y tratamientos postfundición. In *El Campaniforme en la Península Ibérica y su Contexto Europeo*, M.A. Rojo-Guerra, R. Garrido-Pena and I. García-Martínez de Lagrán (eds). Valladolid: Junta de Castilla y León, 495–521.

Rovira Llorens, S. and Gómez Ramos, P. 2003. *Estudios Metalográficos.* Las primeras etapas metalúrgicas en la Península Ibérica. Imprenta Taravilla.

Sangmeister, E. 1995. Zambujal. Kupferfunde aus den Grabungen 1964 bis 1973. In *Zambujal. Kupferfunde aus den Grabungen 1964 bis 1973: Los Amuletos de las Campaños 1964 hasta 1973*, E. Sangmeister and M.d.L.C. Jiménez Gómez (eds). Madrider Beiträge. Mainz: Philipp von Zabern, 1–153.

Sangmeister, E. and Schubart, H. 1972. Zambujal. *Antiquity* 46: 191–7, pls XXIX–XXXIII.

Sangmeister, E. and Schubart, H. 1981. *Zambujal. Die Grabungen 1964 bis 1973.* Madrider Beiträge Mainz: Verlag Philipp von Zabern.

Santos Oliveira, J.M., Farinha, J., Matos, J.X. *et al.* 2002. Diagnóstico Ambiental das Principias Áreas Mineiras Degradadas do País. *Boletim de Minas* 39(2): 67–85.

Santos Zalduegui, J.F., García de Madinabeitia, S., Gil Ibarguchi, J.I. and Palero, F. 2004. A lead isotope database: the Los Pedroches – Alcudia area (Spain): implications for archaeometallurgical connections across southwestern and southeastern Iberia. *Archaeometry* 46(4): 625–34.

Soares, A.M.M. 2005. A metalurgia de Vila Nova de São Pedro. Algumas reflexões. In *Construindo a memória. As colecções do Museu Arqueuológico do Carmo*, J.M. Arnaud and C.V. Fernandes (eds). Lisbon: Associação dos Arqueólogos Portugueses, 179–88.

Soares, A.M.M, Araujo, M.d.F. and Cabral, J.M.P. 1994. Vestígios da práctica de metalurgia em povoados Calcolíticos da bacia do Guadiana, entre o Ardila e o Chança. In *Arqueología en el entorno del Bajo Guadiana. Actas del encuentro internacional de arqueologia del suroeste*, J. Campos, J. Perez and F. Gomez (eds). Huelva: Junta de Andalucia, 165–200.

Soares, A.M.M., Araujo, M.d.F., Alves, L. and Ferraz, M.T. 1996. Vestígios metalúrgicos em contextos do Calcolítico e da Idade do bronze no sul de Portugal. In *Miscellanea em Homenagem ao Professor Bairrão Oleiro*, M.J. Maciel (ed.). Lisbon: Edições Colibri, 553–79.

Sousa, A.C., Valério, P. and Araúho, M.D.F. 2004. Metalurgia antiga do Penedo do Lexim (Mafra): Calcolítico e Idade do Bronze. *Revista Portuguesa de Arqueologia* 7(2): 97–117.

Sperl, G. 1981. Untersuchungen von Resten eines Kupfer-Gießplatzes in Zambujal. In Sangmeister and Schubart 1981, 341–5.

Spindler, K. 1969. Die kupferzeitliche Siedlung von Penedo/Portugal. *Madrider Mitteilungen* 10: 45–116, figs 4–27, tables 9–12.

Spindler, K. and Gallay, G. 1973. *Kupferzeitliche Siedlung und Begräbnisstätten von Matacães in Portugal, mit einem Beitrag von A. von den Driesch.* Madrider Beiträge 1. Mainz: Philipp von Zabern.

Stos-Gale, Z., Gale, N.H., Houghton, J. and Speakman, R. 1995. Lead isotope data from the Isotrace Laboratory, Oxford: Archaeometry data base 1, ores from the western Mediterranean. *Archaeometry* 37(2): 407–15.

Tornos, F. and Chiaradia, M. 2004: Plumbotectonic evolution of the Ossa-Morena Zone (Iberian Peninsula): constraints from ore lead isotopes. *Economic Geology* 99: 965–85.

Trincherini, P.R., Barbero, P., Quarati, P. and Domergue, C. 2001. Where do the lead ingots of the Saintes-Maries-de-la-Mer wreck come from? Archaeology compared with physics. *Archaeometry* 43(3): 393–406.

Uerpmann, H.-P. and Uerpmann, M. 2003. *Zambujal. Die Stein- und Beinartefakte aus den Grabungen 1964 bis 1973.* Madrider Beiträge 5, 4. Mainz: Philipp von Zabern.

Wang, Q. and Ottaway, B. forthcoming. Metallographic analyses of copper-based objects excavated at Zambujal, Portugal. In *Kupferzeitliche Metallurgie in Zambujal, in Estremadura, Südportugal und Südwestspanien: Vom Fertigprodukt zur Lagerstätte. Arbeitstagung Alqueva-Staudamm, 27. bis 30. Oktober 2005*, M. Kunst (ed.). Madrider Beiträge. Mainz: Philipp von Zabern.

## Authors' addresses

- Corresponding author: Roland Müller, Curt-Engelhorn-Zentrum Archäometrie gGmbH, An-Institut der Universität Tübingen, C5 Zeughaus, 68159 Mannheim, Germany (rk.mueller@gmx.de)
- Dr. Gert Goldenberg, Institut für Ur- und Frühgeschichte und Archäologie des Mittelalters, Universität Freiburg, Belfortstraße 22, 79098 Freiburg, Germany (gert.goldenberg@ufg.uni-freiburg.de)
- PD Dr. Martin Bartelheim, Institut für Prähistorische Archäologie (Ur- und Frühgeschichte), Freie Universität Berlin, Altensteinstr. 15, 14195 Berlin, Germany (martin.bartelheim@am.tu-freiberg.de)
- PD Dr. Michael Kunst, Deutsches Archäologisches Institut, Madrid, C/ Serrano, 159, E-28002 Madrid, Spain (kunst@madrid.dainst.org)
- Prof. Dr. Ernst Pernicka, Curt-Engelhorn-Zentrum Archäometrie gGmbH, An-Institut der Universität Tübingen, C5 Zeughaus, 68159 Mannheim, Germany (ernst.pernicka@uni-tuebingen.de)

# The use of experimental archaeology/ archaeometallurgy for the understanding and reconstruction of Early Bronze Age mining and smelting technologies

*Simon Timberlake*

*ABSTRACT*   Some 20 years of archaeological investigation have now firmly established the evidence for prehistoric copper mining in Britain; this includes some 12 sites worked for copper sulphide minerals (chalcopyrite) and carbonate ores, most of these between 2100 and 1600 BC. With few artefacts or work areas surviving, interpreting the technology of primitive mining presents a challenge to the archaeologist. Experimental work has been particularly useful here, allowing us to understand the process of firesetting, the use and hafting of stone implements, and to predict the discovery of organic remains such as antler tools within the excavations on Copa Hill (Cwmystwyth, Wales).

In the absence of archaeological evidence, experiments in smelting tin (cassiterite) and copper (malachite and chalcopyrite) ores are now helping us to reconstruct some of the most rudimentary types of furnaces. A recent attempt to smelt chalcopyrite within an open furnace pit at Butser Iron Age Farm appears to show the process of copper prill formation taking place at temperatures of less than 1200 °C under poorly reducing conditions; tantalising evidence perhaps of how copper may have been extracted from a problematical sulphide ore at the beginning of the Early Bronze Age. The scientific examination of the products of these smelting experiments is described in the following paper (pp. 37–45).

*Keywords:* Early Bronze Age, UK, experimental, mining, smelting, tin, copper, chalcopyrite.

## Introduction

Both mining and smelting are inherently destructive operations. Whatever was being sought in terms of ore will, in most cases, have been extracted, then processed, and finally removed from site. Primitive mining, involving the time-consuming hand-crushing, hand-picking and separation of mineral can be a highly efficient operation. So much so in fact, that what might at first sight seem a perfectly straightforward question – for example, whether this was a copper or lead mine – can sometimes prove difficult to answer (Bick 1999; Mighall *et al.* 2000). If it was a copper mine, then was it the sulphides or the carbonate minerals formed from the surface oxidation of these sulphides (or a combination of the two) that was being extracted (Craddock 1995: 11, 32; Timberlake 2003:100–102)?

Mine spoil does survive yet this often has a history of re-deposition and mixing, and the archaeology of these deposits is complex and can be difficult to interpret. Mining tools, in particular stone ones, have a good record of survival, yet these implements are recycled again and again, the fragments of these ending up scattered over the site. Current evidence suggests that little in the way of ritual importance was attached to practical metal mining tools (Timberlake and Prag 2005: 5). Those made of organic materials (Fig. 1a–c) were discarded once broken or worn, and more often than not these were thrown down and used as floor materials within waterlogged areas, or perhaps consigned to the poorly preserving environment of the waste heaps where there was little chance of survival. Alternatively they may have ended up being recycled as fuel within the next fireset hearth (Timberlake 2003: 71). Thus less evidence than one might expect survives of tools, and still less of the mineral sought. What we are in fact dealing with in most cases are just dispersed fragments and negative evidence.

Still more rarely do intact smelting operations survive within the archaeological record. Furnaces are broken down to extract metal from incomplete smelts, components of the furnace walls and tuyeres are recycled, while slags produced as part of earlier and more primitive smelting operations could have been broken up and crushed to release prills of metal, the residue being used as temper within ceramics or refractory materials. Indeed, 'proper' slags may never have been produced in these operations (P.T. Craddock 1994: 75). Paul Craddock has succinctly summarised the situation: 'at best the evidence is enigmatic and at worst non-existent' (Craddock 2003: 8).

With hindsight we might be asking ourselves whether more careful excavation of these sites could have afforded us a better picture of what was going on. If this were the case, we might

**Figure 1** Organic artefacts from the waterlogged anoxic sediments on the floor of the Copa Hill mine: (a) broken withy handle; (b) withy (hazel) tie or rope; (c) red deer antler pick/hammer. (Photos: S. Timberlake)

ology could be used as a predictive tool to help us determine the sorts of things *we should be looking for* in the course of future archaeological excavation.

## Experiments in primitive mining technology

### Firesetting[1]

The first modern experiments investigating firesetting were undertaken at Penguelan Shaft, Cwmystwyth, Wales, in 1987 (Pickin and Timberlake 1988: 165–7), and on numerous occasions since (Timberlake 1990b; 2005: 188) (Fig. 3a). The size of cut-wood fuel used during these experiments has gradually been reduced (from 3 to 0.5 m long split logs mixed with brushwood) as experience has been gained at building more efficient fires. As archaeological evidence from the excavations on Copa Hill has accumulated, it suggests the use of smaller hearths and the exclusive use of oak firewood in the mine (Nayling in Timberlake 2003).

Typically between 100 and 250 kg of firewood was used in each experimental firing, and by the end of this period the wood (fuel)/stone extraction ratio had improved substantially

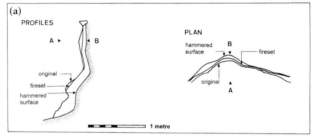

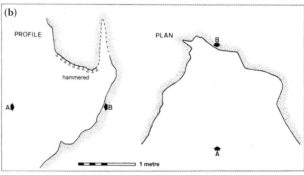

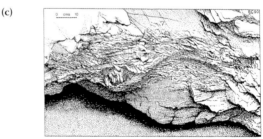

**Figure 2** (a) Plan and profiles of the experimental fireset and hammered rock face, Penguelan, Cwmystwyth (1992); (b) plan and profile of the heading of a Bronze Age mine gallery at Copa Hill, Cwmystwyth (*c.* 1800 BC); (c) drawing of the arched roof of the same gallery with preserved hammer stone batter marks. (Drawings by B. Craddock.)

then have been able to faithfully reconstruct and test these processes by means of a programme of controlled experiments. This is exactly what the two great proponents of experimental archaeology, John Coles (1973) and Peter Reynolds (1999), would have considered to be the ideal. Unfortunately, that same starting point of knowledge would be impossible to achieve in this instance. The partial or even complete absence of archaeological evidence, however, should not necessarily be seen as a reason to invalidate the use of experiment – it just requires a different approach.

The undertaking of well-recorded and repeated experiments in practical mining and smelting can be extremely useful in helping to recognise some of the fragments or artefacts of the various processes witnessed, and to understand the sorts of problems that ancient miners and metallurgists had to grapple with as they developed their technology. A good experimenter will try to enter into the mindset of the miner and smelter to discover the improvements and practical solutions they themselves might have come up with given the limitations of materials to hand and the simple technology available. To take this one step further, experimental archae-

**Figure 3** (a) Experimental firesetting at Penguelan; (b) using stone hammers in experimental hafting on the fireset rock face at Alderley Edge, 1997; (c) using an antler pick to prise out the loosened rock at Penguelan, 1989. (Photos: P.T. Craddock and S. Timberlake.)

from about 1:1 to 1:2. This reflects an increase in skill, more economic use of fuel and a better understanding of the properties of the rock.

Experimentation has provided us with at least one important realisation: that dousing the hot rock with water was not necessary. While 'quenching' might help to shatter the rock surface, this practice did not significantly increase the yield of rock extracted. Two interesting sets of reactions were observed, however, when firesetting shale and quartz veined rocks: first, a decrepitation of the slate brought about by thermal shock as the temperature of the rock face rose rapidly to 800 °C, explosively ejecting debris (< 0.01% of the total mass of extracted); secondly, the splitting open of joints/cleavage by steam as water already present within the pervious rock became superheated. The latter could lead to the collapse of some parts of the rock during firing, but more typically would leave it weakened, allowing for its easy removal in large blocks with hand tools. There may have been other effects of the fire on the rock, such as surface exfoliation and crumbling (a calcination effect as is seen when firesetting is carried out against solid vein quartz (Holman 1927)), but the reaction of trapped moisture in aiding disintegration seemed to outweigh all the rest.

The different sorts of fireset rock waste described above have since been recognised within the various layers of prehistoric spoil from excavation, but archaeology provides no hint of the intentional introduction of water into the workings, or any evidence for its use in quenching to break up the rock face.

### Firesetting at the rock face

The maximum heat penetration of the rock, as suggested by the changed profile of the rock face after mining with stone tools for several hours, seems to have taken place at around 1 m above ground level. This left a prominent rock step or 'heel' at the base of the firesetting hearth (Fig. 2a). These concavities are one of the most distinctive signatures of firesetting (Timberlake 1990a: 49). Although less evident within shale rocks than in granular sandstones and limestones, a good example of a hearth step and the characteristic arched profile of firesetting was observed within the workings on Copa Hill

(Fig. 2b). The Bronze Age miners on Copa Hill, however, managed to sink shafts or vertical trenches on the vein with little more available to them than fire and stone tools. This was also the case at the mines of Zawar, Rajasthan (Willies 1994: 6–7). Recent experiments at Penguelan, Cwmystwyth suggest that the use of smaller fires and the covering of these with heavy wood or stone to reduce airflow not only improve fuel efficiency, but also slow the burn and help direct heat penetration downwards.

### Experimentation and the recognition of fireset mine spoil

Weathered, buried and subsequently exposed mine spoil originating from ancient firesetting activity is often difficult to recognise in the field. Experimentally produced fireset waste from four different occasions (1987, 1989, 1992 and 1995) was separately piled and allowed to weather over a ten-year period. Colour change and soil formation were quantified, suggesting rapid breakdown in the first few years, levelling out thereafter. At the end of this period certain textural similarities were comparable with ancient examples revealed by archaeology (Jenkins and Timberlake 1997: 47–8).

### Firesetting and the dispersion of heavy metals

In 1996, a further experiment was undertaken at the Penguelan site by Jenny Moore of Sheffield University. The purpose of this was to try to reconstruct what might have been the airborne dispersal pattern of microscopic charcoal around a prehistoric opencast mine-working. This charcoal might be picked up within peat cores (the vegetation burning record) along with any geochemical signature with which it might be associated. The origin of this palaeo-pollution record on Copa Hill could be smoke from firesetting on veins, from smelting hearths, or as windblown dust from spoil heaps (Mighall *et al.* 2000, 2002).

Analyses of soot condensates coating the rock face at the Penguelan site have revealed levels of metal (in particular lead and bismuth) up to 40× higher than those within the underlying rock. The analysis of wind-dispersed charcoal particles, however, collected some 20 m away from the fireset, also showed elevated levels of lead (Jenkins and Timberlake 1997: 74–9). Under conditions of turbulent (rotational) wind, particles from the fallout of the smoke plume were collected up to 100 m away from the hearth.[2] The results of such experimental work may yet help to shape the design of future palaeoenvironmental studies.

## Experimental mining using stone and antler tools

### Experiments with stone mining tools

Experimental mining using both hafted and unhafted stone tools and antler picks has been undertaken at Penguelan, Cwmystwyth following all five of the experimental firesets. In 1987, unmodified water-worn cobbles from the floodplain of the river Ystwyth were used experimentally as rock-breaking implements, first hand-held then attached to ropes and swung against the recently fired rock face (Pickin and Timberlake 1988). It was noted that little breakage of these hammers occurred while they were being used against fire-weakened rock. Instead, the characteristic spalling from their ends – the debris of which is so evident among prehistoric mine spoil – began to occur as soon as the sound rock, unaffected by the heat, had been reached. The miners continued to use their tools against this until the rate of breakage and difficulty of progress prompted them to re-fireset. Preserved examples of batter or pounding marks, interpreted as the imprint of the use of hammer stones against a fireset rock, reflect this absolute limit of mining. This was first discovered experimentally, but was then recognised within the roof of a small Bronze Age mine gallery on Copa Hill (Fig. 2c).

Handles made of hazel sticks or twisted willow withies were first used in mining experiments in 1989 (B. Craddock 1990, 1994) (Fig. 3b). Large cobbles could be hafted this way without grooving them, sometimes with little or no notching of the stones required. Experimentally this was a distinct improvement. Hemp twine and leather, later substituted by rawhide strips, were used experimentally for binding and knotting the cobbles in position. These were modelled on the hafted tools recovered alongside the remains of a 2000 year-old Ameri-Indian miner 'Copper Man' from the Restauradora Mine at Chuquicamata, Chile in 1899 (Bird 1979). It did not take long to realise that such hammers could only have been used underarm (Fig. 3b), if the stones were to be retained within the haftings. This type of skill acquisition in experimentation offered simple yet obvious answers to a number of questions regarding the use of hammer stones within the Bronze Age mine: for example, how and why were large numbers of triangular-shaped cobbles used only at the broad end? Quite simply, the use of the tool would force the cobble more firmly back into the ligature of its hafting, while the use of the narrow end, without extensive notching of the cobble, would eject it (Timberlake 2003: 94).

Comparisons made between the use of riverine or beach pebbles as hammer stones, alongside detailed examination of the tools recovered from the excavations, prompted us to change our minds about the sources of the cobbles used. The degree of surface smoothness and polish on these suggested that up to 70% might have been collected from storm beaches some 20 km away on the coast of Cardigan Bay. Experimental hafting of these demonstrated a clear advantage in terms of competency and their survival rate as tools.

The fortunate appearance at the British Museum in 1992 of an intact hammer from Chuquicamata[3] provided Brenda Craddock with a rare opportunity to study this and to suggest a better method of hafting (Craddock et al. 2003). The bruising of the fibres, their twisting, then the looping of a single withy (hazel) handle around the cobble and its fastening with rawhide suddenly seemed obvious. As a result the strength and efficiency of the experimental tools improved. Larger cobbles could now be hafted, while accurate work could still be achieved using smaller hammers (less than 1 kg); in some cases these have dislodged up to a ton of rock with only minimal repairs to the haftings. The facets and spalling surfaces produced on these cobbles have since been examined with an eye to recognising the same types of wear among the tools from the excavations. Similar sorts of bilateral notches, pecked or ground into the sides of experimental cobbles for hafting, have now been recognised from excavated examples, as have facets on the flat surfaces of pebbles for the insertion of wedges, or for the reuse of these tools as ore-crushing anvils.

As a result of working on these experimental tools and from the examination of broken bindings, Brenda Craddock was immediately able to recognise one half of a withy handle seen lying within a waterlogged area of the excavations (Fig. 1a). At the same time, she predicted the find of the second some centimetres away from the first; both of these were found in situ, at the spot where each was discarded when the hammer haft failed some 4000 years ago (Timberlake 2002: 345; 2003: 72).

During the course of experimental archaeology weekends held at Alderley Edge in Cheshire in 1997 and 1998, locally sourced erratic cobbles were fully grooved with a quartz pecking stone and then hafted to reconstruct the implements used within the Bronze Age mines of the Edge (Fig. 3b). Somewhat surprisingly, the complete transverse grooving of a 10 cm diameter cobble to a depth of 1–10 mm with a groove 15–20 mm wide, took rather less than an hour in the hands of a skilled operator. Given the long use histories of some of those tools recovered from the archaeological levels, this would seem to represent a good economy of effort (Timberlake and Craddock 2005).

### The discovery of antler tools

Picks of red deer antler were first used in mining experiments at Cwmystwyth in 1990, primarily as a means to test their effectiveness against hard rocks. These tools were used in a very different way to metal picks, functioning as quite effective mallets to knock out freshly fireset rock or alternatively as levers to prise away blocks after the rock had been fractured and loosened up by hammer stones (cf. the mode of antler pick use in Neolithic flint mines). Approximately 1.5 tons of fireset rock have been removed with the tool shown being used in Figure 3c. This alternation between stone and antler tools proved such an effective combination that we suggested picks could have been used within the Bronze Age mines (Timberlake 1990b). Examples of these were discovered the following year on Copa Hill: one was a broken pick and another reused on the crown as a mallet (Fig. 1c), in the same way as antlers had been used in the experiments. Many small fragments of antler have since been found, most of these indifferently preserved. It has been estimated that each pick could have assisted in the removal of between 15 and 25 tons of rock, thus it is conceivable that between 100 and 300 antlers may have been brought up to site (Timberlake 2003: 84). It is useful to compare this sort of estimate with the evidence recovered from some of the Neolithic flint mines where antler picks were the main tools of extraction. At Grimes Graves in Norfolk, several hundred antler picks were found per shaft (Mercer 1981).

## Smelting of tin oxide and copper carbonate ores

### Bowl furnace smelting of cassiterite

While there is circumstantial evidence for prehistoric tin production in southwest England (Penhallurick 1986), little evidence of smelting has come to light. Some cassiterite and a globule of tin were found within Bronze Age round huts on Dean Moor, Dartmoor (Penhallurick 1986:117), plus glassy tin slag from Early Bronze Age barrows at Caerloggas, St Austell, Cornwall (Tylecote 1986: 43), yet it seems the hearths may have been ephemeral structures, which have left little in the way of any archaeological record.

Our approach to understanding this has been quite different to that of previous experiments which were carried out on a much reduced scale, employing essentially 19th-century Cornish techniques of assaying using a blowpipe and crucible (Earl 1986; Yener *et al.* 2003). We began by smelting cassiterite concentrate (approx. 65% $SnO_2$) with charcoal in clay-lined bowl furnaces using a single tuyere and bag bellows. Experiments conducted at Flag Fen in 1994, in the Netherlands and at Lamorna, Penzance (Cornwall) (the latter filmed for Channel 4's *Time Team*), produced 'prill' tin as well as lumps of pooled metal which solidified beneath the base of the hearth (Timberlake 1994). Successful smelts could be achieved at temperatures a little over 1000 °C, charging the furnace with balls of powdered ore mixed with charcoal, wood shavings or animal dung. At the time this was an intuitively followed procedure, yet it seems to have had some ethnographic parallel, as with copper smelting in Sikkim (Craddock 1995:152). Between 50 and 87% of the maximum theoretical yield of tin was obtained fairly easily this way within the space of about an hour's smelting time.

The metal produced was fairly pure with a low iron content typical of smelting under poorly reducing conditions (see following chapter, p. 39). No slag was formed (only a small amount of glassy clinker which would have been crushed to recover available metal). This may have been the norm at the beginning of the metalworking period in Britain; the smelting of high grade ores within small, low-temperature, poorly reducing open hearths.

### *The smelting of malachite and azurite ores from Alderley Edge: the process of copper prill formation*

Similar experiments were carried out in 1997 and 1998 using bowl furnaces to smelt copper carbonate ores collected from the same ore horizons within the Engine Vein Formation exploited by the prehistoric miners of Alderley Edge (Timberlake and Prag 2005: 198–210). Impure malachite sandstone and pisoliths of azurite collected from a mudstone layer were crushed and moulded into golf-ball sized pellets with charcoal dust, sawdust and animal dung (1:1:1) (Fig. 6a). These were then dried and smelted for between 1½ and 2 hours at 800–1100 °C. On all occasions the part-smelted ore balls were recovered intact at the end of each smelt. Breaking

**Figure 4** Smelting and casting copper at Butser Iron Age Farm (2004–05): (a) smelting malachite within an earth furnace using a paired bag bellows; (b) experimentally smelted metal being cast in a flat-axe mould (scale bar: 10 cm). (Photo (a): J. Bourne; photo (b): J. Graig Argyle.)

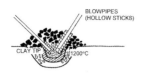

CRUCIBLE AND BLOWPIPES

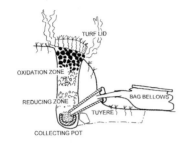

BANK FURNACE

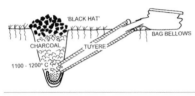

POST-HOLE FURNACE

**Figure 5** Diagrams of the simple furnace designs experimented with at Butser: (top) bowl furnace with a crucible smelt; (middle) bank shaft furnace; (bottom) post-hole furnace. (Drawings: S. Timberlake and B. Craddock).

**Figure 6** (a) Crushing azurite nodules and preparing the ore and fuel balls for the furnace, near Engine Vein, Alderley Edge (1997); (b) communal blowpipe smelt of a malachite/azurite ball within a small pit, Alderley Edge (1998). (Photos: P.T. Craddock.)

some of these open revealed what appeared to be copper prills. The products of these experiments and their significance are described in the following paper (pp. 37–45).

At most, only 5.7% of the copper within any of the balls had been smelted. Insufficient temperature and duration of smelt and/or reducing conditions resulting from the wrong ratio of ore/fuel within the balls had meant that, in effect, we ended up with a series of failed smelts. Yet the net result was a useful insight into what was a credible process. If the smelting could have been brought to its conclusion, the open structure of these balls would have collapsed under the surface tension of molten metal as both small and large prills and the outer skin coalesced and the metal sank to the base of the furnace (see pp. 38–9). The actuality of early smelting, however, may have been quite different.

Changes taking place within the ore and fuel during the smelting process have rarely been examined before. Using techniques of processing that might have been employed in the Bronze Age, this work supports the notion that the earliest copper production may have involved the extraction and re-melting of copper prills.

### Copper smelting and casting experiments at Butser Iron Age Farm (2004–2005)

Together with Paul and Brenda Craddock, I have run week-long courses in experimental archaeology for archaeology students[4] at Butser Iron Age Farm in Hampshire. These have combined two useful roles: training students in the discipline of experimental archaeology and undertaking experiments to discover the simplest workable smelting technology. In one sense we were trying to reinvent the wheel, quite relevant perhaps if some of the practices of metallurgy in Britain at the beginning of the Early Bronze Age had developed quite independently of Europe (Craddock 1992b: 25).

To help demonstrate the viability of short-lived smelting campaigns, students had to build and construct their own furnaces using local clay, sand, straw and dung. They moulded and fired in an open fire their own refractories such as tuyeres and crucibles. They sewed bag bellows then crushed, prepared and smelted copper ore, finally collecting the metal and re-melting it to cast into ingot or axe moulds (Fig. 4b). One concession here was the use of high-grade malachite ore, but otherwise these experiments were all conducted in open conditions under simple shelters. The results were quite astonishing given that for some this was their first attempt at metallurgy.

It was found that some of the benefits of shaft furnaces – such as the maintenance of reducing conditions beneath a 'black hat' of charcoal, greater heat insulation, higher temperatures and longer smelts using smaller amounts of fuel – could all be achieved using earth-bound hearths (Fig. 4a). One of these hearths was simply a reused conical-shaped post-hole some 30 cm deep, while another consisted of a narrow vertical shaft dug 50 cm into the side of an earth bank and provided with an air supply to the base and a pot or crucible to collect metal (Fig. 5a–b). Temperatures of between 1000 and 1200 °C could easily be maintained for up to 2 hours and good yields of raw copper were obtained. This could then be re-melted and refined within a crucible in a bowl hearth; although this made it easier to direct heat onto the metal in order to melt it for pouring and casting, considerably more fuel was used in the process. Crucible smelts were also attempted in a bowl furnace using blowpipes (Fig. 6b), similar to the experiments carried out by Merkel at Batan Grande in Peru (Shimada and Merkel 1991), though these were less successful than those carried out using a tuyere and bellows.

While some of the furnaces have since been reused, others have been left to weather or have been filled in. It is hoped that some of these will be excavated at some future date in order to see what evidence of this activity has survived within the archaeological record. Virtually no evidence could be seen of an experimental bowl furnace site at Flag Fen some 18 months after its last use (Timberlake 1994:127). Poorly fired refractories and furnace lining seem to disintegrate rapidly if they are not properly slagged or vitrified. This sort of evidence has important implications for archaeology.

## Primitive smelting of sulphide ores

Some 20 years of excavation appear to confirm that the earliest miners at Parys Mountain and in mid-Wales were extracting primary sulphide ores – the presumed goal of their operation being chalcopyrite (copper-iron sulphide), the only significant copper ore present (Timberlake 2002; 2005: 101–2). Micro-excavation of areas of the Copa Hill tips has revealed the presence of crushing platforms and fines containing relict chalcopyrite (<2 mm), presumably material lost to processing. This contrasts with the larger lumps of intentionally discarded galena (Mighall *et al.* 2000). Experimental beneficiation of the Copa Hill chalcopyrite using hand-crushers and an anvil stone appears to show similar levels of loss.

Chalcopyrite and other iron-copper-sulphide ores were also being extracted from the Blue Hole workings on Ross Island, Eire, during the Beaker period operation of the mine. It seems that, along with the arsenical tennantite ore, these sulphide ores were being roasted and smelted using a non-slagging process (O'Brien 2004: 472). Even at the Pentrwyn smelting site on the cliffs of the Great Orme, analysis of the copper prills now suggests that chalcopyrite was being smelted (Chapman 1997).[5] So how were they doing it? Billy O'Brien and colleagues successfully smelted tennantite within a reconstructed furnace pit hearth at Ross Island. He also referred to the experiments of Rostoker *et al.* (1989) which showed how roasting and direct reduction of chalcopyrite to copper was possible. A schematic reconstruction of this process is shown in Figure 7.

A first attempt at roasting then crucible-smelting chalcopyrite within an earth pit furnace partially covered with turf was carried out by the author at Roewen, North Wales in 2004. A copper sulphide/copper-iron sulphide matte was obtained which when investigated by scanning electron microscopy (SEM) revealed small growths of copper metal (see pp. 41–4).

In April 2005, a further roasting and smelting experiment was undertaken at Butser Iron Age Farm using the same ore, but within a reconstruction of one of the Ross Island type 1

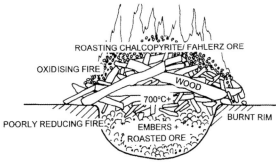

ROASTING / SMELTING PIT

**Figure 7** Diagrammatic section through 'Ross Island-type' furnace pit burning wood fuel (Drawing: S. Timberlake and B. Craddock).

**Figure 8** Roasting and smelting chalcopyrite at Butser Iron Age Farm (2005): (a) chalcopyrite ore charge for roasting, left in direct contact with the charcoal, and right in small crucibles; (b) smelting in pit using paired bag bellows. (Photo (a): C. Denis.)

furnace pits (Fig. 7) (O'Brien 2004: 467). The cleaned and concentrated ore was first roasted over charcoal for about 2½ hours at a temperature of 750 °C, then for a further 2 hours, the final roasting taking place within the interior of the fire (Fig. 8a). This was then smelted under a thick bed of charcoal for 2¼ hours with the aid of a pair of bag bellows (Fig. 8b), the temperature of the furnace reaching over 1300 °C, but averaging about 1180 °C for the duration of the smelt. The tip of the tuyere was extensively slagged and vitrified with red-coloured glassy areas containing copper either as an extremely fine precipitate or dissolved in the glassy surface. SEM examination of the smelted chalcopyrite recovered from the base of this furnace also revealed the presence of tiny copper cones known as 'moss copper' growing within the gas voids of the matte. The quantity of metal produced within this incomplete smelt was probably only a few grams out of an initial charge of 430 g chalcopyrite, most of this remaining as iron-copper-sulphide matte. (A full description of the analysis and formation of this moss copper is discussed in the following paper, pp. 37–45.)

## Discussion

The above experiments are both examples of a smelting process 'frozen' in its initial stages, yet the implications of this as a potential method for smelting a problematical sulphide ore within a firepit-type furnace are interesting. With repeated roastings and smelts, and with a greater volume of sulphide ore to assist in the combustion, as witnessed within the chalcopyrite roasting heaps at Rio Tinto and on Parys Mountain (Rowlands 1966: 43–5; Salkield 1987), copper metal might gradually be recovered from such an ore, bit by bit, without the production of a slag. Even if quartz (sand) had been added, it seems very unlikely that the conditions would have been right for slag formation. Whatever the improvement to the smelting process, these early metallurgists would still have needed to recover all the tiny prills of copper through careful crushing, washing and panning of the heat-treated ore. While this process seems to be suggested by the presence of fine washing residues close to the furnace pits on Ross Island, no associated furnace debris (either crushed or uncrushed) or copper prills have ever been found (O'Brien 2004: 467–70). In fact, the sum total of evidence for the successful smelting of ore to metal at this site amounts to little more than the recovery of a few droplets of melted copper with a high arsenic content, the latter suggesting an early stage in the production of metal prior to its re-melting, refining and casting.

Given the time-consuming nature of this process, it is interesting to speculate on the clearance of woodland for mining and farming activities and the creation of large open fires burning for days at a time; fires which might be used for roasting and smelting sulphide ores. It seems more credible now that wood rather than charcoal was being used in this process. Once exposed to the fierce oxidising conditions of a large fire, followed by reducing conditions under a heap of embers, a significant proportion of the copper present in such an ore could have been reduced to metal. The development of a bowl furnace technology using crucibles and bellows would then be required to recover usable metal. The use of furnace pits, perhaps even the controlled burning of very large tree boles into which prepared ore had been carefully placed, such as carried out by the 19th-century prairie lead smelters in North America (Goodway 1987), are ideas worthy of experimental reconstruction.

The earliest mining of chalcopyrite ores in Europe may well have taken place within the alpine areas where glaciation had stripped the copper deposits of much of their oxidised ores. Assuming a rudimentary level of smelting technology, processing these ores would have made it more difficult to achieve the level of metal yield possible from oxidised ores. One response to this may have been a much increased level of prospection for new deposits; this is reflected perhaps in the flurry of Beaker/Early Bronze Age interest in the metal resources of western Britain and Ireland at the end of the 3rd to the beginning of the 2nd millennium BC.

## Conclusions

Experimental archaeology carried out over the last 17 years has proved invaluable in interpreting primitive mining and smelting technologies. This has been used as a tool to help predict the presence of artefacts and working debris, to reconstruct the tool kits of ancient miners, and to understand skills and techniques of which little evidence now remains. Archaeological experiment can guide excavation strategy while its data will formulate the research questions of the future.

In lieu of the virtual absence of smelting evidence from the British Bronze Age, experiments have enabled us to propose and test some of the simplest credible models for furnace design, allowing us opportunities for intuitive discovery and an understanding of what we should be looking for within the archaeological record. Smelting experiments undertaken out-of-doors using authentic materials can prove useful in other ways. For example, most experimental replication of ancient alloys seems to have been fabricated under controlled laboratory conditions (Merkel 1990; Tylecote and Boydell 1978). While this enables the conditions of the experimental production to be controlled exactly and a specific product to be obtained, the thermodynamics and oxidising/reducing conditions of ancient furnaces were actually highly variable, thus the net composition and type of metal produced within the first smelting stage may have been quite different from that produced by modern laboratory experiments. For instance, a comprehensive ore provenancing study might include the experimental smelting of ores of distinctly differing composition from a variety of known sources. Smelting these within the laboratory using an electric or gas burning furnace may provide a better yield of metal than could be achieved by using a clay-lined bowl or shaft furnace and bag bellows, but which technique is likely to produce metal most closely approximating the composition of that originating from these sources in prehistory? We learn much more about the process of when using credible technologies and materials.

To date, the continuing experiments backed up by metallurgical analyses owe much to Paul Craddock's own interest and enthusiasm for this work. What is needed now is properly funded research that could help support field and laboratory-based PhD projects in experimental archaeology and archaeometallurgy. This challenge still waits to be taken up.

## Acknowledgements

Brenda Craddock and other members of the Early Mines Research Group have been integral to the success of many of the mining and early smelting experiments. Jenny Moore and Tim Mighall have provided useful commentary on the distribution of heavy metals and charcoal. CADW and the Crown Estates gave permissions for experimental work undertaken at Penguelan, Cwmystwyth; the National Trust for work undertaken on Alderley Edge. The latter work was supported by the Manchester Museum and the Leverhulme Trust. Dave Chapman of Roewen assisted with the first chalcopyrite smelt, the second undertaken at Butser Iron Age Farm was carried out by Christophe Denis under the direction of the author. Paul Craddock has worked with the author on all the smelting experiments undertaken at Butser Iron Age Farm. Birkbeck College funded the Butser experiments as a module of its MA Archaeology course. We would like to thank Christine Shaw and volunteers at Butser Iron Age Farm and all the Birkbeck students who took part in the courses for their help in making this possible. This paper is dedicated to the memory of Peter Reynolds.

## Notes

1. A good background to the historic record for the use of firesetting in mining can be found in Timberlake (1990a), Craddock 1992a and 1995 (33–7), and Willies (1994).
2. J. Moore, pers. comm.
3. On loan from Dr Wray.
4. Birkbeck College, University of London, MA degree course.
5. Peter Northover, pers. comm.

## References

Bick, D.E. 1999. Bronze Age copper mining in mid Wales – fact or fantasy? *Historical Metallurgy* 33(1): 7–12.

Bird, J.B. 1979. The 'Copper Man': a prehistoric miner and his tools from northern Chile. In *Pre-Columbian Metallurgy of South America*, E.P. Benson (ed.). Washington, DC: Dumbarton Oaks, 105–32.

Chapman, D. 1997. Great Orme smelting site, Llandudno. *Archaeology in Wales* 37: 56–7.

Coles, J. 1973. *Archaeology by Experiment*. London: Hutchinson.

Craddock, B. 1990. The experimental hafting of stone mining hammers. In Crew and Crew 1990, 58.

Craddock, B. 1994. Notes on stone mining hammers. In Ford and Willies 1994, 28–30.

Craddock, B., Cartwright, C., Craddock, P.T. and Wray, W.B. 2003. Hafted stone mining hammer from Chuquicamata, Chile. In Craddock and Lang 2003, 52–68.

Craddock, P.T. 1992a. A short history of firesetting. *Endeavour* 16(2): 145–50.

Craddock, P.T. 1992b. Copper production in the Bronze Age of the British Isles. *Bulletin of the Metals Museum* 18: 3–28.

Craddock, P.T. 1994. Recent progress in the study of early mining and smelting in the British Isles. *Journal of Historical Metallurgy* 28(2): 69–84.

Craddock, P.T. 1995. *Early Metal Mining and Production*. Edinburgh: Edinburgh University Press.

Craddock, P.T. 2003. Introduction. In Craddock and Lang 2003, 1–3.

Craddock, P.T. and Lang, J. (eds) 2003. *Mining and Metal Production through the Ages*. London: British Museum Press.

Crew, P. and Crew, S. (eds) 1990. *Early Mining in the British Isles*. Maentwrog: Plas Tan y Bwlch Snowdonia Study Centre.

Earl, B. 1986. Melting tin in the west of England: Part 2. *Historical Metallurgy* 20(1): 17–32.

Ford, T.D. and Willies, L. (eds) 1994. Mining before powder. *Bulletin of the Peak District Mines Historical Society* 12(3).

Goodway, M. 1987. More on wooden smelting furnaces. *Journal of Field Archaeology* 14: 383.

Holman, B.W. 1927. Heat treatment as an agent in rock-breaking. *Transactions of the Institute of Metallurgical and Mining Engineers* 36: 219–62.

Jenkins, D.A. and Timberlake, S. 1997. Geoarchaeological research into prehistoric mining for copper in Wales. Unpublished report to the Leverhulme Trust. Bangor: University of Wales.

Mercer, R.J. (ed.) 1981. *Grimes Graves Excavations 1971–2*, vol. 1. London: HMSO.

Merkel, J.F. 1990. Experimental reconstruction of Bronze Age copper smelting based on archaeological evidence from Timna. In *The Ancient Metallurgy of Copper*, B. Rothenberg (ed.). London: IAMS, 78–122.

Mighall, T.M., Timberlake, S., Grattan, J.P. and Forsyth, S. 2000. Bronze Age lead mining at Copa Hill, Cwmystwyth – fact or fantasy? *Historical Metallurgy* 34(1): 1–12.

Mighall, T.M., Abrahams, P.W., Grattan, J.P., Hayes, D., Timberlake, S. and Forsyth, S. 2002. Geochemical evidence for atmospheric pollution derived from prehistoric copper mining at Copa Hill, Cwmystwyth, mid-Wales, UK. *Science of the Total Environment* 292: 69–80.

O'Brien, W. 2004. *Mining, Metal and Society in Early Ireland*. Bronze Age Studies 6. Galway: Department of Archaeology, University of Galway.

Penhallurick, R.D. 1986. *Tin in Antiquity*. London: Institute of Metals.

Pickin, J. and Timberlake, S. 1988. Stone hammers and firesetting: a preliminary experiment at Cwmystwyth Mine, Dyfed. *Bulletin of the Peak District Mines Historical Society* 10(3): 165–7.

Reynolds, P.J. 1999. The nature of experiment in archaeology. In *Experiment and Design in Archaeology: Essays in Honour of John Coles*, A.F. Harding (ed.). Oxford: Oxbow Books.

Rostoker, W., Pigott, V.C. and Dvorak, J.R. 1989. Direct reduction to copper metal by oxide-sulfide mineral interaction. *Archaeomaterials* 3(1): 69–87.

Rowlands, J. 1966. *Copper Mountain I*. Llangefni: Anglesey Antiquarian Society.

Salkield, L.U. 1987. *A Technical History of the Rio-Tinto Mines*. London: Institution of Mining and Metallurgy.

Shimada, I. and Merkel, J.F. 1991. Copper alloy metallurgy in ancient Peru. *Scientific American* 256(1): 62–75.

Timberlake, S. 1990a. Review of the historical evidence for firesetting. In Crew and Crew 1990, 49–52.

Timberlake, S. 1990b. Firesetting and primitive mining experiments, Copa Hill, Cwmystwyth. In Crew and Crew 1990, 53–5.

Timberlake, S. 1994. An experimental tin smelt at Flag Fen. *Journal of the Historical Metallurgy Society* 28: 122–8.

Timberlake, S. 2002. Ore prospection during the Early Bronze Age in Britain. In *The Beginnings of Metallurgy in the Old World*, M. Bartelheim, E. Pernicka and R. Krause (eds). *Archaometrie – Freiberger Forschungen zur Altertumswissenschaft 1*. Freiberg, 327–57.

Timberlake, S. 2003. *Excavations on Copa Hill, Cwmystwyth (1986–1999): An Early Bronze Age Copper Mine within the Uplands of Central Wales*. BAR British Series 348. Oxford: Archaeopress.

Timberlake, S. 2005. In search of the first melting pot. *British Archaeology* 82: 32–3.

Timberlake, S. and Craddock, B. 2005. Experimental mining on Alderley Edge. Part 2: The manufacture of stone mining hammers: the record of a communal experiment on Alderley Edge. In Timberlake and Prag 2005, 192–7.

Timberlake, S. and Prag, A.J.N.W. (eds) 2005. *The Archaeology of Alderley Edge: Survey, Excavation and Experiment in an Ancient Mining Landscape.* BAR British Series 396. Oxford: John and Erica Hedges.

Tylecote, R.F. 1986. *The Prehistory of Metallurgy in the British Isles.* London: Institute of Metals.

Tylecote, R.F. and Boydell, P.J. 1978. Experiments on copper smelting based on early furnaces found at Timna. In *Archaeo-Metallurgy.* IAMS Monograph no. 1. London: IAMS, 27–49.

Willies, L. 1994. Firesetting technology. In Ford and Willies 1994, 1–8.

Yener, K.A., Adriaens, A., Earl, B. and Ozbal, H. 2003. Analysis of metalliferous residues, crucible fragments, experimental smelts, and ores from Kestel Tin Mine and the tin processing site of Goltepe, Turkey. In Craddock and Lang 2003, 181–97.

## Author's address

Simon Timberlake, 19 High Street, Fen Ditton, Cambridge CB5 8ST, UK (simon.timberlake@btinternet.com; www.earlyminesresearch-group.org.uk)

# On the Edge of success: the scientific examination of the products of the Early Mines Research Group smelting experiments

*Paul Craddock, Nigel Meeks and Simon Timberlake*

*ABSTRACT*    This paper records some of the scientific work carried out on the products of the smelting experiments described in the previous paper, and discusses some of the implications of the results. The products of some of the experiments which were only partially successful reveal stages in the process that are not normally preserved in more successful smelts. The sparcity of smelting remains from the British Isles in the Bronze Age means that it is often necessary to argue from negative evidence; indeed making metal while leaving no durable evidence was one of the principal objectives of the experiments. The smelting of the oxidised copper and tin ores were successful in this, but the smelting of the more prevalent mixed copper-iron-sulphide ores has proved more difficult. This suggests that very different extraction strategies may have been used in the British Isles, at least during the Bronze Age.

*Keywords:* smelting, experiment, copper, bronze, tin, Bronze Age, Britain, Ireland, sulphide, oxide, smelting, slag, matte, analysis.

## Introduction

As outlined in the preceding paper (pp. 27–36), the excavations of the Early Mines Research Group (EMRG) and other groups in the British Isles have uncovered the remains of Copper and Bronze Age copper mines but with little or no tangible evidence of smelting, with smelting sites identified, necessarily tentatively, at Ross Island in Ireland (O'Brien 2004) and at Great Orme in Wales (Chapman 1997). Specifically, little or no slag was found, slag being defined here as the once molten waste materials from the smelting process. A slagging process is defined as one in which the formation of a slag was deliberate in order to separate the forming metal from the waste materials in the furnace. The contemporary bronzes from Britain and western Europe generally have very low iron contents, typically below 0.05%, suggestive of rather primitive smelting under conditions that were only moderately reducing (Craddock and Meeks 1987).

The EMRG has initiated experiments to determine if it is possible to smelt both oxide and sulphidic copper ores leaving little or no durable debris and producing copper with a similar composition to that found in prehistoric bronzes. This paper describes the scientific examination of the products of smelting experiments carried out by the EMRG at various locations over the past few years as outlined in the previous paper. The examination of the products of the Alderley Edge, Cheshire, experiments smelting oxidised copper ores are fully described in Craddock *et al.* (2005), and the examination of the materials from the 2004 Butser experiments, including the smelting of both oxidised copper and tin ores, are briefly described in Craddock and Timberlake (2004) and Timberlake (2005), but the examination of the products of the more recent attempts to smelt chalcopyrite ores have not been previously reported.

## The techniques used

All the samples taken from the experiments were examined by optical microscopy and the ores and their various smelted product were analysed by X-ray diffraction (XRD) to establish the minerals present. The balls from the Alderley Edge experiments were radiographed. Sections cut from a selection of the balls and their prills, as well as from the matte obtained from the chalcopyrite smelts, were examined by scanning electron microscopy (SEM). Energy-dispersive X-ray microanalysis (EDX) analyses were obtained of some of the areas of metallic copper that had formed.

Analyses have already been published (Craddock *et al.* 2005) on a series of samples of the smelted products from the Alderley Edge and Butser experiments by atomic absorption spectrometry (AAS) to determine the amounts of copper metal produced. Further quantitative analysis was carried out by X-ray fluorescence (XRF) on the copper, tin and bronze metals and also by inductively coupled plasma spectrometry (ICP) to determine the trace elements associated with the copper ore and the metal to establish how these were modified by the smelting process.

## The smelting of malachite and azurite ores from Alderley Edge, Cheshire

Experimental copper smelting was carried out at Alderley Edge using pisoliths of azurite collected from a mudstone layer in the locality, adjacent to those that had been worked in antiquity.

The ores were crushed and moulded into egg-sized balls with wet charcoal dust, sawdust and animal dung following traditional smelting practice for non-ferrous metals in the Middle East and South Asia (Craddock 1995: 161; see also Timberlake, this volume, Fig. 6a, p. 32). After drying, these were smelted for between 1½ and 2 hours at 800–1100 °C variously in a bowl furnace, shaft furnace and in a blowpipe hearth (see Timberlake, this volume, Fig. 5, p. 32). On smelting, the organic materials burn out, creating the observed open structure through which the reducing gases could freely circulate (Fig. 3). On all occasions the part-smelted ore balls were recovered intact at the end of the smelt, suggesting that the temperatures had not been high enough and the smelting times too short. Some of the balls broke open, however, revealing droplets or prills several millimetres in diameter (Fig. 1) that had the appearance of being of copper (but see Fig. 3). Radiography suggested that many such 'prills' had formed within several of the balls (Fig. 2).

As the balls were quite robust it was decided to cut some of them in half with a diamond-wheel saw, preserving one piece intact for further study. The examination showed that in every instance some of the copper mineral had been reduced to metal prills. This suggests that their formation was not unusual, but rather the product to be expected from smelting under somewhat primitive conditions in the first stages of the operation. It was observed that the prills were concentrated towards the periphery of the balls with few in the centre, again suggesting that the firing time had been too short for their formation throughout the ball. This was especially noticeable on the blowpipe-smelted ball, which had by far the lowest overall content of metallic copper in the prills. One ball was gently broken up and the prills released revealing a variety of shapes, textures, colours and sizes, ranging from what appear to be droplets of metal to irregular lumps of fused material, from peppercorn to pea in size.

One of the balls was sectioned for SEM examination in order that the prills could be studied in the matrix in which they were forming when the firing ceased (Fig. 3). One of the pea-sized copper prills was also sectioned and examined to study its formation. It would seem that pieces of the ground ore shrank somewhat during the firing, bringing the individual particles of reduced copper into contact. These formed into a coherent skin defining the prill (Figs 4 and 5) which, together with some molten copper silicate, began to contract because of surface tension, leaving a space around the forming prill approximately 300–500 μm in width (Fig. 3). In a number of places the quartz fragments can be observed protruding from the skin, suggesting that the copper skin had moved over the quartz which remained immobile. Some of the copper inside the defining copper skin of the prills was reduced *in situ* while still solid and still in the original regular arrangement of the mineral. If the reaction had proceeded to its conclu-

**Figure 1** A broken clay ball from Alderley Edge revealing a copper 'prill'.

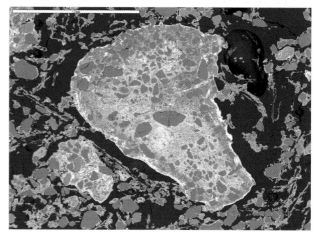

**Figure 2** Radiograph of a smelting ball: the copper shows bright within the clay matrix (S. La Niece/British Museum).

**Figure 3** SEM backscattered electron image of ball 3 after firing, in cross-section. Note the very open structure (black areas). The central light-coloured feature is compact, partly reduced ore mixture (termed 'prill' for convenience), containing quartz grains (grey), on which a skin of copper (bright) has formed (scale bar: 2 mm) (N. Meeks/British Museum).

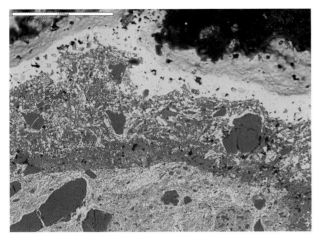

Figure 4 SEM backscattered electron image detail of the forming copper skin (bright layer, top) (scale bar: 200 μm) (N. Meeks/British Museum).

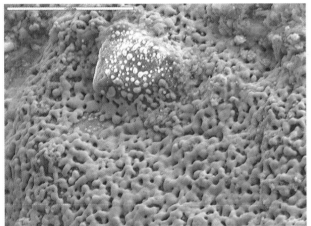

Figure 5 SEM backscattered electron image of the surface detail of the forming copper skin, showing the early stages of nucleation and growth (scale bar: 100 μm) (N. Meeks/British Museum).

sion the contracting copper skin would have picked up and melted the solid copper metal in its path, thereby increasing in thickness.

X-ray diffraction analysis of samples from the prills showed metallic copper, cuprite and silica to be present. In addition, a glassy phase visibly present in the balls probably contains copper silicates. Qualitative XRF analysis indicated that the iron content of the ore was very low compared to most copper ores. If iron had been present, the formation of at least some of the mineral delafossite ($CuO \cdot FeO$) would have been expected, such as has been found at several early smelting sites (Craddock 1995: 128), but none was detected here. The addition of iron oxides as a flux would have suppressed the formation of copper silicate, but would not have created a satisfactory liquid slag to remove the waste materials unless the temperature of the operations had been maintained at several hundred degrees higher than was actually achieved (under 1000 °C). The copper silicate was a problem, not so much because it stopped the reduction of copper oxide to metal, but rather because it hindered the metal from coalescing. Its formation here is probably due to the particular ore chosen at Alderley Edge, which was virtually iron free. Copper silicates would not really be expected to be regularly produced on

early smelting sites and anyway would be unlikely to survive long-term burial.

As the prills in all the balls were similar, it is possible that they had contracted as much as feasible under the prevailing smelting conditions. The bulk of the copper was still present as copper oxide but given more time and probably more reductant within the balls, the copper mineral would have continued to be reduced to copper and form into prills. The appearance of the majority of the copper metal shows it had only been molten very intermittently, if at all in some instances, with little or no chance to coalesce. This suggested that a higher temperature above the melting point of copper, more consistently maintained for longer periods, would have been necessary for the metal to coalesce. Given that some copper oxide was reacting with the silica throughout the balls, it is surprising that the edges of the fragments of crushed quartz from the ore are so crisp and angular, showing little evidence of attack by copper oxide even where the fragments are coated with metallic copper. The incomplete nature of the Alderley Edge experimental smelts effectively froze the process and enabled the actual formation of the copper droplets to be recorded and studied.

In 2004 and 2005 the experiments were repeated at Butser using high-grade malachite ore from Zambia in more or less the same range of smelting units and blowing arrangements but with more charcoal. This time, however, the experiments extended over five days instead of two and were operated by postgraduate students rather than members of the general public. Higher temperatures were maintained for longer periods resulting in successful smelts in which the balls disintegrated and runs and prills of copper were formed, some weighing several hundred grams. Samples were taken of four of the pieces from one of the smelting experiments for analysis by XRF. Three of the samples had approximately 0.01% iron but the fourth had 0.55%, reflecting differing conditions within the smelting hearth. In only one of the samples was arsenic detected (0.04%, detection limit 0.01%).

## Tin smelting at Butser, Hampshire 2004 and 2005

The EMRG had previously conducted some simple tin-smelting experiments at the Historical Metallurgy Society annual general meeting at Flag Fen, Peterborough in 1994 (Timberlake 1994) and this activity formed part of the programme of smelting experiments conducted at Butser in 2004 and 2005. This was partly to provide the material to make bronze but also, as tin is an easy metal to smelt, it acted as a failsafe to ensure that all the students did at least successfully smelt some metal during the course. The tin ore used was cassiterite washed from Cornish workings and contained between 10 and 20% iron and 5–10% arsenic, figures that are probably quite typical of the ore used in prehistory after beneficiation. The ore was smelted both in the dung balls similar to those used for some of the copper-smelting experiments and directly spread dampened onto the charcoal. Temperatures were typically in the range of 1000–1300 °C and the smelts lasted for between 2 and 3 hours. Metallic tin was successfully produced in all the experiments with little slag formation except for very minor amounts of tin silicates and tin-iron-calcium silicates coating some of remaining fuel and tuyeres together with the hearth bottoms and sides. These silicates would probably not survive burial for long periods.

In some of the experiments there was considerable volatilisation of the tin evidenced by white tin oxide deposits on the sides of the smelting units and on the tuyere. One tuyere developed a lip of vitrified tin silicate. The tuyere tip might survive burial and the presence of tin should still be evidenced as heavy metal contamination of the soil for millennia. In the tin-producing areas of southwest England this would be of little significance but elsewhere in Britain it would be an important indicator of tin smelting.

The product of the smelt was tin with about 2–5% arsenic and 5–10% iron. X-ray diffraction analysis showed that the iron was present as the intermetallic compound, FeSn, well known to tin producers in recent centuries as hardhead (Earl 1986, 1994). The purification of the tin is relatively simple, either by exposure of the molten tin to an oxidising atmosphere where the hardhead breaks down to tin and iron oxide or by simply re-melting. After re-smelting the metal smelted at Butser under more oxidising conditions the iron was all but eliminated but the arsenic was largely retained. Thus it is possible that the tin was actually supplying a small amount of arsenic to the Bronze Age alloys, in this case about 0.2–0.5% in a typical 10% tin-bronze. The persistence of arsenic in ancient copper alloys even after tin had replaced it as the main alloying agent has been noted (Craddock 1979).

It might seem strange that tin, which is easier to smelt than copper, should pick up any associated iron while copper does not. This is because the smelting process is a function not only of the free energy requirements but also of the melting point of the metal. Thus copper oxides are easy to reduce under conditions where the iron remains as a mineral, and as shown in the free energy diagram (Fig. 6), the copper has already been reduced to metal while still in the solid state but requires increased temperature before it can melt, by which time it is physically separate from the iron minerals. Conversely tin requires more exacting reducing conditions, very similar to those of iron, but the forming tin metal is already several hundred degrees above its melting point and can dissolve the iron forming at the same time. On cooling the iron is shed from solution as hardhead.

No tin ingots have been recognised from the Early Bronze Age in Britain but most early ingots from other parts of the world, notably the eastern Mediterranean, are remarkably pure. Significantly, hardhead has apparently not been encountered in the early ingots. One reason for this could have been that the very simplest non-slagging processes were neither hot nor reducing enough for hardhead to form. The processes carried out at Butser, however, could not have been simpler and certainly produced no recognisable slag, but the resulting tin contained quantities of hardhead. Thus it appears that the production of hardhead was more or less inevitable but it probably broke down during the re-melting of the freshly smelted metal to form more regular cast ingots.

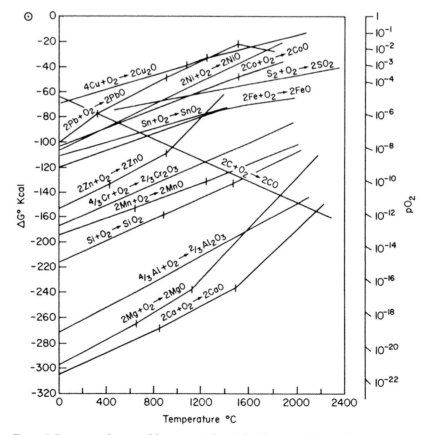

**Figure 6** Free energy diagram of the main metal–metal oxide system showing the main parameters necessary to reduce the oxides to metal. Note copper is reduced under considerably less reducing conditions than tin or iron (from Gilchrist 1967).

Our knowledge of the production of tin in prehistoric western Europe is sparse and only tiny quantities of tin metal have been found, especially when compared to the quantities held in the surviving bronzes. This led Jim Charles (1975) to suggest that perhaps tin metal was not normally produced at all, but instead the beneficiated tin ore could have been co-smelted with the copper ore or added straight to the molten copper and an alloy produced directly rather in the manner that brass was to be made somewhat later in antiquity (Craddock and Eckstein 2003). In 2005 at Butser the experiment was carried out of co-smelting the pure Zambian malachite with the Cornish cassiterite. Bronze was formed with approximately 10% tin (average of four XRF analyses). The metal also contained between 0.8 and 0.9% arsenic and 0.21–0.23% iron (cf. above with the much purer copper when the malachite was smelted by itself). Thus, the experiments conducted here strongly suggest that the iron, which the tin ore almost invariably contains, would have become incorporated in the bronze if the tin ore had been used directly. Thus the bronzes so formed would have had a substantial trace of iron, as in the Butser experiments, something which is not regularly encountered in the prehistoric bronzes from western Europe. From this we conclude that Bronze Age alloys are likely to have been made by mixing copper and tin metals.

## Primitive smelting of sulphidic ores

Although in general it is believed that the early copper smelters concentrated on the oxidised ores, most ore deposits consist predominantly of the primary sulphidic ores. Furthermore, some of the investigated deposits in western Europe with evidence of early exploitation would seem to have only contained sulphidic deposits, with chalcopyrite and lesser amounts of bornite as the usual minerals. This is well exemplified by the Copper Age mines at Ross Island, Ireland (O'Brien 2004) where the ores were of chalcopyrite with some of the copper arsenic sulphide mineral, tennantite, as well as at Cwmystwyth (Timberlake 2003) and Parys Mountain in Wales, where the main ore mined and smelted is believed to have been chalcopyrite. In particular at Ross Island, the careful excavation of the occupation areas immediately adjacent to the mine cautiously concluded that in the Copper Age the ores were likely to have been smelted there in simple pits, yet there were no traces of slag. The excavators at Ross Island attempted some simple smelting experiments, but chose to smelt the arsenical copper ore, tennantite, rather than the mixed copper-iron-sulphide ore, chalcopyrite.

It is well known that copper metal was often found in the heaps where the chalcopyrite ores had been roasted under oxidising conditions (Hofman 1914). Roger Doonan's (1994) roasting experiments also produced minute plebs of copper metal in the small quantity of slag that had formed. Rostoker et al. (1989) describe the smelting of copper-iron-sulphide ores by simple oxidising processes, although these produced quantities of durable slag. Attempts at smelting chalcopyrite ores by non-slagging processes were undertaken by one of the authors (ST) at Roewen, North Wales in 2004 with the ore in a crucible, and again as part of the Butser programme in

2005 with a replication of the Ross Island putative smelting pits (see Timberlake, this volume, pp. 27–36 for details). No flows of molten copper were produced but a sample of the ore together with the smelting products of these experiments was examined to ascertain what had been produced.

The ore used in both experiments was of chalcopyrite from a lode on Copa Hill, Cwmystwyth near to the ancient workings on the Comet lode (Timberlake 2003). A section of the ore was examined by SEM–EDX, which showed it to contain a homogenous copper-iron-sulphide mineral of uniform composition with a few cracks containing lead-rich mixed sulphides. The main product of the two smelting experiments was a dark dense mass that solidified in the bottom of the crucible in the Roewen experiment and in the base of the hearth in the Butser experiment.

The whole sample from Roewen was examined uncoated in the SEM using secondary and backscattered electron (BSE) imaging. This showed that the sample was a mixture of solid material with the appearance of having been molten or viscous,

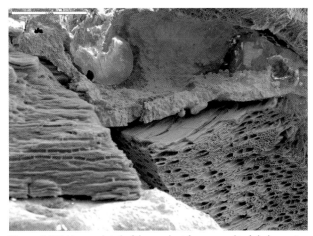

**Figure 7** SEM image detail of the product from a smelt of chalcopyrite ores by non-slagging processes, with the ore in a crucible, undertaken by Simon Timberlake at Roewen, North Wales in 2004. This detail shows (from top) a layer of partly reduced sulphide (matte) lying on two fragments of charcoal (scale bar: 1 mm) (N. Meeks/British Museum).

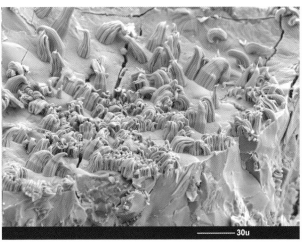

**Figure 8** SEM image of the surface of the matte in Figure 7, showing a field of copper cones produced from the reduction process (500×) (scale bar: 30 μm) (N. Meeks/British Museum).

41

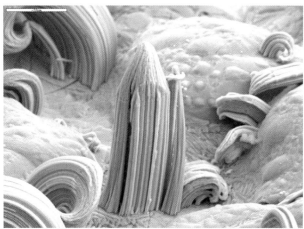

**Figure 9** SEM detail of Figure 8 showing linear growth of the copper cones, superficially with the appearance of having been extruded from the surface of the matte (1200×) (scale bar: 20 μm) (N. Meeks/British Museum).

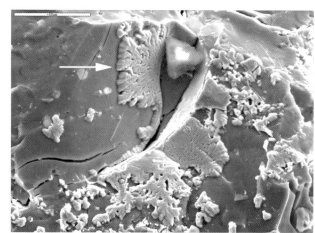

**Figure 11** SEM image of the fracture surface of the matte exposing copper 'dendrites'. At the edges (arrow), these low-profile dendrites have incipient grooves similar to those observed on the cones, suggesting the cones were built up by a similar process. Note how the dendrites have grown in the crack lower left centre (900×) (scale bar: 50 μm) (N. Meeks/British Museum).

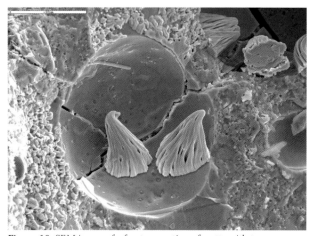

**Figure 10** SEM image of a fracture section of matte with two copper cones on the curved surface of a gas bubble. Note at top right, the flat base of a copper cone that has separated from the matte indicating that the cone formed on the matte surface and was not extruded from within (600×) (scale bar: 50 μm) (N. Meeks/British Museum).

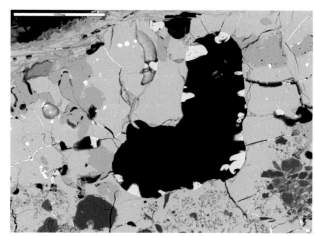

**Figure 12** SEM backscattered electron image (at 180×) of a polished section of matte cutting through a large gas pore (black) and copper cones (white) inside the pore (scale bar: 200 μm) (N. Meeks/British Museum).

containing trapped charcoal fragments (Fig. 7). Energy-dispersive X-ray microanalysis showed the material to be a mixture of copper and iron sulphides, the familiar matte of more recent copper-sulphide smelting. On the surface of the matte are fields of fluted copper cones (Fig. 8). At high magnification the copper cones show parallel vertical striations indicating linear growth (Fig. 9). The growth of some of the cones following the initial nucleation produced a fairly uniform shaft giving the appearance of extrusion. Surface analysis showed the cones to contain about 95% copper and 5% iron.

Examination of the products of the Butser experiments showed comparable features. Examination of the surfaces of the fractured specimens and on the mounted polished samples of the cindery matte fragments revealed some internal gas pores. At high magnification tiny grooved cones are visible no more than 0.3 mm in diameter (Fig. 10). They are similar to those of the crucible smelt but are generally shorter, perhaps caught at an earlier stage. Essentially they are the apexes of the cones of the previous experiment. The bases

of the cones are flat at the point of attachment and where they had detached from the matte they showed that there were no 'feeders' or 'roots' from within the matte (Fig. 10). Others appear to have grown dendritically in two dimensions having been constrained between cracks in minerals where they grew (Fig. 11). At their edges (arrow, Fig. 11), these dendritic spreads have the same grooved appearance at the point where they grow from the surface as the cones, indicating a similar growth process.

Polished samples were made to examine the relationship between the cones and the matte. Figures 12–14 show that the cones have grown on the walls of gas pores. Some copper formation was constrained in cracks in mineral grains (Fig. 13 left) as already shown by the fracture section (Fig. 11). The form of the cones is not the usual rounded blebs and irregular dendrites found in the slags etc. from early smelting operations. In section, some cones have gaps which are the residual interdendritic porosity (Fig. 13). The deposited copper metal must have been forming dendritically at temperatures very

**Table 1** EDX analysis and compounds identified in the false-coloured phases shown in Figure 15. Analysis of chalcopyrite ore from a lode near the ancient workings on Copa Hill, Cwmystwyth is given for comparison.

| | Atomic % | | | | | |
| --- | --- | --- | --- | --- | --- | --- |
| | Cu | Fe | S | Mg | Si | Type |
| Brown = quartz grains | | | | | 100 | oxide |
| Red = $Cu_{2.5} \cdot FeS_2$ | 45 | 17 | 38 | | | sulphide |
| Green = FeS | 1.5 | 50 | 48.5 | | | sulphide |
| Blue = Fe/Mg oxides in the ratio of 5 Fe/1 Mg | 0.5 | 84 | | | 16 | oxide |
| Light green = Fe/Mg oxides in the ratio of 2 Fe/1 Mg | 1.5 | 60 | 0.5 | 37 | | oxide |
| Chalcopyrite ore = $Cu\,Fe\,S_2$ | 27 | 27 | 46 | | | sulphide |

close to the melting point of the copper-iron alloy as the matte cooled (discussed below).

Analysis of the cross-section (Figs 14 and 15) allowed the approximate atomic % and stoichiometrical composition to be determined. From these the main mineral phases were identified (Table 1). The cones were of copper with approximately 1% iron (average of five analyses).

The chalcopyrite has transformed into matte, that is, two phases: copper-iron-sulphide with enhanced copper, and iron sulphide with a little copper. There are two Fe-Mg oxide phases of different composition. The magnesium was not in the chalcopyrite ore, but must have been in a significant quantity in the furnace charge judging by the amount in the samples examined.

## Discussion

The cones superficially gave the impression of having been extruded from the matte and it is surely significant that they

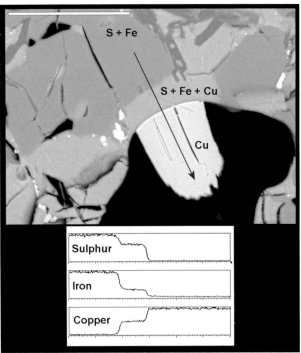

**Figure 14** SEM element line-scan across a polished section of matte passing through two mineral phases and into a copper cone. The mineral phases can be interpreted from the steps corresponding to changes in composition: dark grey Fe+S, light grey Cu+Fe+S, white Cu (1000×) (scale bar: 20 µm) (N. Meeks/British Museum).

have developed only on the surface of the regions containing copper sulphide, not on those of iron sulphides alone (Fig. 14). It is clear however, when viewed in cross-section that the cones grew on the surface of the matte – they had not been extruded as metal from within the matte. Also analytical line-scans in the SEM across the interface between the cones and underlying sulphide ore show no significant diffusion zones of changing copper concentrations. This suggests that the reaction zone is right at the interface between the forming copper

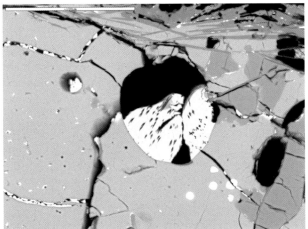

**Figure 13** SEM backscattered electron image detail (at 500×) of a polished section of matte cutting through a small gas pore with two cones. The copper cones have been sectioned and clearly show that they developed on the surface and were not extruded from the matte on which they sit. Note the interdendritic porosity exposed in the sectioned cones. The five small white globules lower right are lead-rich inclusions from the original chalcopyrite. Note also the tiny cones growing in the cracks on the left of the image, seemingly from both sides of the cracks (scale bar: 100 µm) (N. Meeks/British Museum).

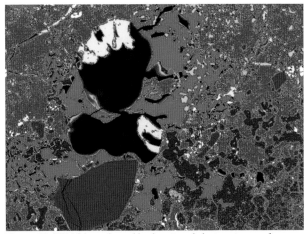

**Figure 15** SEM false-coloured backscattered electron image of a section of the matte. The phases were analysed by quantitative EDX analysis (see Table 1) and the false colour identifies the distribution of the various phases present: yellow = copper; brown = quartz grains; red = $Cu_{2.5} \cdot FeS_2$; green = FeS; blue = mixed Fe/Mg oxides in ratio of 5 Fe/1 Mg; light green = mixed Fe/Mg oxides in ratio of 2 Fe/1 Mg; white inclusions = Pb, Cu, Fe, Mg sulphides (400×). Width of view: 300 µm. (N. Meeks/British Museum.)

and the sulphide matte. The cones have nucleated and grown in the solid state out from the surface into the voids while growing wider at the base in a three-dimensional manner. The grooves in the cones appear to be the gaps between 'linear dendrites' which start to form as nucleating clusters around the base. In a way this mimics directional single crystal growth, more common in modern metallurgical processes. An explanation for the formation of the cones is that such copper as was produced during the smelt dissolved in the molten cuprous sulphide only to be shed as the mass cooled. Copper with a few per cent of iron and cuprous sulphide has approximately the same melting range, around 1100 °C, and a eutectic at about 1070 °C. If this melt cooled slowly, then, as the copper was shed from solution in the still molten sulphide, it would have time to form the dendritic cones and fluted columns observed on the surfaces.

Similar phenomena were noted in the matte smelting of copper in the 19th century, where the copper was known as moss copper. Percy (1861: 359–60) described how filaments of copper had formed in the cavities or on the underside of the matte where it had been cast onto sand. He further reported that 'Under the microscope the filaments present numerous minute, parallel and longitudinal or grooves as though they had consisted of bundles of extremely delicate fibres.' This would be a fair description of the cones in Figure 9. Furthermore Percy reported analyses that showed the copper to be relatively pure with approximately 0.4% sulphur and only traces of iron. He made the valid point that if the copper had been precipitated at a temperature above its melting point then the molten droplets would certainly have picked up other metals present in the matte and the copper would have been much more impure. The cones studied here are more impure, possibly due to a longer gestation or higher temperature.

## Conclusions

The scientific examination of the products of the EMRG smelting experiments suggests that it is possible to smelt oxidised copper ores without creating durable waste products, notably iron silicate molten slags. The few thin skins of alkali-fluxed silicate vitrification were similar to those found by Chapman (1997) and, under most conditions, are unlikely to survive for many years. The tuyeres, although very roughly made of clay dung and straw (see previous paper), did survive surprisingly well and should be durable during burial, especially the partially vitrified ends that had been exposed to high temperatures in the furnace. A complete but used tuyere has been exposed to the elements on the surface for three years now and shows no sign of weathering. The absence of tuyeres from mines in the British Isles rather suggests that either the smelting processes did not use tuyeres or the smelting took place away from the mines. The survival of a complete but apparently unused tuyere from a Middle Bronze Age burial at Ewanrigg in Cumbria shows that tuyeres were known in Bronze Age Britain, if only rarely attested (Bewley et al. 1992).

The composition of the copper produced in the oxide ore smelting experiments is commensurate with that produced in Bronze Age Britain with erratic but generally low iron contents. The experiments showed that tin is a more likely potential source of iron in bronze, as predicted by the free energy diagram. Unless the tin was refined there is a strong possibility that it would contribute substantial traces of iron to the bronze. There is no observed rise in the iron content of the copper-based metals in the Bronze Age following the introduction of bronze, strongly suggesting that bronze was made by mixing copper and tin metals, not by co-smelting copper and tin ores or adding tin ore to molten copper.

It seems that sulphidic ores were exploited from the inception of copper production in western Europe, although the general absence of slag heaps coupled with the low iron content of the contemporary copper alloys suggests the ores were smelted under generally poor reducing conditions during the European Bronze Age (Craddock 1999). There is increasing evidence for the smelting of sulphidic ores in processes that did produce a slag from the Copper Age (Bourgarit and Mille 2001; Bourgarit, this volume, pp. 3–14) although such processes seem not to have become prevalent until the end of the Middle Bronze Age in the Alpine region.

The EMRG attempts at smelting chalcopyrite ores reached the first stage of the process, a matte with some microscopic cones of moss copper. Even though this represents the simplest non-reducing stages in the process under only moderately reducing conditions, the two attempts made so far still had 0.5% and c. 5% of iron respectively in the copper. Furthermore it is difficult to see how the process could proceed further without producing a durable slag and copper with at least a substantial trace of iron.

It would seem that there are some basic misunderstandings in our perception of the production of copper in prehistory, at least as far as the British Isles are concerned. Possibly the ores selected really were exclusively oxides and the sulphides were rejected. It could be argued that the ore left at the workings was that which was not required, although this is refuted by the presence of sulphidic minerals in the waste from the ore-crushing operations at Ross Island and Copa Hill, Cwmystwyth. It is possible that all of the ore was removed to distant smelting sites that have not yet been recognised: John Jackson (1980) argued that the ore from the Early Bronze Age copper mines at Mt. Gabriel, Co. Cork could have been transported some hundreds of metres to the west-facing slopes and smelted there in windblown furnaces. This idea is apparently refuted by the smelting sites at Ross Island and at Pen Trwyn, Great Orme, which are not well sited to pick up the wind, and at Cwmystwyth, where the mine-workings are located at the top of a steep slope facing the prevailing wind, ideally suited for wind-assisted furnaces but there is no surviving evidence of smelting. Even if there are as yet unlocated furnaces, it does not address the problem of the low iron content of the contemporary copper.

It does seem most likely that, as at Ross Island for sulphidic ores and at Pen Trwyn, Great Orme for oxidised ores, a significant portion was smelted at the mines using processes which produced little durable debris. The EMRG experiments have demonstrated a number of different processes for the smelting of oxide ores that conform to the archaeological evidence, but the smelting of the more prevalent mixed copper-iron-sulphide ores remains problematic.

## References

Bewley, R., Longworth, I.H., Browne, S., Huntley, J.P. and Varndell, G. 1992. Excavation of a Bronze Age cemetery at Ewanrigg, Maryport, Cumbria. Appendix on the tuyere by P.T. Craddock. *Proceedings of the Prehistoric Society* 58: 325–54.

Bourgarit, D. and Mille, B. 2001. La transformation en métal de minerais de cuivre a base de sulfures: et pourquoi pas dès le Chalcolithique. *Revue d'Archéometrie* 25: 145–55.

Chapman, D. 1997. Great Orme smelting site, Llandudno. *Archaeology in Wales* 37: 56–7.

Charles, J. 1975. Where is the tin? *Antiquity* 49: 19–24.

Craddock, P.T. 1979. Deliberate alloying in the Atlantic Bronze Age. In *The Origins of Metallurgy in the Atlantic Bronze Age, Proceedings of the 5th Atlantic Colloquium*, M. Ryan (ed.). Dublin: Stationery Office, 369–85.

Craddock, P.T. 1995. *Early Metal Mining and Production*. Edinburgh: Edinburgh University Press.

Craddock, P.T. 1999. Paradigms of metallurgical innovation in prehistoric Europe. In *The Beginnings of Metallurgy*, A. Hauptmann, E. Pernicka, Th. Rehren und Ü. Yalçin (eds). *Der Anschnitt* Beiheft 4. Bochum: Deutsches Bergbau-Museum, 75–92.

Craddock, P.T. and Meeks, N.D. 1987. Iron in ancient copper. *Archaeometry* 29(2): 187–204.

Craddock, P.T. and Eckstein, K. 2003. The production of brass in antiquity by direct reduction. In *Mining and Metal Production through the Ages*, P.T. Craddock and J. Lang (eds). London: British Museum Press, 216–30.

Craddock, P.T. and Timberlake, S. 2004. Smelting experiments at Butser. *HMS News* 58: 1–3.

Craddock, P., Cowell, M. and Meeks, N. 2005. Appendix 1: Report on the scientific examination of materials from an experimental copper smelt at Alderley Edge. In *The Archaeology of Alderley Edge: Survey, Excavation and Experiment in an Ancient Mining Landscape*, S. Timberlake and A.J.N.W. Prag (eds). BAR British Series 396. Oxford: John and Erica Hedges, 211–15.

Doonan, R. 1994. Sweat, fire and brimstone: pre-treatment of copper ore and the effects on smelting techniques. *Journal of the Historical Metallurgy Society* 28(2): 84–97.

Earl, B. 1986. Melting tin in the west of England: Part 2. *Journal of the Historical Metallurgy Society* 20(1): 17–32.

Earl, B. 1994. Tin from the Bronze Age smelting viewpoint. *Journal of the Historical Metallurgy Society* 28(2): 117–20.

Gilchrist, J.D. 1967. *Extractive Metallurgy*. London: Pergamon.

Hofman, H.O. 1914. *The Metallurgy of Copper*. New York: McGraw Hill.

Jackson, J. 1980. Bronze Age copper mining in counties Cork and Kerry. In *Scientific Studies in Early Mining and Extractive Metallurgy*, P.T. Craddock (ed.). British Museum Occasional Paper 20. London, 9–30.

O'Brien, W. 2004. *Mining, Metal and Society in Early Ireland*. Bronze Age Studies 6. Galway: Department of Archaeology, University of Galway.

Percy, J. 1861. *Metallurgy 1: Fuel; Fire-clays; Copper; Zinc; Brass*. London: John Murray.

Rostoker, W., Pigott, V.C. and Dvorak, J.R. 1989. Direct reduction to copper metal by oxide-sulfide mineral interaction. *Archaeomaterials* 3(1): 69–87.

Timberlake, S. 1994. An experimental tin smelt at Flag Fen. *Journal of the Historical Metallurgy Society* 28: 122–8.

Timberlake, S. 2003. *Excavations on Copa Hill, Cwmystwyth (1986–1999): An Early Bronze Age Copper Mine within the Uplands of Central Wales*. BAR British Series 348. Oxford: Archaeopress.

Timberlake, S. 2005. In search of the first melting pot. *British Archaeology* 82: 32–3.

## Authors' addresses

- Paul Craddock, 56 St. Margaret's Street, Rochester, Kent ME1 1TU, UK (pcraddock@thebritishmuseum.ac.uk)
- Nigel Meeks, Department of Conservation, Documentation and Science, The British Museum, London WC1B 3DG, UK (nmeeks@thebritishmuseum.ac.uk)
- Simon Timberlake, 19 High Street, Fen Ditton, Cambridge CB5 8ST, UK (simon.timberlake@btinternet.com; www.earlymines-researchgroup.org.uk)

# Towards a functional and typological classification of crucibles

*Justine Bayley and Thilo Rehren*

*ABSTRACT*   Two approaches to crucibles classification are outlined. The first is based on technical attributes such as form, fabric and thermal properties. The second is based on functional categories: namely cementation, assaying and metal melting. In both classifications there is considerable variability within each of the defined groups – much of it due to technological and cultural choices. The identification of technical attributes can often be carried out in the field or museum, while identification of function frequently requires more invasive instrumental analysis. Despite their differences in approach, both typologies end up with similar groupings, reflecting a strong relationship between functional requirements and technical attributes of crucibles.

*Keywords:* crucible, metalworking, ceramic, melting, cementation, assaying.

## Introduction

Crucibles are a major and varied group of ceramic vessels. They can be defined as potentially movable reaction vessels in which high-temperature transformations take place, but with no permanent unidirectional airflow; it is the latter condition that separates them from furnaces (Rehren 2003). Crucibles have been used for thousands of years all over the world wherever high-temperature processes were carried out; they are thus truly cross-cultural artefacts. The examples given here all relate to metals, but other materials such as glass (Bayley 2000; Rehren 1997a) or artificial pigments (Heck *et al.* 2003) are also made or worked using crucibles. Common crucible processes are physical changes such as melting but a wide range also involves chemical reactions. It is the uses to which the vessels are put rather than specific material properties or stylistic attributes that define them as crucibles; indeed some crucibles were made as domestic pottery.

Investigation of crucibles provides information in three separate but related areas:

- their stylistic attributes such as shape, size and other design features that reflect their cultural context and date;
- the ceramic fabric, the choice and preparation of which may be culturally determined but also needs to be 'fit for purpose';
- and the technical function of the crucible.

The first two aspects are summarised as technical attributes, while the third one places crucibles in their functional categories.

Crucibles and the processes carried out in them can be very diverse. To increase our understanding of crucibles and to highlight the link between attributes and function, we propose a typology into which individual finds can be placed. In this way past work and experience can readily be applied to new material, bringing order out of the apparent chaos as well as deducing, at least tentatively, specific functions without necessarily having to employ scientific and often invasive analysis.

Archaeologists tend to classify objects first by shape – and then ask how they were used. As archaeological scientists, our view is different. While form has relevance, it is not the best basis for classifying crucibles; our research has shown that function is the prime factor of archaeological importance. The properties of crucibles and crucible materials limit their functionality but within these technical constraints a considerable range of culturally determined design solutions are found. We have developed two models with different starting points, which at first sight may appear contradictory but complement each other – and incidentally end up with similar groupings. These approaches are based on the technical attributes of the crucibles themselves, and on the nature of the processes carried out in them.

## Technical attributes

The first approach to classifying crucibles is based on functional requirements which are common to all cultures and periods because they are technically determined. For a crucible to function, it needs a combination of physical and chemical properties that are interrelated and affect vital process characteristics such as thermal shock resistance, melting speed, redox control and ease of manipulation. The variables that can be controlled are fabric, form and how the crucible is heated. These are not independent, however: for instance, even a poor fabric can be used provided the crucible form and the way it is heated are appropriate.

## Crucible fabrics

Crucibles must be strong enough to hold the weight of the metal they contain, at high temperatures as well as when cold. In general, the fabric must be sufficiently refractory so it does not soften at high temperatures, which usually means a ceramic high in silica and/or alumina and low in iron, alkalis, and alkali earth elements that act as fluxes. It must also be relatively inert otherwise it will react with metal oxides (especially lead oxide) and alkalis from the fuel ash. The selection of specific clays for crucible fabrics is not seen regularly before the Roman period; until then, the required properties were achieved by adding different tempers or choosing specific shapes.

The main temper chosen for crucible fabrics throughout the Bronze Age was organic material, such as straw or chaff. This increased thermal insulation due to the voids left when it burnt out, as did walls several centimetres thick (Fig. 1). This was important in crucibles that were heated from above as the underside of the crucible would then remain relatively cool. In these cases, less refractory clays could be used as the cooler outer zone provided the required strength even if the inside had been melted or fluxed. The transition from heating from above to heating from the side or below is characterised by a change to thinner walls and to mineral temper, mostly quartz or igneous rock fragments, often together with selecting lower iron (white-firing) clays (Fig. 2). The thin walls and mineral temper facilitate heat transfer through the fabric and the temper increases its refractoriness.

A number of specialist fabrics were also used for metalworking including bone ash, which was used to make cupels (see below), graphitic clays, and even pure graphite (Hochuli-Gysel and Picon 1999). Graphite is a highly refractory mineral form of carbon with excellent thermal conductivity; in addition, it acts in a similar way to inclusions of vegetable matter – both help produce reducing conditions within the crucible.

## Crucible form

The overall crucible form is determined by the refractoriness of its fabric and hence the way in which it was heated. As noted above, crucibles that were not refractory had shallow open forms and were heated from above. More refractory fabrics allowed heating from below, and the vessel form (and fabric) was modified to increase heat transfer from below while reducing the loss of heat from the upper surface of the melt. This meant the crucible diameter was smaller relative to its height, and the shape became more closed – in extreme cases with only a narrow opening near the top (Fig. 3d). Additional features that may or may not be present affect the ease of use of crucibles of all types. Examples are the provision of knobs or handles (Fig. 3 and d–e), pouring lips or spouts (Figs 3a–b and 4b), which can aid the manipulation of the vessel when pouring molten metal into a mould – though it is notable how

**Figure 1** Cross-section through the rim of a Late Bronze Age bronze-melting crucible from Qantir-Piramesses, Egypt. It is made from non-refractory Nile silt and heated from above/inside, resulting in extensive vitrification and bloating of the ceramic. Wall thickness *c.* 20 mm.

**Figure 2** Early medieval bronze-melting crucible from Novgorod, Russia. It is made from refractory clay tempered with quartz to improve its thermal performance and has been heated from below/outside. Wall thickness 6–10 mm.

few crucibles carry the imprint of the tools that must have been used to manipulate them in the fire.

Most Roman to medieval crucibles have rounded or pointed bases as these have good stability when placed in a heap of red-hot charcoal as well as thermal shock resistance (Figs 3 and 4). Most top-fired crucibles were flat-bottomed vessels as the fire was above them. More sophisticated hearths or furnaces with an iron grate to support the crucibles required flat bases for stability (Fig. 5); thus the crucible form must suit that of the hearth in which it is to be heated.

The control of redox conditions within vessels is also important as oxidation leads to the loss of metal into massive crucible slags, which is usually not desired. Deep rather than shallow forms and luted-on lids (for example Fig. 3a) increase reducing conditions within the crucible and the use of carbon-

**Figure 3** Roman and early medieval crucibles from the British Isles, mostly used for melting copper alloys, not to the same scale: (a) reconstruction of lidded type, height *c.* 55mm (Dinas Powys, after Alcock 1963); (b) open form, diameter 65 mm (Doncaster, this example used for melting silver); (c) hand-made thumb pots, diameter 45 mm (York); (d) with handle and rim pinched together, length *c.* 75 mm (Ipswich); (e) handled, used for gold melting, height 25 mm (Dunadd); (f) wheel-thrown, diameter 84 mm (Snodland); (g) wheel-thrown with added outer layer, partly broken away, height *c.* 120 mm (Dorchester).

**Figure 4** Viking Age (10th century) melting crucibles from (a) Haithabu, Germany (height 35–40 mm); (b) York (diameter *c.* 80 mm); (c) Dublin, Ireland (height 34 mm). Those from Haithabu were used to melt copper or brass; the other types were used for melting both copper alloys and silver.

**Figure 5** Flat-bottomed 16th-century crucibles from the Tower of London, probably used for melting copper alloys: (a) coarse refractory fabric, rim diameter 67 mm; (b) finer fabric, rim diameter 108 mm.

rich fabrics also helps to counteract oxidation. Despite this, many crucibles used for metal melting had shallow open forms. Heating was effected by radiation from the fire above, and the red-hot charcoal would have limited the air that reached the metal. In these cases the position of the crucible, relative to the tuyere or blowing hole in the hearth, was critical as the air blast introduced oxygen, making the fuel burn more fiercely and increasing the temperature but also potentially oxidising the metal in the crucible.

For some processes, such as scorification and cupellation, oxidising conditions are essential, and in these cases the vessel form is always a shallow open one (Fig. 6). Conversely, a lid is essential for cementation processes that require reaction of the charge with a vapour phase.

### Thermal properties

Crucibles had to allow the temperature of their contents to be raised relatively easily, while also providing thermal insulation to retain the heat and keep the charge or melt at the desired temperature. The early top-fired vessels therefore had a wide opening and were relatively shallow, typically not deeper than

hemispherical, and their fabrics were designed to be insulating. The thermal insulation of the ceramic was crucial both to maximise heat retention in the vessel and to guarantee its mechanical integrity at high temperatures, as discussed above. Crucibles heated in this way normally have a distinctive pattern of surface vitrification – mainly on and inside the rim (Figs 1 and 7).

For vessels heated from below, the ratio of surface area to volume was particularly important as the heat needed to drive transformations within the volume was gained from the outer or lower surface of the crucible. The volume of a body increases as the cube of its dimensions, while its surface area only increases as the square of its size. Furthermore, heat transfer into the crucible increases with increasing temperature difference between the inside and the outside. These constraints mean that most crucibles heated from below are relatively small in size, are made from more refractory fabrics, tend to be relatively taller, and have narrow rather than wide

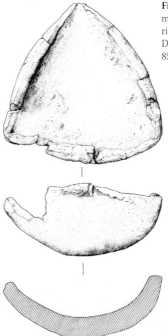

**Figure 7** Iron Age bronze-melting crucible with vitrified rim from Gussage All Saints, Dorset. Maximum width 85 mm (after Wainwright 1979).

**Figure 6** Examples of the four main groups of late 16th-century bone ash cupels from Oberstockstall, Austria. The smallest measure 30–35 mm across, the largest 85–100 mm.

mouths – to reduce heat loss through the surface and prevent oxidation of the charge. In these cases the vitrification is on the outside, mainly on the base (Fig. 3d–g).

In the context of early crucibles, refractoriness is a relative term, so defining a ceramic as refractory is best done by comparison with domestic pottery from the same area and period. The technical ceramic used for a crucible should only be called 'refractory' if it is more heat-resistant than the local domestic wares, either because of the added temper or through choice of particular clays. Some crucible processes require more highly refractory ceramics than others.

Many thin-walled, externally heated crucibles have an added outer layer of less refractory clay which is often deeply vitrified (Figs 3g and 8). This sacrificial layer had several functions. It provided an insulating layer of highly viscous material, distributing the heat more evenly and reducing thermal shock. This insulating effect may appear to contradict the requirement that externally heated crucibles are thermally conducting, but the glassy nature of the added layer means it is relatively conducting, so on balance their overall effect is positive. The added layer would also seal any cracks in the crucible proper during use thereby reducing the likelihood of breakage and metal loss. It protected the crucible from being fluxed by the fuel ash in the fire, so helping maintain its structural integrity and strength, and it increased the crucible's thermal capacity, helping to keep the melt liquid for longer during casting.

The discussion above has shown how many features can be seen either as constraining the crucible design or resulting from the technical choices made. No one attribute can therefore be considered in isolation; indeed particular combinations of features are characteristic of specific uses. Although most of the examples are metal-melting crucibles, the same technical attributes are also relevant for other crucible processes, which leads to the second model.

## Functional categories

The second model starts from the premise that crucibles are reaction vessels whose contents undergo a variety of chemical and/or physical changes. Three main groups of processes are considered: cementation, assaying and melting. It is the nature of these processes that fundamentally determines the vessel form and the ceramic fabric, and is used to classify the crucibles (Rehren 2003). Cultural traditions then modulate the crucible design within the technically determined classes.

### Cementation

This general term is applied to a group of processes where chemically active vapour phases produce chemical transformations. They all produce metal of a specific composition, in some cases the process is effectively alloying, in others refining. Cementation processes are usually reactions of solids with a vapour phase, which require carefully controlled temperatures and atmospheres so the vapour phase stays in contact with the other ingredients. The crucibles must thus be essentially closed vessels, though unless the fabric is sufficiently porous a small opening is essential to relieve any build-up of pressure. The processes involved are typically endothermic, requiring a constant supply of heat. In order to maximise the ratio of surface area (= heat input) to volume (= heat use), these vessels are usually relatively small and/or tubular rather than

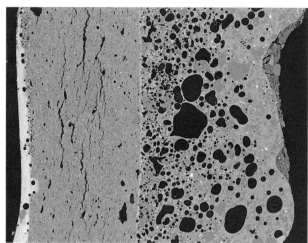

**Figure 8** Backscattered electron image of a cross-section through a late Roman bronze-melting crucible from Xanten (Colonia Ulpia Trajana), Germany. It is made from refractory clay and has a thin lead-rich slag layer on the inner surface (left). The outer surface is covered in a layer of less refractory clay which is heavily vitrified and bloated (right). Image width *c.* 20 mm.

**Figure 9** Roman domestic lidded pot from London used as a parting vessel. Lid diameter 150 mm.

spherical in shape; in the case of parting, the relatively low temperatures used mean the vessel shape is not important. The processes used as examples here are parting gold from silver, brass-making and crucible steel-making.

The parting vessels from Sardis, dating to the 6th century BC, are probably the earliest true cementation vessels. In these, finely dispersed native argentiferous gold was embedded in a salt-rich matrix and heated to a sufficient temperature to facilitate the reaction of chloride from the salt with the silver in the gold to form volatile silver chloride, leaving behind the now porous, but pure, gold particles (Ramage and Craddock 2000). In Britain, examples of parting vessels are known from the mid-1st century AD onwards (Bayley 1991, 2001) but all come from small-scale urban contexts and seem to have been used to purify precious metal that was being recycled. The process continued in use into the medieval period (for example Bayley 1992a: 751–4) but was eventually superseded by a wet process using nitric acid in the later Middle Ages/Renaissance (Bayley 1996).

The rarity of parting vessels in the archaeological record is partly due to lack of recognition. Even domestic pots could be used (Ramage and Craddock 2000; Marsden 1975: 100; Fig. 9) as a highly refractory fabric was not required because the process took place at relatively low temperatures; the vitrified deposits that are typical of many crucible processes are often absent. The key diagnostic feature is a bleached appearance to the inside of the vessels, often combined with a pink or purple colour due to the presence of specular hematite, a by-product of the reaction with salt (Bayley 1991; Fig. 9). Many vessels were probably crushed to extract the silver they contained by smelting, which is a further reason for the scarcity of archaeological finds.

Brass-making on a considerable scale developed in the Roman world in the 1st century BC. The lidded crucibles contained finely divided copper mixed with crushed calamine ore ($ZnCO_3$) and charcoal and were heated at around 900 °C, below the melting point of the copper; once again a large surface area of solid metal was essential so the zinc vapour could diffuse in. These crucibles were usually small in size (for example Bayley 1984; Rehren 1999a) as this was beneficial for the energy balance of the process. Increased production used larger numbers of vessels, rather than increasing the size of the individual vessels (Rehren 1999a) though there are exceptions (Picon *et al.* 1995). After a hiatus of several hundred years, brass-making in central Europe started again in the medieval period, initially in open vessels (Rehren *et al.* 1993). Improvements in ceramic technology and furnace designs, together with a better understanding of the metallurgy involved, led to a gradual increase in crucible sizes and the re-emergence of lidded vessels (Martinon-Torres and Rehren 2002; Rehren 1999b). The temperature regime changed, too, from the Roman solid-state process; once the cementation reaction had finished the temperature was raised to melt and homogenise the charge within the same vessel.

Diagnostic features of many of these brass-making crucibles are their poorly refractory fabrics, particularly during the Roman period, and the high levels of zinc detectable in all of them; frequently, the clay used for the cementation vessels is less refractory than that used for brass-melting crucibles in the same workshop (Martinon-Torres and Rehren 2002). The

highly coloured vitreous deposits that are typical of metal-melting crucibles are usually absent from these cementation vessels.

Pre-industrial crucible steel-making was concentrated in central and south Asia; in both regions, large-scale production sites are known with very large numbers of crucible fragments. Rehren and Papachristou (2003) provide a compilation of the available archaeological and technological data. The thermal requirements for cementation crucibles for steel-making are very different from those for parting or brass-making. Steel melts at about 1400–1550 °C, temperatures outside the typical thermal stability range of archaeological ceramics. The two steel-making areas had contrasting ways to achieve the required thermal stability of the crucibles. The central Asian ones are relatively big with a volume of up to one litre, made from a very dense, almost white-firing fabric with a deliberately produced opening in the lid (Fig. 10, left and centre). Those from south India and Sri Lanka are highly porous, small or medium in size, tightly closed and black (Fig. 10, right).

The ceramic material used in central Asia is based on rich kaolinitic clay, with typically less than 5% oxides other than alumina and silica. The crucibles were test-fired in the 1960s and found to be stable up to 1650 °C (Abdurazakov and Bezborodov 1966), well above the required temperature. This allowed the production of large, thin-walled crucibles, based on the availability of this highly refractory clay.

The clays available in the south of the Indian subcontinent contain high levels of iron oxide and are less rich in alumina than those from central Asia. There is evidence that specific clays were being selected for these crucibles but their iron oxide content was too high to make them sufficiently refractory despite their reasonably high alumina content. To compensate, large amounts of rice husk were added to the clay which on firing reduced the iron oxide to iron metal, finely dispersed in the ceramic matrix and no longer acting as a flux (Freestone and Tite 1986). In addition, the rice husk contributed large quantities of free silica which further increased the refractoriness of the clay. The ceramic became suitable for crucible steel production, but only just, and only for relatively small vessels.

Overall, there emerges a clear relationship between the functionality of a cementation crucible and the technical attributes required. Key among these is the closed shape necessary to maintain the particular gaseous atmosphere inside the vessels. The fabric, on the other hand, only matters for the hotter cementation processes, in particular for steel-making; at the lower temperature end, normal domestic pottery could do the job. It must be stressed though that not every closed crucible is from a cementation process. A number of closed melting and casting crucibles are also known, primarily as a protective measure to prevent oxidation of the contents (for example Bachmann 1976).

## Assaying

Fire assay, the practice of testing unknown substances by subjecting samples of them to a series of chemical or metallurgical operations carried out in small crucibles, probably developed

**Figure 10** Typical steel-making crucibles from 11th-century Akhsiket, Uzbekistan (left and centre) and 19th-century Sri Lanka (right); both are lidded. The Uzbek crucibles are made from white-firing dense kaolinitic clay tempered with quartz sand, are *c.* 300 mm tall and *c.* 80 mm wide internally, and have a small opening in the lid. The Sri Lankan crucibles are about 200–250 mm total length and *c.* 30 mm wide internally. Their ceramic is black and porous due to a high amount of rice husk temper, and their lids are pierced with a few hair-fine vents.

during the Middle Ages in the context of mining and coin production (Rehren 1997b). By the 16th century, fire assay was standardised across Europe, with almost identical remains known from Norway to Portugal to central Europe and into Turkey. The main groups of assaying vessels are scorifiers, which are flat dishes for oxidising operations, crucibles, often triangular at the top and circular near the bottom, and bone-ash cupels for the final determination of the gold and silver content of a sample, concentrated in lead bullion.

One of the key objectives of assaying, as spelled out by Georgius Agricola (Hoover and Hoover 1950), is to emulate on a small scale the metallurgical operations typically done in smelting furnaces. The emphasis, therefore, is again on chemical transformations rather than simple melting. The ceramic materials used for fire assay vessels had to meet specific demands for both thermal and chemical refractoriness; the latter required withstanding attack by liquid metal oxides and aggressive fluxes at high temperatures. In addition, the vessels had to be suitable for quantitative processes, so there was virtually no loss of material during the operations.

These functional requirements are reflected in the technical attributes of the three main assaying vessel types. The scorifiers are shallow, open and often rather thick but not particularly refractory; they are designed for general oxidising processes where some reaction between the slag that forms and the ceramic is acceptable. The triangular crucibles are more typically used for reducing operations or those involving particular atmospheres; hence, they are deep, often closed with a lid or cover, and of highly refractory ceramic material (Martinon-Torres 2005). Their walls are thin to promote heat transfer and their rim shape facilitates decanting the contents. The most characteristic assaying vessels are known as cupels; they are small, shallow dishes used to separate noble metals from a surplus of lead metal which is oxidised to lead oxide, while the noble metals remain metallic. For this process they have to be open, to promote oxidation, and have a highly absorbent fabric, to soak up the liquid lead oxide while retaining the liquid metal at the surface (Fig. 6). Silica-rich materials are not suitable for this as the silica reacts with the lead oxide to form a viscous melt which will stop further absorption. Throughout the Bronze Age and into the Roman period, people used open hearths lined with calcareous materials such as clay marls, crushed shells and/or bone ash for this process (Bayley and Eckstein 2006). From late- and post-medieval times cupels were made of finely ground bone ash. The advantage of this material was that it absorbed the lead oxide through capillary action without chemically reacting with it (Rehren and Klappauf 1995).

In summary, assaying vessels have very particular shapes (wide, shallow dishes for scorifiers and cupels, with no pouring lip) and materials, for instance bone ash cupels, or triangular crucibles of either Hessian or graphitic ware (Martinon-Torres and Rehren 2005). Of the assaying vessels, only the triangular crucibles could also be used for either cementation or melting. The technical attributes of scorifiers and cupels are totally determined by their function; they are highly specific in their use, and their shape and fabric are thus diagnostic of assaying.

## *Melting*

Melting is the third and most common functionally defined process. The key requirements of metal-melting crucibles are to hold the charge, to contain the heat, and to maintain a neutral to reducing atmosphere. They are the most widely found types of crucibles, with examples known from the inception of metallurgy in the Late Neolithic up to the present day. Melting is essentially a physical transformation, though sometimes metals are alloyed while being melted, for example tin can be added to molten copper to make bronze. Once molten, the metal is usually poured into a mould.

These functional aspects – heating the charge to fusion and then casting it – determine the necessary technical attributes. The need for casting often leads to the presence of a spout and sometimes a handle (Figs 3 and 4). It also requires the vessels to be small and strong enough that they can be moved and tilted with the liquid metal inside. Therefore, crucible volumes

rarely exceed one litre or about 10 kg of metal prior to the industrial revolution (Bayley 1992b). The need for heating is met in two fundamentally different ways: either from above through direct contact of the heat source with the charge, or indirectly by heating the crucible from underneath until its contents become molten. The former leads to open vessels with thick insulating fabrics, the latter to more closed shapes with thin, thermally conductive fabrics. The different technical attributes required to meet these demands, and their developments over time, have already been discussed in more detail above.

## Discussion

Studying crucibles provides both technical and cultural information, about the crucibles themselves, their function, and the intentions, skills and activities of the workmen using them. The technical attributes and traces of use provide mostly technical information. Understanding the functional constraints on the one hand, and the particular design solutions such as the quality of the fabrics relative to the local domestic or contemporary technical wares on the other, gives a wealth of culturally relevant information. Based on this wider picture, one can discuss independent development of technical solutions (Rehren 1999b) or adaptation to particular environmental situations (Rehren and Papachristou 2003).

European metal-melting crucibles can be used as an example that draws together the common themes from the two different approaches outlined above. It demonstrates the relevance of technical attributes to the subdivision of this functionally defined class of crucibles. Table 1 summarises a crucible classification based on technical attributes or design features. Note that these technically determined groups have a chronological significance; often there is increasing technical sophistication with time. The attributes in the left-hand column are those commonly found in early crucibles; in British and European contexts these are prehistoric ones – dating to the Bronze and Iron Ages. The centre column shows the features that then develop, the italics indicating changes that are found on Roman and earlier medieval crucibles across Europe. The right-hand column shows further changes, again italicised, which are typical of later medieval and post-medieval crucibles in the same region.

Within each of these three groups there are culturally determined variants. Figure 3 shows some of the considerable variability in form and fabric that is found within Roman and early medieval crucibles from Britain. They all share the

**Table 1** Technical attributes of crucibles.

| small size | small size | *larger size* |
|---|---|---|
| shallow form | *deep form ± lid* | deep form |
| ceramic not refractory | *refractory ceramic* | refractory ceramic |
| thick walls | *thin walls* | thin walls |
| heated from above | *heated from below* | heated from below |
| round/pointed base | round/pointed base | *flat base* |

⟶ Time

attributes in the central column of Table 1 and, although some of the variability is due to the range of processes carried out in them, most is culturally determined – so like many other archaeological finds, they can be used to identify the date and cultural affinities of their users. Understanding process requirements allows us to identify which attributes are technically required and which are culturally determined, through 'technological styles', available raw materials, and economic frameworks of workshop associations and governance.

Examining the examples in Figure 3 further we can identify some of the variables that would not have significantly affected their function and can therefore only be attributed to cultural choices. Most of the Roman crucibles are wheel-thrown (Fig. 3f–g) (although some simple hand-made forms continue in use). This reflects the widespread use of the wheel for making domestic pottery; indeed some of the crucibles are of forms also common in domestic assemblages. The post-Roman period is the first time that lids and handles are found regularly on metal-melting crucibles in the British Isles (Fig. 3a–c). They are known in the Roman period, but mainly for specific applications such as brass-making crucibles or parting vessels, i.e. on cementation vessels where they are functionally required (for example Bayley 1984: figs 1 and 2; 1992a: fig. 320).

In the Viking world a range of crucible types is also found. The tall, thin thimble-shaped crucibles (Fig. 4a) are typical of the Scandinavian homelands but are only occasionally found, presumably as imports, in other lands conquered by the Vikings. At this period the dominant crucible types in northern England, and to a limited extent in the south, were made from Stamford Ware, a fine-grained and highly refractory fabric used to make a range of bi-conical and bag-shaped forms (Bayley 1992a: fig. 322; Fig. 4b). In Viking Dublin a few crucibles of both these types are found, but the majority are hand-made crucibles with a pointed base and a D-shaped or triangular rim (Fig. 4c). These are similar to pre-Viking crucibles from sites such as Garranes (Ó Ríordáin 1942: fig. 25). All these forms are found in a range of sizes (for example Bayley 1992a: fig. 322).

Figures 3 and 4 show that within the one functional group there are many forms. All are culturally determined design solutions – but function comes before form.

## Conclusions

Starting from two different approaches, we have shown that similar groupings arise from both. We have come a long way towards a classification system for crucibles. It should however be noted that there are grey areas, so in some places no hard and fast divisions can be made. We have shown that to be useful, any crucible typology must be technically based but can focus primarily either on technical attributes or on functional classes. The identification of technical attributes can often be carried out in the field or museum, while identification of function often requires more invasive instrumental analysis. It is important in this context to realise that technical and functional typologies are simply two facets of the same complex relationship between cultural and technological traditions, practical requirements, individual know-how, economic

conditions and available materials. The technical attributes directly reflect the functional requirements; hence, these are not independent systems but closely linked projections of the same multidimensional entity. Therefore, both approaches lead to similar groupings because specific technical attributes are necessary if the crucible is used for a particular process, and individual processes demand specific attributes of the crucibles used. Generally, cultural affinity shows itself in variations in properties that do not affect functionality – such as shape, or in the choice of a particular design solution.

## Acknowledgements

We would like to thank all the excavators who have allowed us to study their crucibles and other metalworking finds. Figures 3b–3g, 4b, 5a, 5b and 9 are all © English Heritage. Figures 1, 2, 4a, 4c, 6, 8 and 10 are © the authors.

## References

Abdurazakov, A. and Bezborodov, M. 1966. *Medieval Glasses from Central Asia*. Tashkent: Academy of Sciences.

Alcock, L. 1963. *Dinas Powys*. Cardiff: University of Wales Press.

Bachmann, H.-G. 1976. Crucibles from a Roman settlement in Germany. *Historical Metallurgy* 10(1): 34–5.

Bayley, J. 1984. Roman brass-making in Britain. *Historical Metallurgy* 18(1): 42–3.

Bayley, J. 1991. Archaeological evidence for parting. In *Archaeometry '90*, E. Pernika and G. A. Wagner (eds). Basel: Birkhäuser Verlag, 19–28.

Bayley, J. 1992a. *Non-Ferrous Metalworking at 16–22 Coppergate*. The Archaeology of York 17/7. London: Council for British Archaeology.

Bayley, J. 1992b. Metalworking ceramics. *Medieval Ceramics* 16: 3–10.

Bayley, J. 1996. Innovation in later medieval urban metalworking. *Historical Metallurgy* 39(2): 67–71.

Bayley, J. 2000. Glass-working in early medieval England. In *Glass in Britain and Ireland AD 350–1100*, J. Price (ed.). British Museum Occasional Paper 127. London, 137–42.

Bayley, J. 2001. Precious metal refining in Roman Exeter. *Proceedings of the Devon Archaeological Society* 59: 141–7.

Bayley, J. and Eckstein, K. 2006. Roman and medieval litharge cakes: structure and composition. In *Proceedings of the 34th International Symposium on Archaeometry, 3–7 May 2004, Zaragoza, Spain*, J. Pérez-Arantegui (ed). Zaragoza: Institución 'Fernando el Católico', 145–53. (http://www.dpz.es/ifc/libros/ebook2621.pdf)

Freestone, I. and Tite, M. 1986. Refractories in the ancient and preindustrial world. In *High-Technology Ceramics: Past, Present and Future*, D. Kingery and E. Lense (eds). Westerville, OH: American Ceramic Society, 35–63.

Heck, M., Rehren, Th. and Hoffmann, P. 2003. The production of lead-tin yellow at Merovingian Schleitheim (Switzerland). *Archaeometry* 45: 33–44.

Hochuli-Gysel, A. and Picon, M. 1999. Les creusets en graphite découverts à Avenches/Aventicum. *Bulletin de l'Association Pro Aventico* 41: 209–14.

Hoover, H.C. and Hoover, L.R. (trans.) 1950. *Georgius Agricola's De Re Metallica*. New York: Dover Publications.

Marsden, P. 1975. The excavation of a Roman palace site in London, 1961–1972. *Transactions of the London and Middlesex Archaeological Society* 26: 1–102.

Martinon-Torres, M., 2005. *Chymistry and Crucibles in the Renaissance Laboratory: An Archaeometric and Historical Study*. PhD dissertation, Institute of Archaeology, University College London.

Martinon-Torres, M. and Rehren, Th. 2002. Agricola and Zwickau: theory and practice of Renaissance brass production in SE Germany. *Historical Metallurgy* 36(2): 95–111.

Martinon-Torres, M. and Rehren, Th. 2005. Ceramic materials in fire assay practices: a case study of 16th-century laboratory equipment. In *Understanding People through their Pottery: Proceedings of the 7th European Meeting on Ancient Ceramics (Emac '03)*, I. Prudêncio, I. Dias and J.C. Waerenborgh (eds). Lisbon: Instituto Português de Arqueologia (IPA), 139–48.

Ó Ríordáin, S.P. 1942. The excavation of a large earthen ring-fort at Garranes, Co Cork. *Proceedings of the Royal Irish Academy* 47(c): 77–150.

Picon, M., le Nezet-Celestin, M. and Desbat, A. 1995. Un type particulier de grands récipients en terre réfractaire utilisés pour la fabrication du laiton par cémentation. *Société Française d'Étude de la Céramique Antique en Gaule, Actes du Congrès de Rouen*, 207–15.

Ramage, A. and Craddock, P. 2000. *King Croesus' Gold*. London: British Museum Press.

Rehren, Th. 1997a. Ramesside glass colouring crucibles. *Archaeometry* 39: 355–68.

Rehren, Th. 1997b. Metal analysis in the Middle Ages. In *Material Culture in Medieval Europe*, G. De Boe and F. Verhaeghe (eds). Zellik: Instituut voor het Archeologisch Patrimonium, 9–15.

Rehren, Th. 1999a. Small size, large scale: Roman brass production in Germania Inferior. *Journal of Archaeological Science* 26: 1083–7.

Rehren, Th. 1999b. The same … but different: a juxtaposition of Roman and medieval brass-making in central Europe. In *Metals in Antiquity*, S.M.M. Young, M. Pollard, P. Budd and R. Ixer (eds). BAR International Series 792. Oxford: Archaeopress, 252–7.

Rehren, Th. 2003. Crucibles as reaction vessels in ancient metallurgy. In *Mining and Metal Production through the Ages*, P. Craddock and J. Lang (eds). London: British Museum Press: 147–9, 207–15.

Rehren, Th. and Klappauf, L. 1995. … ut oleum aquis. Vom Schwimmen des Silbers auf Bleiglätte. *Metalla* 2: 19–28.

Rehren, Th. and Papachristou, O. 2003. Similar like white and black: a comparison of steel-making crucibles from Central Asia and the Indian subcontinent. In *Man and Mining*, T. Stöllner, G. Körlin, G. Steffens and J. Cierny (eds). *Der Anschnitt* Beiheft 16. Bochum: Deutsches Bergbau-Museum, 393–404.

Rehren, Th., Lietz, E., Hauptmann, A. and Deutmann, K.H. 1993. Schlacken und Tiegel aus dem Adlerturm in Dortmund: Zeugen einer mittelalterlichen Messingproduktion. In *Montanarchäologie in Europa*, H. Steuer and U. Zimmermann (eds), 303–14. (*Archäologie und Geschichte - Freiburger Forschungen zum ersten Jahrtausend in Südwestdeutschland*, 4.)

Wainwright, G.J. 1979. *Gussage All Saints*. Department of the Environment Archaeological Reports 10. London: HMSO.

## Authors' addresses

- Justine Bayley, English Heritage, Fort Cumberland, Fort Cumberland Road, Eastney, Portsmouth PO4 9LD, UK (justine.bayley@english-heritage.org.uk)
- Thilo Rehren, Institute of Archaeology, University College London, 31–34 Gordon Square, London WC1H 0PY, UK (th.rehren@ucl.ac.uk)

# Records of palaeo-pollution from mining and metallurgy as recorded by three ombrotrophic peat bogs in Wales, UK

*T.M. Mighall, Simon Timberlake, S. Singh and M. Bateman*

*ABSTRACT*   This paper presents geochemical data from three ombrotrophic peat bogs located close to two former lead mines, Craig y Mwyn and Nantymwyn in North and South Wales, UK respectively. The research objective was to reconstruct a record of the pollution generated by activities associated with mining preserved in each bog. Radiocarbon dates from each site confirm that the peat provides a record of pollution since prehistoric times. Small lead peaks appear in the peat record at dates corresponding to prehistoric and possibly the Roman and Dark Ages at Nantymwyn, but mining did not commence on a continuous basis until the 18th century AD. At Craig y Mwyn, lead pollution occurs during the Roman period and, hitherto, provides the best estimate for the start of mining at the site. These examples demonstrate the usefulness of palaeo-pollution records from peat bogs to reconstruct the origins and history of metal mining in Britain.

*Keywords:* peat, palaeo-pollution, lead, zinc, mining, Wales.

## Introduction

Despite the major advances made in the fields of mining archaeology and archaeometallurgy, there is still a lot of debate about the origin and history of many mines, particularly their possible use in antiquity. Indeed, while some have questioned the antiquity of prehistoric mines per se (Briggs 1991), others have challenged the widely held assertion that prehistoric mines were exploited for lead and silver rather than copper (Bick 1999). Detailed surveys and excavation have not always been able to decipher the earliest phases of mining and/or metalworking at a site and answer such challenging questions. The majority of metal mines are multi-period, some having been exploited repeatedly, yet later activity and a lack of datable artefacts can make it difficult to build a chronological sequence of events using archaeological evidence alone. Reworking can destroy or alter a site stratigraphy and/or hide evidence of earlier activities. These are problems that archaeology can highlight but not necessarily answer. Thus, alternative approaches must be applied to such problems. For example, separate phases of mining and/or smelting can be determined and dated using palaeo-pollution records in peat bog archives, thus providing important information about the timing and longevity of metal mining and any associated metallurgy, especially when archaeological evidence is confused or lacking.

It has also long been recognised that trace element analysis can provide a useful insight into any metallurgical events in the history of an archaeological site, including provenancing metal ores and objects as well as the identification of slags resulting from the working of metals (Jenkins 1988; Rohl and Needham 1998), yet archives of chemical data stored in peat bogs have not been fully exploited by archaeologists as part of their research strategy. Peatlands store signals (e.g. in the form of deposition of metals, microfossils) over time resulting from the flux of another reservoir, the atmosphere. Peat bogs are generally undisturbed by human action and therefore provide continuous archives of environmental and climatic changes. As such, they represent ideal databases by which to understand both natural and human-induced changes on natural ecosystems such as mining and metallurgy (Martínez Cortizas and Weiss 2002). Moreover, peat deposits are well distributed across the UK's metalliferous orefields.

Analytical procedures used to reconstruct metal pollution histories have undergone rigorous scientific testing so the methods employed here are well established. Studies now confirm the reliability of these archives for certain heavy metals, such as lead and copper, and processes operating internally in the peat bog are no longer thought to induce serious post-depositional re-mobilisation of metals (Mighall *et al.* 2002a; Shotyk *et al.* 1997). Thus, the use of peat bogs as archives of past pollution is now established and atmospheric pollution histories have been reconstructed for past metallurgical activities on local and regional scales (Rosen and Dumayne Peaty 2001; West *et al.* 1997) including pollution from prehistoric, Roman and medieval workings (Mighall *et al.* 2002a, 2002b; Timberlake and Jenkins 2000). Such studies have shown that peat bogs located close to mines can preserve a specific geochemical signature of mining and/or smelting.

However, there has been no comprehensive study of the use of records of metal deposition contained within peat bogs to answer archaeological-based questions. Records of atmospheric pollution from peat bog archives can produce plausible and challenging data to pose and isolate problems for the archaeologist. Chemical records in peat profiles can offer potential information with regard to aspects of a metal mining and working site that may have so far eluded archaeologists. Peat bog archives can provide evidence to resolve some of these problems. To really establish the importance of metal exploitation in the British Isles requires a more extensive investigation at other known or suspected early mine sites. Given the recent debate, a critical examination of these issues is timely. It is fair to suggest that much of the evidence for the earliest exploitation of metals within the UK will be found in Wales (Timberlake 2002). For this reason targeted research at potential sites should yield evidence to answer some of the questions frequently posed by archaeologists and archaeometallurgists interested in mining and metalworking.

This paper seeks to provide a record of geochemical data from two peat bogs located close to the mines of Nantymwyn near Rhandirmwyn, Carmarthenshire, South Wales, and one close to Craig y Mwyn, Llanrhaeadr-ym-Mochnant, Powys, North Wales, in order to demonstrate that peat bog metal archives can determine the origins and history of mining at each site.

## Site details and archaeology

Craig y Mwyn and Nantymwyn (Figs 1 and 2a) were chosen because both mines appear to have had a long history of lead mining, yet the origin of each mine is unknown. A summary of what is known of their archaeology and history is provided here while more detailed accounts can be found in Timberlake (2003, 2004) and Jones *et al.* (2004).

### *Craig y Mwyn Mine, Llanrhaeadr-ym-Mochnant, Powys (SJ 074 285)*

The Craig y Mwyn lead mine is located in the district of Llangynog (Fig. 1), on the north-facing slopes of Y Clogydd

at an elevation of *c.* 500 m (Jones *et al.* 2004). Here lead ore has been extracted from galena and sphalerite-rich veins within the Ordovician Age shales, slates and tuffs (Jones *et al.* 2004). Although the detailed history of this mine before 1749 is largely unknown, there are references to leases and exploration in this area going back to 1656, while in 1751, O'Connor refers to the site as an old and formerly rich mine (Bick 1978). Hushing channels were thought to date from the mid-17th century or earlier (Jones *et al.* 2004); for instance, a more recent shaft close to the 'ancient' opencast was dated to around 1747, while the foundations for a rectangular stone building which appears to postdate one of the hushing ponds has now been linked to a house and mine smithy recorded here in 1751 (Jones and Frost 1996). Several ventures worked the mine with mixed success from the mid-to-late 1800s right up until the time it was abandoned in 1911 (Jones *et al.* 2004).

While there are no specific claims for Roman mining, there are the remains of an ancient building and 'Roman' lead smelting site ('the cubil') close by at Cwm Glanafon (Tyler 1982). On recent examination, however, Williams (1985) considered this to be early post-medieval. Davies in 1810 considered the mine itself to be 'an ancient work of great note' which bore 'marks of very ancient mining by the assistance of fire and water' (Williams 1985). Recent examinations of some of the small workings here support this assertion of firesetting (Timberlake 2004). Moreover, phases of hushing were probably integral to the excavation of the massive, ancient opencast at Craig y Mwyn, a practice similar to that employed in hydraulically assisted mining carried out at Cwmystwyth (Hughes 1994). At the latter site this has been ascribed to the late medieval/ early post-medieval period. This method of hushing, however, is also rather similar to the Roman technique described by Pliny (Timberlake 2004). The location of tanks and leats are characteristic of those attributed to the Romans at Dolaucothi (Lewis and Jones 1969).

A peat infill of one of the leats supplying water to what was probably the last phase of hushing practised on-site has provided an age of 520 ± 60 years BP (cal AD 1300–1460),

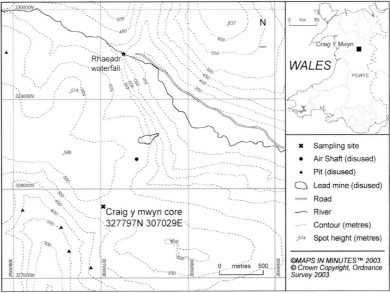

**Figure 1** Location of Craig y Mwyn lead mine and the sampling site.

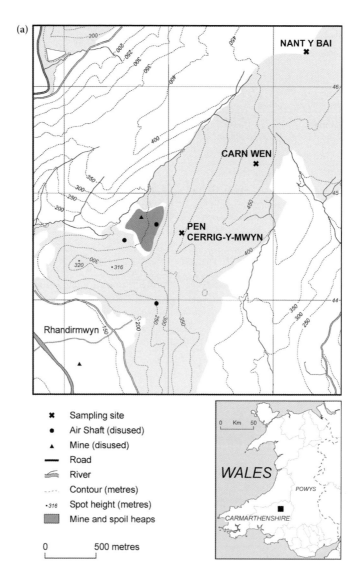

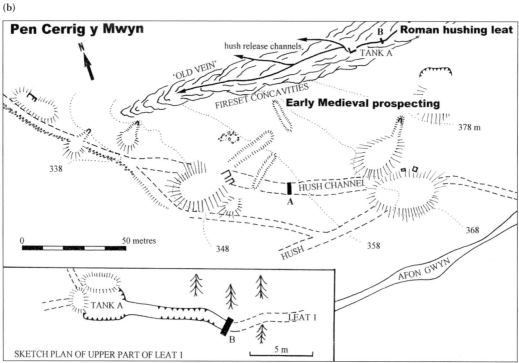

**Figure 2** (a) Location of the Nantymwyn lead mines and the sampling sites at Nant y Bai (NYB) and Carn Wen (CW). (b) A plan of early mining features at Pen Cerrig y Mwyn. Black rectangular blocks mark the position of the sections cut during the excavation.

suggesting that hushing may have continued into the medieval period, yet implying that some of the opencast workings and leat/pond/hush systems could be earlier in age (Timberlake 2004). By the mid-to-late 1700s the richest ore was more or less exhausted.

## Rhandirmwyn (Pen Cerrig y Mwyn and Nantymwyn Mine(s)), Carmarthenshire (SN 789 443)

The extensive workings at Nantymwyn are located east of Rhandirmwyn village, c. 10 km north of Llandovery (Fig. 2a). The main group of galena-rich lodes strike NNE and outcrop in the Nant y Bai valley at the foot of the steep upper slopes of Pen Cerrig y Mwyn. Sphalerite (zinc blende) formed a small but significant by-product during the 1930s reworking of the mine (Hall 1971: 95). Cerussite (lead carbonate) and arsenic-bearing pyromorphite have been recorded within the upper oxidised zones of the Old Vein and other mineralised outcrops on the Pen Cerrig y Mwyn (Bevins 1994).

The various workings here form part of the largest lead mine in South Wales, employing around 400 miners at its peak in 1791, while the extent of the older spoil heaps suggests that a very sizable amount of ore was extracted hundreds of years before 18th-century reworking. The earliest recorded mining dates back to 1530 (Hughes 1992). Claims of antiquity remain unsubstantiated although in 1996 the Early Mines Research Group revealed some evidence for very early medieval prospection at Pen Cerrig y Mwyn (Fig. 2b). This consisted of primitive-looking fireset hollows excavated into the side of a cliff. Charcoal recovered from here provided a date of 1080 ± 60 years BP (AD 865–1035) (Timberlake 1998; Timberlake and Craddock 1996).

Subsequently, archaeological investigation has uncovered evidence for what could be Roman prospecting activity, possibly for gold rather than lead. This activity is sealed by peat resting upon a thin layer of weathered shale gravel within the base of a proposed hushing leat/channel situated on top of the ridge outcrop at Pen Cerrig y Mwyn. A radiocarbon date of 1450 ± 60 years BP (cal AD 460–680 or cal AD 520–680) was obtained from this peat providing an age that clearly postdates this phase of mineral prospection (Timberlake 2004) (Fig. 2b). This is probably the strongest indication so far of Roman hushing at a lead mine in Wales. Therefore it is plausible that mining could have commenced even earlier, making Rhandirmwyn an ideal site for further investigation.

## Methods

At Rhandirmwyn, samples were collected from peat deposits within a kilometre of mining activity. A 1.48 m core was collected using a Russian corer (a specially designed peat corer that encloses the sampled core to prevent it from compacting and contamination (Moore *et al.* 1991)) from a blanket peat that covers Pen y Darren, at the head of the Nant y Bai valley, c. 462 m OD (grid reference SN 79929 46268). A 1 m deep monolith was also cut from a freshly exposed and cleaned

section of blanket peat that occupies a small summit known as Carn Wen, c. 442 m OD (grid reference SN 80043 45519), less than 1 km from the previous site. At Craig y Mwyn, a 1.2 m core was obtained from a blanket peat located at the top of the pass between the Rhaeader and Tanat valleys, overlooking the hillfort at Craig Rhiwarth.

Each core or monolith was cut into contiguous 1 cm slices, which were oven dried at 40 °C and homogenised. Between 0.5 and 0.6 g of each sample was digested in a Mars 5 microwave using 2 ml deionised water, 6 ml nitric acid (AR $HNO_3$) and 1 ml hydrogen peroxide (AR $H_2O_2$) per sample. Lead (Pb), copper (Cu) and zinc (Zn) were determined using a Unicam 939 flame atomic absorption spectrometer (FAAS) while aluminium (Al), calcium (Ca) and magnesium (Mg) were determined using an inductively coupled plasma-atomic emission spectrometer (Perkin Elmer Plasma 400 model) housed in the chemistry department at Coventry University. An estimation of the efficiency of the digestion method and of the accuracy of the analytical measurements was obtained by use of replicate subsamples, spiked blanks and certified reference materials (CRMs) (Ebdon *et al.* 1998). Spiked samples of known concentration (10.0 mg/L), for the metals have been used to test the efficiency of the acid, microwave digestion. Two certified reference materials were also used: sphagnum energy peat (NJV 94-2) and carex energy peat (NJV 94-1, Swedish University of Agricultural Sciences, Department of Agricultural Research for Northern Sweden, Laboratory for Chemistry and Biomass). Standards of known metal concentrations were used to calibrate the FAAS and to ensure optimum performance (Holler *et al.* 1996). Total metal concentrations for each sample are expressed in μg g⁻¹ dried (40 °C) peat and are calculated as enrichment ratios (ERs) following Shotyk (1996).

## Results

Two horizons make up the stratigraphy of the Craig y Mwyn core. The basal 15 cm is light grey clay which is overlain by herbaceous peat. The stratigraphy for the Nant y Bai (NYB) core consisted of sphagnum peat down to 45 cm and then herbaceous peat until a basal mineral clay layer at 1.48 m. The Carn Wen monolith was 90 cm deep and consisted primarily of herbaceous peat with a sphagnum layer between 20 and 23 cm depth.

### Radiocarbon dating

Radiocarbon dates from each site are shown in Table 1. The radiocarbon dates are calibrated using CALIB 4.1 radiocarbon calibration program and IntCal98 (Stuiver and Reimer 1993; Stuiver *et al.* 1998) and the calibration age ranges to 2σ are also shown. The radiocarbon date of 145 ± 30 years BP from NYB appears to be relatively young given the depth of the peat. The lead peak coincides with historical evidence for a major phase of mining (see below) suggesting that it is reliable, however the possibility of contamination or a break in peat accumulation cannot be ruled out.

**Table 1** Radiocarbon dates from Craig y Mwyn (CYM) and Rhandirmwyn (CW and NYB).

| Laboratory no. | Depth | Uncalibrated date | Calibrated age range (2σ) |
|---|---|---|---|
| **CYM** | | | |
| B-199412 | 62–66 cm | 1720 ± 40 | Cal AD 230–410 |
| B-199414 | 98–102 cm | 3810 ± 60 | Cal BC 2460–2120 & Cal BC 2100–2040 |
| **CW** | | | |
| B-199411 | 38–40 cm | 400 ± 60 | Cal AD 1420–1640 |
| **NYB** | | | |
| Poz-14668 | 58–62 cm | 145 ± 30 | Cal AD 1668–1953 |
| B-199415 | 122–128 cm | 2340 ± 70 | Cal BC 770–380 |

## Peat chemistry

Lead, zinc and copper concentrations and enrichment ratios are shown for both Craig y Mwyn (CYM) and the two sites at Rhandirmwyn (Carn Wen (CW) and Nant y Bai (NYB)) are shown in Figures 3–5. Notwithstanding the occasional outlier, the percentage recovery of the CRMs was within 10% of the certified value. The spiked sample results produced a percentage mean recovery rate over 90%. These results suggest that metal recovery using the microwave digestion is very efficient and the FAAS is providing reliable and accurate data.

To help separate natural variations in metal concentrations and those caused by human activity, enrichment ratios have been calculated. Aluminium has been used as a conservative element known to have been derived by crustal weathering. The ERs of lead and zinc indicate the degree to which the metal is enriched in a sample compared with the abundance of that element in crustal rocks (derived from Wedepohl 1995) normalised to the conservative element (Shotyk 1996). In order to establish that the peat is ombrotrophic (i.e. the bog is raised above the ground surface and may be fed only by rainwater) and therefore the elements supplied to the bog are atmospheric, calcium/magnesium molar ratios have been calculated (Shotyk 1996) (Figs 3–5). When they remain below 1, the peat is most likely to be ombrotrophic.

## Craig y Mwyn

Lead concentrations remain low from the base of the core to 62 cm (Fig. 3). They then increase to peaks exceeding 150 μg g$^{-1}$ at 60 cm, 40 cm and 22 cm before gradually decreasing to 60 μg g$^{-1}$. Zinc is characterised by low concentrations of less than 50 μg g$^{-1}$ below 62 cm before rising to 200 μg g$^{-1}$ in the uppermost 30 cm. Copper concentrations remain below 40 μg g$^{-1}$ throughout the profile and, notwithstanding the occasional reversal, decrease from the base of the core to the bog surface. Only the broad peak at 20 cm correlates with lead. The ERs follow a similar pattern to the concentrations for both lead and zinc, although the peaks at 44 cm, 30 cm and at the top of the core are more pronounced. Copper is totally different when compared to the concentration profile and shows a broadly similar pattern to lead and zinc ERs, suggesting that the source of copper is related to mining in the top 62 cm of the core.

## Rhandirmwyn

### Site 1: Nant y Bai

Lead concentrations fall from the very base of the core to very low concentrations at 140 cm (Fig. 4). They remain low except for a short-lived peak at 120 cm until they increase rapidly from 62 to 60 cm. They remain high; peaking between 44 cm and 36 cm with concentrations exceeding 5000 μg g$^{-1}$, remain fairly stable until 16 cm before decreasing in the top part of the profile. Zinc and copper follow a very similar pattern to lead with a small peak around 120 cm and a dramatic rise at 62 cm, with concentrations exceeding 900 μg g$^{-1}$ and 350 μg g$^{-1}$ respectively. The ERs follow a similar pattern to the concentrations for both lead and copper although the decrease in lead ER is more pronounced than the concentrations. Zinc ER is totally different when compared to the concentration profile, peaking at 60 cm and then falling in value gradually to the bog surface. High calcium/magnesium ratios are determined at 122 and 136 cm.

### Site 2: Carn Wen

Concentrations for lead, copper and zinc remain low until 40 cm (Fig. 5). The first evidence of pollution occurs at 40 cm when concentrations and ERs increase for all three metals. The increases are slightly staggered with first lead increasing most rapidly, followed by copper and then zinc. All peak at 30 cm although lead concentrations also peak between 36 and 34 cm, while zinc has a second peak at 26 cm. After a rapid decline, a second peak in lead and copper occurs around 12 cm but zinc concentrations remain low. The ERs for lead, copper and zinc are in broad agreement with the concentrations.

## Interpretation and discussion

The absence of any metal pollution peaks below 62 cm at Craig y Mwyn suggests that the site was not exploited during prehistory (Fig. 3). The rise in lead and zinc from 62 cm is dated to 1720 ± 40 years BP, confirming that mining commenced here during the Roman period. Low aluminium concentrations and the gradual increase in concentrations and ERs suggest that the metals supplied to the bog are anthropogenic in origin and the three peaks at 60 cm (Roman), c. 40 cm (estimated age 1127 years BP) and 22 cm (estimated age 620 years BP) represent periods of more intense mining. If the latter estimated ages are reasonably accurate, this suggests that significant phases of lead mining occurred well before documentary records and the suggestion by Jones et al. (2004: 29) that 'the earliest workings appear to have been relatively muted' deserves to be re-evaluated.

The slightly elevated but declining concentrations in the basal peat at NYB are the result of increased clay underlying the peat as reflected in the high aluminium concentrations (Fig. 4). These elevated concentrations are not shown by the ERs and thus represent a natural flux of metals. The first evidence of possible pollution related to lead mining occurs at 128 cm (374 μg g$^{-1}$) and from 124 cm to 122 cm when lead concentrations peak at 934 μg g$^{-1}$ and ERs reach a value of 3585. Copper and zinc also show elevated concentrations and

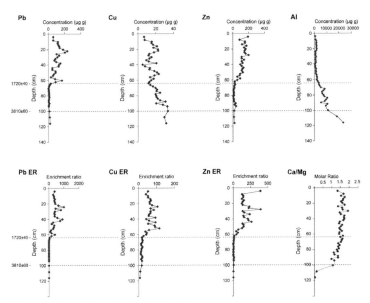

**Figure 3** Chemical profiles for Craig y Mwyn.

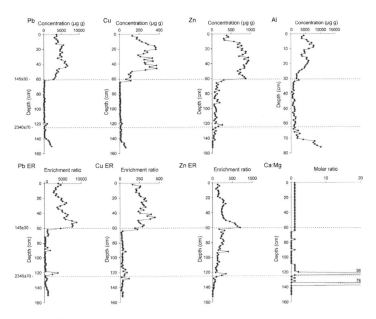

**Figure 4** Chemical profiles for Nant y Bai.

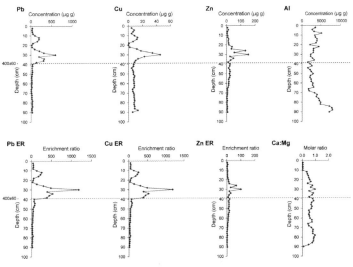

**Figure 5** Chemical profiles for Carn Wen.

ERs. The peaks may represent an early, possibly experimental, small-scale phase of lead and/or copper mining and/or metallurgy. A radiocarbon date suggests that this activity took place during the Late Bronze Age or Early Iron Age.

There is a limited amount of artefact evidence for the use of lead within the British Isles prior to 1500 BC. For example, Hunter and Davis (1994) describe a discovery of a lead bead necklace in an Early Bronze Age Beaker grave in southeast Scotland. Lead was also intentionally being alloyed with copper and tin in bronze by the end of the Middle Bronze Age and again at the beginning of the Late Bronze Age (Rohl and Needham 1998), while small pieces of lead have been found in Late Bronze Age occupation sites (e.g. Needham and Hook 1988). Craddock (1994) suggests that trench mines at Charterhouse on Mendip, in Somerset, may be much earlier in origin than previously thought. Todd (1996) discovered a pre-Roman enclosure associated with lead-smelting activity here, and there is evidence to suggest that lead isotope ratios within some of the LBA Wilburton Phase metalwork may also be associated with Mendip ore sources (Rohl 1995). Le Roux et al. (2004) used stable lead isotopes to reconstruct successfully the atmospheric lead pollution history from Lindow Moss, near Manchester, and reveal evidence of lead pollution extending back to the Late Iron Age.

The evidence for Nant y Bai must, however, be treated with caution as the peak in lead corresponds with a spike in calcium/magnesium molar ratios at 122 cm (Fig. 4). This suggests that the peat may not be completely ombrotrophic and may have received additional material from groundwater sources, which could account for the higher values of zinc and copper. It is unlikely that human activity alone can account for the zinc since zinc ore was not intentionally exploited at this time, although it was introduced into copper well before Roman times in other parts of the world (Craddock 1995). Sphalerite (zinc sulphide), like galena (lead sulphide), invariably accompanies copper, and the crushing of this to free the attached galena or chalcopyrite may liberate large amounts of this within the immediate environment of the mine. This would rapidly oxidise upon the exposed spoil tips, be leached out, or could even be blown some distance in the form of dust by gusting wind (Mighall et al. 2000). Some zinc could have been released in the smelting of impure lead ores although there is no evidence for smelting at the site.

A second phase of possible exploitation occurs between 90 and 86 cm when lead concentrations and ERs peak (Fig. 4). The first peak coincides with higher copper and zinc. Aluminium concentrations remain low and this suggests that the metals deposited onto the bog surface are of anthropogenic origin. If the radiocarbon date obtained for samples between 58 and 62 cm is contaminated and therefore too young, an extrapolated age (assuming the peat bog surface is 0 BP) for the start of this pollution is 1680 years BP and therefore Roman. There is evidence for Roman lead activity elsewhere in Wales and elevated pollution signals are common in peat bogs in Britain (e.g. Lee and Tallis 1973; Martin et al. 1979; Mighall et al. 2002b; West et al. 1997) and across Europe (e.g. Görres and Frenzel 1997; Martínez Cortizas et al. 1997). The concentrations here are much lower and it is likely that no significant mining activity for lead, copper or zinc occurred at Rhandirmwyn and that the activity described

by Timberlake (2003, 2004) was small scale and insufficient to generate a widely distributed pollution signal. It is possible that this was a phase of prospecting activity for gold rather than for lead (Timberlake 2003, 2004) as the Roman gold mine and settlement at Dolaucothi lies only 10 km to the west. If the radiocarbon date is reliable, this peak is dated to approximately 1450 years BP and may represent a period of small-scale mining during the Dark Ages, which still predates the earliest recorded phase of mining (Hughes 1992). Alternatively if there is a break in peat accumulation between 62 and 60 cm and part of the record is missing, it is difficult to establish an age for this phase of lead enrichment.

The first evidence of pollution at Carn Wen occurs at 40 cm (Fig. 5). A radiocarbon date of 400 ± 60 years BP suggests this pollution is caused by mining from possibly as early as the 14th century AD. A sharp peak in lead ERs at 26 cm is tentatively dated to the AD 1500s and it is plausible that this rise coincides with the earliest documented phase of mining during the early 15th century AD. The rise in lead at Carn Wen is earlier than that recorded at Nant y Bai. At NYB the main phase of pollution occurs from 62 cm to the surface of the bog. Lead, copper and zinc concentrations and ERs rise rapidly. Lead concentrations are an order of magnitude greater than concentrations previously recorded in peat bogs in Britain and remain well above 2600 µg g$^{-1}$ and peak at 6680 µg g$^{-1}$ at 42 cm (Fig. 4). A radiocarbon date of 145 ± 30 years BP dates this rise to between AD 1668 and 1953. An examination of the relative area of probability of the calibrated age ranges at 2σ suggests that it is more likely that the true age of the date lies somewhere between AD 1717 and 1781 or 1797 and 1891, which coincides with a major phase of lead extraction between AD 1775 and 1797 (Owen 1999). A second phase of lead and copper pollution, between 18 and 10 cm, occurs in the last 200 years at Carn Wen and this coincides with a sizable period of mining activity when Nant y Mwyn was the largest lead mine in South Wales (Owen 1999).

Notwithstanding the limitations caused by a lack of radiocarbon dates for both cores, the peaks at CW do not always coincide with NYB. It is possible that CW is recording a localised source of pollution as its position is sheltered from the main area of mining within the bottom and on the southern slopes of the Nant y Bai valley. Alternatively, the differences between these two sites might also be explained by a break in peat accumulation at NYB. Further dating, particularly lead[210] dating, would resolve this problem and provide a firm chronology with which to compare the historical pollution record more effectively with documentary archives of mining at Rhandirmwyn. The low concentrations and ERs from 40 cm to the base of the monolith suggest that no earlier activity occurred locally. Zinc does increase gradually from approximately 66 cm but this rise is more likely to be the result of downward translocation of zinc. Zinc is thought to be mobile in peat and similar trends have been described elsewhere (e.g. Mighall et al. 2002b).

The NYB and CW lead concentrations and ERs are also much higher than those for copper and zinc (Fig. 4), and this suggests that lead was the primary target for the miners. All three metals sustain high concentrations until the top 16 cm at NYB when all fall gradually, suggesting a possible cessation of activity. Despite the abandonment of the mines, the concentrations and ERs remain relatively high and do not return to background values, which are recorded in the basal section of the core. Copper ERs remain elevated except for a reversal between 32 and 24 cm. This period of higher copper concentrations does, however, coincide with increased aluminium concentrations, which also rise gradually from 60 cm onwards, to peak at *c.* 15 cm. Thus, some of the contribution of metals to the bog must be the result of wind dispersal of eroded contaminated soils and finely crushed mine tailings blown off of the abandoned and exposed spoil tips as well as from pollution actively generated by mining. The position of NYB, at the head of the valley and directly downwind of the main area of historical mining, is ideally situated to receive windblown material. This phenomenon is not observed at Carn Wen and the cessation of mining is clearly seen as lead concentrations and ERs fall to close to zero in the top 10 cm.

## Conclusions

Peat bog archives offer tremendous potential to increase our understanding of the origins and history of mining and metalworking. The results presented in this paper strongly suggest that lead mining at Craig y Mwyn commenced during the Roman period. The interpretation of the chemical records at Nant y Mwyn is more problematic because there is evidence for a break in peat accumulation. The results presented here, however, suggest that two phases of early exploitation, one of which is dated to the prehistoric period, occurred before a more intensive period of historical mining.

## Acknowledgements

We would like to dedicate this paper in memory of David Bick who sadly passed away in January 2006, a colleague and sparring partner of Paul Craddock: two of the main inspirations for us pioneers into early mining research in the UK. We thank Susan La Niece and an anonymous reviewer for suggesting improvements to this paper. We are grateful to the Cambrian Archaeological Association for funding the radiocarbon dates. This research was completed during the tenure of a Leverhulme Trust Grant reference number F/00/732/C.

## References

Bevins, R.E. 1994. *A Mineralogy of Wales.* Geological Series no. 16. Cardiff: National Museum of Wales.

Bick, D. 1978. *The Old Metal Mines of Mid-Wales – Part 5: Aberdovey, Dinas Mawddwy and Llangynog.* Newent: The Pound House.

Bick, D.E. 1999. Bronze Age copper mining in Wales – fact or fantasy? *Historical Metallurgy* 33: 7–12.

Briggs, C.S. 1991. Early mines in Wales: the date of Copa Hill. *Archaeology in Wales* 31: 5–7.

Craddock, P.T. 1994. Recent progress in the study of early mining and metallurgy in the British Isles. *Historical Metallurgy* 28: 69–84.

Craddock, P.T. 1995. *Early Mining and Metal Production.* Edinburgh: Edinburgh University Press, 362.

Ebdon, L., Evans, E.H., Fisher, A.A. and Hill, S.J. 1998. *An Introduction to Analytical Atomic Spectrometry.* Chichester: John Wiley and Sons, 193.

Görres, M. and Frenzel, B. 1997. Ash and metal concentrations in peat bogs as indicators of anthropogenic activity. *Water, Air and Soil Pollution* 100: 355–65.

Hall, G.W. 1971. *Metal Mines of Southern Wales.* Westbury on Severn: Griffin Publications.

Holler, F.J., Skoog, D.A. and West, D.M. 1996. *Fundamentals of Analytical Chemistry,* 7th edn. Philadelphia, PA: Saunders College Publishing.

Hughes, S. 1994. The hushing leats at Cwmystwyth: mining before powder. *Bulletin of the Peak District Mines Historical Society* 12: 8–53.

Hughes, S.J.S.H. 1992. Nantymwyn Mine, Northern Mines Research Society. *British Mining* 49: 87–110.

Hunter, F. and Davis, M. 1994. Early Bronze Age lead: a unique necklace from south-east Scotland. *Antiquity* 68: 824–30.

Jenkins, D.A. 1988. Trace element analysis in the study of ancient metallurgy. In *Aspects of Ancient Mining and Metallurgy, Bangor, 1986,* J. Ellis-Jones (ed.). Bangor: University College, 95–105.

Jones, N.W. and Frost, P. 1996. *Powys Metal Mines Survey.* Clwyd-Powys Archaeological Trust, Report no.111.1.

Jones, N.W., Walters, M. and Frost, P. 2004. *Mountains and Orefields: Metal Mining Landscapes of Mid and North-east Wales.* Council for British Archaeology (CBA) Research Report 142, 180.

Le Roux, G., Weiss, D., Grattan, J. *et al.* 2004. Identifying the sources and timing of ancient and medieval atmospheric lead pollution in England using a peat profile from Lindow Bog, Manchester. *Journal of Environmental Monitoring* 6: 502–10.

Lee, J.A. and Tallis, J.H. 1973. Regional and historical aspects of lead pollution in Britain. *Nature* 245: 216–18.

Lewis, P.R. and Jones, G.D.B. 1969. The Dolaucothi gold mines I: the surface evidence. *Antiquity* XLIX: 244–72.

Martin, M.N., Coughtrey, P.J. and Ward, P. 1979. Historical aspects of heavy metal pollution in the Gordano valley. *Proceedings of the Bristol Naturalists Society* 37: 91–7.

Martínez Cortizas, A. and Weiss, D. 2002. Peat bog archives of atmospheric metal deposition. *Science of the Total Environment* 292: 1–5.

Martínez Cortizas, A., Pontvedra-Pombal, X., Nóvoa Muños, J.C. and García-Rodeja, E. 1997. Four thousand years of atmospheric Pb, Cd and Zn deposition recorded by ombrotrophic peat bog of Penido Vello (northwestern Spain). *Water, Air and Soil Pollution* 100: 387–403.

Mighall, T.M., Timberlake, S., Grattan, J.P. and Forsyth, S. 2000. Bronze Age lead mining at Copa Hill: fact or fantasy? *Historical Metallurgy* 34: 1–12.

Mighall, T.M., Abrahams, P.W., Grattan, J.P., Hayes, D., Timberlake, S. and Forsyth, S. 2002a. Geochemical evidence for atmospheric pollution derived from prehistoric copper mining at Copa Hill, Cwmystwyth, mid-Wales, U.K. *Science of the Total Environment* 282: 69–80.

Mighall, T.M., Grattan, J.P., Lees, J.A., Timberlake, S. and Forsyth, S. 2002b. An atmospheric pollution history for lead-zinc mining from the Ystwyth valley, Dyfed, mid-Wales, UK as recorded by an upland blanket peat. *Geochemistry: Exploration, Environment, Analysis* 2: 175–84.

Moore, P.D., Webb, J.A. and Collinson, M.E. 1991. *Pollen Analysis.* Oxford: Blackwell.

Needham, S. and Hook, D.R. 1988. Lead and lead alloys in the Bronze Age: recent finds from Runnymede Bridge. In *Science and Archaeology: Applications of Scientific Techniques to Archaeology, Glasgow, 1987,* E.A. Slater and J.O. Tate (eds). BAR 196. Oxford: Archaeopress, 259–74.

Owen, D. 1999. *Rhandirmwyn: A Brief History* [private publication].

Rohl, B.1995. *Application of Lead Isotope Analysis to Bronze Age Metalwork from England and Wales.* Unpublished D.Phil thesis, Oxford University.

Rohl, B. and Needham, S. 1998. *The Circulation of Metal in the British*

*Bronze Age: The Application of Lead Isotope Analysis*. British Museum Occasional Paper 102. London, 234.

Rosen, D. and Dumayne-Peaty, L. 2001. Human impact on the vegetation of South Wales during late historical times: palynological and palaeoenvironmental results from Crymlyn Bog NNR, West Glamorgan, Wales, UK. *The Holocene* 11: 11–24.

Shotyk, W. 1996. Peat bog archives of atmospheric metal deposition: geochemical evaluation of peat profiles, natural variations in metal concentrations, and metal enrichment factors. *Environmental Reviews* 4: 149–83.

Shotyk, W., Norton, S.A. and Farmer, J.G. 1997. Summary of the workshop on peat bog archives of atmospheric metal deposition. *Water, Air and Soil Pollution* 100: 213–19.

Stuiver, M. and Reimer, P.J. 1993. A computer program for radiocarbon age calibration. *Radiocarbon* 35: 215–30.

Stuiver, M., Reimer, P.J., Bard, E. *et al.* 1998. INTCAL98 radiocarbon age calibration 24,000–0 cal BP. *Radiocarbon* 40:1041–83.

Timberlake, S. 1998. Survey of early metal mines within the Welsh Uplands. *Archaeology in Wales* 38: 79–81.

Timberlake, S. 2002. Medieval lead-smelting boles near Penguelan, Cwmystwyth. *Archaeology in Wales* 42: 45–59.

Timberlake, S. 2003. An archaeological examination of some early mining leats and hushing remains in upland Wales. *Archaeology in Wales* 43: 33–44.

Timberlake, S. 2004. Early leats and hushing remains: suggestions and disputes of Roman mining and prospection for lead. *Mining History: Bulletin of the Peak District Mines Historical Society* Special Issue 15: 64–76.

Timberlake, S. and Craddock, B. 1996. Pen Cerrig y Mwyn Mine. *Archaeology in Wales* 36:104.

Timberlake, S. and Jenkins, D.A. 2000. Prehistoric mining: geochemical evidence from sediment cores at Mynydd Parys, Anglesey. In *Archaeological Sciences 97. Proceedings of the Conference held at the University of Durham, 2–4 September 1997*, A. Millard (ed.). BAR International Series 939. Oxford: Archaeopress, 193–9.

Todd, M. 1996. Ancient mining on Mendip, Somerset: a preliminary report on recent work. *Mining History* 13: 47–51.

Tyler, A.W. 1982. *Prehistoric and Roman Mining for Metals in England and Wales.* Unpublished PhD thesis, University of Wales (Cardiff).

Wedepohl, K.H. 1995. The composition of the continental crust. *Geochimica et Cosmochimica Acta* 59(7): 1217–32.

West, S., Charman, D.J., Grattan, J.P. and Cherburkin, A.K. 1997. Heavy metals in Holocene peats from south west England: detecting mining impacts and atmospheric pollution. *Water, Air and Soil Pollution* 100: 343–53.

Williams, R.A. 1985. *The Old Mines of the Llangynog District.* British Mining No. 26. Sheffield: Northern Mine Research Society.

## Authors' addresses

- T.M. Mighall, Department of Geography and Environment, School of Geosciences, University of Aberdeen, Elphinstone Road, Aberdeen AB24 3UF, UK (geo515@abdn.ac.uk)
- Simon Timberlake, Cambridge Archaeological Unit, Department of Archaeology, University of Cambridge, 34A Storeys Way, Cambridge, CB3 0DT, UK
- S. Singh, School of Process, Environmental and Materials Engineering, Houldsworth Building, University of Leeds, Clarendon Road, Leeds, West Yorkshire, LS2 9JT, UK
- M. Bateman, Department of Geography, Environmental Sciences and Disaster Management, Faculty of Business, Environment and Society, Coventry University, Priory Street, Coventry, CV1 5FB, UK

# Copper, tin and bronze

# Prehistoric copper production at Timna: thermoluminescence (TL) dating and evidence from the East

*Andreas Hauptmann and Irmtrud Wagner*

*ABSTRACT*   A major question in archaeometallurgy concerns the field evidence for the beginnings of metallurgical techniques. It has been suggested that at Timna, Israel, the earliest evidence for copper production is represented by the Chalcolithic site 39 and the Late Neolithic site F2 (6th millennium BC). This model is compared with other examples of social and spatial patterns of prehistoric metallurgy. Results of thermoluminescence (TL) dating are presented for site F2 which show in fact that it dates to the Late Bronze Age. Evidence for extensive copper smelting in the middle of the 3rd millennium BC comes from two sites near Aqaba: Tell Magass and Tell Hujayrat al-Ghuzlan; (self-fluxing) copper ores were imported from Timna and smelted inside these villages. The hinterland of these sites was probably Egypt.

*Keywords:* prehistoric, Timna, copper, smelting, thermoluminescence dating, TL, mining, metallurgy, Tell Magass, Tell Hujayrat al-Ghuzlan, Yotvata, Israel, Egypt.

## Introduction

A fundamental question in archaeometallurgical research concerns the incipient stages of metallurgy. The search for the answer to this question is contemporary with the beginnings of modern industrial archaeology, which in the 1970s, after many years of studying metal objects, developed into interdisciplinary research into ancient smelting sites. One of the most important examples of this was the pioneering research in the copper ore district of Timna (Rothenberg 1973), which is reported on here. In a series of ground-breaking research activities, the researchers primarily investigated the Late Bronze Age and Iron Age smelting sites and slag at Timna. These remains, several thousand tons in total, had accumulated close to the numerous copper ore mines. Some of these smelting sites, such as the New Kingdom site of Timna 30, were fortified by a wall.

Possible models for the early organisation of metal production were suggested, perhaps influencing the ideas of Levy and Shalev (1989). The main points are summarised as follows:

1. Smelting of ores took place close to the ore deposits.
2. Metal production was connected with the production of large amounts of slag.
3. Metal production was connected with fuel problems.
4. Metallurgical activities inside settlements were limited to secondary processes such as casting, smithing, alloying etc.

Such a model, however, is static – it does not allow for any chronological, structural and spatial changes in mining and metal production. These changes occur everywhere, from the very beginning of the use of metals until the ages of mass production such as during the Bronze Age and later periods (Stöllner 2003). These changes are dependent on numerous factors: the geological structure and mineral content of an ore deposit, fuel and water supply, agriculture, trade, markets and the hinterland, as well as the development of innovations.

The earliest evidence for the utilisation of metal comes from Anatolia and dates back to the 8th millennium BC Pre-Pottery Neolithic (PPN) period. As shown by finds from Çayönü Tepesi (Maddin *et al.* 1999) and from Ashikli Höyük (Yalçin and Pernicka 1999), native copper was utilised to form beads and small tools by cold-working and annealing. This first metallurgy, however, is unlikely to have had any major social and cultural impact. It was not until a few thousand years later, during the Late Neolithic period, that the knowledge of copper smelting, i.e. the extraction of metal from ore by smelting processes, spread across the Near and Middle East and southeast and central Europe. In Late Chalcolithic levels, in a period of rich cultural development and significant technological craftsmanship, remains of smelting processes such as slags, crucibles and other metallurgical installations are more frequently found.

What is the situation at Timna? When was copper first smelted in this ancient copper district? Was it smelted close to the mines, as suggested by Levy and Shalev (1989)? No native copper occurs at this ore deposit, unlike in Anatolia, and therefore it probably excludes metallurgical activities during the PPN. Even a pendant made of (native?) copper discovered in Halaf levels of Tell Ramad (Damascus) does not point to a provenance from the Wadi Arabah: its context with objects of obsidian rather points to it being an import from the north.

The abundance of oxidic copper ores at Timna provided the basis for a simple one-stage smelting process. Does Timna match the organisational pattern of metallurgy known in the Near and Middle East? To elucidate the last question we will present some examples of the most ancient copper-smelting activities and compare their archaeological evidence with that of Timna. We will further present results of thermoluminescence (TL) dating of a prehistoric slag from Timna and include results from excavations of two mid-4th millennium BC settlements near Aqaba, located at a distance some 30 km from Timna and less than 10 km from copper mineralisations south of Timna.

## Copper production in settlements: organisational patterns at the very beginnings of extractive metallurgy

The maxim formulated by Levy and Shalev (1989), that ore was smelted close to the ore deposit, is very difficult to prove for early examples of the extractive process. The application of this pattern was perhaps one of the reasons why it was not possible to identify the first steps of metallurgy for a long time. This is exemplified by the investigation of the early Neolithic copper mine of Rudna Glava, which was excavated by Borislav Jovanovic and which was dated by [14]C to the second half of the 6th millennium BC (Weisgerber and Pernicka 1995). It was suggested that the mine was the earliest supplier of ore for metal objects in the Balkan Peninsula. No slag was found however in the neighbourhood of the mine that could prove the hypothesis of ore smelting. Lead isotope analysis could not prove a possible origin of early Neolithic metal artefacts in the Balkans from Rudna Glava copper ores, nor was it possible to verify that copper ore from the mine at Ai Bunar, dated to the end of the 5th millennium BC, was used to produce the earliest copper objects from Bulgaria (Pernicka et al. 1993). As suggested by Gale et al. (1991), Ai Bunar copper ore was traded a considerable distance from the mine, to unknown places.

It seems likely that another rule, applicable across all regions, must have existed to explain the organisational beginnings of metallurgy. This can be deduced from the tradition of the Neolithic trade in exotic materials such as carnelian, obsidian, native copper in Anatolia, and of 'greenstones' that mainly consisted of green secondary copper ores. These were transported from sources to settlements, often over distances of a hundred kilometres and more. There are numerous examples that indicate that ores were smelted inside these settlements of the 5th/4th millennium BC. If villages were located within a catchment area of more than one ore deposit, sometimes, as was observed at Arslan Tepe in eastern Anatolia (Palmieri et al. 1993), ores were imported for metal production from more than one source. Transport posed no problem. At the beginnings of metallurgy, metal items had a prestige object status: in the southern Levant at least, they were produced in limited quantities. The use of metal for tools for daily use was an exception. Such metallurgical practices required other forms of social organisation than existed later during the later Bronze Age and Iron Age, when metal was produced on the

scale of many thousands of tons, and slags and other metallurgical remains, as a rule, are to be found in the direct neighbourhood of the mines. The transportation of ore over greater distances was naturally seen as untenable.

For a long time the remains of metallurgical finds inside settlements were only understood as belonging to the further treatment of metal such as casting, smithing and alloying in crucibles. This appears to have been a misinterpretation. Today, apart from the sites mentioned above, there are many examples known which confirm the pattern of a 'domestic mode production' of metal inside settlements in the 5th/4th millennium BC.

Copper ores from Feinan were smelted, for example at the cultural centres of the Beersheba basin at Abu Matar (Shugar 2000) and at Shiqmim (Golden et al. 2001), and were found further to the west at the Early Bronze Age I site of Asqelon-Afridar (Segal et al. 2004). Settlements with remains of 5th/4th millennium BC metal production in eastern Anatolia are especially known from the Altinova area on the upper Euphrates. Details from Arslantepe, Degirmentepe, Norsuntepe, Tepecik, Tülintepe are compiled in Müller-Karpe (1994).

In the Alpine region, copper settlement production using ores from partly remote sources was found at Maria Hilfsbergl near Innsbruck (Bartelheim et al. 2002), at Pfyn (Switzerland) and Götschenberg near Salzburg (compiled by Ottaway 1999). Copper was also smelted inside at the village of Selevac in Serbia (Glumac and Tringham 1990).

## Prehistoric copper production at Timna

The southern Negev and the area around the Gulf of Aqaba are rich in prehistoric sites (Avner 2006), but few settlements are dated to the 4th millennium BC. Two sites of copper smelting at Timna were suggested to be archaeological key sites of the Late Neolithic and Chalcolithic period (Merkel and Rothenberg1999). The Chalcolithic site is exemplified by the site of Timna site 39 (Rothenberg 1978), while site F2 is suggested to date to the Qatifian period, i.e. the 6th/5th millennium BC (Merkel and Rothenberg 1999; Rothenberg and Merkel 1995) (Fig. 1). In the following we will focus on these two sites and the longstanding ongoing debate on the dating and technology of copper production at Timna from these periods (Adams 1998; Avner 2002, 2006; Craddock 1993, 2000; Muhly 1984; Rothenberg 1978; Weisgerber 2006). In addition, a 4th-millennium BC site at Yotvata, showing remains of metallurgical activities, should be mentioned here.

### Timna, site 39

Timna 39 is located at the eastern side of the semicircular ore district of Timna, at the opening to the Wadi Arabah. Timna 39A consists of a small habitation site which seems to be isolated. No larger settlement from this period exists nearby. At a distance of c. 130 m, some scattered slags were found at Timna 39B and a smelting furnace was excavated. It is suggested that this furnace, a hole in the ground, was operated by tuyeres, and the slags seemed to have been produced

**Figure 1** Satellite image of the area between Timna and Yotvata in the north and the Gulf of Aqaba. Indicated are archaeological sites mentioned in the text.

using fluxing agents. This 'Chalcolithic' smelting furnace has become established in a large part of archaeometallurgical literature. In most cases it is forgotten, however, that there are no parallels of such furnace constructions at such an early date known in the Old World, and that the $^{14}$C measurements of charcoal from Timna 39B gave a date between 400 BC and AD 450 (BM 1116; see also Hauptmann 2000; Weisgerber 2006). Another $^{14}$C date of the 5th millennium BC (4355–4327 BC, OxA 7632; see Rothenberg and Merkel 1998) comes from the habitation site Timna 39A, and not from the slag scatter of Timna 39B.

### Timna, site F2

As pointed out by Merkel and Rothenberg (1999), it was of major interest to discover that Timna site F2 is located in the close vicinity of extensive primitive (but undated) pit-mining activities at the western half of the erosion circle; not only was very primitive metallurgical debris found at the site, but also working tools, flint and pottery. The site was excavated in 1976. Based upon field evidence, and supported by petrographic observations of the rather primitive character of the pottery, the authors dated the site to the Qatifian period (the late 6th millennium BC). In a previous publication, Rothenberg (1990) had already suggested that some tuyeres found among the nut-sized fragments of slags at F2 would prove the very earliest use of such integral parts in furnace construction. The proposed dating of this smelting site not only would predate the beginning of extractive metallurgy in the Levant by a few millennia, it also would raise the know-how of earliest metallurgical techniques in the entire Old World to a hitherto unknown level of sophistication, unparalleled by any other findings.

Segal *et al.* (1998) stressed the difficulty of directly dating the metallurgical activities of Timna site F2 and pointed out that, due to the lack of inclusions of charcoal, no physical dating of the slags from F2 could be carried out as yet. However, [14]C dating was performed on charcoal from a bowl-shaped pit below layer 3 (Burleigh and Matthews 1982; BM 1368) which, after Merkel and Rothenberg (1999), gave an age of 1080 ± 50 BC, i.e. Iron Age. Craddock (1993, 2000 and oral communication), who supervised the excavation of the site, pointed out that the majority of the flint, ceramic and metallurgical finds, tuyere fragments and slags, were found lying on the surface together with other material of all ages from the Palaeolithic to the present. The three identified archaeological layers were extremely thin, typically not more than 5 cm in total. Claims for Neolithic copper metallurgy at Timna have been rejected by Adams (1998).

### Yotvata, site 44

Yotvata is located *c.* 10 km north of Timna. On a hill close to the Kibbutz Yotvata, an Early Iron Age stronghold (site 44) was excavated by Meshel (1993). The core of a rampart built to the west of the fortress was dated by [14]C to 4360–4240 BC. Another [14]C date from an ash and slag layer was 3520–3360 BC (Avner 2002). The author suggests that Chalcolithic/Early Bronze Age I smelting activities on the Yotvata hill represent a widespread phenomenon of slag dumps found on almost every hilltop in the southern Arabah. This phenomenon shows similarities with the copper district of Feinan, where numerous slag heaps are to be found on top of hills. But these are exclusively dated to the Early Bronze Age II–IV (Hauptmann 2000). Rothenberg *et al.* (2006) propose, based upon a comparative slag typology with Timna 39 and Timna F2, a tentative dating of copper smelting at Yotvata to the Late Pottery Neolithic (6th to 5th millennium BC) and to the Chalcolithic period (5th to 4th millennium BC). Unfortunately no radiocarbon dating for this classification is available.

**Figure 2** Heavily slagged fragment of a tuyere from Timna, site F2. Note the opening of the nozzle which is *c.* 2 cm in diameter (see Fig. 3). Width of the sample: 2.7 cm.

### Thermoluminescence dating of Timna, site F2

A slag sample[1] with adhering host rock and inclusions, from Timna site F2, was analysed by thermoluminescence (TL) to verify the proposed archaeological dating of this site and to provide evidence for or against such an early beginning of metallurgy in the southern Levant. The then proposed age of the slag was the Qatifian period, i.e. the late 6th millennium BC.

The $2 \times 2.5 \times 3.3$ cm sample is a heavily slagged fragment of a tuyere's nozzle made of coarse-grained, sand-tempered pottery. It is obviously the part of a tuyere which was inserted into a furnace and reacted with the charge and/or the liquid slag. The surface shows a black layer of porous slag, partly coloured by red inclusions of cuprite ($Cu_2O$) (Fig. 2). This points to rather oxidising conditions during the formation of the slag and seems to be in accord with a general oxygen surplus in a furnace close to the tuyeres. The diameter of the tuyere is roughly 2 cm (Fig. 3). The typology of our sample closely

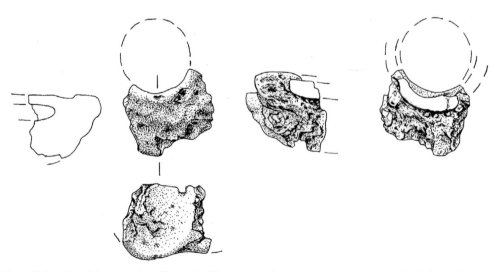

**Figure 3** Drawing of the tuyere from Timna, site F2, investigated in this study. Note the size of the hole which is *c.* 2 cm in diameter (see Fig. 2). (Drawing: A. Weisgerber.)

resembles the fragments of slags and tuyeres from site F2, as shown in Rothenberg (1990) and Merkel and Rothenberg (1999) which were said to come from 6th-millennium levels of site F2.

## Sampling and measurement

Due to the small amount of material, an ordinary procedure of TL dating could not be performed, and, hence, a simplified variant was carried out which usually is applied for authenticity testing of archaeological ceramic objects (Aitken 1985). This seemed to be reasonable as the crucial question of dating concerned a relatively long time span, i.e. 6th millennium BC as proposed by Rothenberg as against 12th/11th century BC as suggested by previous [14]C dating. We therefore could easily accept a TL age with a relatively large error. The larger error of TL authenticity tests, in comparison to TL dating is caused by: (1) the unknown environmental dose rate of the sample, (2) simplified sampling and sample-preparation procedure ignoring the disproportionate distribution of uranium, thorium and potassium in temper and matrix minerals, (3) a less precise determination of the archaeodose and the k-value, and (4) the unknown moisture content.

After removing the light-exposed surface, a powdered sample of about 250 mg was taken by drilling from the inner part of the slag under strictly subdued illumination at the Forschungsstelle Archäometrie at the Max-Planck-Institute. It was treated with diluted hydrochloric acid, and the grain size fraction of 4–11 μm was separated by sedimentation in acetone. Ten subsamples were prepared. Thermoluminescence was measured in a nitrogen atmosphere at a heating rate of 10 °C/s using an ELSEC 7188 TL-reader with an EMI 9635 QA photomultiplier and a Corning 7-59 filter. The natural thermoluminescence was measured on three subsamples; another four subsamples were irradiated by a [90]Sr-ß-source with 22.1 Gy, 25.23 Gy, 31.53 Gy and 62.99 Gy. To determine the specific k-value, three subsamples were irradiated by a [241]Am-α-source with 212.55 Gy, 221.9 Gy and 426.4 Gy respectively. During measurement of the first α-irradiated subsample there was an instrumental irregularity hence this glow curve was not suitable for further interpretation of data. The intercept correction was done by using the three subsamples already taken for natural luminescence analysis and the three α-irradiated ones already taken for k-value determination. They again were irradiated by 15.79 Gy, 31.49 Gy and 47.20 Gy. The plateau test was applied to determine the temperature range of the glow curves utilised for age calculation. We chose the range between 360 and 420 °C and calculated all the values presented in Table 1. Testing of anomalous fading was not performed.

**Table 1** Measurement results of the natural dose ($D_E + I_0$) and the k-value.

|  | $D_E$ | $I_0$ | $D_E + I_0$ | k-value |
|---|---|---|---|---|
| 360–420 °C | 17.57 ± 2.57 Gy | 2.52 ± 0.84 Gy | 20.09 Gy | 0.132 |

## Natural dose rate

To determine the natural dose rate, both the ceramic and the slaggy part of the sample were α-counted. The internal dose rate was determined by calculating first the α-dose rate using the α-count rate of the ceramic part and the k-value. The ß-dose rate of uranium and thorium was then calculated by assuming a U/Th-ratio of 1/3.

The potassium concentration for sample and surroundings was estimated to be 2 ± 1.8 wt%. The external dose rate was estimated by deriving the γ-dose rate of uranium and thorium from the α-count rate of the slaggy part of the sample taking the U/Th-ratio given by the 'pairs' technique of α-counting and adding the potassium concentration. The cosmic dose rate was estimated with 0.3 ± 0.1 mGy/a because the sample was found near the surface. We assumed that the water content was negligible in both sample and environment due to the arid conditions at Timna and we therefore assumed a pore volume of 20% with a saturation level 1 ± 1%. All (calculated and estimated) values for the dose rate determination are given in Table 2. Conversion factors were taken from Adamaiec and Aitken (1998).

## Results and discussion

From the natural dose and dose rate data, given in Tables 1 and 2, a TL age of 2846 years with 25.8% as 1σ-error is calculated. This corresponds to 850 BC ± 735 years. Depending on the required probability, the following ranges are obtained:

- 1σ-range (= 68.3% probability): 1585–115 BC
- 2σ-range (= 95.4% probability): 2320 BC–AD 620
- 3σ-range (= 99.7% probability): 3055 BC–AD 1355

This result definitively confirms the younger of the two proposed ages of the smelting site. Even the 3σ-range of probability of our result does not support an age of 6th millennium. As already mentioned, the [14]C dating of charcoal (Burleigh and Matthews 1982) excavated from the base of a pit (layer 3) at the site (Craddock 2000: fig. 2b) gave a date of 3030 ± 50 BP. According to the calibration curve by Stuiver et al. (1998), the calibrated date is 1394–1219 BC (1σ-range) and, herewith, corresponds with the Late Bronze Age, and not

**Table 2** Determination of the dose rate. Pore volume and water saturation of sample and environment were estimated as 20% resp. 1%.

|  | α-counting rate (ks$^{-1}$) | Potassium (wt.%) | k-value | $dD_{\alpha,humid}/dt$ (mGy/a) | $dD_{\beta,humid}/dt$ (mGy/a) | $dD_{\gamma,humid}/dt$ (mGy/a) | $dD_{humid}/dt$ (mGy/a) |
|---|---|---|---|---|---|---|---|
| ceramic | 14.50 | 2 | 0.132 | 2.44 | 2.64 |  | 7.06 |
| slag | 15.34 | 2 |  |  |  | 1.98 | |

the Iron Age. Considering the archaeological setting of site F2, not only the slag investigated here, but also the charcoal utilised for $^{14}$C dating are obviously proof for Late Bronze Age metallurgical activities. We conclude that, for the time being, there is no convincing dating available yet for any Qatifian metallurgy at Timna.

Finally, we should add an observation concerning the occurrence of tuyeres at site F2 which clearly supports a later date. The diameter of the hole in our sample and those published by Rothenberg is *c.* 2 cm They are easily distinguished from any of the clay tips that are found in metal workshops all over the Middle East and Europe in the early stages of metallurgy, which were utilised both for smelting and melting operations. Such clay tips were applied to blowpipes made perhaps of reed to protect them from the glowing charcoal bed. The diameter of such clay tips ranges between 2 and 6 mm (Roden 1988). A nozzle hole of 2 cm necessarily needs an air supply from mechanically operated bellows. Even if the earliest bellows were constructed exclusively of organic material and are all lost, it remains to be mentioned that such clay tuyeres have apparently not been found on any of the most primitive smelting sites, which may indicate that bellows were unknown (Craddock 1995: 135). Archaeological evidence of hand- or foot-operated bellows, and of tuyeres adapted to bellows, does not exist before the late 3rd or early 2nd millennium BC (Davey 1988; Müller-Karpe 1994). The suggested dating of a Qatifian origin of the slag/tuyere from Timna, site F2 would predate the use of bellows by at least 3000 years.

## Late Chalcolithic settlements at Aqaba: Tell Magass and Tell Hujayrat al-Ghuzlan

In the late 1980s, and subsequently during the joint project Archaeological Survey and Excavations at the Wadi al-Yutum and Magass Area al-Aqaba (ASEYM), under the leadership of L. Khalil and R. Eichmann, two major settlements were found on the gravel fan of the Wadi al-Yutum, directly north of the modern city of Aqaba, Jordan. These are Tell Hujayrat al-Ghuzlan (*c.* 150 × 180 m) and Tell Magass (*c.* 30 × 100 m). They are dated by $^{14}$C to between *c.* 4000 and 3600 BC, i.e. the Late Chalcolithic period (Görsdorf 2002). Excavations have revealed a variety of metallurgical remains such as copper ores, many kilograms of slags, fragments of (s)melting crucibles, grinding and hammer stones, and finally several dozen oval- and rectangular-shaped casting moulds. In addition some ingots, lumps and small objects were retrieved (Khalil 1988; Khalil and Riederer 1998; Müller-Neuhof *et al.* 2003). These finds clearly indicate that the two settlements were centres of copper production in the southern Wadi Arabah during the 4th millennium BC.

Both at Tell Magass and Tell Hujayrat al-Ghuzlan, copper ores were analysed for their mineralogical, chemical and lead isotope composition to identify their provenance and the nature of the material used for smelting (Hauptmann *et al.* in press; Khalil and Riederer 1998). It was investigated whether the ores were suitable to directly produce metal or whether fluxing agents had to be added to facilitate slag formation. The copper ores derived from a sedimentary ore deposit.

Consequently, the ore district of Timna and surrounding mineralisations to the south, at a distance from the settlement of only 10 to 30 km (Nahal Roded, Wadi Amram), were suggested to be the most likely source among all mineralisations in question. As this region is just one part of the much larger copper ore district of the Wadi Arabah, however, all other mineralisations at the eastern margin, between Wadis Abu Kusheibah and Abu Qurdiyah and the region of Feinan, *c.* 100 km to the north, also had to be taken into consideration.

It has been pointed out (Hauptmann 1989, 2000) that petrographic analyses of the texture and fabric of ores in some cases can be a useful tool to trace archaeological copper ores in the southern Levant back to their sources in the Wadi Arabah. One of the most significant petrographic features of the ores excavated at Tell Magass and Tell Hujayrat al-Ghuzlan are cuprified remains of plants which are partly replaced by malachite and other secondary copper minerals and by limonite (Fig. 4; Hauptmann *et al.*, in press). They are identical to those described by Bartura *et al.* (1980) and Keidar (1984) from the Amir/Avrona Formation of the copper ore district at Timna. This is the formation where the most intensive ancient mining activities were found. Such floral remains have been observed exclusively at Timna; they occur neither at Feinan nor at any other mineralisations in the Wadi Arabah. For the time being, the occurrence of cuprified floral remains seems to provide the only discriminating and secure feature to pinpoint the origin of these ores to Timna. Otherwise, due to the common genesis, it is not possible to clearly distinguish between ores originating from Timna and those from Feinan – either by ore mineralogy, by geochemistry or by lead isotope analysis. The identification of ores from Timna at Tell Magass thus indirectly proves mining activity at this copper deposit during the 4th millennium BC with the ores being smelted at settlements of the prehistoric area of Aqaba.

We can demonstrate further that high grade and partly self-fluxing copper ores, i.e. ores that could be smelted to

**Figure 4** Sample JD-35/41b, Tell Hujayrat al-Ghuzlan, F5, Locus 1. Micrograph of a piece of high-grade copper ore consisting of malachite, (par-)atacamite and chalcocite (black). Note the cellular texture of plant remains, overgrown by secondary copper minerals. They are characteristic for copper mineralisations at the western margin of the Wadi Arabah, and are extensively described from Timna (Bartura *et al.* 1980). Comparable ores were also found at Tell Magass. Transmitted light, scale bar: 500 μm.

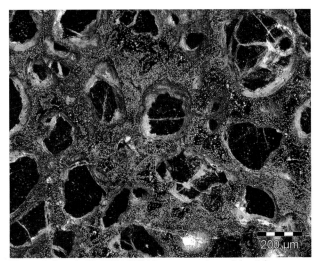

**Figure 5** Sample JD-35/16, Tell Magass, area A, sq. 6, loc 0 (top soil). Micrograph of a piece of 'tile ore': sandstone impregnated and cemented by a mixture of cuprite and limonite (red). Quartz grains are surrounded by a rim of malachite and others. The chemical composition of a similar ore from Tell Magass ($SiO_2$ = 32 wt%, $Fe_2O_3$ = 30 wt%, Cu = 14 wt%) is characteristic for the ore deposit of Timna and mineralisations in the vicinity. It is close to that of a self-fluxing ore. Transmitted light, scale bar: 200 μm.

metal directly without adding any fluxing agents for slag formation, were found at the two sites mentioned above. They agree in their material composition and texture with ores from Timna. Such ore, termed 'tile-ore', consists of a fine-grained, intensively intergrown mixture of iron-(hydr-)oxides, such as limonite and hematite, with changing amounts of secondary copper minerals (cuprite and malachite) and of quartz (Fig. 5). There is no reason to assume that the ores found and smelted at Tell Magass and Tell Hujayrat al-Ghuzlan came from sources other than the Wadi Arabah.

## Conclusions

Chalcolithic copper production at Timna, as suggested by Rothenberg, can be supported in relation to the exploitation of copper ores from the ore mineralisations in the ore district and surroundings. The excavations at Tell Magass and Tell Hujayrat al Ghuzlan clearly indicate that the social organisation pattern of metal production – as it is known from the eastern Mediterranean, the Balkans and from Europe – can be applied to the prehistoric southern Arabah also. Ore was transported from the mine to the settlement and smelted there. At Tell Hujayrat al-Ghuzlan, metal was produced on a scale that exceeded the level of domestic production. The excavation of the settlements underlines the larger than regional importance as a possible gateway for the trade to the Red Sea in the Late Chalcolithic period, even as far as the Nile Delta (Müller-Neuhof *et al.* 2003). The identification of Timna copper ores from the two sites also indicates a larger scale mining activity than was presumed previously. This is supported by a large number of club-like hammer stones made of local rocks from Timna that suggests active mining activities possibly as early as the 4th millennium BC (Weisgerber 2006). Wherever smelting took place, high-grade

oxidic copper ores were available which, due to their composition, in theory could be smelted in a single-stage metallurgical operation. Mixed copper-iron ores with high amounts of quartz also point to smelting operations without any need of adding fluxes for slag formation.

As stressed in a number of publications (e.g. Avner 2002; Hauptmann 2000), there is no conclusive evidence of metallurgical activities in the southern Levant before the late 5th/4th millennia BC. Thermoluminescence dating of a slag with adhering host rock from Timna site F2 does not support a Late Neolithic age. It is in agreement with previous radiocarbon measurements from this site which indicate metallurgical activities during the Late Bronze Age.

## Appendix

Some comments regarding the thermoluminescence date are appropriate here. The TL method provides the date of the last heating of a sample, thus a reheating of the tuyere, e.g. by a camp fire, would result in a younger age. Due to the small amount of material available, no test for anomalous fading could be performed. If this phenomenon is observed, the determined age has to be interpreted as the minimum age. In our case, however, we would exclude the possibility that the true age due to anomalous fading would be higher than the age covered by the 3σ-range – contamination of the sample by light-exposed material would lead to lower TL ages. This cannot be ruled out with 100% certainty. But if indeed the sample had had an age exceeding 7000 years, a mixture with light-exposed material (= TL age zero) of more than 50% would be necessary in order to obtain an apparent TL age of 3000 years. This, however, is extremely unlikely.

## Acknowledgements

We are very much obliged to Prof. Dr Beno Rothenberg for allowing us to perform thermoluminescence dating of a sample from Timna, site F2. We gratefully acknowledge permission from Prof. Dr Lutfi Khalil, University of Jordan, Amman, and from Prof. Dr Klaus Schmidt, Deutsches Archäologisches Institut, to analyse copper ores from Tell Magass and Tell Hujayrat al-Ghuzlan. The study was supported by the Deutsche Forschungsgemeinschaft.

## Note

1. The sample was received from Prof. Dr Beno Rothenberg on 24 October 1992, during a workshop entitled 'Archaeometallurgy in the Arabah, the Negev and Sinai' at the Institute of Archaeology, University of London (Hauptmann and Merkel 1992).

## References

Adamaiec, G. and Aitken, M. 1998. Dose-rate conversion factors: update. *Ancient TL* 16(2): 37–50.

Adams, R.B. 1998. On early copper metallurgy in the Levant: a response to claims of Neolithic metallurgy. In *The Prehistory*

of Jordan II: Perspectives from 1996, H.G. Gebel, Z. Kafafi and
G. Rollefson (eds). Berlin: Ex Oriente, 651–5.

Aitken, M. 1985. *Thermoluminescence Dating*. London: Academic
Press.

Avner, U. 2002. *Studies in the Material and Spiritual Culture of the
Negev and Sinai Populations during the Sixth–Third Millennia BC*.
PhD dissertation, Institute of Archaeology, Hebrew University.

Avner, U. 2006. Settlement patterns in the Wadi Arabah and the
adjacent areas: a view from the Eilat region. In *Crossing the Rift:
Resources, Routes, Settlement Patterns and Interaction in the Wadi
Arabah*, P. Bienkowski and K. Galor (eds). Levant Supplement
Series 3, 51–74.

Bartelheim, M., Eckstein, K., Huijsmans, M. *et al.* 2002. Kupferzeitliche
Gewinnung in Brixlegg, Österreich. In *Die Anfänge der Metallurgie
in der Alten Welt*, M. Bartelheim, E. Pernicka and R. Krause (eds).
Forschungen zur Archäometrie und Altertumswissenschaft 1.
Rahden: Marie Leidorf, 33–82.

Bartura, Y., Hauptmann, A. and Schöne-Warnefeldt, G. 1980. Zur
Mineralogie und Geologie der antik genutzten Kupferlagerstätte
im Timna-Tal. In *Antikes Kupfer im Timna-Tal*, H.G. Conrad and
B. Rothenberg (eds). *Der Anschnitt* Beiheft 1. Bochum: Deutsches
Bergbau-Museum, 41–56.

Burleigh, R. and Matthews, K. 1982. British Museum radiocarbon
measurements XIII. *Radiocarbon* 24(2): 165.

Craddock, P.T. 1993. Review of *The Ancient Metallurgy of Copper* by
B. Rothenberg. *Journal of Archaeological Science* 20: 589–94.

Craddock, P.T. 1995. *Early Metal Mining and Metallurgy*. Edinburgh:
Edinburgh University Press.

Craddock P.T. 2000. From hearth to furnace: evidences for the earliest metal smelting technologies in the eastern Mediterranean.
*Paléorient* 26(22): 151–65.

Davey, C. 1988. Tell edh-Dhiba'i and the southern Near Eastern
metalworking tradition. In *The Beginning of the Use of Metals
and Alloys. Papers from the Second International Conference on
the Beginning of the Use of Metals and Alloys, Zhengzhou, China,
21–26 October 1986*, R. Maddin (ed.). Cambridge, MA: MIT Press,
63–8.

Gale, N.H., Stos-Gale, Z.-A., Lilov, P. *et al.* 1991. Recent studies of
eneolithic copper ores and artefacts in Bulgaria. In *La Découverte
du Metal*, J.P. Mohen and C. Éluère (eds). Paris: Société des amis du
Musée des antiquités nationales et du Château de Saint-Germain-
en-Laye, 49–75.

Glumac, P. and Tringham, R. 1990. The exploitation of copper minerals. In *Selevac: A Neolithic Village*, R. Tringham and D. Kristić
(eds), *Monumenta Archaeologica* 15. Los Angeles: University of
California, 549–66.

Golden, J., Levy, T.E. and Hauptmann, A. 2001. Recent discoveries
concerning chalcolithic metallurgy at Shiqmim, Israel. *Journal of
Archaeological Science* 28(9): 951–63.

Görsdorf, J. 2002. New ¹⁴C-datings of prehistoric settlements in the
south of Jordan. In *Ausgrabungen und Surveys im Vorderen Orient
I*, R. Eichmann (ed.). Orient-Archäologie, Band 5. Rahden: Marie
Leidorf, 333–9.

Hauptmann, A. 1989. The earliest periods of copper metallurgy
in Feinan. In *Old World Archaeometallurgy*, A. Hauptmann,
E. Pernicka and G.A. Wagner (eds). *Der Anschnitt* Beiheft 7.
Bochum: Deutsches Bergbau-Museum, 119–35.

Hauptmann, A. 2000. *Zur frühen Metallurgie des Kupfers in Fenan,
Jordanien*. *Der Anschnitt* Beiheft 11. Bochum: Deutsches
Bergbau-Museum.

Hauptmann, A. and Merkel, J.F. 1992. IAMS Workshop Series:
Archaeo-metallurgy in the ancient Near East. *Institute for Archaeo-
Metallalurgical Studies Newsletter* 18: 14–15.

Hauptmann, A., Pernicka, E., Rehren, Th. und Yalçin, Ü. (eds) 1999.
*The Beginnings of Metallurgy*. *Der Anschnitt* Beiheft 9. Bochum:
Deutsches Bergbau-Museum.

Hauptmann, A., Khalil, L. and Schmitt-Strecker, S. in press. Evidence
for Late Chalcolithic/Early Bronze Age I copper production

from Timna ores at Tall Magass, Aqaba. In *Prehistoric Aqaba*,
R. Eichmann, L. Khalil and K. Schmidt (eds). Rahden: Marie
Leidorf.

Keidar, Y. 1984. *Mineralogy and Petrography of Copper Nodules in the
Upper White Nubian Sandstone in the Timna Valley*. MSc thesis,
Ben Gurion University of the Negev, Beer Sheba.

Khalil, L. 1988. Excavation at Magass-Aqaba, 1985. *Dirasat* XV(7):
71–109.

Khalil, L. and Riederer, J. 1998. Examination of copper metallurgical
remains from a Chalcolithic site at el-Magass, Jordan. *Damaszener
Mitteilungen* 10: 1–9.

Levy, T. and Shalev, S. 1989. Prehistoric metalworking in the southern Levant: archaeometallurgical and social perspectives. *World
Archaeology* 20: 350–72.

Maddin, R., Muhly, J.D. and Stech, T. 1999. Early metalworking at
Çayönü. In Hauptmann *et al.* 1999, 37–44.

Merkel, J.F. and Rothenberg, B. 1999. The earliest steps to copper
metallurgy in the western Arabah. In Hauptmann *et al.* 1999,
149–65.

Meshel, Z. 1993. Yotvata. In *The New Encyclopedia of Archaeological
Excavations in the Holy Land*, vol. 4. Jerusalem: Israel Exploration
Society/Carta and New York: Simon and Schuster, 1517–20.

Muhly, J.D. 1984. Timna and King Salomon. *Bibliotheca Orientalis*
XLI: 275–92.

Müller-Karpe, A. 1994. *Altanatolisches Metallhandwerk*. Offa-Bücher,
Bd. 75. Neumünster: Wachholtz.

Müller-Neuhof, B., Schmidt, K., Khalil, L. *et al.* 2003. Warenproduktion
und Fernhandel vor 6000 Jahren. *Alter Orient* 4: 22–5.

Ottaway, B. 1999. The settlement as an early smelting place for copper.
In *The Fourth International Conference on the Beginnings of the
Use of Metals and Alloys (BUMA-IV), Shimane, Japan, 1998*,
165–72.

Palmieri, A.M., Sertok, K. and Chernykh, E. 1993. From Arslantepe
metalwork to arsenical copper technology in eastern Anatolia.
In *Between the Rivers and Over the Mountains. Archaeologica
Anatolica et Mesopotamica Alba Palmieri Dedicata*, M. Frangipane,
H. Hauptmann, M. Liverani *et al.* (eds). Rome: Universita 'La
Sapienza', 573–99.

Pernicka, E., Begemann, F., Schmitt-Strecker, S. *et al.* 1993. Eneolithic
and Early Bronze Age copper artefacts from the Balkans and their
relation to Serbian copper ores. *Prähistorische Zeitschrift* 68(1):
1–54.

Roden, C. 1988. Blasrohrdüsen. Ein archäologischer Exkurs zur
Pyrotechnologie des Chalkolithikums und der Bronzezeit. *Der
Anschnitt* 40(3): 62–82.

Rothenberg, B. 1973. *Das Tal der biblischen Kupferminen*. Bergisch-
Gladbach: Lübbe.

Rothenberg, B. 1978. Excavations at Timna Site 39, a Chalcolithic
copper smelting site and furnace and its metallurgy. In
*Archaeometallurgy: Chalcolithic Copper Smelting*, B. Rothenberg
(ed.). Institute for Archaeo-Metallurgical Studies, Monograph 1.
London: IAMS, 1–15.

Rothenberg, B. 1990. Copper smelting furnaces, tuyeres, slags,
ingot-moulds and ingots in the Arabah: the archaeological data.
In *Researches in the Arabah 1959–1984. Vol. II: The Ancient
Metallurgy of Copper*, B. Rothenberg (ed.). London: Institute for
Archaeo-Metallurgical Studies, 1–77.

Rothenberg, B. 1999. Archaeometallurgical researches in the southern
Arabah 1959–1990. Part I: Late Pottery Neolithic to Early Bronze
Age IV. *Palestine Exploration Quarterly* 131: 68–89.

Rothenberg, B. and Merkel, J.F. 1995. Late Neolithic copper smelting in the Arabah. *Institute for Archaeo-Metallurgical Studies
Newsletter* 15/16: 1–8.

Rothenberg, B. and Merkel, J.F. 1998. Chalcolithic, fifth millennium
BC, copper smelting at Timna. *Institute for Archaeo-Metallurgical
Studies Newsletter* 20: 1–3.

Rothenberg, B., Segal, I. and Khalaily, H. 2006. Late Neolithic and
Chalcolithic copper smelting at the Yotvata oasis (south-west

Arabah). *Institute for Archaeo-Metallurgical Studies Newsletter* 24: 17–28.

Segal, I., Rothenberg, B. and Bar-Matthews, M. 1998. Smelting slag from prehistoric sites F2 and N3 in Timna, SW Arabah, Israel. In *Metallurgica Antiqua: In Honour of Hans-Gert Bachmann and Robert Maddin*, Th. Rehren, A. Hauptmann and J.D. Muhly (eds). *Der Anschnitt* Beiheft 8. Bochum: Deutsches Bergbau-Museum, 223–34.

Segal, I., Halicz, L. and Kamenski, A. 2004. The metallurgical remains from Ashqelon, Afridar – Areas E, G and H. *Atiqot* 45: 311–30.

Shugar, A. 2000. *Archaeometallurgical Investigation of the Chalcolithic Site of Abu Matar, Israel: A Reassessment of Technology and its Implications for the Ghassulian Culture*. PhD dissertation, Institute of Archaeology, University of London.

Stöllner, T. 2003. Mining and economy: a discussion of spatial organisations and structures of early raw material exploitation. In *Man and Mining*, T. Stöllner, G. Körlin, G. Steffens and J. Cierny (eds). *Der Anschnitt* Beiheft 16. Bochum: Deutsches Bergbau-Museum, 415-46.

Stuiver, M., Reimer, P.J., Bard, E. *et al.* 1998. INTCAL98 Radiocarbon age calibration, 24.000–0 cal. BP. *Radiocarbon* 40(3): 1041–83.

Weisgerber, G. 2006. The mineral wealth of ancient Arabia and its use I: copper mining and smelting at Feinan and Timna – comparison and evaluation of techniques, production, and strategies. *Arabian Archaeology and Epigraphy* 17: 1–30.

Weisgerber, G. and Pernicka, E. 1995. Ore mining in prehistoric Europe. In *Prehistoric Gold in Europe*, G. Morteani and J.P.S. Northover (eds). Dordrecht/Boston/London: Kluwer, 159–82.

Yalçin, Ü. and Pernicka, E. 1999. Frühneolithische Metallurgie von Ashikli Höyük. In Hauptmann *et al.* 1999, 45–54.

## Authors' addresses

- Andreas Hauptmann, Deutsches Bergbau-Museum, Forschungsstelle Archäologie und Materialwissenschaften, 44791 Bochum, Germany (andreas.hauptmann@bergbaumuseum.de)
- Irmtrud Wagner, Forschungsstelle Archäometrie der Heidelberger Akademie der Wissenschaften, Max-Planck-Institut für Kernphysik, 69117 Heidelberg, Germany

# On the origins of metallurgy in prehistoric Southeast Asia: the view from Thailand

*Vincent C. Pigott and Roberto Ciarla*

*ABSTRACT*   Research over the last 30 years has markedly improved our understanding of metallurgical developments in prehistoric Thailand. The chronology of its earliest appearance, however, remains under debate. Current evidence suggests that tin-bronze metallurgy appeared rather abruptly as a full-blown technology by the mid-2nd millennium BC. Questions also continue to arise as to the sources of the technology. Current arguments no longer favour an indigenous origin; researchers are increasingly pointing north into what is today modern China, linking metallurgical developments to the regions of the Yangtze valley and Lingnan and their ties to sophisticated bronze-making traditions which began during the Erlitou (*c.* 1900–1500 BC) and the Erligang (*c.* 1600–1300 BC) cultures in the Central Plain of the Huanghe. In turn, links between this early 2nd-millennium BC metallurgical tradition and the easternmost extensions of Eurasian Steppe cultures to the north and west of China have been explored recently by a number of scholars. This paper assesses broadly the evidence for 'looking north' into China and eventually to its Steppe borderlands as possible sources of traditions, which, over time, may be linked to the coming of tin-bronze in Thailand/Southeast Asia.

*Keywords:* China, Eurasian Steppes, Khao Wong Prachan valley, Lingnan, Southeast Asia, Thailand, archaeometallurgy, copper, prehistory, tin-bronze.

## Introduction

It is generally accepted that copper/bronze metallurgy first appears in the region of mainland Southeast Asia sometime after *c.* 2000 BC. Exactly when this occurs in Thailand continues to be debated.[1] As discussed below, the gist of this paper pivots on increasing evidence that metallurgy may have come to Southeast Asia from the region of modern China; based on current evidence, copper-base metallurgy was earlier, i.e. 3rd millennium BC in the realm of northern modern China (for example Higham 1996; Linduff *et al.* 2000) and the Eurasian Steppes (for example Linduff 2004), than in Thailand/Southeast Asia (see Figs 1 and 2). Specifically, current arguments suggest that the adoption of metallurgy in Southeast Asia was stimulated through interactions with the southern tributaries of the Yangtze via Lingnan (Ciarla 2000; Higham 1996, 2004, 2006) (Fig. 3). This region of China in turn can be linked technologically to the Central Plain of China where in the early 2nd millennium BC the sophisticated bronze-casting metallurgical traditions of the Erlitou-Erligang cultures had evolved.

Over the last decade the question of links between Steppe metallurgy and metallurgical developments in the Central Plain of China has been the subject of considerable discussion. At the end of the 3rd millennium BC, the western and northern borders of northern China were in regular contact with bronze-using Steppe cultures, in particular, the Andronovo and Afanasievo (Anthony 1998; Mei 2000: 58–71) (Fig. 4). Based on archaeological research at sites dating to the 2nd millennium BC in northwestern China, particularly along the Gansu corridor, a pattern of artefact distribution has been

seen to emerge as Steppe-related copper/bronze artefacts co-occur with copper/bronze artefacts from further east (Ke 1998; Mallory and Mair 2000:327, figs 173–4; Mei 2000: figs 3.1–3.17; 2003a).

Scholars have long debated the question of west to east trending contact between the Eurasian Steppes and China (see Jettmar 1985) and Southeast Asia (see Ciarla 1994; Chiou-Peng 1998). A new wave of scholars (e.g. Fitzgerald-Huber 1995, 2003; Li 2003: 16–18; Linduff *et al.* 2000; papers in Mair 1998; Mei 2000, 2003a, 2003b) argue more in favour than against the suggestion that among the possible consequences of this stimulating interaction between 'the Steppe and the Sown' was ultimately a melding of metallurgy and indigenous craft traditions in the early 2nd millennium BC under the mantle of the socially complex Xia and Shang dynasties in the Zhongyuan, the Central Plain of northern China. Regardless of which side one takes in the discussion, what resulted was one of the Old World's most sophisticated metallurgical traditions, typified by the use of a complex piece-mould casting technology responsible for the production of ritual bronzes. These items, in life and in death, served as markers of status and rank for a powerful ruling dynastic elite. 'Social hierarchy was manifest in differential burial customs, in the prescriptive process of bronze designing and casting' (Linduff 1998: 621; also Franklin 1983). Yet this unique state-centred tradition of production did not, to any significant extent, extend its reach technologically much beyond the boundaries of the Central Plain (Fig. 5).[2]

This sophisticated metallurgical tradition which, for the sake of brevity we will term the 'Shang tradition', did not

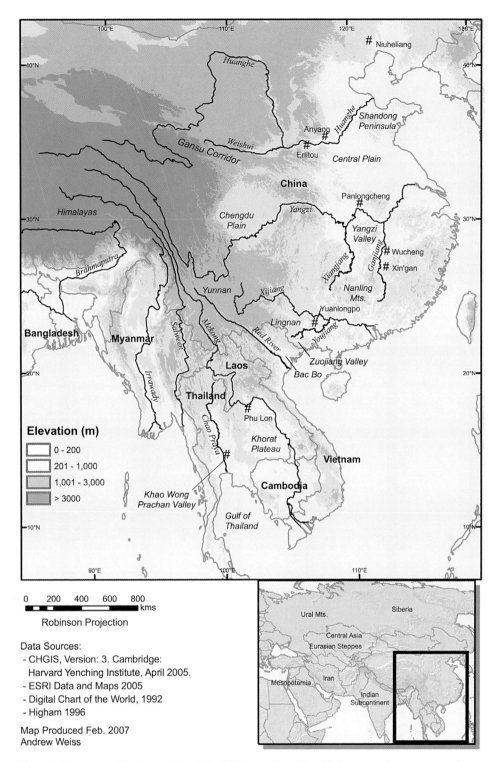

**Figure 1** The topography of greater East Asia which exerted a profound influence on the movements of peoples and on the development of settlement patterns over the millennia.

develop in isolation, but was part of an interaction sphere which, at least from the 4th millennium BC linked many different advanced cultures spread across what is modern China and beyond (Chang 1986). Looking at the general picture of the artefact inventories which typify the three main cultural periods of the Central Plain Bronze Age, that is Erlitou (c. 1900–1600 BC), Shang-Erligang (c. 1600–1300 BC) and Shang-Yin (c. 1300–1045 BC), as well as from references contained in the oracle bones inscriptions of the Shang-Yin

period, we can clearly discern exchange networks which brought highly desired raw materials and goods to the Shang elite centres: jade, turquoise and other semi-precious stones, cowries, turtle carapaces, copper and tin, 'slaves' and gold (Chang 1980: 149–57; see also Higham 2006). The search for and exchange of these goods and raw materials, implying direct and indirect contacts with quite distant sources, activated an ever increasing and complex chain of interlocked regional interaction spheres (we might term 'the Far Eastern

Interaction Sphere') through which copper/bronze casting technology eventually reached Southeast Asia.

Consistent with the new scholarship on eastward trending movements and interactions on the Eurasian Steppes as well as a continually improving understanding of metallurgical developments in this region and in ancient China, there is little evidence to suggest that the copper/bronze metallurgical tradition, which appears in 2nd-millennium BC Thailand was of indigenous origin. Unfortunately, among the reasons for the paucity of evidence surrounding this significant technological transition is that much of Southeast Asia remains an archaeological *terra incognita*. While much work is being

done in Vietnam (e.g. Nguyen *et al.* 2004), it remains difficult to gain access to its publications and language barriers persist (Glover and Bellwood 2004: 205). Thus, Thailand provides the bulk of the information upon which the observations presented here are based. This is, to be sure, an imperfect vantage point given the geographical enormity and ancient cultural diversity of Southeast Asia. Nevertheless, almost 20 years ago, White (1988: 179) argued against independent invention of metallurgy in Southeast Asia:

There is mounting indirect evidence ... that Asian people were interacting during [the 3rd millennium BC]

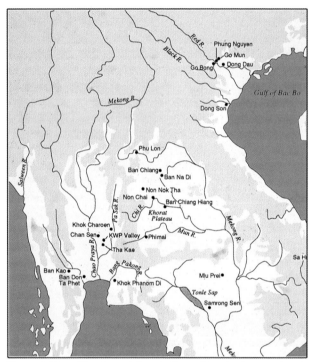

**Figure 2** Map of Southeast Asia including prehistoric sites mentioned in the text.

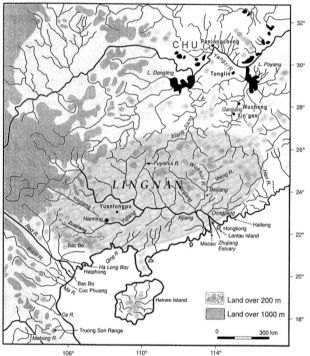

**Figure 3** Map of southeast China including Lingnan with sites, river and regions mentioned in the text.

**Figure 4** Distribution of Andronovo and Afanasievo cultures on the Eurasian Steppes at the end of the 3rd millennium BC (after Mei 2000: 161, fig. 6.1).

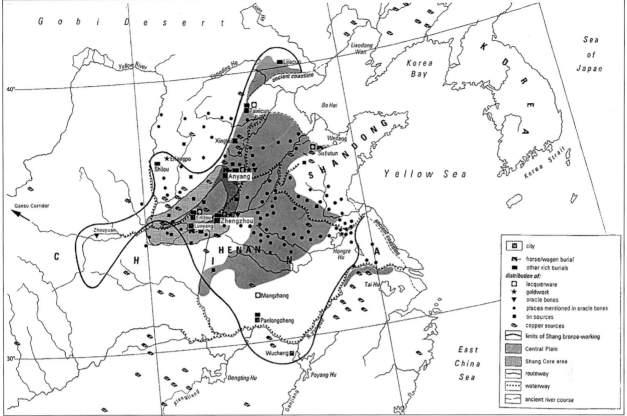

**Figure 5** Map of Shang China indicating limits of its cultural distribution. (© 1999 Chris Scarre. Permission to publish this map courtesy of HarperCollins Publishers Ltd. Copyright in the customized version vests in Archeotype Books, Oxford.)

... although independent invention may be an appropriate concept when discussing the emergence of New World versus Old World agriculture, it has come to the point of naiveté to use it in third millennium Asia ... the best working assumption for Southeast Asia during the third millennium BC is that the region ... interacted on some level ... with other societies on the continent. When more archaeological research is conducted in Burma, south China, central Asia and the coastal regions of Southeast Asia, we may be able to specify the interaction.

We now have the benefit of the results of the past 20 years of abundant archaeological research in central and eastern Asia. This has specified in even greater detail the extent of Asian interactions – in particular those pertinent to the question of the origins of Southeast Asian copper/bronze metallurgy. Many questions, however, remain unanswered.

Looking broadly at the evidence from Thailand, it includes prehistoric copper mines, at least one significant production centre and local community-based copper and bronze processing distributed selectively across the landscape. Items produced locally are typologically rather limited and major categories include socketed implements and distinctive bangles. At the same time, however, Thailand has yet to offer indications of a gradually emerging tradition of copper/bronze production. To presage the discussion to follow, this lack of a developmental stage may link to 'a rapid and broad spread of bronze technology in Asia east of the Urals dating from about 2000 BC' (White 1997: 104).

## Potential external source areas for metallurgy

### South Asia

The current evidence for tin-bronze metallurgy offers indications that it 'arrived' in Thailand as a full-blown technology, that is conducted by people who knew how to find copper and tin, and mine, smelt and alloy them when so desired. Only two directions are available in which to look for an external source area for copper/bronze technology, namely west towards India and north into China. In looking westward, across Burma and Bangladesh to the Indian subcontinent, in comparison to Southeast Asia, metallurgical developments in South Asia are unquestionably earlier by millennia (c. 5th millennium BC) with Harappan Phase (c. 2300–1900 BC) metallurgy among its apogees (Kenoyer and Miller 1999; see also Agarwal 2000). The strength of evidence for cultural influences filtering eastwards from the subcontinent, however, is not recognised archaeologically in Southeast Asia until late in the 1st millennium BC (Bellina and Glover 2004; Rispoli 1997b, in press a, in press b). Thus, given the current lack of adequate information, it is not possible at this time to consider South Asia as a source area. The only alternative direction in which to look, therefore, is to the north into modern China, to those locales where early metallurgy has been securely dated and documented early in the later 3rd/early 2nd millennium BC (An 1992, 1993; Higham 1996; Linduff et al. 2000).

## China

By the mid-to-late 3rd millennium BC, evidence of copper-base metallurgy becomes consistently distributed across northern China (see Linduff *et al.* 2000: 10–22 with maps), from the Shandong Peninsula in the east to the upper Huanghe valley during the Longshan period (for example Shao 2002; Thorp 2006: 54), which some term the 'Chalcolithic' (i.e. Yan 1984; Zhang and Wei 2004). By the end of the 3rd to the beginning of the 2nd millennium BC, two main centres of intentional copper/bronze metallurgy had emerged in northern China. The earliest is the Qijia culture (*c.* 2200–1600 BC/Early Bronze Age) (An 1992; Debaine-Francfort 1995; Yan 1984: 38–40) distributed in the upper drainage of the Weishui-Huanghe system (northwest China), which was based on agriculture and a pig- and horse-raising economy; the second is the Erlitou culture (*c.* 1900–1600 BC), which had highly ranked social contexts and was centred between the Yi-Luo drainage and the Huanghe middle valley (Thorp 2006: 21–61; Zhao 2003). While it cannot be ruled out that the first tradition might have stimulated the metallurgical developments of the second, the differences between these two metallurgical traditions are particularly striking. The Qijia metal artefacts were, in fact, tools, weapons (non-socketed and socketed) and simple ornaments (rings and bracelets) mostly of pure copper (Cu 96% on average), followed by tin and tin-lead alloys (*c.* 5%), cast in open or bivalve stone moulds; finishing by hammering was not the rule. Erlitou bronze casting, although later than Qijia, was instead based on the preferential use of a ternary alloy (copper-tin-lead) cast into ceramic piece-moulds to produce vessels of complex shapes derived from earlier ceramic prototypes for the ritual use of an already well-established elite. No socketed tools or weapons (with the exception of a knife and an axe), or rings/bracelets similar to northwestern types have been found at the site of Erlitou. What deserves note

concerning Qijia is that it was instrumental in establishing the basis of the flourishing of the 'Northern Zone Complex' during the 2nd millennium BC, 'a broad area in which different people shared certain common traits, in particular their bronze inventory. This metallurgical tradition typifies the north and marks the cultural boundary with the civilization of the Central Plain' (Di Cosmo 1999: 893).

From the 2nd millennium BC on, the several cultures of the Northern Zone Complex, linked, on the one hand, the agricultural society of the Central Plain of China to the Steppe populations, while on the other, it participated integrally in the exchange networks and migration routes along the Steppe and Forest Belt, from Siberia through central Asia to eastern Europe, and concurrent with one of the Old World's major tin belts (Fig. 6) (see also Parzinger 2000: 67, fig. 1). The continuing availability of this normally scarce resource facilitated the production and use of copper/bronze as Steppe peoples moved either east/west or, in other cases, south towards China's borders, corresponding largely to modern northern Shanxi, Ningxia, Gansu, Qinghai provinces. The basic metallurgical tradition shared by these Steppe cultures, including the Northern Zone Complex, utilised the small furnace and crucible for smelting/melting, bivalve and open moulds for casting and cold and hot hammering for working tin-bronze. Tools and weapons were typically socketed (Chernykh 1992; Linduff 2004).

In contrast, piece-moulding, one of the Old World's most sophisticated, certainly state-centred, metallurgical traditions, rooted in the Erlitou culture, is attested only in the agricultural plains of the Huanghe, of the Yangtze and of their tributaries. This technique appears fully mastered not only by the bronzesmiths of the Erligang culture (or early Shang dynasty period *c.* 1600–1300 BC) which, centred in the mid-Huanghe valley, reached the Yangtze with an outpost excavated at Panlongcheng (Wuhan, Hubei province), but also by other bronze-producing regional cultures strictly linked to Erligang, at least from the technological point of view. The Sanxingdui-Jinsha culture of the upper Yangtze, in the Chengdu Plain (Sichuan province), whose earliest manifestations were possibly contemporary to Erlitou, flourished from the eve of the Erligang period to the early Western Zhou period (*c.* 1045–770 BC) (Thorp 2006: 234, 249–63). This poorly understood culture, whose astonishing jade, bronze and gold production manifested a cultural matrix separate from the realm of the Central Plain, was centred in the Chengdu Plain, thus commanding the hub of narrow river valleys looping southwards from eastern Tibet towards Yunnan and northern Southeast Asia. Most logically, it should be through these north–south narrow corridors that, according to van Driem (2002, 2005), a branch of Tibeto-Burmans peopled the southern and southeastern Himalayas, moving from their Gansu-Qinghai homeland during the Majiayao Neolithic period (*c.* 3500–2000 BC). This same hub, crossing the rich copper deposits of Malong and the tin belt of Gejiu, continued to act during the 1st millennium BC in the spread of 'animal-style' elements from the Steppes to Yunnan and the Red river/Song Hon valley in northern Vietnam (Chiou-Peng 1998; Ciarla 1994; Orioli 1994).

Moving downstream along the Yangtze from the Chengdu Plain, another strategic communication hub formed by the middle Yangtze and its tributaries flowing from the north and

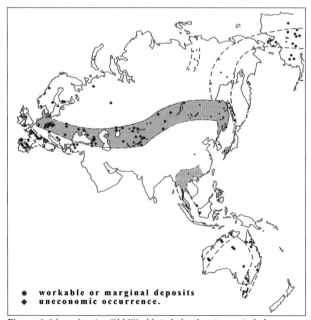

**Figure 6** Map of major Old World tin belts showing main belt crossing Eurasia into East Asia where it bifurcates north into Inner Mongolia/Manchuria and south in Southeast Asia (after de Jesus 1978: 36, fig. 3).

south, was the seat of several local bronze-producing cultures, under-investigated but evidently in close contact with the Erligang centre of the Central Plain. Of particular relevance to our discussion are the Wannian and the Wucheng/Xin'gan archaeological cultures in modern Jiangxi province. These are characterised by the hybridisation of local elements with Shang-derived bronze technology and artefact types, and a chronological span covering the Erligang period (*c.* 1600–1300 BC) (Thorp 2006: 107–16). Both cultures are located in the valley of the Ganjiang, which flows from the south into Poyang Lake, one of the three main siphons of the mid-Yangtze, where the copper-mining complex at Tongling, in one of the richest copper regions of China, is located. Attempts to control the Tongling copper might explain the southern expansion of the Erligang culture and the flourishing of the local Wannian-Wucheng cultures. Moreover, we cannot rule out a westwards destination for Tongling copper heading towards the Chengdu Plain (Sanxingdui-Jinsha culture). Slightly to the west of the Ganjiang is the Xiangjiang, the main southern Yangtze tributary, which flows into Dongting Lake, where a few 14th/13th-century BC archaeological sites have been excavated. One of these sites, at Zaoshi (Shimen county, Hunan province), contained fragments of stone bivalve moulds and a poorly preserved furnace, and some slag possibly associated with copper smelting (Hunan sheng Wenwu Kaogu Yanjiusuo 1992: 191, 216).

Both the Xiang and the Gan rivers, particularly the valley of the Xiangjiang, have been suggested by Rispoli (2004) as the most probable route of the earliest Neolithic expansion (by the 5th/4th millennium BC) towards the northern Southeast Asian regions (also Higham 2002b). These rivers connect the middle Yangtze to the Guangxi-Guangdong region south of the Nanling Mountains (Lingnan means south of the Nanling).

## Cultural expansion south from Lingnan

This proposed route of Neolithic expansion implies a network which was active over the centuries, and one which later on moved highly desired goods ultimately from Southeast Asia along the river valleys (see Higham 2006 for suggested goods). Regional geography had to play a key role in the movements of peoples and ideas within China and into Southeast Asia. These two major regions are inextricably linked by nine of the world's great rivers: the Brahmaputra, Irrawady, Salween, Chao Phraya, Mekong, Red/Song Hon, Huanghe, Xijiang and the Yangtze. Significantly eight of 'These rivers flow along radial lines and arcs centred on Yunnan' (Takaya 1987: 3, fig. 2) (Fig. 7). It has long been understood that, other than along the seacoasts, the conduits of human movement lay along the rivers and associated valleys which radiate through the region in a fashion peculiar to China itself, where many of these rivers originate, and to Southeast Asia through which several of them flow. With the movement of peoples and information along these 'lines of least resistance' (Higham 1996: 337) could have come the knowledge and practice of new technologies.

Over time, as exchange and contact progressed, there ensued an ever-widening pattern of cultural expansion southwards by the 3rd/2nd millennium BC. Based on her ceramic

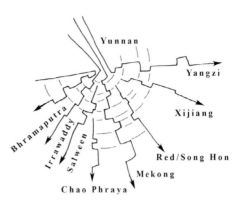

**Figure 7** Diagram of direction of flow of eight major rivers of East and Southeast Asia which are centred on Yunnan, China (after Takaya 1987: 3, fig. 2).

research, Rispoli (2004) has argued that this process involved a 'cultural package' with its roots in the Chinese Neolithic, marked by a characteristic ceramic tradition (see also Rispoli 1997a), rice and millet agriculture, pig/dog/cattle husbandry, polished stone tools (mainly adzes), shell-reaping knives, and the manufacture of marine shell and stone ornaments. This 'package', it has been argued, moved along the river valleys radiating, with a ripple-like effect, into northern Vietnam, into Yunnan, and along the Mekong, running its course across Southeast Asia, including along the Chao Phraya river valley down the centre of Thailand (Higham 1996; 2002a: 108–11; 2004, 2006; Rispoli 2004). Thus the suggestion is that cultural influences filtered out of the Chinese heartland over time. They moved south into and along the Yangtze valley via networks of exchange and spread further south into the region of Lingnan and the Bac Bo Plain (northern Vietnam) during the period of major momentum in the Erlitou-Erligang cultural growth in the centuries astride the mid-2nd millennium BC (Ciarla 2000; Higham 2004, 2006). Thus, both Ciarla and Higham would argue that following this same path of expansion came copper/bronze metallurgy, which was ultimately to reach Thailand. Just how early this happened is, as mentioned, the subject of current debate.

In piecing together a picture of cultural development from the Yangtze valley to the Gulf of Thailand between the late 3rd and the end of the 1st millennium BC, Higham (1996: 338; see also 2006) noted 'the possibility that the Southeast Asian tradition of copper-base metallurgy was initially stimulated by the spread of ideas and goods down the very rivers which introduced the rice farming groups a millennium or so earlier' (late 3rd millennium BC). Moreover, Higham argues from the archaeological record of the Lingnan region in southeastern China, that this region comprises the most suitable candidate for the initial local development of the aforementioned 'ideas and goods' that were to have spread south by the mid-2nd millennium BC.

Ciarla (2000) discussed Lingnan as a physiographically quite distinct region, linked not only to the Yangtze valley through a system of rivers flowing northward from the Nanling range, but also open to the Bac Bo and the Red river/Song Hon valley through the high Zuojiang valley and along the coast. The region he indicated, as has been demonstrated by Allard (1994, 1995), was directly exposed via exchange to northern ideas and goods. Exchange networks among Late Neolithic communities of Lingnan fostered contact with sophisticated bronze-producing centres, such as those of the Wannian and Wucheng cultures in the Ganjiang valley or the lesser known, still unnamed, regional cultures of southern Hunan along the Xiangjiang valley. He concurs that Lingnan is a likely candidate to have acted as the conduit transmitting southwards the knowledge necessary to transform metal-bearing ores into molten copper/bronze. Needless to say, this hypothesis concerning the transmission of metallurgical stimuli from the Yangtze valley into Lingnan and ultimately into Southeast Asia is one which remains in need of rigorous testing.

## Copper and bronze metallurgy in Thailand

Copper/bronze metallurgy, it can be argued based on current evidence, has all the appearances of having reached Thailand from an external northern source(s) by a pattern of movement along the radiating river valleys mentioned above. Particularly apropos to this discussion is the prehistoric copper-mining complex at Phu Lon (Natapintu 1988; Pigott 1998; Pigott and Weisgerber 1998) perched on the south bank of the Mekong in northernmost Loei province, well upstream from Ban Chiang and related sites on the Khorat Plateau (northeast Thailand) (see Fig. 2). While the majority of Phu Lon's radiocarbon dates cluster at c. 1000 BC and after, one early date of c. 1700 BC (lacking a direct cultural association) suggests possible early activity at the mine (Pigott and Weisgerber 1998: 151, fig. 24). This early date could complement the evidence for contemporaneous metallurgy on the Khorat Plateau. It has been argued elsewhere that mining expeditions from settlements such as Ban Chiang and related sites on the plateau exploited the mine's significant copper reserves on a seasonal basis, probably using the Mekong as one route of access (Pigott 1998; White and Pigott 1996). Several of these prehistoric Khorat settlements (for example Ban Chiang, Non Nok Tha and Ban Na Di) share a 2nd into 1st millennium BC tradition of copper/bronze metallurgy marked by strong similarity in metals-related material culture (for example crucible type and manufacture, casting in bivalve stone and ceramic moulds, and commonly socketed implements and ornaments/bangles). Peoples moving along the Mekong also could have split off from groups following the river and headed south into Thailand along the valley of the Chao Phraya which divides Thailand down the middle along a zone of contact between two small pieces of continental plate. This zone, due to its metallogeny, is rich in ore minerals, including copper. Research in central Thailand has documented a copper production locus at the major regional centre of Tha Kae (Ciarla 1992, 2005), as well as in the copper-rich Khao Wong Prachan valley where an industry dated c. 1500 BC and later focused considerable effort on copper (apparently not tin-bronze) production as revealed by a long-term programme of archaeological fieldwork under the auspices of the Thailand Archaeometallurgy Project (Natapintu 1987, 1988, 1991; Pigott and Natapintu 1988; Pigott et al. 1997) as well as by a range of detailed analytical studies on production remains (Bennett 1988a, 1988b, 1989; Pigott 1999; Rostoker

et al. 1989; Wang et al. 1998). Thus, as will be outlined below, what one sees in both metal and metallurgical artefacts in central and northeast Thailand is a basic similarity in simple implement production based on casting in bivalve moulds, a tradition common to the Central Plain of northern China, southeastern China and to the Eurasian Steppes as well.

## Linking prehistoric Thailand to China and the Eurasian Steppes

As for Rispoli's (2004) purported 'cultural package' for the Neolithic expansion from the Yangtze towards the river plains of Southeast Asia, in the process of hypothesis testing and to explore the nature of tenuous links between China and Thailand, one would have to begin by looking individually at the items included.

### Copper/bronze production

Within Thailand, prehistoric copper-base metal production shares the same basic technology, though some regional variation is clearly in evidence. We see a tradition of mining copper ores by shaft and gallery, followed by ore crushing and beneficiation at or not far from the mines. Ores are smelted in crucibles or shallow bowl furnaces. Bivalve moulds (stone and ceramic) are used to cast a somewhat limited repertoire of items in copper/bronze, for example small, frequently socketed implements, fishhooks, ornaments (often bangles) and some weaponry (spear points).

In the production of socketed implements common at prehistoric sites in Thailand, mould plugs around which molten metal is cast would have been used between mould halves to form the socket. Such plugs (ceramic) are known from the Khao Wong Prachan valley. There also exists a tradition of marking bivalve moulds (and other mould types as well) with geometric and curvilinear designs (noted at Phu Lon and in the Khao Wong Prachan valley), which may serve as makers' marks or marks of ownership. Evidence from the Khao Wong Prachan valley sites suggests, on the basis of stratigraphic evidence, that the marked bivalve moulds from central Thai sites of Non Pa Wai and Nil Kham Haeng, are 1st millennium BC in date. The Phu Lon marked mould also dates to this same period.

Of particular importance is possible evidence for the persistent tradition of what, on the Eurasian Steppes (east of the Urals) (Chernykh 1992: 222, fig. 76), are termed 'founders' burials' (as in foundry related) (see examples in Chernykh 1992: 80–1, 83, pls 7, 8; 135, pl. 12; and in Tylecote 1976: 17, fig. 8). Such burials, also termed 'smith's graves', contain items related to metal production, often bivalve moulds and/ or crucibles. Burials of this type have also been found in both northeast Thailand at Non Nok Tha (Bayard 1984) and Ban Non Wat,[3] and central Thailand at Non Pa Wai (Pigott et al. 1997: 147, figs 7–9), Nil Kham Haeng (White and Pigott 1996: 166, fig. 13.9). In the northeast, crucibles and bivalve moulds were interred, while in central Thailand burials contained bivalve moulds for large socketed axes (Fig. 9a) and

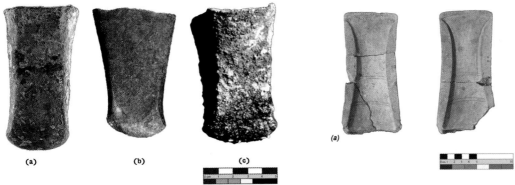

**Figure 8** (a) Socketed copper/bronze axe/adze from Miaobu Locus North site (*c.* 1300–1200 BC) at Yinxu (Anyang) China (Zhongguo Shihui Kexueyuan Kaogu Yanjiusuo 2006: fig. 7.1, table IV.5). (b) Socketed copper/bronze axe/adze (TK93: TNo.1904) from 'Bronze Age layer' (late 2nd millennium BC) at the site of Tha Kae (Lopburi province), central Thailand. (c) Socketed copper/bronze axe/adze from Period 2A burial (*c.* 1500 BC) at the site of Non Pa Wai, Khao Wong Prachan valley, central Thailand.

**Figure 9** (a) Pair of ceramic bivalve moulds for casting a large socketed axe/adze from the 'Grave of the Metalworker', Non Pa Wai Period 2A, *c.* 1500 BC, Khao Wong Prachan valley, central Thailand (Pigott *et al.* 1997: 147, fig. 7). (b) A single stone bivalve mould for the casting of a large socketed axe/adze, from the site of Tangxiahuan (fourth layer), late 2nd millennium BC, Zhuhai city, Guangdong province, China (Guangdong sheng Wenwu Kaogu Yanjiusuo 1998: 14, fig. 13.6).

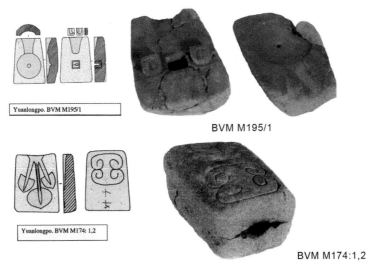

**(a)   Yuanlongpo, Lingnan, southeast China**

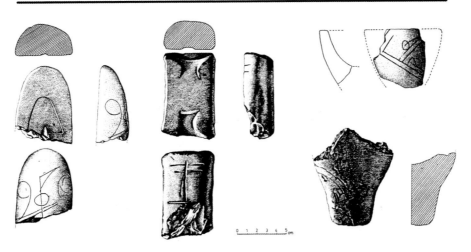

**(b)   Marked moulds from the Khao Wong Prachan valley**

**Figure 10** (a) Two pairs of ceramic bivalve casting moulds from burials M195 and M174 at the site of Yuanlongpo, Lingnan, southeast China, *c.* 1100–800 BC. Note markings on exterior of lower mould pair (Guangxi sheng Wenwu Kaogu Yanjiusuo 1988: figs 28, 29). (b) Two ceramic bivalve moulds and one ceramic conical mould from sites in the Khao Wong Prachan valley, central Thailand, *c.* mid-1st millennium BC. Note markings incised in the moulds' external surfaces.

furnace chimneys (Fig. 12). On the Eurasian Steppes, founders' burials with bivalve moulds have also included mould plugs (Chernykh 1992: 81, pl. 80). Thus, on the one hand, given the above evidence there is little to suggest a direct link to metallurgical production typical of the Erlitou-Erligang tradition in China from the early 2nd millennium BC onward, on the other hand, links to Steppe traditions appear to be more consistent. Joyce C. White (1988), in characterising the so-called 'southern metallurgical tradition', emphasised the distinctions between traditions of working copper/bronze in the Chinese heartland and in mainland Southeast Asia. These include use of the socketed implements, which she argues was rare in the Chinese heartland, but very common in Southeast Asia. The last 20 years of research in China, however, now indicate that socketed tools (for example, ploughshares, several types of spade blades, adzes, chisels and weapons) begin to appear regularly in the Shang inventory from the Erligang period (c. 1600–1300 BC) onwards, including the hybrid Wannian/Wucheng contexts in the Ganjiang valley. Apparently, it is piece-mould casting that typifies bronze production in the areas of the Shang and Zhou cultures of China more than a tradition based on bivalve mould casting which, however, was known and practised for the production of tools and weapons. In contrast, in Southeast Asia, production is characterised more by bivalve mould casting as well as hammering of metal to shape, and the piece-mould tradition is not used (or at least not until quite late – mid-to-late 1st millennium BC – perhaps in the casting of 'Dongson drums').

In the north, weaponry is omnipresent from the Erligang period (c. 1600–1300 BC) onward, but in Southeast Asia it becomes much more common only in the 1st millennium BC. Bangles as personal ornament are more common on the Eurasian Steppes and in Southeast Asia than in the Central Plain where they are practically absent (Franklin 1983: 95), whereas, stone/shell bangles, with D- and T-shaped cross-sections, and 'open ring' earrings are typical of the Yangtze world immediately to the south and extending into Southeast Asia from the Neolithic onward.

Even in the Shang-Zhou Huanghe heartland, a utilitarian copper/bronze metallurgy as practised for implements, weapons, some ornaments, and from c. 1200 BC horse-and-chariot harnesses, bears marked similarities with that of the Steppes and the Southern Tradition as well. To take one example, the socketed axe, an artefact known earliest from the 3rd-millennium BC Eurasian Steppes. In China, the socketed axe in Figure 8a is from grave 60 (M60:29) excavated at Huayuanzhuang Locus East at Yinxu (Anyang) (Zhongguo Shihui Kexueyuan Kaogu Yanjiusuo 2006: fig. 7.1, table IV.5) and is dated to the Yinxu phase 1 (c. 1300–1200 BC). Compare this Anyang implement to those from central Thailand, from Tha Kae's 'Bronze Age level' (Fig. 8b) and Non Pa Wai (beginning of Period 2A, mid-2nd millennium BC) (Fig. 8c); also *in situ* (see Pigott *et al.* 1997: 146, fig. 5). These latter two implements could have been cast in ceramic bivalve moulds basically similar to those found in the two metalworkers' graves mentioned above from Non Pa Wai Period 2A dated

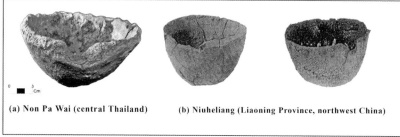

**(a) Non Pa Wai (central Thailand)**     **(b) Niuheliang (Liaoning Province, northwest China)**

**Figure 11** (a) Ceramic smelting crucible from early Period 2 Non Pa Wai, later 2nd millennium BC, Khao Wong Prachan valley, central Thailand. (b) Ceramic vessels which resemble that seen in Figure 11a and which may represent crucibles from the site of Yuanlongpo, Lingnan, southeast China, *c.* 1100–800 BC (Guangxi sheng Wenwu Kaogu Yanjiusuo 1988: figs 16–17).

**(a) Nil Kham Haeng (central Thailand)**     **(b) Yuanlongpo (Lingnan, southeast China)**

**Figure 12** (a) Ceramic furnace chimney reconstructed from fragments from a burial at the site of Nil Kham Haeng, early 1st millennium BC, Khao Wong Prachan valley, central Thailand (see burial photo in White and Pigott 1996: 166, fig. 13.9). (b) What appears to be a comparable chimney from a smelting installation excavated from a late Lower Xiajiadian context (*c.* 2300–1600 BC) at the site of Niuheliang in Liaoning province, northwest China (Li *et al.* 1999: fig. 10).

at its earliest *c*. 1500 BC (Fig. 9a). A second firm comparison with moulds in the burial from Non Pa Wai mentioned above can be found at sites in Lingnan, as well as for examples at Tangxiahuan fourth layer (Zhuhai city, Guangdong province), where one half of a stone bivalve mould and a similar fragment were excavated and dated to the end of the 2nd millennium BC (Fig. 9b) (Guangdong sheng Wenwu Kaogu Yanjiusuo 1998). Fragments or entire specimens are plentiful in Lingnan (particularly Hong Kong), northern Vietnam and Thailand), and are largely dated to the same chronological range (*c*. 1500–800 BC).

Two final examples merit mention here, again related to bivalve moulds. First, burial M203 (12th–11th century BC) contained a pair of ceramic bivalve moulds for casting an unsocketed *ge*-halberd from the Miabu Locus North site (a foundry site) in Anyang-Yinxu (Zhongguo Shihui Kexueyuan Kaogu Yanjiusuo 1987). Thus even in the Central Plain there is an inkling of the persistence of the tradition of interring metalworking accoutrements, a tradition that could have its roots in the Eurasian Steppes. As a second example, Ciarla (2000) and Higham (1996, 2004, 2006) have recognised the Lingnan region in southeastern China as the most suitable candidate for the initial local development of the cultural complex that was to spread south with relative rapidity by the mid-2nd millennium BC, and with it copper/bronze metallurgy. Additional support for this argument, as discussed above, comes from Rispoli's (2004) contribution highlighting the earlier role of Lingnan in the transmission of the Neolithic 'package' southward. Over time, Lingnan has clearly had an influential role to play, serving as a continuous link with southernmost copper/bronze-using communities from early on in this metal's appearance.

While there are various key sites in this scenario, one site was focused on by both Ciarla and Higham, namely Yuanlongpo, a cemetery site in Lingnan, near Nanning in Guangxi (Guangxi sheng Wenwu Kaogu Yanjiusuo 1988) which provided several [14]C dates ranging from the late 2nd to the mid-1st millennium BC (Zhongguo Shihui Kexueyuan Kaogu Yanjiusuo 1991: 220–21). The site yielded some 110 copper/bronze artefacts from 350 graves, among them two ritual bronze vessels of the late Western Zhou dynasty (1050–771 BC): a *you* (M147:1) and a *pan* (M33:3). These are certainly imports from the north, while the other 108 artefacts comprised weapons and implements. The socketed artefacts included 21 lance heads (4 types), 15 *dui* (a kind of rectangular lance head), 11 *yue* battle-axes, and 23 *fu* (axes); and the unsocketed artefacts included three spoons, ten arrow heads, 12 knives, five items of unknown use, one chisel, two needles and seven bells. All of these items (except for the bells) appear to have been cast locally in bivalve moulds as indicated by the presence of six complete pairs of sandstone bivalve moulds and numerous fragments. What is particularly important about these moulds is that at least one of them (M174:3) bears geometric motifs incised on the exterior surface. Significantly, on the exterior surface is an incised motif resembling two 'Cs' opposed at the open side (Fig. 10a) (Guangxi sheng Wenwu Kaogu Yanjiusuo 1988: 13, figs 29, 31). As mentioned earlier, a marked bivalve mould was excavated at 1st-millennium BC Phu Lon and a great number from probable mid-1st millennium BC contexts at sites in

the Khao Wong Prachan valley (Fig. 10b). Are we looking at a shared tradition of mould marking or simply coincidental behaviour? Ciarla (2000) also suggests that at least two bowl types from Yuanlongpo burials (M4:3 and M119:4) – defined respectively by the excavators as 'type I bowl' (*I shi bo*) and 'type II bowl' (*II shi bo*) (Guangxi sheng Wenwu Kaogu Yanjiusuo 1988: 6, figs 16–17), in terms of shape, paste, proportions, volume and rim breakage – appear comparable to similar containers excavated at Non Pa Wai and which are undoubtedly copper-smelting crucibles (Fig. 11a, b). One final item of smelting equipment that may provide an additional link between north and south is the so-called 'furnace chimney'. A number of these were excavated in central Thailand at Non Pa Wai (mid-2nd/early 1st millennium BC) as fragments and at Nil Kham Haeng (later 1st millennium BC) among the grave goods in several burials (Fig. 12a). A Chinese publication (Li *et al*. 1999) shows what appears to be a comparable chimney from a smelting installation excavated from a late Lower Xiajiadian context (*c*. 2300–1600 BC) at the site of Niuheliang in Liaoning province (northwest China) (Fig. 12b), in the context of a local metallurgical tradition with strong links with the Steppe world. The two chimneys do bear a reasonable resemblance to one another, although they are not from contemporaneous contexts. This single comparative example suggests connections between the north and south; however, there are no indications that such chimneys are part of the Steppe tradition of metallurgy. Further research on excavated collections in China and from the Eurasian Steppes may well provide additional links among the metallurgical artefacts.

## Conclusions

In summary, therefore, we would argue that the above evidence offers further suggestions of north–south connections and interaction including the south trending flow of technological innovation and information which may well have resulted in the introduction of copper/bronze metallurgy into Thailand/Southeast Asia. The 'southern metallurgical tradition' with its socketed implements, bivalve moulds, bangles and founders' burials has more in common with the Eurasian Steppes than Erlitou-Erligang China. The piece-mould technology of the Shang-Zhou dynastic elites did not lend itself well to small-scale, community-based production among less socially complex cultural groups who, over time, were apparently mobile enough to expand out of southeastern China into new territories to the south.

However, if there is a demonstrable link to be made here between the movement of peoples and technological traditions north and south, it is only going to be confirmed through further archaeological research in China and Southeast Asia facilitated by markedly improved information exchange. The preceding discussion pursues a hypothesis concerning the interactions of peoples and technology which is in need of continued testing and revision. It is hoped that this ongoing process will serve to raise to new levels the dialogue between the scholarly communities who focus on these two deeply intertwined culture areas.

## Acknowledgements

We are pleased to have been given the opportunity to participate in the conference and this volume in honour of Paul Craddock. We look forward to his continued and substantive contributions to the field of archaeometallurgical research. We wish to thank, in particular, Surapol Natapintu (Silpakorn University, Bangkok), who serves as Thailand Archaeometallurgy Project (TAP) and Thai-Italian Lopburi Regional Archaeology Project (LoRAP) co-director, for his continued support of our ongoing research.

We wish to acknowledge the following organisations for their support of the TAP research: the National Science Foundation, the National Geographic Society, the American Philosophical Society and the University of Pennsylvania Museum.

We are grateful to the following individuals for their assistance with the illustrations and computer graphics: at the Institute of Archaeology, Helena Coffey and Stuart Laidlaw and in Seattle, WA, Andrew D. Weiss who produced Figure 1. Elisabeth A. Bacus, Fiorella Rispoli and Joyce C. White offered insightful comments on earlier drafts of this paper. We are indebted to them for their assistance.

We also wish to thank HarperCollins Publishers Ltd. for permission to use the map in Figure 5.

## Notes

1. For those who wish to wade into the chronological, stratigraphic and typological complexities of this debate, we refer them to the following: Early chronology: for example White 1986; 1997: 104; in press; also Onsuwan 2000; Eyre 2006; Spriggs 1996–97. Late chronology: for example Higham 1996: 5–16; 2002a:113, 129; 2004: 51–2; 2006.
2. The following changes or additions were made to the original map utilised in Figure 5: Map of Shang China © 1999 Chris Scarre. It was converted from colour to b/w; Shang Core area and Xiangjiang and Ganjiang rivers were added, white boxes were placed around Anyang, Erlitou, Luoyang and Zhengzhou, and a black arrow pointing to the Gansu Corridor was added.
3. H. Cawte, pers. comm.

## References

Agarwal, D.P. 2000. *Ancient Metal Technology and Archaeology of South Asia: A Pan-Asian Perspective*. New Delhi: Aryan Books International.

Allard, F. 1994. Interaction and social complexity in Lingnan during the first millennium BC. *Asian Perspectives* 33: 309–26.

Allard, F. 1995. *The Interaction and Social Complexity in Lingnan during the Late Neolithic and the Bronze Age*. PhD dissertation, University of Pittsburgh.

An, Z.m. 1992. The Bronze Age in eastern parts of central Asia. In *History of Civilizations of Central Asia. The Dawn of Civilization: Earliest Times to 700 BC*, vol. 1, A.H. Dani and V.M. Masson (eds). Paris: UNESCO Publishing, 319–36.

An, Z.m. 1993. Shilun Zhongguo de zaoqi tongqi [On the early bronze in China]. *Kaogu* 12: 1110–19 (trans. in Linduff *et al.* 2000: 29–46).

Anthony, D. 1998. The opening of the Eurasian Steppe at 2000 BC. In Mair 1998, vol. 1, 94–113.

Bayard, D. 1984. Rank and wealth at Non Nok Tha: the mortuary evidence. In *Southeast Asian Archaeology at the Fifteenth Pacific Science Congress*, D. Bayard (ed.). Dunedin: Department of Anthropology, University of Otago, 87–128.

Bellina, B. and Glover, I. 2004. The archaeology of early contact with India and the Mediterranean world, from the fourth century BC to the fourth century AD. In Glover and Bellwood 2004, 68–88.

Bellwood, P. and Renfrew, C. 2002. *Examining the Farming/Language Dispersal Hypothesis*. Cambridge: McDonald Institute for Archaeological Research.

Bennett, A.1988a. *Copper Metallurgy in Central Thailand*. PhD thesis, Institute of Archaeology, University College London.

Bennett, A.1988b Prehistoric copper smelting in Thailand. In Charoenwongsa and Bronson 1988, 125–35.

Bennett, A.1989 The contribution of metallurgical studies in Southeast Asia. *World Archaeology* 20(3): 324–51.

Chang, K.c. 1980. *Shang Civilization*. New Haven/London: Yale University Press.

Chang, K.c. 1986. *The Archaeology of Ancient China*, 4th edn. New Haven: Yale University Press.

Charoenwongsa, P. and Bronson, B. (eds) 1988. *The Stone and Metal Ages in Thailand*. Bangkok: Thai Antiquity Working Group.

Chernykh, E.1992. *Ancient Metallurgy in the USSR*. Cambridge: Cambridge University Press.

Chiou-Peng, Th. 1998. Western Yunnan and its steppe affinities. In Mair 1988, vol. 1, 280–304.

Ciarla, R. 1992. The Thai-Italian Lopburi Regional Archaeological Project: preliminary results. In *South East Asian Archaeology 1990*, I. Glover (ed.). Centre for South-East Asian Studies. Hull: University of Hull, 111–28.

Ciarla, R. 1994. Pastoralism and nomadism in south-western China: a long debated question. In Genito 1994, 73–86.

Ciarla, R. 2000. Rethinking Yuanlongpo: the case for technological links between the Lingnan (RPC) and central Thailand in the Bronze Age. Paper presented at the Eighth International Conference of the European Association of South-east Asian Archaeologists, Sarteano, Italy, 2–6 October 2000 (submitted to *East and West* 58, 2008).

Ciarla, R. 2005. The Thai-Italian 'Lopburi regional archaeological project': a survey of fifteen years of activities. In *La Cultura Thailandese e le Relazioni Italo-Thai, Atti del Convegno Italo-Thai: La cultura Thailandese e le relazioni Italo-Thai (Torino, Italia, 20–21 May 2004)*, I. Piovano (ed.). Torino: CeSMEO, 77–104.

Ciarla, R. and Rispoli, F. (eds). 1997. *South-East Asian Archaeology 1992. Proceedings of the Fourth International Conference of the European Association of South-East Asian Archaeologists, Rome, 28 September–4 October, 1992*. Rome: Istituto Italiano per l'Africa e l'Oriente.

Debaine-Francfort, C.1995. *Du Néolithique à l'Âge du Bronze en Chine du Nord-Ouest. La culture de Qijia et ses connexions*. Paris: Editions Recherche sur les Civilisations/Association pour la diffusion de la pensée Française.

De Jesus, P S. 1978. Considerations on the occurrence and exploitation of tin sources in the ancient Near East. In *The Search for Ancient Tin*, A.D. Franklin, J.S. Olin and T.A. Wertime (eds). Washington, DC: US Government Printing Office, 33–8.

Di Cosmo, N. 1999. The northern frontier in pre-Imperial China. In *The Cambridge History of Ancient China: From the Origins of Civilization to 221 BC*, M. Loewe and E.L. Shaughnessy (eds). Cambridge: Cambridge University Press, 885–966.

Eyre, C. Onsuwan 2006. *Prehistoric and Proto-Historic Communities in the Eastern Upper Chao Phraya River Valley, Thailand: Analysis of Site Chronology, Settlement Patterns, and Land Use*. PhD dissertation, Department of Anthropology, University of Pennsylvania.

Fitzgerald-Huber, L.G. 1995. Qijia and Erlitou: the question of contacts with distant cultures. *Early China* 20: 17–67.

Fitzgerald-Huber, L.G. 2003. The Qijia culture: paths East and West. *Bulletin of the Museum of Far Eastern Antiquities* 75: 55–78.

Franklin, U. 1983. The beginnings of metallurgy in China: a comparative approach. In *The Great Bronze Age of China: A Symposium*, G. Kuwayama (ed.). Los Angeles: Los Angeles County Museum of Art, 94–9.

Genito, B. (ed.) 1994. *Archeologia delle Steppe: Metodi e Strategie di Lavoro*. Naples: Istituto Universitario Orientale.

Glover, I. and Bellwood, P. (eds) 2004. *Southeast Asia: From Prehistory to History*. London: Routledge Curzon.

Guangdong sheng Wenwu Kaogu Yanjiusuo [Institute of Cultural Relics and Archaeology of Guangdong Province] 1998. Zhuhai Pingsha Tangxiahuan yizhi fajue jianbao [Short report on the excavation of Tangxiahuan site at Pingsha near Zhuhai]. *Wenwu* 7: 4–16.

Guangxi sheng Wenwu Kaogu Yanjiusuo [Institute of Cultural Relics and Archaeology of Guangxi Province] 1988. Guangxi Wumin Matou Yuanlongpo muzang fajue jianbao [Short report on the excavation of the Yuanlongpo graveyard in Matou county, Wumin district, Guangxi province]. *Wenwu* 12: 1–13.

Higham, C. 1996. *The Bronze Age of Southeast Asia*. Cambridge: Cambridge University Press.

Higham, C. 2002a. *Early Cultures of Mainland Southeast Asia*. Bangkok: River Books.

Higham, C. 2002b. Languages and farming dispersal: Austrasiatic languages and rice cultivation. In Bellwood and Renfrew 2002, 223–32.

Higham, C. 2004. Mainland Southeast Asia from the Neolithic to the Iron Age. In Glover and Bellwood 2004, 41–67.

Higham, C. 2006. Crossing national boundaries: southern China and Southeast Asia in prehistory. In *Uncovering Southeast Asia's Past*, E.A. Bacus, I.C. Glover and V.C. Pigott (eds). Singapore: Singapore University Press.

Hunan sheng Wenwu Kaogu Yanjiusuo [Institute of Cultural Relics and Archaeology of Hunan Province] 1992. Hunan Shimen Zaoshi Shangdai yizhi [The Shang period site at Zaoshi, Shimen County, Hunan]. *Kaogu Xuebao* 2: 185–219.

Jettmar, K. 1985. Cultures and ethnic groups west of China in the second and first millennium BC. *Asian Perspectives* 24(2): 145–62.

Ke P. 1998. The Andronovo bronze artefacts discovered in Toquztara County in Ili, Xinjiang. In Mair 1998, vol. 2, 573–80.

Kenoyer, M. and Miller, H. 1999. Metal technologies of the Indus valley tradition in Pakistan and western India. In *The Archaeometallurgy of the Asian Old World*, V.C. Pigott, (ed.). MASCA Research Papers in Science and Archaeology, vol. 16. Philadelphia: University Museum, University of Pennsylvania.

Li, S.c. 2003. Ancient interactions in Eurasia and northwest China: revisiting Johan Gunnar Andersson's legacy. *Bulletin of the Museum of Far Eastern Antiquities* 75: 31–54.

Li Y.x., Han R.b., Bao W.b. and Chen T.m. 1999. Niuheliang yetong hubi canbian yanjiu [Researches on a copper-smelting furnace fragments from Niuheliang]. *Wenwu* 12: 44–51.

Linduff, K. 1998. The emergence and demise of bronze-producing cultures outside the Central Plain of China. In Mair 1998, vol. 2, 619–43.

Linduff, K. 2004. *Metallurgy in Ancient Eastern Eurasia from the Urals to the Yellow River*. Lampeter: Edwin Mellen Press.

Linduff, K., Han, R.b. and Sun, S.y. (eds) 2000. *The Beginnings of Metallurgy in China*. Lewiston: Edwin Mellen Press.

Maddin, R. (ed.) 1988. *The Beginning of the Use of Metals and Alloys. Papers from the Second International Conference on the Beginning of the Use of Metals and Alloys, Zhengzhou, China, 21–26 October 1986*. Cambridge, MA: MIT Press.

Mair, V.H. (ed.) 1998. *The Bronze Age and Early Iron Age Peoples of Eastern Central Asia*, vols 1 and 2. Washington, DC: The Institute for the Study of Man and the University of Pennsylvania Museum Publications, Philadelphia.

Mallory, J.P. and Mair, V.H. 2000. *The Tarim Mummies*. London: Thames and Hudson.

Mei, J. 2000. *Copper and Bronze Metallurgy in Late Prehistoric Xinjiang: Its Cultural Context and Relationship with Neighboring Regions*. British Archaeological Reports International Series 865. Oxford: Archaeopress.

Mei, J. 2003a. Qijia and Seima-Turbino: the question of early contacts between northwest China and the Eurasian Steppe. *Bulletin of the Museum of Far Eastern Antiquities* 75: 31–54.

Mei, J. 2003b. Early copper technology in Xinjiang, China: the evidence so far. In *Mining and Metal Production through the Ages*, P. Craddock and J. Lang (eds). London: British Museum Press, 111–21.

Natapintu, S. 1987. Current research on prehistoric copper-based metallurgy in Thailand. *SPAFA Digest* 8(1): 27–35.

Natapintu, S. 1988. Current research on ancient copper-base metallurgy in Thailand. In Charoenwongsa and Bronson 1988, 107–24.

Natapintu, S. 1991. Archaeometallurgical studies in the Khao Wong Prachan valley, central Thailand. *Bulletin of the Indo-Pacific Prehistory Association* 11: 153–9.

Nguyen Khac Su, Pham Minh Huyen and Tong Trung Tin 2004. Northern Vietnam from Neolithic to the Han period. In Glover and Bellwood 2004, 177–208.

Onsuwan, C. 2000. *Excavation of Ban Mai Chaimongkol, Nakhon Sawan Province, Central Thailand: A Study of Site Stratigraphy, Chronology and its Implications for the Prehistory of Central Thailand*. Unpublished MA thesis, Department of Anthropology, University of Pennsylvania.

Orioli, M. 1994. Pastoralism and nomadism in South-West China: a brief survey of the archaeological evidence. In Genito 1994, 87–108.

Parzinger, H. 2000. The Seima-Turbino phenomenon and the origin of the Siberian animal style. *Archaeology, Ethnology and Anthropology of Eurasia* 1: 66–75.

Pigott, V.C. 1998. Prehistoric copper mining in the context of emerging community craft specialisation in northeast Thailand. In *Social Approaches to an Industrial Past: The Archaeology and Anthropology of Mining*, A.B. Knapp, V.C. Pigott and E.W. Herbert (eds). London: Routledge, 205–25.

Pigott, V.C. 1999. Reconstructing the copper production process as practiced among prehistoric mining/metallurgical communities in the Khao Wong Prachan valley of central Thailand. In *Metals in Antiquity*, S.M.M. Young, M. Pollard, P. Budd and R. Ixer (eds). BAR International 792. Oxford: Archaeopress, 10–21.

Pigott, V.C. and Natapintu, S. 1988. Archaeological investigations into prehistoric copper production: the Thailand Archaeometallurgy Project 1984–1986. In Maddin 1988, 156–62.

Pigott, V.C. and Weisgerber, G. 1998. Mining archaeology in geological context: the prehistoric copper mining complex at Phu Lon, Nong Khai province, northeast Thailand. In *Metallurgica Antiqua: In Honour of Hans-Gert Bachmann and Robert Maddin*, Th. Rehren, A. Hauptmann and J.D. Muhly (eds). *Der Anschnitt* Beiheft 8. Bochum: Deutsches Bergbau-Museum, 135–62.

Pigott, V.C., Weiss, A.D. and Natapintu, S. 1997. The archaeology of copper production: excavations in the Khao Wong Prachan valley, central Thailand. In Ciarla and Rispoli 1997, 119–58.

Rispoli, F. 1997a. Late third/early second millennium BC pottery traditions in central Thailand: some preliminary observations in a wider perspective. In Ciarla and Rispoli 1997, 59–97.

Rispoli, F. 1997b. *Ad Occidente è l'India. Alla ricerca delle radici del processo di indianizzazione nella Thailandia Centrale*. Doctoral dissertation, Istituto Universitario Orientale, Napoli (National Library of Florence – Doctoral Dissertation No. TDR.1997.003737).

Rispoli, F. 2004. Looking for the root of Neolithization in Southeast Asia. Paper presented at the Tenth International Conference of the European Association of Southeast Asian Archaeologists, British Museum, London, 13–17 September 2004 (submitted to *East and West* 57(1–4), 2007).

Rispoli, F. 2005. To the West and India. In *Studi in onore di Maurizio Taddei*, P. Callieri, and A. Filigenzi (eds), *East and West* 55(1–4): 243–61.

Rispoli, F. in press b. Terracotta ear-studs and skin-rubbers: looking for the roots of Indianization in central Thailand. In *La Cultura Thailandese e le Relazioni Italo-Thai, Atti del Convegno Italo-Thai: La cultura Thailandese e le relazioni Italo-Thai (Torino, Italia, 20–21 May 2004)*, I. Piovano (ed.). Torino: CeSMEO.

Rostoker, W., Dvorak, J. and Pigott, V.C. 1989. Direct reduction to copper metal by oxide/sulfide interaction. *Archeomaterials* 3: 69–87.

Scarre, C. (ed.) 1999. *The Times Archaeology of the World*. London: Times Books/Harper Collins Publishing.

Shao, W.p. 2002. The formation of civilization: the interaction sphere of the Longshan period. In *The Formation of Chinese Civilization: An Archaeological Perspective*, K.c. Chang and P.f. Xu (S. Allan (ed.)). New Haven: Yale University Press, 85–123.

Spriggs, M. 1996–97. The dating of Non Nok Tha and the 'Gakushiun factor'. In *Ancient Chinese and Southeast Asian Bronze Age Cultures*, vol. 2, D. Bulbeck (ed.).Taipei: SMC Publishing Inc., 941–8.

Takaya, Y. 1987. *Agricultural Development of a Tropical Delta*. Honolulu: University of Hawaii Press.

Thorp, R. 2006. *China in the Early Bronze Age: Shang Civilization*. Philadelphia: University of Pennsylvania Press.

Tylecote, R.F. 1976 *A History of Metallurgy*. London: Metals Society.

Van Driem, G. 2002. Tibeto-Burman phylogeny and prehistory: languages, material culture and genes. In Bellwood and Renfrew 2002, 233–49.

Van Driem, G. 2005. Tibeto-Burman vs. Indo-Chinese: implications for population geneticists, archaeologists and prehistorians. In *The Peopling of East Asia: Putting Together Archaeology, Linguistics and Genetics*, L. Sagart, R. Blench and A. Sanchez-Mazas (eds). London and New York: Routledge-Curzon, 81–106.

Wang, D.N., Pigott, V.C. and Notis, M.R. 1998. The archaeometallurgical analysis of copper-based artefacts from prehistoric Nil Kham Haeng, central Thailand: a preliminary report. Bethlehem: Lehigh University (www.lehigh.edu/~inarcmet/papers/thailand.pdf).

White, J.C. 1986. *A Revision of the Chronology of Ban Chiang and its Implications for the Prehistory of Northeast Thailand*. PhD dissertation, Dept. of Anthropology, University of Pennsylvania.

White, J.C. 1988. Early East Asian metallurgy: the southern tradition. In Maddin 1988, 175–81.

White, J.C. 1997. A brief note on new dates for the Ban Chiang cultural tradition. *Bulletin of the Indo-Pacific Prehistory Association* 16: 103–6.

White, J.C. in press. Dating early bronze at Ben Chiang, Thailand. In *Proceedings of the 11th International Conference of the European Association of Southeast Asian Archaeologists, Bourgon, France*, J.-P. Patreu *et al.* (eds).

White, J.C. and Pigott, V.C. 1996. From community craft to regional specialization: intensification of copper production in pre-state Thailand. In *Craft Specialization and Social Evolution: In Memory of V. Gordon Childe*, B. Wailes (ed.). Philadelphia: University of Pennsylvania Museum, 151–75.

Yan W.m. 1984. Lun Zhongguo de tongshi bingyong shidai [On the Chalcolithic Age of China]. *Shiqian Yanjiu* 1: 36–44.

Zhang J.k. and Wei J. 2004. *Xinshiqi Shidai Kaogu* [The archaeology of the Neolithic period], 211–26. Beijing: Wenwu Chubanshe [Cultural Relics Publishing House].

Zhao Z.q., 2003. Xiadai qianqi wenhua zonglun [A comprehensive discussion on the early Xia culture], *Kaogu Xuebao* 4: 459–82.

Zhongguo Shihui Kexueyuan Kaogu Yanjiusuo [Institute of Archaeology of the Chinese Academy of Social Sciences] 1987. *Yinxu fajue baogao 1958–1961* [Report on the 1958–1961 excavations at Yinxu), 34, 231, table VIII.2. Beijing: Wenwu Chubanshe [Cultural Relics Publishing House].

Zhongguo Shihui Kexueyuan Kaogu Yanjiusuo [Institute of Archaeology of the Chinese Academy of Social Sciences] 1991. *Zhongguo Kaoguxue zhong tan shisi niandai shujuji 1965–1991* [Radiocarbon dates in Chinese archaeology 1965–1991]. Beijing: Wenwu Chubanshe [Cultural Relics Publishing House].

Zhongguo Shihui Kexueyuan Kaogu Yanjiusuo [Institute of Archaeology of the Chinese Academy of Social Sciences] 2006. Henan Anyang Yinxu Huayuanzhuang dongdi 60 hao mu [Tomb n. 60 at Huayuanzhuang Locus East within the Yin Ruins in Anyang, Henan]. *Kaogu* 1: 7–18.

## Authors' addresses

- Vincent C. Pigott, Institute of Archaeology, University College London, London WC1H 0PY, UK (vcpigott@aol.com). (Also Research Associate in the Museum Applied Science Center for Archaeology (MASCA) at the University of Pennsylvania Museum, Philadelphia, Pennsylvania, USA.)

- Roberto Ciarla, Istituto Italiano per l'Africa e l'Oriente (IsIAO), Via U. Aldrovandi, 1600195 Rome, Italy (rispoli.ciarla@flashnet.it). (Also Conservator in Chief of the Far Eastern archaeological and art collections at the 'Giuseppe Tucci' National Museum of Oriental Art in Rome, Italy.)

# Coals to Newcastle, copper to Magan? Isotopic analyses and the Persian Gulf metals trade

*Lloyd Weeks*

*ABSTRACT*   This paper reviews the most recent evidence for metal production and exchange systems in the Bronze Age Persian Gulf. In particular, the evidence for the use of non-Omani metal (tin and tin-bronze) is addressed, and the implications for our understanding of the development of local alloying and exchange are discussed. Subsequently, the more unexpected possibility that copper ingots were being imported into copper-rich Oman in the Bronze Age is addressed. Upon consideration of the evidence, it seems possible that the putative foreign copper ingots are in fact local products, although there is a deal of ambiguity in relation to these claims.

*Keywords:* copper, tin, exchange, Persian Gulf, Mesopotamia, lead isotope analysis.

## Introduction

This paper presents a brief overview of the history of archaeometallurgical research in the Persian Gulf (Fig. 1), related particularly to the evidence for the location of the copper-producing land of Magan known from Bronze Age Mesopotamian sources. More specifically, the results of recent archaeometallurgical research in the region are discussed. As suggested by the title, these analyses have challenged a number of longstanding beliefs about the nature of metal exchange systems in the Persian Gulf, and have outlined (albeit in a most preliminary manner) a metal production and exchange system of hitherto unexpected complexity.

## Fieldwork in Oman and the location of Magan

The history of archaeometallurgical research in the Persian Gulf region has been discussed at length in a number of publications and will be addressed only briefly here (see Potts 1990; Weeks 2004, for a full discussion and bibliography). Interest in the topic can be traced to the study of Bronze Age Mesopotamian cuneiform documents, whose predominant concentration on economic matters has yielded an abundance of information on raw materials production, exchange and administration in ancient western Asia. Beginning in the late 4th or early 3rd millennium BC and continuing through to the early 2nd millennium BC, the acquisition of copper by various institutions and individuals in southern Mesopotamia, a land devoid of metallic resources, was achieved through exchange with a number of areas denoted by specific toponyms, in particular the lands of Dilmun and Magan (Potts 1990). Assyriological and archaeological research has played a critical role in determining the location of these lands, and

it is now established beyond reasonable doubt that Dilmun is to be located on the southern shores of the Persian Gulf, specifically on the island of Bahrain by the later 3rd millennium BC (Crawford 1998: 1–8; Heimpel 1987). Dilmun copper is mentioned as early as the Proto-Literate texts from Uruk (*c.* 3100 BC) and copper from Dilmun is sporadically mentioned over the course of the 3rd millennium BC in individual quantities of a hundred kilograms or so (Potts 1990: 182). It is in the early 2nd millennium BC, however, in the Isin-Larsa and Old Babylonian periods that the copper trade between Dilmun and southern Mesopotamia reaches its peak. At this time, texts from the house of a Mesopotamian copper merchant named Ea Nasir attest to multiple shipments of Dilmun copper totalling approximately 18 tonnes (Leemans 1960: 50).

Significantly, geological and archaeological research is clear in indicating that no copper deposits exist in Dilmun.

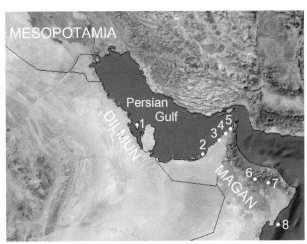

**Figure 1** Ancient toponyms and sites mentioned in the text. Sites: (1) Saar (2) Umm an-Nar Island (3) Al Sufouh (4) Tell Abraq (5) Unar1 and Unar 2 (6) Wadi Bahla/Al Aqir (7) Maysar 1 (8) Masirah Island.

Thus, Dilmun could only have been the proximate source of the copper used in Bronze Age Mesopotamia. As a result, the search for the ultimate sources of the Dilmun copper has focused to a great extent on locating the land of Magan. Magan is mentioned in cuneiform sources over a much shorter period than Dilmun, first appearing in the Sargonic period (*c.* 2350 BC) and reaching a peak of importance in the Ur III period in the final century of the 3rd millennium BC (Heimpel 1988). As attested in the cuneiform texts at this time, set portions of the copper obtained on trips to Magan were deposited as tithes in Mesopotamian temples. As the tithed amounts themselves are often as large as a few hundred kilograms (Leemans 1960: 23–36), the total amount of copper obtained in these individual ventures may have been on the order of thousands of kilograms.

In contrast to Dilmun, Magan itself seems to have been a copper-*producing* land, rather than one that engaged merely in the exchange of metal. The generally accepted explanation for the patterns observable in the cuneiform evidence outlined above has regarded Magan as the ultimate source of almost all the copper that was traded to Mesopotamia through the Persian Gulf throughout the Bronze Age (for example Cleuziou and Méry 2002: 282). The prominence of Dilmun as a copper supplier in the early/mid-3rd millennium BC and the early 2nd millennium BC can be seen as simply a reflection of the monopoly on direct trade with Mesopotamia that Dilmun merchants were able to exercise at specific points in history.

Although Magan seems a likely ultimate source for much of the copper reaching southern Mesopotamia, the cuneiform evidence for the geographical location of Magan is more difficult to interpret than that for Dilmun. It is very likely that Magan was located on the southern shores of the Persian Gulf, in southeastern Arabia and the areas occupied by the modern countries of the United Arab Emirates (UAE) and the Sultanate of Oman (Potts 1990: 133–49). It is possible, however, that areas across the Straits of Hormuz, on the northern side of the Persian Gulf in modern-day Iran, may also have been part of the (predictably hazy) Mesopotamian conception of Magan (Abdi 2000). The delineation of the location of Magan has therefore relied, to a greater extent than for Dilmun, on the buttressing of the fragmentary cuneiform record with archaeological and archaeometallurgical evidence from the region.

The earliest scientific research on the question of the location of Magan was the study by H. Peake (1928), undertaken for the Sumerian Copper Committee. Peake analysed a number of copper-based objects from southern Mesopotamia and the Persian Gulf, in addition to a copper ore sample from Oman, and famously drew a connection between Mesopotamian objects and Omani copper ores based upon higher than usual concentrations of nickel. Very little subsequent archaeological research into the location of Magan was undertaken until the 1970s, when the publication of Geoffrey Bibby's book *Looking for Dilmun* (1970), which discussed the results of the early Danish archaeological research in the Persian Gulf and the issue of copper from Dilmun and Magan, inspired renewed geological surveys for copper deposits in Oman. At that time, geologists recorded the remains of a number of old mine workings and at least 44 pre-modern smelting sites in Oman, some of which were very large (Goettler *et al.* 1976:

43–4). Subsequent archaeological surveys in the mid-1970s by American and Italian teams (Hastings *et al.* 1975; Tosi 1975) were able to provide a firmer chronological context for some of the extractive activity, and to suggest that copper smelting extended back as far as the 3rd millennium BC. Nevertheless, little in the way of analytical evidence was published in these studies conclusively supporting an association between Magan and the Oman Peninsula.

Intensive investigation of early copper extraction in southeastern Arabia can really be said to have begun with the arrival in Oman in 1977 of a team from the German Mining Museum, Bochum. Over a series of field seasons in the late 1970s and early 1980s, the German team located and recorded more than 150 mining and smelting sites (Weisgerber 1980a: 68; 1980b: 115), and their detailed fieldwork was able to provide the first comprehensive body of data supporting discussions of the chronology, scale and technological development of copper production in southeastern Arabia (e.g. Hauptmann 1985: 108–9; Weisgerber 1981: fig. 4). In brief, the findings of the German team in relation to the question of the location of Magan were as follows: they recorded 20 sites with evidence for 3rd-millennium copper smelting, with slag deposits ranging in size from *c.* 100 to 4000 tonnes. Based upon reconstructed extraction and smelting processes and the total volume of Bronze Age slag recorded, in addition to the likely destruction of much of the Bronze Age evidence by later (especially Islamic) smelting operations, it was estimated that 2000–4000 tonnes of copper had been produced in southeastern Arabia in the 3rd millennium BC (Hauptmann 1985: 108). In short, the archaeological discoveries made by the German Mining Museum and their predecessors provided strong circumstantial evidence supporting the hypothesis that the copper of Magan came from southeastern Arabia.

At the same time that archaeometallurgical investigations into the primary extraction of copper in Oman were taking place, analytical studies of early copper-based artefacts from southeastern Arabia were initiated (Berthoud 1979; Craddock 1981, 1985; Frifelt 1991; Hauptmann 1987; Hauptmann *et al.* 1988). While the overall number of Arabian artefacts analysed was small, the analyses were relatively consistent in suggesting conservative fabrication and alloying technologies in the region: the composition of copper ingots and copper-based artefacts was regarded as indicating the predominant use in southeastern Arabia of local copper (identifiable by its substantial impurities of arsenic and nickel), rarely if ever alloyed with tin before the late 2nd millennium BC. The statistical analyses of compositional data on ores and artefacts from Mesopotamia, Iran and the Persian Gulf undertaken by Berthoud (1979; Berthoud *et al.* 1980) were further regarded as supporting the use of southeastern Arabian copper in Mesopotamia and southwest Iran by the mid-3rd millennium BC (although see Seeliger *et al.* 1985 for questions over the reliability of Berthoud's interpretation).

In general, the conclusions of the various studies undertaken in this initial phase of research into Persian Gulf metallurgy were largely consistent, and suggested a straightforward reconstruction of early metal production and exchange systems. First, artefact and ore analyses and the evidence for large-scale copper production in southeastern Arabia in the 3rd millennium BC suggested that the copper traded in the

Bronze Age Gulf (i.e. the copper of both Dilmun and Magan) came from Oman, and that Oman was thus the location of the ancient land of Magan. Secondly, analytical studies suggested that the copper-based artefacts found throughout the Bronze Age Gulf were made from the copper of Magan, and that this copper was rarely alloyed with foreign metals such as tin, perhaps due to its improved mechanical properties resulting from high levels of arsenic (As) and nickel (Ni) impurities (Hauptmann 1987: 217).

### Recent archaeometric studies of Bronze Age Persian Gulf material

This reconstruction of Bronze Age metal production, exchange and use has been challenged, in certain critical respects, by the results of archaeometallurgical research conducted in the Persian Gulf over the course of the last decade. In particular, the series of compositional and lead isotope analyses of ores, ingots and finished artefacts undertaken by M.K. Prange from the German Mining Museum (Prange 2001; Prange *et al.* 1999) and those by the present author (Weeks 2004; Weeks and Collerson 2004, 2005). In this section the various analytical programmes are introduced and their implications for our understanding of the Persian Gulf metals trade summarised.

### *Early alloy use in the Persian Gulf*

Beginning in 1996, the author conducted compositional analyses on 83 copper-based finished artefacts from four Umm al-Nar tomb assemblages on the northern coast of the UAE, which could be closely dated within the period from *c.* 2450 to 2000/1900 BC. In addition, a series of 16 finished artefacts, 3 copper ingots and 20 samples of metalworking debris was analysed from the Dilmun period settlement of Saar on Bahrain (dated *c.* 2050–1700 BC). These compositional studies predominantly utilised the technique of proton-induced X-ray emission analysis (PIXE), although four samples from Saar were analysed using energy-dispersive spectrometry on a scanning electron microscope (EDX–SEM) (Weeks 2004; Weeks and Collerson 2005). In addition, lead isotope analysis was undertaken on a subset of 55 of the compositionally analysed samples (Weeks and Collerson 2004, 2005). The results of these analyses have been published in detail elsewhere and will only be briefly outlined here.

In contrast to the previous analyses of finished objects from the region that were undertaken up to the early 1990s, the PIXE analyses of the Umm al-Nar period copper-based artefacts showed a much greater prevalence of alloying, especially with tin. Objects made of relatively pure copper, and of copper with significant impurities of arsenic and nickel were also recorded, but in the latest Umm al-Nar period tomb assemblages at the sites of Unar2 (at Shimal, Ra's al-Khaimah) and Tell Abraq (Sharjah), tin-bronze was the dominant metal recorded, accounting for 50–60% of all analysed artefacts (Weeks 2004: fig. 4.28). The proportion of tin-bronze used at Saar was somewhat lower than observed for the sites in

the UAE, although five tin-bronzes were found among the 16 analysed finished artefacts from the site (Weeks and Collerson 2005). In general, these analyses challenged the existing understanding of the Gulf metals trade by indicating that alloying was more widespread than previously thought, with the more important implication that, as tin deposits are not found anywhere in eastern Arabia, a significant amount of non-Omani metal must have been traded to and through the Persian Gulf. The availability of foreign metallic tin in the Persian Gulf was supported by the discovery in the tomb at Tell Abraq of a tin ring (Weeks 2004: fig. 3.10).

However, lead isotope analysis of the material from the UAE and Bahrain has added an extra degree of complexity to this picture, by indicating that not only metallic tin but also pre-alloyed tin-bronze must have been traded through the Persian Gulf. A number of the isotopically analysed artefacts from southeastern Arabia show anomalous or radiogenic lead isotope signatures indicative of an origin outside the Persian Gulf region. These are all tin-bronzes (Weeks 2004: fig. 7.1). Moreover, when the lead isotope distribution of the non-radiogenic samples is examined in comparison to isotope measurements on ores from Oman (Fig. 2), it is apparent that many, especially tin-bronzes but also some copper artefacts, are incompatible with an Omani origin.

As it is likely (although not certain) that the lead isotope signature of the tin-bronzes is controlled by lead contributions from the copper used in their production rather than the tin (Begemann *et al.* 1999; Pernicka *et al.* 1990), these objects are likely to have been imported in their entirety. That is, they

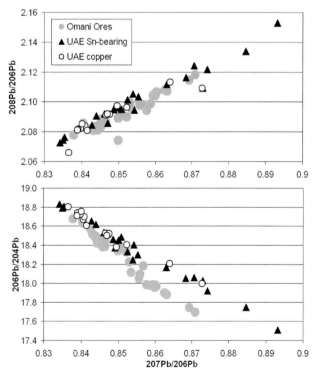

**Figure 2** Lead isotope data for non-radiogenic artefacts from four Umm al-Nar period tomb assemblages in the UAE (data from Weeks and Collerson 2004) in comparison to copper ores from Oman (data from Calvez and Lescuyer 1991; Chen and Pallister 1981; Prange 2001; Stos-Gale *et al.* 1997). 'UAE Sn-bearing' objects are copper-based alloys with more than 1% Sn; 'UAE copper' objects contain less than 1% Sn. Absolute 2σ analytical errors are smaller than the size of the individual data points.

were not produced in the Persian Gulf by adding foreign tin to local Omani copper. For a number of reasons, it is probable that such pre-alloyed tin-bronze (and copper) was arriving in the Persian Gulf in the form of finished artefacts, and there are indeed occasional cuneiform sources attesting to the exchange of finished metal alloy artefacts through the Gulf (Weeks 2004: ch. 8). This conclusion is supported by compositional analyses of the small-scale metallurgical production debris from Saar (Weeks and Collerson 2005). The complete absence of even minor levels of tin in these samples indicates that while tin-bronze artefacts were present at Saar, tin-bronze was not worked there to any great extent.

### Foreign copper ingots in Magan?

The extensive analytical programme conducted by Prange (2001) incorporated compositional analyses of 91 copper ores and 250 Bronze Age/Iron Age slags, ingots, and finished objects from Oman using inductively coupled plasma optical emission spectroscopy (ICP–OES), in addition to analyses of Bronze Age artefacts from the UAE (n=17) and Bahrain (n=43) using the same technique. The compositional data were statistically investigated using hierarchical cluster analysis for the purposes of comparison and grouping. Finally, lead isotope analysis was undertaken on a subset of 37 of these samples, including 15 ores, 3 slags, 2 copper inclusions in slag, 11 copper ingots and 6 finished artefacts, all from Oman (Prange 2001: table 30). Numerous important conclusions regarding early metal production and exchange have been drawn from this comprehensive study, but I choose here to focus upon only one major aspect of the research; specifically, the compositional and lead isotope analyses of plano-convex or 'bun-shaped' copper ingots from the sites of Maysar 1 and Al Aqir.

The ingots found at the settlement of Maysar 1 were of particular significance. Maysar 1 produced abundant evidence for the smelting of copper from local ores (Weisgerber 1980a,b, 1981), and the discovery of a cache of plano-convex copper ingots in one area of the site was particularly important as it suggested that such ingots were the standard end products of Umm al-Nar period copper-smelting operations in southeastern Arabia, destined for external exchange. A large hoard of typologically similar ingots was discovered at Al Aqir near Bahla (Weisgerber and Yule 2003). These ingots differ from the Maysar 1 examples, both in their composition (possessing slag-filled cores) and their context of deposition, such that it has been suggested that they may have been religious offerings deposited during the construction of a Bronze Age dam (Weisgerber and Yule 2003: 50–51). Regardless of these distinctions, however, the discovery of the Maysar 1 and Al Aqir ingots seemed to support the notion of indigenous production of plano-convex copper ingots in Bronze Age southeastern Arabia. The typology of the ingots from Maysar 1 and Al Aqir was readily linked with similar copper ingots reported from further to the north in the UAE (on Umm an-Nar Island, Frifelt 1995), and in Bahrain at Qala'at al-Bahrain (Højlund and Andersen 1994), Saar (Killick and Moon 2005), and Nasariyah (Lombard and Kervran 1989), providing a series of 'footprints' identifying the route used to transport ingots of Magan copper from southeastern Arabia

north to the borders of Mesopotamia that directly verified the Mesopotamian cuneiform accounts of the Persian Gulf copper trade.

It was understandably surprising then, that a major finding of Prange's recent research was that:

> With but one exception … neither the bun shaped copper ingots from the Bronze Age hoards at al Maysar, nor from Al Aqir match any of the copper ores and metal objects from Oman in their chemical composition. This is supported by lead isotope analysis: copper ingots from al Maysar form a distinct group and are different from ores and metal objects (Prange 2001: 102).

Compositionally, the Maysar 1 and Al Aqir ingots form a relatively homogeneous group that is distinguishable from contemporary southeastern Arabian ores and objects by low levels of selenium (generally <25 ppm Se) and relatively high concentrations of silver (c. 300–3000 ppm Ag). Additionally, it is claimed that the Maysar 1 and Al Aqir ingots do not compositionally match *any* of the analysed finished objects from the central or southern Persian Gulf, although their chemical composition is matched by some objects from Susa in southwest Iran (Prange 2001: 102). Isotopically, moreover, most of the Maysar 1 ingots exhibit unusual lead isotope ratios that are relatively depleted in their thorogenic lead component in comparison to most ores and finished artefacts from southeastern Arabia (Prange 2001: figs 81 and 82). As illustrated in Figure 3, a similar isotopic signature has been recorded for one

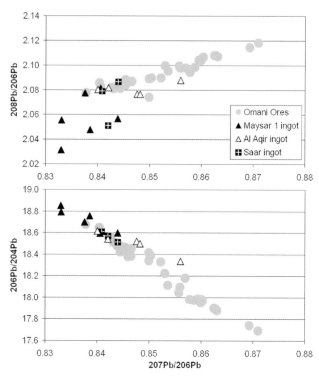

**Figure 3** Lead isotope data for Bronze Age plano-convex copper ingots from Maysar 1 and Al Aqir analysed by M.K. Prange (data from Prange 2001: table 30) in comparison to contemporary plano-convex copper ingots from Saar (data from Weeks and Collerson 2005) and Omani copper ores (data from Calvez and Lescuyer 1991; Chen and Pallister 1981; Prange 2001: table 30; Stos-Gale *et al.* 1997). Absolute 2σ analytical errors are smaller than the size of the individual data points.

of the plano-convex copper ingots from the Saar settlement analysed by Weeks and Collerson (2005).

Together, the findings of the compositional and lead isotope studies have been regarded as evidence that the ingots from Maysar 1 and Al Aqir were made of non-Omani metal (Prange 2001: 102). Needless to say, the conclusion that copper ingots were being imported into Oman, the supposed 'copper mountain of Magan', presented a radical challenge to existing reconstructions of Gulf metal sources and exchange routes. Nevertheless, these findings were not seen to negate the equation of ancient Magan with southeastern Arabia, which was still regarded as the source of most of the metal used in the region due to the compositional similarity between Bronze Age finished artefacts from the Persian Gulf and ores from Oman (Prange 2001: 103; Prange *et al.* 1999).

### Archaeometry and ambiguity

I would like to suggest that, while it is possible that the Maysar 1 and Al Aqir ingots are indeed of non-Omani metal, there are a number of aspects of the research that allow an alternative, more ambiguous interpretation and the possibility of a *local* Omani origin for the ingots from Maysar 1 and Al Aqir. This discussion should in no way be seen as critical of the data collection or the analytical programme conducted by Prange. On the contrary, the data appear to be accurate, precise and highly reliable. Nevertheless, where archaeometric precision meets messy archaeological 'reality', there commonly arise possibilities for competing interpretations of even high-quality data.

In the first instance, the conclusion of a foreign origin for the Maysar 1 ingots can be challenged by more closely scrutinising the actual data. The first possible test derives from the fact that the ingots from Maysar 1 and Al Aqir do not all fall within the one statistical cluster. For example, while the majority of Maysar 1 ingots form one compositional cluster (Prange 2001: table 23 cluster 3) when compared against all the analysed ingots and artefacts from Oman, Maysar 1 ingots DA4028, 4259, 4840U fall into cluster 4.1.1, and Al Aqir ingots DA7405/1, 7410/1, 7410/2 fall into cluster 2. Three additional ingot fragments (from Al Aqir – DA7415; and two Maysar ingots DA4033, 4840D) fall in cluster 4.2.1. These nine Maysar and Al Aqir ingots *do* group compositionally with late 3rd to early 2nd millennium BC artefacts from Oman (Prange 2001: table 23).

Secondly, the main cluster of Maysar 1 and Al Aqir ingots diverges in only two minor/trace elements from the majority of copper-based artefacts in Oman. In all other respects, especially in their arsenic and nickel concentrations, the ingots are compositionally compatible with the remaining Omani Bronze Age copper-based objects. The internal compositional and mineralogical variability of copper deposits is well known, and it would perhaps not be justified to expect even that copper ingots produced at different times from the same ore deposit would be compositionally identical in their concentrations of all trace and minor elements. They are generally very similar in their composition to other Omani copper artefacts, and their high silver and low selenium levels might just be the result of the particular deposit (or part thereof) exploited for their production.

Moreover, the hierarchical cluster analyses that differentiated the Maysar 1 and Al Aqir ingots from other Omani artefacts depend on the amount of compositional variation within the group being statistically analysed. There is relatively little compositional variation in the group of ores and ingots from Oman analysed by Prange, meaning that slight differences in minor/trace element concentrations are enough to statistically isolate the ingots from Maysar 1 and Al Aqir. In contrast, when the ingots are included within a group with more compositional diversity, such as the copper-based objects from the greater Gulf region, the similarity of the Maysar 1 and Al Aqir ingots with other artefacts from Oman, the UAE and Bahrain is immediately apparent, and they are statistically grouped together (Prange 2001: table 29, cluster 3.3.1.1).

To these issues in the interpretation of the compositional and statistical analyses can be added concerns raised by the lead isotope data. In particular, it can be seen that the reconstructed compositional and isotopic groups do not fully replicate one another, i.e. objects from the same compositional group are isotopically distinct and objects from separate compositional groups are isotopically identical. For example, ingots DA4028 and 4259 are compositionally grouped together with a series of Bronze Age artefacts from Oman (Prange 2001: table 23 cluster 4.1.1) and yet one ingot (DA4028) has an anomalous lead isotope signature like many ingots from Maysar 1 while the other has a more common lead isotope signature consistent with contemporary finished artefacts (and ingots) from Oman (Prange 2001: fig. 81).

These specific details and questions raise the possibility that, rather than being examples of metal imported into Oman, the copper ingots from Maysar 1 and Al Aqir might equally have been manufactured in southeastern Arabia from locally collected ores. The lack of known isotopic matches with Omani ores is a stumbling block to this hypothesis. Even though Prange's analytical programme was well planned and extensive, it is still possible that the ingots were produced from as-yet unsampled Omani copper deposits.

Masirah Island, off the southern coast of Oman, initially appeared to be a potential source for such ores, as one copper mine on the island has produced a calibrated radiocarbon date contemporary with the later 3rd millennium BC occupation at Maysar 1 (Yule in press). Moreover, the Masirah Island copper deposits are hosted in an ophiolite sequence unrelated to the mainland Semail Ophiolite (Weeks 2004: 12–14), and might thus be expected to produce copper compositionally similar to but isotopically distinct from that produced on the Omani mainland. It now appears, however, that the ores from Masirah compositionally analysed by Prange (2001: tables 31, 36) do not isotopically match the ingots from Maysar 1 and Al-Aqir[1] and so this possibility must be ruled out.

If we move beyond the data and analyses presented by Prange, the case for an external (non-Omani) origin for the Maysar ingots would be strengthened if there were any ore or artefact data from nearby regions that showed parallels with the unusual lead isotope characteristics of the Omani ingots. As yet, there are no Bronze Age finished metal objects from the greater Persian Gulf region that show lead isotope characteristics matching those of the thorogenic lead-depleted ingots from Maysar 1 and Al Aqir (and Saar). Furthermore, isotopically matching ores have yet to be discovered in India or Iran (Weeks 2004: figs 7.16–17), although relevant lead isotope data are extremely limited thus far. One region that *has* produced copper ores with similar isotope ratios, however, is Saudi Arabia, where deposits in the Arabian Shield show isotope ratios relatively depleted in their thorogenic lead component in a manner similar to the Maysar 1 ingots (Weeks 2004: fig. 7.18), but no archaeological evidence supporting the Bronze Age exploitation of these deposits is currently available. There are sporadic references to 'ancient' or pre-Islamic workings at copper ore bodies within Saudi Arabia and Yemen (for example Windsor 1981) and ongoing archaeological research in Yemen is rapidly expanding the number of known local copper-based artefacts that might be dated back to the 2nd (Crassard and Hitgen 2007), 3rd (Giumlia-Mair *et. al.* 2002), and even 4th millennium BC (Steimer-Herbet *et al.* in press). Needless to say, archaeometric data supporting copper exchange across mainland Arabia would be extremely interesting, however on present evidence it is impossible to assess whether and to what extent central and southern Arabian deposits were worked in prehistory.

Thus, ambiguity commonly arises in archaeometric research through the inevitable 'gaps in the evidence' from limited laboratory analyses and fieldwork. In each of the instances discussed above, further field and laboratory work will be required to adequately address the uncertainties in the preliminary research. The implication of this ambiguity, however, is that the import of foreign copper ingots into southeastern Arabia is possible but by no means certain given the preliminary nature of much of the isotopic research.

## Conclusions

The last decade of archaeometallurgical research in the Persian Gulf has demonstrated the ability of reliable programmes of archaeometric analysis to provide novel information that challenges existing hypotheses, and the conclusions of the research have substantially complicated the archaeological conception of Bronze Age metal production and exchange systems in the region. It is clear that explaining the more comprehensive and complex body of information now available will require the compositional and isotopic data to be fully integrated and contextualised with the range of archaeological evidence.

In particular, new explanations must be developed that can incorporate the evidence for extensive use of foreign artefacts of copper, tin and especially of tin-bronze in a land that produced massive amounts of its own metal. Such complexity is not unique in the Bronze Age world. To list but a few examples, the use of foreign copper in regions with abundant local sources (and even local production) is known to have characterised to a greater or lesser extent the metalworking traditions of Sardinia, the Aegean and Egypt (e.g. Begemann *et al.* 2001; Ogden 2000; Pernicka *et al.* 1990; Shortland 2006).

In these more complex explanatory frameworks, the commonplace use of imported copper and tin-bronze artefacts on the northern coast of the UAE in the later 3rd millennium BC can be understood in relation to southeastern Arabian participation in the Persian Gulf trade and the nature of that exchange system (Weeks 2004: ch. 8), which existed not only

as a series of economic interactions but also as a web of interpersonal relationships and obligations. In this formulation, anthropological considerations suggest that the exchange of finished metallic artefacts could have cemented the interpersonal relationships that underlay all exchange events, and the artefacts themselves may have acted as critical material tokens of participation by individuals in the Persian Gulf exchange system. The role of elites in such systems has received considerable attention (e.g. Earle 1997), and there is little doubt that our ability to understand the first appearance of various metal artefacts and alloys in the Persian Gulf region will ultimately depend as much upon understanding the social contexts of their exchange and use as upon the improved working properties of new alloys.

The possibility that not only finished artefacts but also unalloyed copper ingots were imported *into* southeastern Arabia is perhaps more surprising and difficult to explain. I have argued here that critical ambiguities in the archaeometric data allow for a local origin for the copper ingots found in abundance at Maysar 1 and Al Aqir. Nevertheless, even if the metal *is* eventually proved to be from outside Oman, its presence in Oman can potentially be understood if the exchange is situated within its related technological, cultural and economic contexts. As a final note, it should be mentioned here that there are other factors supporting the possibility of large-scale trade of non-Omani copper in the Bronze Age Persian Gulf, especially in the early 2nd millennium BC. In particular, ongoing fieldwork in Oman has yet to clearly identify copper-smelting sites that date to the early 2nd millennium BC (see Carter 2001; Weeks 2004: ch. 2; forthcoming). Obviously, this lack of evidence has serious implications for our understanding of the metal sources supplying the Dilmun copper trade with Mesopotamia in the Isin-Larsa and Old Babylonian period. The issue is far from resolved – it is possible, for example, that the 'Bronze Age' smelting sites so far identified in fact span the transition from the 3rd to the 2nd millennium BC. Nevertheless, the continued existence of such lacunae in our evidence is indicative of the potential for further radical reconceptions of the Persian Gulf exchange system with future archaeological and archaeometric research.

## Acknowledgements

It is a sincere pleasure to be able to offer this contribution in honour of Paul Craddock, whose general works introduced the author, as a student, to the broad scope and interests of archaeometallurgy, and whose work on material from the United Arab Emirates (Craddock 1981, 1985) has proved important for understanding the development of alloying technologies in the region. The author would like to thank Dr Cameron Petrie (Cambridge University) for reviewing a draft of this paper. Funding for the author's research was provided by the Australian Institute for Nuclear Science and Engineering, the British School of Archaeology in Iraq, and the American School of Prehistoric Research. Collaborative laboratory work was undertaken with Professor Ken Collerson (lead isotope analysis – Dept. of Earth Sciences, University of Queensland) as well as Dr G. Bailey, P. Johnson, E. Stelcer and Dr R. Siegele (PIXE – Australian Nuclear Science and Technology Organisation, Lucas Heights, NSW). Archaeological samples were kindly provided by Dr Sabah Jasim (Sharjah Museum, UAE), Christian Velde and Dr Derek Kennet (Ras al-Khaimah Museum, UAE), Dr Hussein Qandil (Dubai Museum, UAE), Professor

Dan Potts (University of Sydney), and Dr Jane Moon and Dr Robert Killick (London Bahrain Archaeological Expedition).

## Note

1. A. Hauptmann, pers. comm.

## References

Abdi, K. 2000. Review of Potts, T.F., *Mesopotamia and the East: An Archaeological and Historical Study of Foreign Relations 3400–2000 B.C. Journal of Near Eastern Studies* 59: 277–84.

Begemann, F., Kallas, K., Schmitt-Strecker, S. and Pernicka, E. 1999. Tracing ancient tin via isotope analyses. In *The Beginnings of Metallurgy*, A. Hauptmann, E. Pernicka, Th. Rehren and Ü. Yalçin (eds). *Der Anschnitt* Beiheft 9. Bochum: Deutsches Bergbau-Museum, 277–84.

Begemann, F., Schmitt-Strecker, S., Pernicka, E. and Lo Schiavo, F. 2001. Chemical composition and lead isotopy of copper and bronze from Nuragic Sardinia. *European Journal of Archaeology* 4: 43–85.

Berthoud, T. 1979. *Etude par l'analyse de traces et la modélisation de la filiation entre minerais de cuivre et objets archéologiques du Moyen Orient.* PhD dissertation, University of Paris VI.

Berthoud, T., Bonnefous, S., Dechoux, M. and Françaix, J. 1980. Data analysis: toward a model of chemical modification of copper from ores to metal. In *Proceedings of the XIXth Symposium on Archaeometry*, P.T. Craddock (ed.). British Museum Occasional Paper 18. London, 87–102.

Bibby, T.G. 1970. *Looking for Dilmun.* New York: Knopf.

Calvez, J.Y. and Lescuyer, J.L. 1991. Lead isotope geochemistry of various sulphide deposits from the Oman mountains. In *Ophiolite Genesis and the Evolution of the Oceanic Lithosphere*, Tj. Peters, A. Nicolas, and R.G. Coleman (eds). Sultanate of Oman: Ministry of Petroleum and Minerals, 385–97.

Carter, R.A. 2001. Saar and its external relations: new evidence for interaction between Bahrain and Gujarat during the early second millennium BC. *Arabian Archaeology and Epigraphy* 12: 183–201.

Chen, J.H. and Pallister, J.S. 1981. Lead isotopic studies of the Samail Ophiolite, Oman. *Journal of Geophysical Research* 86(B4): 2699–708.

Cleuziou, S. and Méry, S. 2002. In-between the great powers: the Bronze Age Oman Peninsula. In *Essays on the Late Prehistory of the Arabian Peninsula*, S. Cleuziou, M. Tosi and J. Zarins (eds). Rome: Istituto Italiano per L'Africa e L'Oriente. Serie Orientale Roma XCIII, 273–317.

Craddock, P.T. 1981. Appendix V. Report on the scientific investigation of metallurgical samples from the prehistoric site at Umm an-Nar, Abu Dhabi (submitted by W.Y. Al-Tikriti via the Department of Western Asian Antiquities). In *Reconsideration of the Late Fourth and Third Millennium B.C. in the Arabian Gulf with Special Reference to the United Arab Emirates*, W.Y. al-Tikriti, 242–3. PhD dissertation, University of Cambridge.

Craddock, P.T. 1985. Technical appendix 1. The composition of the metal artifacts. In Prehistoric tombs of Ras al-Khaimah, P. Donaldson. *Oriens Antiquus* 24: 85–142.

Crassard, R. and Hitgen, H. 2007. From Cãfer to Bãlhãf: rescue excavations along the Yemen LNG pipeline route. *Proceedings of the Seminar for Arabian Studies* 37: 43–59.

Crawford, H.E.W. 1998. *Dilmun and its Gulf Neighbours.* Cambridge: Cambridge University Press.

Earle, T.K. 1997. *How Chiefs Come to Power: The Political Economy in Prehistory.* Stanford, CA: Stanford University Press.

Frifelt, K. 1991. *The Third Millennium Graves, the Island of Umm an-Nar 1.* Aarhus: Jutland Archaeological Society Publications No. 26/1.

Frifelt, K. 1995. *The Third Millennium Settlement, the Island of Umm an-Nar 2.* Aarhus: Jutland Archaeological Society Publications No. 26/2.

Giumlia-Mair, A., Keall, E.J., Shugar, A.N. and Stock, S. 2002. Investigation of a copper-based hoard from the Megalithic site of al-Midamman, Yemen: an interdisciplinary approach. *Journal of Archaeological Science* 29: 195–209.

Goettler, A., Firth, N.H. and Huston, C.C. 1976. A preliminary discussion of ancient mining in the Sultanate of Oman. *Journal of Oman Studies* 2: 43–56.

Hastings, A., Humphries, J.H. and Meadow, R.H. 1975. Oman in the third millennium BCE. *Journal of Oman Studies* 1: 9–56.

Hauptmann, A. 1985. *5000 Jahre Kupfer in Oman. Band 1: Die Entwicklung der Kupfermetallurgie vom 3. Jahrtausend bis zur Neuzeit. Der Anschnitt* Beiheft 4. Bochum: Deutsches Bergbau-Museum.

Hauptmann, A. 1987. Kupfer und Bronzen der Südostarabischen Halbinsel. *Der Anschnitt* 39: 209–18.

Hauptmann, A., Weisgerber, G. and Bachmann, H.G. 1988. Early copper metallurgy in Oman. In *The Beginning of the Use of Metals and Alloys*, R. Maddin (ed.). Cambridge, MA: MIT Press, 34–51.

Heimpel, W. 1987. Das Untere Meer. *Zeitschrift für Assyriologie* 77: 22–91.

Heimpel, W. 1988. Magan. *Reallexikon der Assyriologie* 7: 195–9.

Højlund, F. and Andersen, H. 1994. *Qala'at al-Bahrain Vol 1. The Northern City Wall and the Islamic Fortress.* Moesgaard: Jutland Archaeological Society Publications 30:1.

Killick, R. and Moon, J. 2004. *The Early Dilmun Settlement at Saar.* London: London-Bahrain Archaeological Expedition, Institute of Archaeology, UCL.

Leemans, W.F. 1960. *Foreign Trade in the Old Babylonian Period.* Leiden: E.J. Brll.

Lombard, P. and Kervran, M. 1989. *Bahrain National Museum Archaeological Collections*, vol. 1. Bahrain: Directorate of Museums and Heritage, Ministry of Information.

Ogden, J. 2000. Metals. In *Ancient Egyptian Materials and Technology*, P.T. Nicholson and I. Shaw (eds). Cambridge: Cambridge University Press, 148–76.

Peake, H. 1928. The copper mountain of Magan. *Antiquity* 2: 452–7.

Pernicka, E., Begemann, F., Schmitt-Strecker, S. and Grimianis, A.P. 1990. On the composition and provenance of metal objects from Poliochni on Lemnos. *Oxford Journal of Archaeology* 9: 263–98.

Potts, D.T. 1990. *The Arabian Gulf in Antiquity*, vol. 1. Oxford: Clarendon Press.

Prange, M.K. 2001. 5000 Jahre Kupfer in Oman. Band II. Vergleichende Untersuchungen zur Charakterisierung des omanischen Kupfers mittels chemischer und isotopischer Analysenmethoden. *Metalla* 8: 1–126.

Prange, M.K., Gotze, H.-J., Hauptmann, A. and Weisgerber, G. 1999. Is Oman the ancient Magan? Analytical studies of copper from Oman. In *Metals in Antiquity*, S.M.M. Young, M. Pollard, P. Budd and R. Ixer (eds). BAR International Series 792. Oxford: Archaeopress, 187–92.

Seeliger, T.C., Pernicka, E., Wagner, G.A. *et al.* 1985. Archäometallurgische Untersuchungen in Nord- und Ostanatolien. *Jahrbuch der Römisch-Germanisches Zentralmuseum Mainz* 32: 597–659.

Shortland, A.J. 2006. Application of lead isotope analysis to a wide range of Late Bronze Age Egyptian materials. *Archaeometry* 48: 657–69.

Steimer-Herbet, T., Saliege, J.-F., Sagory, T., Lavigne, O., As-Saqqaf, Machkour, M. and Guy, H. 2007. Rites and funerary practices at Rawk during the fourth millennium BC (Wadi 'Idim, Yemen). *Proceedings of the Seminar for Arabian Studies* 37: 281–94.

Stos-Gale, Z.A., Maliotis, G., Gale, N.H. and Annetts, N. 1997. Lead isotope characteristics of the Cyprus copper ore deposits applied to provenance studies of copper oxide ingots. *Archaeometry* 39(1): 83–123.

Tosi, M. 1975. Notes on the distribution and exploitation of natural resources in ancient Oman. *Journal of Oman Studies* 1: 187–206.

Weeks, L.R. 2004. *Early Metallurgy of the Persian Gulf: Technology, Trade, and the Bronze Age World*. Boston: American School of Prehistoric Research and Brill Academic.

Weeks, L.R. forthcoming. Iran and the Bronze Age metals trade in the Persian Gulf. In *Essays on the Archaeology and History of the Persian Gulf Littoral*, K. Abdi (ed.). Oxford: BAR International.

Weeks, L.R. and Collerson, K.D. 2004. Lead isotope data from the Gulf. In Weeks 2004, 145–63.

Weeks, L.R. and Collerson K.D. 2005. Archaeometallurgical studies. In *The Early Dilmun Settlement at Saar*, R. Killick and J. Moon (eds). London: London-Bahrain Archaeological Expedition, Institute of Archaeology, UCL, 309–24.

Weisgerber, G. 1980a. …und Kupfer in Oman - Das Oman-Projekt des Deutschen Bergbau-Museum. *Der Anschnitt* 32: 62–110.

Weisgerber, G. 1980b. Patterns of early Islamic metallurgy in Oman. *Proceedings of the Seminar for Arabian Studies* 13: 115–25.

Weisgerber, G. 1981. Mehr als Kupfer in Oman - Ergebnisse der Expedition 1981. *Der Anschnitt* 33: 174–263.

Weisgerber, G. and Yule, P. 2003. Al-Aqir near Bahla – an Early Bronze Age dam site with planoconvex 'copper' ingots. *Arabian Archaeology and Epigraphy* 14: 24–53.

Windsor, G.W. 1981. The other minerals. *Saudi Aramco World* 32: 25–30.

Yule, P. in press. Review of *The Early Dilmun Settlement at Saar* by Killick, R. and Moon, J. *Journal of the American Oriental Society*.

## Author's address

Lloyd Weeks, Department of Archaeology, University of Nottingham, University Park NG7 2RD, Nottingham, UK (lloyd.weeks@nottingham.ac.uk)

# The first use of metal on Minoan Crete

*J.D. Muhly*

ABSTRACT   A review of new evidence for the beginnings of copper metallurgy in eastern Crete, based particularly upon results from the study of excavated remains at the sites of Hagia Photia and Chrysokamino. The latter site remains our only copper-smelting site from Early Minoan Crete. There is now growing evidence for extensive Aegean copper and silver metallurgy during the Final Neolithic period, reflecting contemporary developments in the Balkans.

*Keywords:* Minoans, metallurgy, Chrysokamino, Hagia Photia, copper smelting, Final Neolithic.

In comparison with developments in the north, in central Europe, the Balkans, and northern Greece, especially Macedonia, the use of metal came quite late to the island of Crete. A recent survey of the beginnings of copper metallurgy in southeastern Europe and the Carpathian Basin identified 15 sites with relevant copper finds dating in the 6th millennium BC (or 5800–5200 BC), a period now called Copper Horizon 1a (Kalicz 1992). This is thought to represent an independent southeast European metallurgical development, despite the fact that the first use of copper in nearby Anatolia can now be placed in the late 9th millennium BC, at the site of Çayönü Tepesi, while at least three additional copper-using sites can be dated in the 8th millennium BC (Yalçin 2000, especially chronological table on p. 19). In other respects, especially in burial customs, Bulgarian archaeologists are quite prepared to recognise the importance of Anatolian connections (Bǎčvarov 2003), but metal usage is another matter.

There is still much discussion regarding the independent invention of metallurgical technology in various parts of the ancient world (for the Aegean see Zachos and Douzougli 1999). All go back to the remarkable paper published by Colin Renfrew in the *Proceedings of the Prehistoric Society* for 1969, certainly one of the most influential papers in the history of archaeometallurgical studies over the past century (Renfrew 1969). I do not want to enter the lists and fight yet another engagement in a seemingly never-ending battle. At this point I only want to call attention to the recent pronouncement by Cyprian Broodbank, certainly no pan-diffusionist, that, in the Neolithic period, 'the Aegean formed part of a belt of generically similar societies extending from Anatolia to the Balkans' (Broodbank 2000: 283). I certainly agree (Muhly 2006: 158), and it is within such a context that I feel we must place the early use of metal in the Aegean world.

The remarkable series of gold ring-idol pendants now attested in Greece, known mainly from the antiquities market but also from excavation contexts, have clear, indisputable connections with similar pendants from the Balkans (Makkay 1989: 101–3), especially from several very rich graves within the Varna cemetery complex (Bailey 2000: 218–22; Muhly

1996). The most recent discovery of such a pendant, made by a woman hiking in the vicinity of the town of Ptolemaida, southwest of Thessaloniki, was reported in the Greek press in February 2006 (Kantouris 2006). What has not been properly appreciated is that, from the Aegean world, there are three comparable pendants made of silver. These all come from cave contexts, from the so-called cave of Euripides on the island of Salamis, from a small burial near the Amnisos cave on Crete and from the Alepotrypa cave in Lakonia (Demakopoulou 1998: nos. 62–64). A Balkan background for this 5th-millennium use of gold and silver seems quite convincing, however the exact circumstances are yet to be explained (for problems of dating see Maran 2000: 185–9). What is surprising about the European evidence is the general absence of silver, apart from the Caucasus and the Kuban region (Mallory and Huld 1984).

With the use of precious metals one always finds evidence for the use of base metals as well, especially copper. During the period known as Karanovo VI or the Gumelnitsa culture, as well as gold and silver the inhabitants of the Balkans were also producing elaborate, massive shaft-hole axes of copper as well as a series of flat axes or celts. The shaft-hole axes, or axe-adzes, first examined by Jim Charles in a study published as an appendix to the Renfrew paper mentioned above (Charles 1969), are not known from the Aegean but, in 1923–24, excavating in his Neolithic House A at Knossos, in what he described as 'a pure Neolithic element', Evans found an example of such a flat axe (Evans 1928: 14 and fig. 3:f; see also Glotz 1925: 402; Renfrew 1972: fig. 16.2: lower right). His interpretation of the significance of this find is worth quoting:

> That copper was worked in Crete at this time is highly improbable and it is more likely that the axe had reached the site of Knossos either from Egypt or, owing to some coast-wise drift of commerce from Cyprus, where copper implements seem to have been fabricated from a very remote period (Evans 1928: 14–15).

I doubt that anyone today would derive such a find from Egypt or Cyprus.

97

This axe from Neolithic Knossos is often regarded as the oldest copper object discovered in Crete (Sakellarakis 1973: 145). Unfortunately its context seems not to be secure (J.D. Evans was willing to accept it in 1971: 115; but rejected it in 1994: 18) and the axe itself seems to have disappeared. Yet this axe is unique only for Crete. Examples are fairly common in Final Neolithic contexts from northern Greece, at sites such as Sesklo and Pefkakia (Phelps *et al.* 1979; for a detailed discussion of such axes from Greece and the islands, especially the Zas Cave on Naxos, see Zachos and Douzougli 1999: 962–3).

Not an auspicious beginning for the story of the early use of metals on the island of Crete. Nor, as far as we can tell, did this early use of copper on Crete have any impact upon what was to follow. The real development of metal technology on the island took place in the Final Neolithic period, especially in the eastern part, the Bay of Mirabello. One of the reasons why the Final Neolithic period has achieved such prominence in recent years is because of the greater number of sites producing evidence for some sort of Final Neolithic occupation. Increase in evidence leads to an increase in interest and vice versa. In 2002 the Polish scholar Krzysztof Nowicki published an article 'The end of the Neolithic in Crete' (Nowicki 2002), in which he presented a catalogue of 76 Final Neolithic sites on the island. In a recent lecture he delivered in Athens that number had already increased to 90 sites (Nowicki 2005). This phenomenon is not confined to the island of Crete. Major Final Neolithic sites are being found all over the Greek world, especially in Thessaly and Macedonia. Long-established excavations are now producing Final Neolithic sherds. As far as Aegean archaeology is concerned, the past few years can be regarded as the Age of the Final Neolithic. Even Archaic sites in Crete such as Azoria, near Kavousi, are now producing Final Neolithic sherds (Haggis *et al.* 2004: 390).

This is the period that saw the initial occupation of Crete, as far as the Bay of Mirabello is concerned. We have done enough work in eastern Crete by now to say that there is simply no evidence for the human habitation of the region prior to Final Neolithic (Hayden 2003: 396–7) but Tomkins *et al.* (2004: 56) report Early Neolithic sherds from the modern village of Kavousi. Well and good, but what is meant by Final Neolithic and what is the date of the Final Neolithic period? Just to raise these questions is to introduce the basic problem in the archaeology of the period. For archaeologists working in the Balkans and in northern Greece, the designation Final Neolithic covers a long period, extending for as much as *c.* 1000 years or 4500–3500 BC (Johnson 1999). This is the Balkan Copper Age or Chalcolithic, the time of Karanovo VI, the Gumelnitsa culture, the cemetery of Varna and the settlement at Durankulak (Bailey 2000: 203–22). The period represents, among other things, the beginnings of stone architecture in the Balkans (Boyadžiev 2004). In Macedonia this is the period of Sitagroi III (Elster and Renfrew 2003). In the recent publication of the excavations of Jean Deshayes at Dikili Tash, conducted from 1961 to 1975, this period is designated Chalcolithique Ancien (Treuil 2004), as French scholars have never used the term Final Neolithic.

In the Aegean world, however, a date of *c.* 3500 BC for the end of Final Neolithic creates a serious gap between it and the beginning of EH/EM I, placed at *c.* 3100 BC (Nowicki 2002: 64). For Nowicki, therefore, his 90 Final Neolithic sites are to be placed in the second half of the 4th millennium BC (Nowicki 2002: 65). This confusion is certainly, at least in part, created by what has come to be known as 'the missing 4th millennium BC'. This can best be seen at the site of Mandelo, in western Macedonia, where a 5th-millennium Final Neolithic occupation, with early metal finds including a crucible (Papanthimou and Papasteriou 1993: 1209 and 1213, fig. 2) is followed by a 3rd-millennium EBA occupation, with nothing in between (Kostakis *et al.* 1989: 684, fig. 3; Renfrew 1989: 677). Related to this problem is the possible hiatus in occupation at the site of Sitagroi, between periods III and IV, and an even longer hiatus at Dikili Tash (Treuil 2004: 263, table 4.14). Being unable to solve this problem at the present time we have to accept here that Final Neolithic is an attractive designation of uncertain date and comparative context.

Another reason for this confusion is that we still lack a proper Final Neolithic settlement from Crete. We have bits and pieces of material, usually pottery, almost always out of context. One of our major deposits of Final Neolithic pottery comes from a well, where the shapes in use seem to represent the specialised jars used in drawing water (Manteli 1992, esp. p. 107). Nowicki has much to say about the remote site of Monastiraki Katalimata (his site no. 1, Nowicki 2002: 16–20), but the total amount of pottery from this site amounts only to 2 kg. New work at Phaistos holds out great promise, and everyone is awaiting publication of the excavations at Kephala Siteias, being excavated by Yiannis Papadatos (Papadatos *et al.* 2004). A further complication is the growing conviction that, in the latter part of this period which Nowicki calls Final Neolithic II, there was a major influx of new inhabitants, of refugees or settlers who came to Crete from western Anatolia (Hood 1990; Warren 1974) and chose to live on defensible hilltops, usually commanding the highest elevation in the area (Nowicki 2002: 66–71).

Then, at the very end of the Final Neolithic period, there was a further influx of population, this time from the Cyclades, and it is this Cycladic element that seems to be associated with the introduction of the use of metal, especially in the eastern part of the island (Betancourt 2003). Here I can only discuss two sites that I have been involved with over the past ten years in terms of excavation and/or publication.

The first is that of Hagia Photia Siteias, an Early Minoan I cemetery site excavated by the Greek archaeologist Costis Davaras in 1971 and 1984 (Fig. 1). With its total of 263 excavated burials, it is the largest known Early Minoan cemetery and, as a cemetery, it obviously is not a metalworking or metal-production site. It is of great interest, however, for its early use of metal, both copper and silver. I now have a catalogue of 34 metal artefacts from Hagia Photia, 32 of copper or copper-based alloy, one of silver and one of lead (Fig. 2). In addition there are two silver beads, recently identified and not yet catalogued. The copper/copper-alloy artefacts consist of six daggers, a socketed spearhead (that could very well be LM III), six fishhooks, three small saws, a spatula, six chisels, two awls or reamers, two bracelets, four beads and one zoomorphic pendant. In addition to the finds of copper or copper alloy, we have a zoomorphic pendant of silver, as well as the two silver beads already mentioned, and a pendant made of lead (for this site see Davaras and Betancourt 2004).

**Figure 1** Map of eastern Crete showing the location of Chrysokamino and Hagia Photia (after Davaras and Betancourt 2004: fig. 1, courtesy INSTAP Academic Press).

What are we to make of such a remarkable use of metal on Crete from such an early date? How early? When we get our chronological problems sorted out we might be able to answer that question. At present, I would suggest a date of *c.* 3000 BC. Where did the metal technology displayed here come from? Where were these artefacts manufactured and how did they come to end up as grave goods at a cemetery in northeastern Crete? The outstanding feature of the ceramic assemblage from Hagia Photia has always been its strong Cycladic connections. The excavator, Costis Davaras, called attention to this in a preliminary report published in 1971, the initial year of the excavation itself (Davaras 1971). This observation has been confirmed by all subsequent scholarship dealing with Hagia Photia (Betancourt 2003; Doumas 1976; Watrous 2001: 163–5; more cautious is Broodbank 2000: 301–4).

Back in the mid-1980s the Gales analysed 16 artefacts from Hagia Photia, showing that they were made of arsenical copper

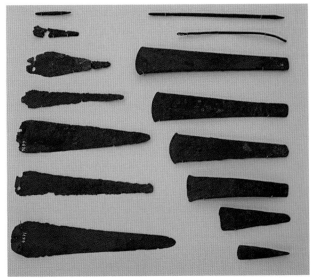

**Figure 2** Photograph of metal finds from Hagia Photia as displayed in Hagios Nikolaos Museum (photo courtesy of the Hagios Nikolaos Museum, Crete).

(Gale 1990: 303, fig. 1), thus supporting the Cycladic connections that were also suggested by the presence of silver and the absence of gold. To appreciate how quickly things changed in eastern Crete one need only turn to the abundance of gold jewellery from EM II Mochlos (Davaras 1975) in a context that was very Minoan, not Cycladic. It would be good to know from where the EM II inhabitants of Mochlos suddenly acquired so much gold (Whitelaw 2005). As for arsenical copper it has to be recognised that, in the mid-3rd millennium, the use of arsenical copper certainly was not confined to the Cyclades. It was, in fact, the dominant copper alloy in the Old World at that time, attested in copper artefacts from the Indus valley to the United Kingdom. There were also two crucibles from Hagia Photia, both of which were found in tombs: one in Tomb 10, the other in Tomb 45 (Davaras and Betancourt 2004: 17–19 and 50–51). The crucibles were of Cycladic type (Blitzer 1995: 502), but earlier than any comparanda known from the Cyclades. The use of crucibles as grave goods is a most unusual phenomenon. From Early Minoan Crete, I know of only one other example, from the plundered tholos tomb at Ayia Kyriaki. In their publication detailing the surviving artefacts from this tomb, Blackman and Branigan mention but do not illustrate this crucible (1982: 40, table 4). There is no reason to designate these two tombs from Hagia Photia as the burials of metalworkers as no metal artefacts were found in either tomb. Still, the crucibles do have clear Cycladic connections.

Far more significant were the lead isotope results derived from the study of 12 copper-alloy artefacts from Hagia Photia (Stos and Gale 2006; preliminary report in Stos-Gale 1993: 122, table 11.1). These results suggest that the copper used at Hagia Photia came from the Cycladic island of Kythnos. It was compatible, at the same time, with the lead isotope signature characteristic of the copper-smelting slags from the site of Chrysokamino. To appreciate the significance of this it is necessary to turn to recent archaeological and analytical work at Chrysokamino itself.

The site known as Chrysokamino (Fig. 3) has been in the archaeological literature for over 100 years, ever since the first reports on the site by Harriet Boyd (Boyd 1901: 156; Hawes *et al.* 1908: 33). Angelo Mosso, in 1910, was convinced not only of the existence of the prehistoric copper-smelting site but also of the existence of a copper mine located in a nearby cave (Mosso 1910: 289–93). Over the intervening century almost all Minoan archaeologists paid visits to the smelting site, making their own collections from the surface accumulations of slag and ceramic furnace fragments. Suggested dates for smelting activity at Chrysokamino ranged from Minoan to medieval (Branigan 1968: 50–51), with even the suggestion that the site was actually a lime kiln (Zoes 1990).

In 1994 Philip Betancourt, together with the present author, decided that it was time for a proper excavation of the site. There was a good chance that the surface exposure would, in fact, turn out to be all that remained, the rest of the site having eroded into the sea, but there was only one way of finding out. Following a detailed topographical and geological survey of the entire surrounding area, together with a mapping of the proposed excavation site in 1995, major excavations were carried out in 1996 and 1997 involving, among other things, dry and water sieving of all the dirt from the excavation, something of a first in the history of Minoan archaeology (preliminary report in Betancourt *et al.* 1999; final report in Betancourt 2006a).

In brief, the pottery goes back to the Final Neolithic and covers the entire Early Minoan period. From EM III there was a hut constructed from wooden posts (with only the holes remaining) and, presumably, some sort of wattle and daub construction. Within this hut we found numerous fragments of a most unusual type of pot bellows (Betancourt and Muhly

2006), the earliest examples of a pot bellows yet attested from the Mediterranean or the ancient Near East (for literature see Davey 1979; Müller-Karpe 1994: 103–7, examples from 17 sites; Tylecote 1981).

The finds from the smelting site itself consist of hundreds of perforated furnace fragments and many tons of copper-smelting slag. Most of the slag had been broken up into tiny pieces, but some of it had been pulverised to a fine black powder. We assume that all of this was done in order to extract the copper prills still embedded in the slag. The scientists at Demokritos, in a research group led by Yannis Bassiakos, have concluded from their detailed analysis of many slag fragments that high-temperature smelting was carried out at Chrysokamino, resulting in a real separation between metallic copper and a tapped slag (Catapotis *et al.* 2004). They also estimate that about 50% of the slag remained inside the furnace and had to be broken up, even crushed, in order to recover the embedded prills of metallic copper (Bassiakos and Catapotis 2006: 349).

Clearly the metalworkers at Chrysokamino were after every tiny piece of copper they could get their hands on – they left virtually nothing behind. We have a limited number of pieces of slag still containing tiny prills of copper, but not a single copper artefact was recovered. Whatever copper was produced was taken away in what seems to have been a most efficient operation. We have no crucibles, no moulds, no metalworking paraphernalia except for one possible tuyere fragment (Ferrence and Koukaras 2006: fig. 9.1, no. 208) that, along with the pot bellows, was part of the smelting operation. Chrysokamino was a single-function site; it existed in order to produce metallic copper through the smelting of imported copper ore. The arsenical copper ore was most likely imported from the Cycladic island of Kythnos. The perforations in the furnace went along with the pot bellows and the windy hillside of the site itself, all designed to bring about maximum draught for the high temperature inside the furnace necessary to achieve the separation of the metallic copper from its slag matrix (Bassiakos and Catapotis 2006: 351–3; Betancourt 2006b: 183–9).

All of this has very close parallels with the Skouries smelting site on the island of Kythnos. Once again we have here a strong Cycladic setting. Indeed the location of the sites of Chrysokamino and Skouries – on an isolated windy hillside – and the separation between smelting and metalworking sites are now both seen as Early Cycladic traditions (Broodbank 2000: 294, 298–9; Hayden 2003: 398). But why Chrysokamino? What was it that brought seasonal metalworkers to this remote promontory of land, next to a cliff and a sheer drop of some 30 m to the rocky coastline below?

Despite repeated reports to the contrary, there are no local sources of copper ore anywhere in the vicinity of Chrysokamino (Farrand and Stearns in Betancourt *et al.* 1999: 352; Stos and Gale 2006: 301–4). This lack of copper deposits seems to be characteristic of the entire island of Crete, although there are still Minoan archaeologists who would strongly disagree with that statement. Keith Branigan goes so far as to argue that 'it seems that the situation of copper sources may have been amongst the factors which determined the siting of the EM settlements in southern Crete' (Branigan 1971: 14). I doubt very much that this was true for southern

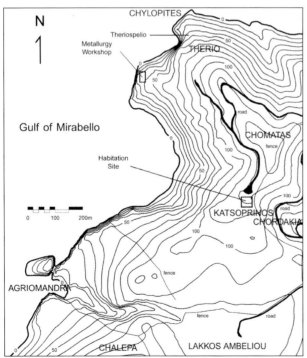

**Figure 3** Topographical map of the territory of Chrysokamino showing the location of the metallurgical site and the habitation site (that could represent the dwelling place of the Early Minoan metalworkers but, as excavated, is predominantly of Late Minoan date) and the small harbour of Agriomandra (where copper ore could have been brought by boat and transported up the ravine to the smelting site) (courtesy Chrysokamino Project).

Crete, and it certainly cannot be argued for the northern part of the island. Cyprian Broodbank has, in fact, an interesting explanation for the practice of shipping ore to a smelting site located some distance from the ore source itself:

> Such localised inter-island transport makes sense in a canoe-based system, given that the rich ores that early miners are likely to have found would have weighed little more than the metals that they yielded, whilst the fuel would have been considerably heavier (Broodbank 2000: 293).

In other words, the ore was brought to the fuel, not vice versa. This implies that the area of Chrysokamino must have been more heavily forested in the Early Bronze Age than it is today (Betancourt and Farrand 2006: 42–3; Grove and Rackham 2001: 318–20).

The excavation of the copper-smelting site at Chrysokamino was certainly a small-scale operation, but the implications of the discoveries made there have been far-reaching. I assumed that the pot bellows found in the EM III hut must have operated like the ones shown in the famous Rekhmire tomb painting, but the details in the design of our pot bellows simply did not fit the Near Eastern model. It was Harriet Blitzer (1995) who finally convinced everyone that what we had to do was to turn our examples upside down (Betancourt and Muhly 2006: 126, fig. 8.1: Rekhmire; 127, fig. 8.2: Blitzer). We finally decided that the perforated furnace fragments, found in hundreds, could not represent the furnace itself. The wall of the cylinder was too thin, the diameter too narrow. They seemed most likely, in fact, to have functioned as some sort of a chimney, placed on top of a bowl furnace that consisted of a clay-lined pit dug in the ground. We now feel that this is the most likely reconstruction (Betancourt 2006b: 184, fig. 14.2), but we found no trace of any bowl furnace on the hillside where the smelting took place. We certainly looked for such evidence, but it was not there (Fig. 4).

Chrysokamino remains the only known Early Minoan copper-smelting site. Many questions have still to be answered, but the impetus to further research that has come from this excavation is most encouraging. My own thoughts regarding the reconstruction of copper-smelting technology at Chrysokamino certainly have evolved over the past ten years. Chrysokamino was something entirely new, unlike any site ever excavated before on the island of Crete. Discussion regarding details of metallurgical operations at the site will, I am sure, continue for many years to come.

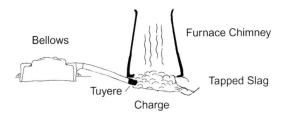

**Figure 4** Reconstruction of the Early Minoan III copper-smelting process at Chrysokamino showing the bowl furnace with bellows and tuyere, and the upright furnace chimney (courtesy Chrysokamino Project).

## References

Băčvarov, K. 2003. *Neolithic Mortuary Practices: Intramural Burials in their Southeast European and Anatolian Context.* Sofia: [in Bulgarian with English summary].

Bailey, D.W. 2000. *Balkan Prehistory: Exclusion, Incorporation and Identity.* London: Routledge.

Bassiakos, Y. and Catapotis, M. 2006. Reconstruction of the copper smelting process based on the analysis of ore and slag. In Betancourt 2006a, 329–53.

Betancourt, P.P. 2003. The impact of Cycladic settlers on Early Minoan Crete. *Mediterranean Archaeology and Archaeometry* 3: 3–12.

Betancourt, P.P. (ed.) 2006a. *The Chrysokamino Metallurgy Workshop and its Territory, Hesperia Supplement 36.* Princeton, NJ: American School of Classical Studies at Athens.

Betancourt, P.P. 2006b. Discussion of the workshop and reconstruction of the smelting practices. In Betancourt 2006a, 179–89.

Betancourt, P.P and Farrand, W.R. 2006. The natural environment. In Betancourt 2006a, 19–44.

Betancourt, P.P. and Muhly, J.D. 2006. The pot bellows. In Betancourt 2006a, 125–32.

Betancourt, P.P., Muhly, J.D., Farrand, W.R. *et al.* 1999. Research and excavation at Chrysokamino, Crete, 1995–1998. *Hesperia* 68: 343–70.

Blackman, D. and Branigan, K. 1982. The excavation of an Early Minoan tholos tomb at Ayia Kyriaki, Ayiofarango, southern Crete. *Annual of the British School at Athens* 77: 1–57.

Blitzer, H. 1995. Minoan implements and industries. In *Kommos I: The Kommos Region and Houses of the Minoan Town*, J.W. Shaw and M.C. Shaw (eds). Princeton, NJ: Princeton University Press, 403–535.

Boyadžiev, Y. 2004. Chalcolithic stone architecture from Bulgaria. *Archaeologia Bulgarica* 8: 1–12.

Boyd, H.A. 1901. Excavations at Kavousi, Crete, in 1900. *American Journal of Archaeology* 5: 156.

Branigan, K. 1968. *Copper and Bronzework in Early Bronze Age Crete.* Studies in Mediterranean Archaeology XIX. Lund: Carl Bloms Boktryckeri A.-B.

Branigan, K. 1971. An Early Bronze Age metal source in Crete. *Studi Micenei ed Egeo-Anatolici* 13: 10–14.

Broodbank, C. 2000. *An Island Archaeology of the Early Cyclades.* Cambridge: Cambridge University Press.

Catapotis, M., Pryce, O. and Bassiakos, Y. 2004. Prehistoric copper smelting at Chrysokamino: archaeological analysis and experimental reconstruction. Abstract of paper presented at the University of Crete Conference on Aegean Metallurgy in the Bronze Age, 19–21 November 2004, 13.

Charles, J.A. 1969. Appendix I. A metallurgical examination of southeast European copper axes. *Proceedings of the Prehistoric Society* 35: 40–42.

Davaras, C. 1971. Protominoikon nekrotapheion Agias Photias Siteias. *Athens Annals of Archaeology* 4: 392–6.

Davaras, C. 1975. Early Minoan jewellery from Mochlos. *Annual of the British School at Athens* 70: 101–14.

Davaras, C. and Betancourt, P.P. 2004. *The Hagia Photia Cemetery I: The Tomb Groups and Architecture.* Prehistory Monographs 14. Philadelphia: INSTAP Academic Press.

Davey, C. J. 1979. Some ancient Near Eastern pot bellows. *Levant* 11: 101–11.

Demakopoulou, K. 1998. *Kosmemata tes Ellenikes Proistorias. O Neolithikos Thesauros.* Athens: Ministry of Culture.

Doumas, C. 1976. Proistorikoi Kykladites sten Krete. *Athens Annals of Archaeology* 9: 69–80.

Elster, E.S. and Renfrew, C. (eds) 2003. *Prehistoric Sitagroi: Excavations in Northeast Greece, 1968–1970. Vol. 2: The Final Report.* Monumenta Archaeologia 20. Los Angeles: Cotson Institute of Archaeology.

Evans, A. J. 1928. *The Palace of Minos at Knossos, Vol II*. London: Macmillan and Co.

Evans, J.D. 1971. Neolithic Knossos: the growth of a settlement. *Proceedings of the Prehistoric Society* 37: 95–117.

Evans, J.D. 1994. The early millennia: continuity and change in a farming settlement. In *Knossos: A Labyrinth of History. Papers in Honour of Sinclair Hood*, D. Evely, H. Hughes-Brock and N. Momigliano (eds). Athens: British School at Athens, 1–20.

Ferrence, S.C. and Koukaras, B. 2006. Miscellaneous ceramic artifacts. In Betancourt 2006a, 133–6.

Gale, N.H. 1990. The provenance of metals for Early Bronze Age Crete: local or Cycladic? In *Proceedings of the 6th Cretological Congress (Chania 1986)*, vol. 1, Chania 1990. Chrysostomes Philogical Society, 299–316.

Glotz, G. 1925. *The Aegean Civilization*. New York: Methuen.

Grove, A.T. and Rackham, O. 2001. *The Nature of Mediterranean Europe: An Ecological History*. New Haven, CT: Yale University Press.

Haggis, D.C., Mook, M.S., Scarry, C.M., Snyder, L.M. and West, W.C. 2004. Excavations at Azoria, 2002. *Hesperia* 73: 339–400.

Hawes, H. Boyd, Williams, B.E., Seager, R.B. *et al.* 1908. *Gournia, Vasiliki and other Prehistoric Sites on the Isthmus of Hierapetra, Crete; Excavations of the Wells-Houston-Cramp Expeditions, 1901, 1903, 1904*. Philadelphia: American Exploration Society, Free Museum of Science and Art.

Hayden, B.J. 2003. Final Neolithic–Early Minoan I/IIA settlement in the Vrokastro area, eastern Crete. *American Journal of Archaeology* 107: 363–412.

Hood, S. 1990. Settlers in Crete *c.* 3000 B.C. *Cretan Studies* 2: 151–8.

Johnson, M. 1999. Chronology of Greece and south-east Europe in the Final Neolithic and Early Bronze Age. *Proceedings of the Prehistoric Society* 65: 319–36.

Kalicz, N. 1992. The oldest metal finds in southeastern Europe and the Carpathian Basin from the 6th to 5th millennia BC. *Archaeologiai Értesitö* 119: 3–14 [in Hungarian with English summary].

Kantouris, C. 2006. Hiker finds 6,500-year-old gold pendant in field. *Athens News* 24 February 2006: 26.

Kostakis, A., Papanthimou-Papaefthimou, A., Pilali-Papasteriou, A. *et al.* 1989. Carbon 14 dates from Mandalo, W. Macedonia. In Maniatis 1989, 670–85.

Makkay, J. 1989. *The Tiszaszölös Treasure*. Budapest: Akadémiai Kiadó.

Mallory, J.P. and Huld, M.E. 1984. Proto-Indo-European 'silver'. *Zeitschrift für Vergleichende Sprachforschung* 97: 1–12.

Maniatis, Y. (ed.) 1989. *Archaeometry. Proceedings of the 25th International Symposium*. Amsterdam: Elsevier.

Manteli, K. 1992. The Neolithic well at Kastelli Phournis in eastern Crete. *Annual of the British School at Athens* 87: 103–20.

Maran, J. 2000. Das Ägäische Chalkolithikum und das erste Silber in Europa. In *Studien zur Religion und Kultur Kleinasiens und des ägäischen Bereichs. Festschrift für Baki Öğün zum 75. Geburtstag*, C. Işık (ed.). Bonn: Rudolf Habelt, 179–93.

Mosso, A. 1910. *The Dawn of Mediterranean Civilisation*. London: Fisher Unwin.

Muhly, J.D. 1996. The first use of metals in the Aegean. In *The Copper Age in the Near East and Europe, XIII International Congress of Prehistoric and Protohistoric Sciences, Colloquium XIX, 10*, B. Bagolini and F. Lo Schiavo (eds). Forlí: Abaco Edizioni, 75–84.

Muhly, J.D. 2006. Chrysokamino in the history of early metallurgy. In Betancourt 2006a, 155–77.

Müller-Karpe, A. 1994. *Altanatolisches Metallhandwerk*. Offa-Bücher, Band 75. Neumünster: Wachholtz Verlag.

Nowicki, K. 2002. The end of the Neolithic in Crete. *Aegean Archaeology* 6: 7–72.

Nowicki, K. 2005. How did the Neolithic end in Crete? Paper presented at the Minoan Seminar, Danish Institute in Athens, 23 March 2005.

Papadatos, G., Tsipopoulou, M., Bassiakos, Y. and Catapotis, M. 2004. The beginning of metallurgy in Crete: new evidence from the FN-EM I settlement at Kephala Petras, Siteia. Abstract of paper presented at the University of Crete conference on Aegean Metallurgy in the Bronze Age, 19–21 November 2004, 36.

Papanthimou, A. and Papasteriou, A. 1993. O proistorikos oikismos sto Mandalo: Nea stoicheia sten proistoria tes Makedonias. *Archaia Makedonia* 5: 1207–16.

Phelps, W.W., Varoufakis, G.J. and Jones, R.E. 1979. Five copper axes from Greece. *Annual of the British School at Athens* 74: 175–84.

Renfrew, C. 1969. The autonomy of the south-east European Copper Age. *Proceedings of the Prehistoric Society* 35: 12–47.

Renfrew, C. 1972. *The Emergence of Civilisation*. London: Methuen.

Renfrew, C. 1989. Theme session: the transition from the Neolithic to Early Bronze Age in the Aegean. In Maniatis 1989, 677–8.

Sakellarakis, Y. 1973. Neolithic Crete. In *Neolithic Greece*, D.R. Theocharis (ed.). Athens: National Bank of Greece, 131–46.

Stos, Z. and Gale, N. 2006. Lead isotope and chemical analyses of slags from Chrysokamino. In Betancourt 2006a, 299–319.

Stos-Gale, Z. 1993. The origin of metal used for making weapons in Early and Middle Minoan Crete. In *Trade and Exchange in Prehistoric Europe*, C. Scarre and F. Healy (eds). Oxford: Oxbow Books, 115–29.

Tomkins, P., Day, P.M. and Kilikoglou, V. 2004. Knossos and the earlier Neolithic landscape of the Herakleion basin. In *Knossos: Palace, City, State. Proceedings of the Conference in Herakleion organized by the British School at Athens and the 23rd Ephoreia of Prehistoric and Classical Antiquities of Herakleion, in November 2000, for the Centenary of Sir Arthur Evans's Excavations at Knossos*, G. Cadogan, E. Hatzaki and A. Vasilakis (eds). Athens: British School at Athens, 51–9.

Treuil, R. (ed.) 2004. *Dikili Tash, village prehistorique de Macedonie orientale. I. Fouilles de Jean Deshayes (1961–1975), Vol. 2, BCH, Supplément 37*. Paris: De Boccard.

Tylecote, R.F. 1981. From pot bellows to tuyeres. *Levant* 13: 107–18.

Warren, P.M. 1974. Crete, 3000–1400 B.C.: immigration and the archaeological evidence. In *Bronze Age Migrations in the Aegean: Archaeological and Linguistic Problems in Greek Prehistory*, R.A. Crossland and A. Birchall (eds). Park Ridge, NJ: Noyes Press, 41–9.

Watrous, L.V. 2001. Review of Aegean prehistory III: Crete from earliest prehistory through the Protopalatial period. In *Aegean Prehistory: A Review*, T. Cullen (ed.). Boston, MA: Archaeological Institute of America, 157–223.

Whitelaw, T. 2005. Alternative pathways to complexity in the southern Aegean. In *The Emergence of Civilization Revisited*, J.C. Barrett and P. Halstead (eds). Oxford: Oxbow Books, 232–56.

Zachos, K.L. and Douzougli, A. 1999. How early and how independent? In *Meletemata. Studies in Aegean Archaeology Presented to Malcolm H. Wiener as he enters his 65th year*, vol. III, P.P. Betancourt, V. Karageorghis, R. Laffineur and W.-D. Niemeier (eds). Liège: Aegaeum 20, 959–68.

Yalçin, Ü. 2000. Anfänge der Metallverwendung in Anatolien. In *Anatolian Metal I. Der Anschnitt* Beiheft 13. Bochum: Deutsches Bergbau-Museum, 17–30.

Zoes, A.A. 1990. Anaskaphi Vasilikes Ierapetras. *Praktika tes Archaiologkies Etaireias* 1990: 340–41.

## Author's address

Professor J.D. Muhly, University of Pennsylvania, American School of Classical Studies, Souidias 54, Kolonaki, Athens 175 61, Greece (jimmuhly@yahoo.com)

# Cross-cultural Minoan networks and the development of metallurgy in Bronze Age Crete

*Noel H. Gale and Zofia Anna Stos-Gale*

*ABSTRACT*   A review is presented of the results of comparative lead isotope analyses of ores and artefacts directed towards the elucidation of the changing ore sources which supplied the metals used in Minoan Crete from the Protopalatial to Postpalatial period. Mention is made in particular of ore sources and Early Bronze Age (EBA) smelting places in the Cyclades, Attica and Crete. Evidence is presented that there are only miniscule ore deposits in Crete and that, with perhaps a minor exploitation of Chrysostomos in the Prepalatial period, they were not exploited by the Minoan civilisation. Evidence is given which supports that Lavrion was an important source of copper, as well as lead and silver, for the Aegean Bronze Age, including Minoan Crete.

*Keywords:* Minoan, metallurgy, ore, slag, Chrysokamino, Crete, Cyclades, Lavrion.

## Introduction

Metal finds of gold, silver, bronze and lead, from 3rd–2nd-millennium BC Crete are of exquisite character and unique workmanship (Hood 1978: 153–208, 233–41), comparable only with contemporary artefacts found in Egypt and the Middle East. The skills required to make gold and silver jewellery and cups, copper-based weapons and tools were developed on Bronze Age Crete in the absence of any significant local metal resources.[1] In fact, we shall see that so far there is minimal evidence that any of the two or three small local sources of copper, and one miniscule source of lead, on Crete were ever exploited in prehistoric times.[2] Moreover Crete had no deposits of tin or gold whatsoever. Further, the lead isotope evidence from the small Early Bronze Age (EBA) copper-smelting operation at Chrysokamino (Stos-Gale 1998; Stos and Gale 2006) on the northeast coast of Crete indicates that even there the ores were brought for smelting from other localities in the Aegean outside Crete. It is therefore quite clear that the development of Minoan metallurgy depended on metals imported through the contemporary intensive cross-cultural seaborne networks that were perhaps fostered by the inhabitants of this island,[3] which presumably also resulted in metallurgical technology transfer. Sir Arthur Evans[4] believed in a Minoan 'Thalassocracy' that developed, among other reasons, in the attempt to find supplies of metals and other exotic materials such as ivory, ornamental stone, etc. This theory is no longer widely accepted (for example Knapp 1993; Starr 1955), but the fact remains that the acquisition of metals from outside the island was essential for Minoan metallurgy to flourish as it did.

It has been a principal aim of research at the Isotrace Laboratory in Oxford to establish Bronze Age trading patterns in the Mediterranean, especially for metals (Gale and Stos-Gale 1982, 1986, 1992). This paper is chiefly concerned with the metals trade which supplied Crete with the metals unavailable in Crete itself, in the light of comparative lead isotope studies of Minoan metal artefacts and ore deposits around the Mediterranean. In common with the majority of contemporary scholars of Bronze Age archaeology in the Mediterranean, we reject the past stultifying tendency to discount the concept of 'trade' in the ancient world, stemming from the influences of 'primitivist' economic historians (Polanyi) feeding into the views of ancient historians (Finley and the Cambridge school), sociologists (Weber), or processual archaeologists, since the evidence of field archaeology and modern archaeological enquiry seems to be against them, as ably argued by Sherratt and Sherratt (1991, 1998).

We shall demonstrate that our lead isotope analyses of over 500 metal artefacts excavated from Bronze Age sites on Crete show a varied pattern for the metal sources used, a pattern which changed through time. Overall about a third of copper-based artefacts are consistent with the copper coming from Cyprus, as are a number of copper ingots of various shapes and their fragments. Approximately 40% of metal indicates an Aegean origin for the copper, but a further 30% of artefacts of specifically Minoan style seem to be made of copper that came from much farther away. Furthermore a number of copper ingots seem to be made of copper originating from far outside the western, central or eastern Mediterranean. A brief critical evaluation will be made of the role of copper sources in Lavrion, Attica.

This paper examines the cross-cultural Minoan supply networks that led to the Cretan hegemony in metallurgy and the sources of the continual supply of much needed metals.

Figure 1 Locations of metal ore showings and of some Minoan sites on Crete.

## Ore sources of metals on Crete

As mentioned, both our own field surveys in Crete and those of others have revealed little in the way of significant metal ore deposits in Crete. There are certainly no sources of either gold or tin, so the Minoan use of each of these metals necessitated a supply of metals brought in from outside the island. There is one minor source of lead at Ano Valsamonero,[5] whose location is shown in Figure 1. Galena has also been found in minute

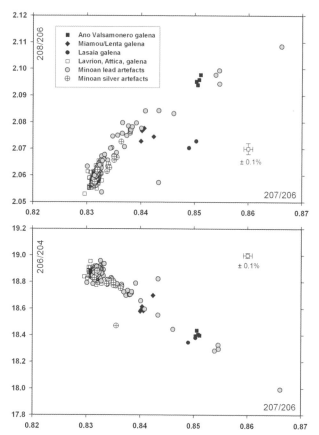

Table 1 Neutron activation analyses of the contents of certain elements in samples of galena from Crete.

| Sample | Locality | Au | Ag | As | Sb | Cu |
|--------|----------|------|------|------|------|------|
| MLB | Lenta B | 0.03 | 1587 | 17 | 113 | 142 |
| LAQ | Lenta A | 0.03 | 986 | 2 | 182 | 41 |
| MJED | Miamou | 0.02 | 982 | 2 | 117 | 34 |
| LAF | Ano Valsomonero | <0.01 | 244 | 267 | 1428 | 112 |
| AN1 | Ano Valsomonero | 0.01 | 157 | 77 | 223 | 22 |
| LASF | Lasaia | <0.05 | 697 | 1531 | 6227 | 505 |

Analytical contents are given as ppm. Instrumental neutron activation analyses using gamma ray spectrometry were made in Oxford (see Stos-Gale 1991) for samples of galena which were neutron irradiated in the Imperial College research reactor in Silwood Park, Berkshire, UK.

amounts by Faure (1966, 1980) and Diallinas at localities in Lenta, Lasaia and Miamou near the south-central coast of Crete (see Fig. 1), but Diallinas[6] has said of these occurrences that they are merely in the float, and were not viable sources of lead ore even in the Bronze Age.

Table 1 presents our instrumental neutron activation analyses of samples of galena from Crete, showing in particular the silver (Ag) contents. The silver contents of galena from the only known ore source in Crete at Ano Valsamonero are too low to have been used in the Bronze Age,[7] though the samples of galena from the float contain silver at exploitable levels. Figure 2 shows lead isotope analyses of Minoan lead and silver artefacts in relation to the lead isotope data for galena samples from Crete and from the large lead/silver/copper deposit of Lavrion in Attica (Marinos and Petrascheck 1956). It is immediately clear from this lead isotope diagram that lead/silver ore sources on Crete are almost completely excluded as sources for the analysed lead and silver Minoan artefacts. On the other hand, Figure 2 shows that a large number, but not all, of the Minoan artefacts are consistent with having been made from ores coming from Lavrion in Attica.

Figure 2 Lead isotope data for Minoan lead and silver objects and from lead ores from Crete and Lavrion, Attica. The analysed lead objects range in date from EMII to LMII and are from the sites of Hagia Triadha, Kastelli, Knossos, Unexplored Mansion Knossos, Royal Road Knossos, Mavro Spilio, Kommos, Mochlos, Palaikastro, Pighi, Pseira, Psychro Cave and Tylissos. The analysed silver objects are from the sites of Hagia Photia, Krasi and Mochlos.

## Ore sources in the Aegean

It is clear that, in view of the complete absence in Crete of ore sources of gold and tin and the virtual absence of sources of copper, silver and lead, the Minoans must largely have pro-

**Table 2** Summary of the chief ore deposits included in the lead isotope and chemical analytical programme.

| Country | No. of occurrences | No. of ore samples |
|---|---|---|
| Greece | 33 | 867 |
| Cyprus | 46 | 700 |
| Turkey | 130 | 290 |
| Near and Middle East | 36 | 226 |
| Italy and Spain | 73 | 610 |
| The Balkans | 20 | 220 |
| TOTAL | 338 | 2,913 |

cured these metals from outside their island. Considering ore sources relatively near to Crete, copper, lead and silver ore sources occur in Spain, Sardinia, Liguria, Tuscany, in the Cycladic islands and mainland Greece (especially Lavrion), in Bulgaria and Anatolia, and in Arabia, while large ore sources of copper occur in Cyprus, Feinan and Timna. Over the past 30 years, surveys of these ore deposits, and collections of ores and metal slags for analysis, have been made by teams principally from Heidelberg, Oxford, the Smithsonian, Tel Aviv, Bochum, Cagliari and Japan. Comparative lead isotope data for these ore deposits were measured chiefly in Oxford and to a lesser extent in Mainz,[8] the Smithsonian[9] and Japan.[10] Table 2 briefly summarises the extent of the sampling of ore sources involved in creating the lead isotope database.

Within Crete itself we have already noted that the copper ore 'sources' suggested by Branigan (1974) and by Faure (1966, 1980) are either miniscule or non-existent, and that modern fieldwork (Becker 1976; Gale 1990; Wheeler *et al.* 1975) has confirmed the paucity of copper ore sources in Crete. Figure 1 shows the locations of the showings of copper ores that do exist on Crete, mostly stains on the host rock, but with two real small copper ore deposits, one at Sklavopoulou in western Crete, and the other at Chrysostomos in central southern Crete. Sklavopoulou (where Davies (1935: 267) found sherds possibly dating to the 5th century BC, but no slag) was mined in the 19th and early 20th centuries AD, while there was exploitation by quarrying of Chrysostomos in 1952 by the Greek firm Psaromarolis. Though both are small deposits, Chrysostomos is the larger, while both Faure (1966: 52) and J.A. Macgillivray have found a very small amount of Minoan sherds near the edge of the Chrysostomos pit; of course none of this is proof of exploitation by the Minoans. The results of our own surveys on Crete, combined with all reliable geological information about mineral deposits on the island, indicate that significant metal extraction (copper, lead and silver) from the local ores in the Bronze Age times is very unlikely. To widen our survey for possible metal extraction slags on Crete we enlisted the help of archaeologist Dr Krzysztof Nowicki of Warsaw, who surveyed large areas of the mountains in Crete. No copper slag heaps have been located, however, other than that at Chrysokamino (Betancourt *et al.* 1999; Stos-Gale 1998), although layers have occasionally been found in the excavations of Minoan sites such as Malia (Pelon 1987; Poursat 1985). It may be worth pointing out here that the structures at such places as Kato Zakros and Hagia Triadha, which some have sought to identify as Bronze Age copper furnaces, clearly are not, as has been agreed by all archaeometallurgical experts

who have examined them (for example Slater, Muhly, Eveley), and proved by the analytical work of Liritzis (1984).

## Lavrion and the Cyclades as sources of copper ore

Before the advent of lead isotope studies for provenancing metals, most Aegean archaeologists would probably have thought of Cyprus as the chief, if not the only, source of copper for the Bronze Age Mediterranean world.[11] This view appeared to be challenged by the application of lead isotope studies to Bronze Age copper alloy artefacts from the Cyclades and Late Minoan Crete (Gale and Stos-Gale 1982), which suggested that a considerable number of these artefacts had been made using copper ores from the Lavrion ore deposit in Attica, Greece. While many Aegean archaeologists (e.g. McGeehan-Liritzis 1996: 119; Niemeier 1998: 36–7) have accepted the idea that Lavrion was indeed a significant source of copper in the Late Bronze Age, others (e.g. Matthäus 1998; Muhly 1983: 217; 1997: 28–9) have cast doubt on this, despite the increasing body of evidence in its favour. This is perhaps understandable given that the Lavrion ore deposits were formerly known to archaeology chiefly as the source of silver (and lead metal) for the Athenian city-state (for example Conophagos 1980; Kraay 1976: 62 ff.) It has long been known, however, that lead ores from Lavrion were smelted much earlier than this, in the MBA and LBA (Stos-Gale and Gale 1982). Recent work has pushed even earlier the exploitation of ores from Lavrion, with the spectacular discoveries of Final Neolithic and EBI silver and lead production at Lambrika (Koropi) and Merenda (Markopoulou, Fig. 3) in the Mesogeia, Attica (Kakavoyianni 2005; Kakavoyianni *et al.* 2007). This work is welcome support for our own finding that three Final Neolithic artefacts from the Alepotrypa cave in Laconia (a silver earring, a silver pendant and a copper axe) were made of silver or copper extracted from Lavrion ores (Gale and Stos-Gale in press). Our own work has also shown that both copper and lead ores from Lavrion were being smelted in the Early Helladic II period at Raphina in Attica (Gale *et al.* 2007).

Since the validity of the hypothesis that Lavrion was a principal copper source in the Bronze Age Aegean world is clearly important for the comprehension of the Bronze Age 'trade' in metals in the Mediterranean region, we believe that a more extensive and critical re-examination of the evidence is needed, which we will attempt in a subsequent paper. Here we will briefly mention some of the evidence supporting this hypothesis. There is good direct evidence on the ground for copper ore deposits in the Lavrion region of southern Attica,[12] especially at Kamareza and Sounion (Fig. 3). The mines in the Sounion and Kamareza regions of Lavrion, which we visited in 1987 with Dr S. Papastavrou[13] of the Greek Institute of Geology and Mineral Exploration and an old miner who had worked for the French Lavrion Company, have abundant thick deposits of mixed rich malachite/azurite constituting a rich source of copper minerals of a type which is very easily smelted with the production of minimal slag. Direct evidence for this is given in Figure 4, which shows in colour both a typical sample of mixed rich azurite/malachite ore from Lavrion and an underground region at Kamareza showing rich oxidised

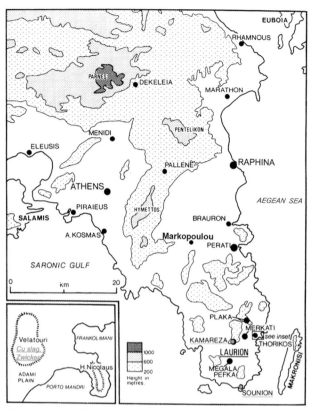

**Figure 3** Map of the Lavrion region, Attica, Greece.

**Figure 4** A guide from the former Compagnie Française des Mines du Laurium shows a large mass of oxidised copper ores exposed in a gallery wall in the Kamareza region in 1987.

**Table 3** Summary of the range of archaeological metal artefacts included in the lead isotope and chemical analytical programme.

| Country | No. of sites | No. of Cu or Cu/Sn artefacts | No. of Pb + Ag artefacts |
|---|---|---|---|
| Greece, Crete, and Aegean islands | 61 | 780 | 690 + 70 |
| Cyprus | 20 | 420 | 70 + 6 |
| Egypt and the Near East | 12 | 200 | 90 + 30 |
| Turkey | 10 | 700 | 125 + 0 |
| The Balkans | 45 | 500 | 2 + 0 |
| TOTAL | 148 | 2,600 | 977 + 106 |

copper ore deposits in the gallery wall. Zwicker[14] has shown us small fragments of copper slag (undated) which he found on the slopes of the Velatouri Hill, Thorikos, in the Lavrion region of Attica (see Fig. 3 for the location).

As for the Cyclades, Gale *et al.* (1985) have reported the presence of copper ore and antique copper slag heaps on the islands of Kythnos and Seriphos, the presence of copper ore on Siphnos and Thera (Gale and Stos-Gale in press), and proved the EBA dates of the Skouries copper slag heap on Kythnos (Gale *et al.* 1985) and the Chrysokamino copper slag heap on Crete (Stos-Gale 1998: 720–21). Subsequent archaeological fieldwork by Philaniotou (2007) has produced archaeological evidence that copper was smelted in the Bronze Age at the copper slag heaps Avyssalos and Kefala on Seriphos, previously reported by Gale *et al.* (1985: 83, map on 94).

## Lead isotope analyses of Minoan artefacts and Aegean ore deposits

Table 3 lists the range of Bronze Age metal artefacts for which lead isotope analyses were made, while Table 2 lists the metal ores and slags analysed for their lead isotope composition. The methods used for discovering the provenance of the metal of particular artefacts has been described by Gale and Stos-Gale (2000). For particular artefact types a useful preliminary sorting, from thousands of lead isotope data, of the relevant ore and slag deposits for detailed comparison is made by using software which computes the Euclidean three-dimensional distances between the lead isotope analyses for artefacts and ores.[15] Final provenance assignments are always made by plotting the comparative lead isotope data on the usual two complementary two-dimensional lead isotope diagrams, searching for one-to-one correspondences between artefacts and ores, as has earlier been illustrated in Figure 2. Lack of space clearly prevents presentation in this paper of all the many lead isotope diagrams prepared to make provenance assignments of the analysed Minoan copper-based metal artefacts. Instead a review will be given of the results of such plots for each of the chief Minoan periods.

### Hagia Photia

The Late EMI (Late Early Minoan I) site of Hagia Photia Siteias in east Crete is the largest Early Minoan (EM) cemetery yet discovered on Crete and was excavated by Davaras in 1971 (Davaras and Betancourt 2004). It is mentioned individually here because it is commonly agreed that this site has a very strong Cycladic element, to such an extent that the site may represent an active influx of settlers from the Cycladic islands in the final phase (Kampos Group) of Early Cycladic I.[16] In the 1980s we analysed 16 metal artefacts from this site (Gale 1990); they proved to be of arsenical copper. Lead isotope analyses, which can now be interpreted again on the basis of the greatly enlarged database, show that these 16 objects were not made of copper smelted from the small ore deposits on Crete at Chrysostomos or Sklavopoulou. Instead the lead isotope evidence links the artefacts from Hagia Photia partly with the slags and ores on Seriphos,[17] but more strongly with copper smelted at the proven EBA smelting site of Skouries on Kythnos and also strongly with copper ores from Aspros Pyrgos on Siphnos. Just two of the Hagia Photia artefacts are compatible with having been made of copper coming ultimately from Lavrion ores, but the lead isotope evidence is that copper ores from Kythnos, Siphnos and Lavrion were all being smelted at EBA Skouries on Kythnos and Chrysokamino on Crete. Consequently the strong evidence for a Cycladic connection for Hagia Photia from the presence of Cycladic marble figurines and Kampos Group pottery is extended by the use at this site of copper originating from Cycladic ore sources.

## Sources of copper used in Prepalatial Crete (*c.* 3100/3000–1925/1900 BC)[18]

Figure 5 summarises the indications of comparative lead isotope analyses for the provenance of copper for 43 copper-based artefacts, chiefly the typical triangular daggers,[19] a few long daggers[20] and small tools from Prepalatial sites in Crete.[21] Branigan, based on the comparative typology of EM copper-based artefacts and his belief in the existence of copper ore deposits on Crete, thought that 'it seems probable that initially its supplies of copper came from within the island, but that from EM III (Early Minoan III) onwards copper was imported'. The source of this copper, and also of the tin needed for bronze production, was the western Mediterranean.[22] In contrast our comparative lead isotope analyses show that in the Prepalatial period very little use was made of copper deposits in Crete and that about 58% of copper-based artefacts are consistent with an origin from ore deposits in the Cycladic islands (Kythnos, Siphnos, Seriphos, possibly Keos). The second largest import, about 26%, seems to have been copper from Cyprus which, being prior to the advent of the oxhide ingots, was imported in an unknown form. About 6% (three artefacts) of copper already came from Lavrion in this period, though it may have been smelted on Kythnos or at Chrysokamino. About 6% of copper came from Anatolia while a small amount of copper (about 2%) is consistent with having come from the ore deposits in the Wadi Arabah (Feinan/Timna); the origin of some 2%

cannot at present be identified. It should be mentioned that about ten artefacts from tholos tombs in the Mesara just about match ores from Chrysostomos, so that there was perhaps very limited use of this indigenous Cretan ore deposit in this early period, a use which was later displaced by other, richer ore sources.

## Sources of copper used in Crete in the Protopalatial period (*c.* 1925/1900–1750/1720 BC)

Figure 6 summarises the evidence from comparative lead isotope analyses for the provenance of copper for 73 copper-based artefacts from Protopalatial sites in Crete.[23] The dominant source of copper (at 66%) remains the Cycladic islands, but the use of copper (11%), lead and silver from the Lavrion deposits increases in this period by roughly a factor of two. The importation of copper from Cyprus in some form continues. The Protopalatial period also sees development of the Minoan procurement of copper from other regions, reaching possibly into the Middle East and into the Taurus Mountains in southern Anatolia. In view of the lead isotope evidence that Lavrion was important as a source of copper, lead and silver for Minoan Crete, we should emphasise that the evidence is that copper metal smelted from Lavrion copper ores came to Crete, and that the quantitative scientific evidence is firmly against the hypothesis that copper came from Cypriot ores but that its lead isotope signature was masked by the addition to the copper of lead metal from Lavrion ores. A similar qualitative hypothesis, involving the suggested addition of Sardinian lead metal to Cypriot copper metal (Kassianidou 2001, 2005, 2006) was advanced in the attempt to explain why nuragic bronzes have lead isotope compositions corresponding to Sardinian ores, yet all the analysed oxhide ingots found on Cyprus have lead isotope compositions corresponding with those of copper ores from the Apliki region on Cyprus. This hypothesis was thoroughly examined and discussed and it was shown that quantitative chemical and lead isotope analyses did not support it in any way (Gale 2006). It was also shown that all empirical evidence supported that ancient copper alloy artefacts containing lead in the range approximately 0.01% to about 4% (or even higher) had lead that originated from the

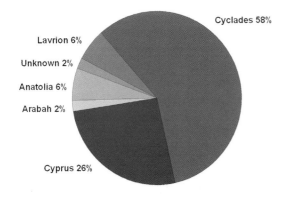

**Figure 5** Sources of copper for the Minoans in the Prepalatial period (*c.* 3100/3000–1925/1900 BC) based on lead isotope analyses of 43 copper-based artefacts.

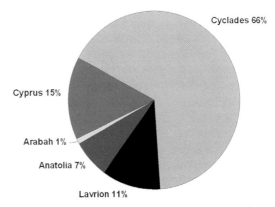

**Figure 6** Sources of copper for the Minoans in the Protopalatial period (*c.* 1925/1900–1750/1720 BC) based on lead isotope analyses of 73 copper-based artefacts.

copper ores from which the copper metal was smelted and refined (Gale 2006). The recent direct evidence that Lavrion copper ores were being smelted as early as the Early Bronze Age at Raphina in Attica (Gale *et al.* 2007) adds to the plausibility that Lavrion copper metal came to Minoan Crete.

## Sources of copper used in Crete in the Neopalatial period (*c.* 1750/1720–1490/1470 BC)

The Neopalatial period (154 analyses, Fig. 7) witnesses a great change in the sources of copper for the Minoans, with the dominant source becoming Lavrion in Attica (42%) while the component from the Cyclades drops off to about 3%. There remains an important supply of copper from Cyprus (17%), perhaps beginning towards the end of this period to arrive in oxhide ingot form,[24] known, for example at Mochlos, from LMIB (Late Minoan IB). There seems to be a remarkable increase in supplies of copper from the Taurus Mountains (21%). This is also the period where the use of tin-bronze greatly increases, with the average tin content in Minoan bronze higher than ever before. This might be relevant to the tin trade routes, since tin probably came from central

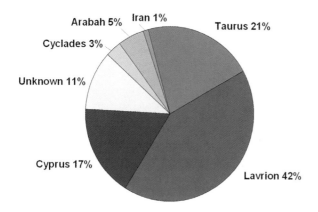

**Figure 7** Sources of copper for Minoan Crete in the Neopalatial period (*c.* 1750/1720–1490/1470 BC) based on lead isotope analyses of 154 copper-based artefacts.

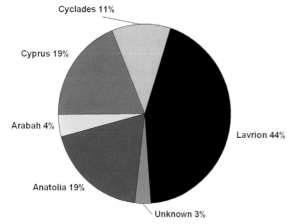

**Figure 8** Sources of copper for Mycenaean Knossos and the Postpalatial period (*c.* 1490/1470–1075/1050 BC) based on lead isotope analyses of 161 copper-based artefacts.

Asian tin ore deposits, perhaps associated with lapis lazuli from Afghanistan.

## Copper sources in Mycenaean Knossos and the Postpalatial period (*c.* 1490/1470–1075/1050 BC)

After *c.* 1450 BC, the copper (and tin?) importing networks seem to remain broadly unchanged (161 artefacts, Fig. 8). The major sources of copper are Lavrion (44%), Cyprus (19%) and the Taurus Mountains (19%), with the relatively minor Cycladic copper deposits at about 11%. There seem to remain quite small amounts (4%) of copper coming from the Arabah. This is of course the period where Cypriot copper was being moved extensively around the Mediterranean in oxhide and planoconvex ingot form, even as far west as Sardinia (Gale 1999). It is also the period where we have direct evidence of copper and tin ingots being transported about the Mediterranean on such ships[25] as *Uluburun* (*c.* 1300 BC) and *Cape Gelidonya* (*c.* 1200 BC).

## Conclusions

In 1980 very little was known about EBA copper sources and copper smelting in the Aegean beyond the bare facts about Raphina (Theocharis 1951, 1952) and the very small-scale operation at Kephala (Coleman 1977) on the Cycladic island of Keos. Now we have considerably larger, proven EBA copper-smelting sites on the Cycladic islands of Kythnos (2) and Seriphos (3) (Gale *et al.* 1985; Philaniotou 2007), and on Crete at Chrysokamino (Betancourt *et al.* 1999; Stos-Gale 1998), with proven smelting of ores from Kythnos, Seriphos, Siphnos and Lavrion at all sites apart from Raphina and Kephala, where solely ores from Lavrion were smelted.

At Chrysokamino, Kythnos, Raphina and Kephala on Keos the typical furnace fragments with holes occur[26] which point to EBA windblown (partially?) furnaces.

The lead isotope analyses show that from the 3rd millennium BC, Minoan copper-based artefacts were made of metal from various ore sources, principally the Cyclades, Lavrion and Cyprus, but not from ores occurring in Crete, apart from a possible minor Prepalatial exploitation of Chrysostomos for artefacts in some of the tholos tombs of the Mesara.

The earliest, EMI, silver and copper-based artefacts from Hagia Photia in east Crete prove to have been made respectively from silver/lead ores from Lavrion and copper ores smelted on the Cycladic islands.

Throughout the later Bronze Age periods, the dominant sources of copper and silver for the Minoans were in the Aegean – first the Cycladic islands and then Lavrion. Cyprus also figures as an important source of copper for the Minoans from the earliest periods through to the Postpalatial period.

Copper ore sources in the Taurus Mountains, in Cilicia, appear to have become an important copper source for the Minoans in later periods, from the Neopalatial onwards.

The earliest copper oxhide ingots discovered anywhere in the Mediterranean remain the Type I ingots excavated from LMIB levels at Hagia Triadha, central southern Crete (Cucuzza *et al.* 2004; Gale and Stos-Gale 1986). Lead isotope analyses show their copper to derive from Precambrian ores not found in Cyprus or the Mediterranean generally[27] – the ultimate origin of the copper oxhide ingots thus remains at present a mystery.

## Notes

1. It used to be averred that there were in Crete sufficient deposits of minerals of copper and lead to supply all the needs of the Minoans (e.g. Branigan 1974; Faure 1966, 1980), but modern surveys of mineral deposits in Crete (by Muhly and Rapp, quoted in Wheeler *et al.* 1975, by Becker 1976 and Gale 1990) show that this is not correct, a view which we shall see is confirmed by comparative lead isotope analyses.
2. Of the two earlier experts who personally surveyed Hellenic mineral deposits, Fiedler (1841), who represented King Otto of Greece, did not visit Crete which was not then part of Greece. Davies (1935: 266–8) examined minor copper deposits in western Crete but found no associated sherds dating earlier than possibly the 5th century BC. Davies summarises his view about the suggestion that copper ores from western Crete were a source for Minoan bronze in the sentence: 'I could discover no foundation for such a theory.'
3. A view supported by Wiener (1987, 1990).
4. Evans 1921–35, I, 283–4; II, 238–52.
5. We were taken to this small ore deposit by Dr K Zervantonakis in 1983.
6. M. Diallinas of Heraklion was well known, as a source of information on the occurrence of ores in Crete, to earlier archaeologists in Crete such as Branigan and Faure. Branigan (1968: 50–52) mentions Diallinas as a source of such information, from whose advice we also have benefited considerably.
7. The lowest content of silver in a lead ore which could be exploited in the Aegean Bronze Age was discussed in Gale *et al.* 1984 and in Pernicka and Wagner 1982.
8. For example Wagner *et al.* 2003 and references therein.
9. For example Sayre *et al.* 2001 and references therein.
10. For example Hirao *et al.* 1995 and references therein.
11. The view that copper from Cyprus dominated the supply to the Late Bronze Age Aegean permeates the literature of the prehistory of this region, for example Knapp 1986, 1998.
12. The major treatise on the ore deposits in Lavrion remains Marinos and Petrascheck (1956), who mention many deposits of copper ores, both oxidised and sulphidic, especially at Kamareza and Sounion. Cambresy, a mining engineer with expert practical knowledge of the Lavrion ores, mentions huge aggregations of malachite in the Lavrion deposits (Cambresy 1889). Fiedler mentions a number of copper deposits within Lavrion (Fiedler 1841: 43–5, 49, 55, 61, 89).
13. Now sadly deceased.
14. Pers. comm. 1983.
15. Use of this software, and newer analyses, has led to the present reappraisal of the Minoan metal sources, which differs in minor ways from that published by Stos-Gale (2001).
16. See Muhly 2007.
17. Note that Philaniotou 2007 has now produced evidence that the Seriphos copper slag heaps were active in EC times.
18. The absolute dates for the Minoan periods remain controversial. Those given in this paper are those suggested in the Dartmouth Classics Department internet site Prehistoric Archaeology of the Aegean: http://projectsx.dartmouth.edu/classics/history/bronze_age/chrono.html#top.
19. See Branigan 1968, figs 5 and 6 for illustrations.
20. See Branigan 1968, figs 1–3 for illustrations.
21. The sites from which the analysed Prepalatial copper-based artefacts originated were chiefly the tholos tombs of Hagia Triadha, Kalathiana, Koumasa, Krasi, Marathokephala and Platanos, and from the cave of Pyrgos. The reuse of these tombs and their looting in the Bronze Age complicates dating of their metal artefacts, which we took from Branigan 1993: 143–52.
22. Branigan 1968: 57. In fact our comparative lead isotope analyses exclude the possibility that Minoan copper-based artefacts of any period were made of copper coming from ore deposits in Tuscany, Sardinia, Liguria or southern Spain.
23. The sites from which the analysed Prepalatial copper-based artefacts originated were chiefly the tholos tombs of Hagia Triadha, Kalathiana, Koumasa, Krasi, Marathokephala and Platanos, and from the cave of Pyrgos.
24. It should not be forgotten that 19 copper oxhide ingots of early typological form (Buchholz Type I) had already arrived in Crete at Hagia Triadha in LMIB, but that they were not made of Cypriot copper – indeed the ore source of their copper has not yet been identified (Cucuzza *et al.* 2004; Gale 1991, 1999), but seems probably to be a Precambrian ore deposit which may perhaps be in central Asia, although Precambrian copper deposits in the southern Sinai (El Shazly *et al.* 1955) have not yet been excluded.
25. For discussions of these Bronze Age ships and other relevant matters see Yalçin *et al.* 2005.
26. The typical furnace fragments with holes are discussed and illustrated by Mosso (1910: 289–93).
27. The 19 Hagia Triadha oxhide ingots have lead isotope compositions which moreover do not match those of the copper ores from Saudi Arabia, Feinan, Timna or the Eastern Desert, Egypt. The Precambrian ores in southern Sinai have not yet been analysed.

## References

Becker, M. 1976. Soft stone sources on Crete. *Journal of Field Archaeology* 3(4): 361–74.

Betancourt, P.P., Muhly, J.D., Farrand, W.R. *et al.* 1999. Research and excavation at Chrysokamino, Crete, 1995–1998. *Hesperia* 68: 343–70.

Branigan, K. 1968. *Copper and Bronzework in Early Bronze Age Crete.* Studies in Mediterranean Archaeology XIX. Lund: Carl Bloms Boktryckeri A.-B.

Branigan, K. 1974. *Aegean Metalwork of the Early and Middle Bronze Age.* Oxford: Clarendon Press.

Branigan, K. 1993. *Dancing With Death.* Amsterdam: Hakkert.

Cambresy, A. 1889. Le Laurium, Chapitre III. *Revue Universelle des Mines* VII(3): 175–217.

Cline, E.H. and Harris-Cline, D. (eds) 1998 *The Aegean and the Orient in the Second Millennium. Aegaeum* 18. Liège/Austin: Université de Liège.

Coleman, J. 1977. *Keos Volume 1, Kephala.* Princeton, NJ: American School of Classical Studies.

Conophagos, C.E. 1980. *Le Laurium Antique et la Technique Grecque de la Production de l'Argent.* Athens: Ekdotike Hellados.

Cucuzza, N., Gale, N.H. and Stos-Gale, Z.A. 2004. Il mezzo lingotto oxhide da Haghia Triada. *Creta Antica* 5: 137–53.

Davaras, C. and Betancourt, P.P. 2004. *The Hagia Photia Cemetery I: The Tomb Groups and Architecture.* Prehistory Monographs 14. Philadelphia: INSTAP Academic Press.

Davies, O. 1935. *Roman Mines in Europe.* Oxford: Clarendon Press.

El Shazly, E.M., Abdel Naser, S. and Shukri, B. 1955. *Contributions to the Mineralogy of the Copper Deposits in Sinai. Paper No. 1.* Cairo: Egyptian Geological Survey.

Evans, A. 1921–35. *The Palace of Minos.* London: Macmillan.

Faure, P. 1966. Les minerais de la Crète antique. *Revue d'Archéometrie* 1: 45–78.

Faure, P. 1980. Les mines du roi Minos. In Πεπραγμένα του Δ Διεθνούς Κρητολογιχού Συνεδρίου 1. Heraklion: 151–68.

Fiedler K.G. 1841. *Reise durch alle Theile des Konigreiches Griechenland,* 2. Leipzig: Friedrich Fleischer.

Gale, N.H. 1990. The provenance of metals for EBA Crete: local or Cycladic? In Πεπραγμένα του ΣΤ Διεθνούς Κρητολογιχού Συνεδρίου 1. Chania: 299–316.

Gale, N.H. 1991. Copper oxhide ingots: their origin and their place in the Bronze Age metals trade in the Aegean. In *Bronze Age Trade in the Mediterranean*, N.H. Gale (ed.). SIMA 90. Jonsered: Paul Åströms Förlag, 197–239.

Gale, N.H. 1999. Lead isotope characterisation of the ore deposits of Cyprus and Sardinia and its application to the discovery of the sources of copper for Late Bronze Age oxhide ingots. In *Metals in Antiquity*, S.M.M. Young, M. Pollard, P. Budd and R. Ixer (eds). BAR International Series 792. Oxford: Archaeopress, 122–33.

Gale, N.H. 2006. Sardinia and the Mediterranean: provenance studies of artefacts found in Sardinia. *Instrumentum* 23: 29–34.

Gale, N.H. and Stos-Gale, Z.A. 1982. Bronze Age copper sources in the Mediterranean: a new approach. *Science* 216: 11–19.

Gale, N.H. and Stos-Gale, Z.A. 1986. Oxhide ingots in Crete and Cyprus and the Bronze Age metals trade. *Annual of the British School at Athens* 81: 81–100.

Gale, N.H. and Stos-Gale, Z.A. 1992. Lead isotope studies in the Aegean (The British Academy Project). In *New Developments in Archaeological Science (Proceedings of the Joint Royal Society/British Academy Conference on Science Based Archaeology)*, M. Pollard (ed.). Oxford: Oxford University Press, 63–108.

Gale, N.H. and Stos-Gale Z.A. 2000. Lead isotope analyses applied to provenance studies. In *Modern Analytical Methods in Art and Archaeology*, E. Ciliberto and G. Spoto (eds). Chemical Analyses Series, vol. 155. New York: John Wiley and Sons, 503–84.

Gale, N.H. and Stos-Gale, Z.A. in press. Changing patterns in metallurgy. In *Proceedings of the Cycladic Colloquium, Cambridge, 2004*, C. Renfrew and N. Brodie (eds).

Gale, N.H., Stos-Gale, Z.A. and Davis, J.L. 1984. The provenance of lead used at Ayia Irini, Keos. *Hesperia* 53(4): 389–406.

Gale, N.H., Papastamataki, A., Stos-Gale, Z.A. and Leonis, K. 1985. Copper sources and copper metallurgy in the Aegean Bronze Age. In *Furnaces and Smelting Technology in Antiquity*, P.T. Craddock and M.J. Hughes (eds). British Museum Occasional Paper 48. London, 81–103.

Gale, N.H., Kayafa, M. and Stos-Gale, Z.A. 2007. Early Helladic metallurgy at Raphina, Attica, and the role of Lavrion. In Tzachili 2007.

Hägg, R. and Marinatos, N. (eds) 1987. *The Function of the Minoan Palaces. Proceedings of the Fourth International Symposium at the Swedish Institute in Athens. Skrifter utgivna av Svenska i Athen 4°, 35.* Stockholm: Svenska institutet i Athen.

Hirao, Y., Enomoto, J. and Tachikawa, H. 1995. Lead isotope ratios of copper, zinc and lead minerals in Turkey. In *Essays on Ancient Anatolia and its Surrounding Civilizations*, H.I.H. Prince T. Mikasa (ed.). Wiesbaden: Harrasowitz, 89–114.

Hood, S. 1978. *The Arts in Prehistoric Greece.* Harmondsworth: Penguin Books Ltd.

Kakavoyianni O. 2005.Εργαστήριο μεταλλουργίας αργύρου της Πρωτοελλαδικής I εποχής στα Λαμπρικά Κορωπίου. *Archeologia kai Technes*, 94: 45-8.

Kakavoyianni, O., Douni, K. and Nezeri, F. 2007. Silver metallurgical findings from the end of the Final Neolithic period to the MBA at the area of Mesogeia, Attica. In Tzachili 2007.

Karageorghis, V. and Stampolidis, N. (eds) 1998. *Proceedings of the International Symposium Eastern Mediterranean: Cyprus-Dodecanese-Crete, Sixteenth–Sixth Centuries BC*, Athens: A.G. Leventis Foundation.

Kassianidou, V. 2001. Cypriot copper in Sardinia: yet another case of bringing coals to Newcastle? In *Italy and Cyprus in Antiquity: 1500–450 BC*, L. Bonfante and V. Karageorghis (eds). Nicosia: Severis Foundation, 97–120.

Kassianidou, V. 2005. Cypriot copper in Sardinia: yet another case of bringing coals to Newcastle? In *Archaeometallurgy in Sardinia*,

F. Lo Schiavo, A. Giumlia-Mair and R. Valera (eds). Monographies Instrumentum 30. Montagnac: Mergoil, 384–96.

Kassianidou, V. 2006. The production, use and trade of metals in Cyprus and Sardinia: so similar and so different. *Instrumentum* 23: 34–7.

Knapp, A.B. 1986. Production, exchange and socio-political complexity on Bronze Age Cyprus. *Oxford Journal of Archaeology* 5: 35–60.

Knapp, A.B. 1993. Thalassocracies in Bronze Age eastern Mediterranean trade. *World Archaeology* 24(3): 332–47.

Knapp, A.B. 1998. Mediterranean Bronze Age trade: distance, power, and place. In Cline and Harris-Cline 1998, 193–207.

Kraay, C.M. 1976. *Archaic and Classical Greek Coins*. London: Methuen.

Liritzis, Y. 1984. Reappraisal of Minoan kilns by thermoluminescence and neutron activation/XRF analysis. *Revue d'Archéometrie* 8: 7–20.

Marinos, G. and Petrascheck, W. 1956. *Laurium*. Institute for Geology and Subsurface Research IV. Athens: Institute for Geology and Subsurface Research.

Matthäus, H. 1998. Discussion. In Karageorghis and Stampolidis 1998, 45–6.

McGeehan-Liritzis, V. 1996. *The Role and Development of Metallurgy in the Late Neolithic and Early Bronze Age of Greece.* SIMA Pocket Book 122. Jonsered: P. Åströms Förlag.

Mosso, A. 1910. *The Dawn of Mediterranean Civilization.* London: Fisher Unwin.

Muhly, J.D. 1983. Lead isotope analysis and the kingdom of Alashia. *Report of the Department of Antiquities Cyprus*, 210–18.

Muhly, J.D. 1997. Metals and metallurgy: using modern technology to study ancient technology. In *Archaia Elliniki Technologia: Praktika tou Diethnes Synedrio*. Thessaloniki, Etaireia Meletis Archaias Ellinikis Technologias. Thessaloniki: Techniko Mouseio Thessalonikis, 27–33.

Muhly, J.D. 2007. Hagia Photia and the Cycladic element in Early Minoan metallurgy. In Tzachili 2007.

Niemeier, W-D. 1998. The Minoans in the south-eastern Aegean and in Cyprus. In Karageorghis and Stampolidis 1998, 29–49.

Pelon, O. 1987. Minoan palaces and workshops: new data from Malia. In Hägg and Marinatos 1987, 269–72.

Pernicka, E. and Wagner, G.A. 1982. Lead, silver and gold in ancient Greece. *PACT, Journal of the European Study Group on Physical, Chemical and Mathematical Techniques Applied to Archaeology* 7: 419–25.

Philaniotou, O. 2007. Early copper production in the Aegean: the case of Seriphos. In Tzachili 2007.

Poursat, J.-C., 1985. Ateliers et artisans minoens. In *Pepragmena tou E' Diethnous Kritologikou Synedriou (Agios Nikolaos, 25 Septemvriou – 1 Oktovriou 1981), Vol. A'* Irakleio, Kritis: Etairia Kritikon Istorikon Meleton, 297–300.

Sayre, E.V., Joel, E.C., Blackman, M.J., Yener, K.A. and Ozbal, H. 2001. Stable lead isotope studies of Black Sea Anatolian ore sources and related Bronze Age and Phrygian artefacts from nearby archaeological sites. *Archaeometry* 43(1): 77–115.

Sherratt, A. and Sherratt, S. 1991. From luxuries to commodities: the nature of Mediterranean Bronze Age trading systems. In *Bronze Age Trade in the Mediterranean*, N.H. Gale (ed.). SIMA 90. Jonsered: Paul Åströms Förlag, 351–86.

Sherratt, A. and Sherratt, S. 1998. Small worlds: interaction and identity in the ancient Mediterranean. In Cline and Harris-Cline 1998, 329–43.

Starr C.G. 1955. The myth of the Minoan thalassocracy. *Historia* 3: 282–91.

Stos, Z.A. and Gale, N.H. 2006. Lead isotope and chemical analyses of slags from Chrysokamino. In *The Chrysokamino Metallurgy Workshop and its Territory, Hesperia Supplement 36*, P.P. Betancourt (ed.). Princeton, NJ: American School of Classical Studies at Athens, 299–319.

Stos-Gale, Z.A. 1991. Neutron activation analysis of copper ores, copper-based metal and slags. In *Neutron Activation and Plasma Emission Spectrometry Analyses in Archaeology*, M.J. Hughes, M.R. Cowell and D.R. Hook (eds). British Museum Occasional Paper 82. London, 227–48.

Stos-Gale, Z.A. 1998. The role of Kythnos and other Cycladic islands in the origins of Early Minoan metallurgy. In *Kea-Kythnos: History and Archaeology*, L.G. Mendoni and A. Mazarakis Ainian (eds). *Meletimata* 27. Athens: National Hellenic Research Foundation, 717–36.

Stos-Gale, Z.A. 2001. Minoan foreign relations and copper metallurgy in MMIII–LMIII Crete. In *The Social Context of Technological Change: Egypt and the Near East 1650–1550 BC*, A.J. Shortland (ed.). Oxford: Oxbow Press, 195–210.

Stos-Gale, Z.A. and Gale, N.H. 1982. The sources of Mycenaean silver and lead. *Journal of Field Archaeology* 9: 467–85.

Theocharis, D.P. 1951. Excavations at Raphina. *Praktika tis en Athenais Archaiologikis Etaireias*: 77–92.

Theocharis, D.P. 1952. Excavations at Raphina. *Praktika tis en Athenais Archaiologikis Etaireias*: 130–33.

Tzachili, I. (ed.) 2007. *Aegean Metallurgy in the Bronze Age*. Heraklion: University of Crete Press.

Wagner, G. A., Wagner, I., Öztunalı, Ö., Schmitt-Strecker, S. and Begemann, F. 2003. *Archäometallurgischer Bericht über Feld-forschung in Anatolien und Blei-isotopische Studien an Erzen und Schlacken. Der Anschnitt* Beiheft 16. Bochum: Deutsches Bergbau-Museum, 475–94.

Wheeler, T.S., Maddin, R. and Muhly, J.D. 1975. Ingots and the Bronze Age copper trade in the Mediterranean. *Expedition* 17(4): 31–9.

Wiener, M. 1987. Trade and rule in palatial Crete. In Hägg and Marinatos 1987, 261–6.

Wiener, M. 1990. The Isles of Crete? The Minoan thalassocracy revisited. In *Thera and the Aegean World* III.1, D.A. Hardy, C.G. Doumas, J.A. Sakellarakis and P.M. Warren (eds). London: Thera Foundation, 128–61.

Yalçin, Ü., Pulak, C. and Slotta, R. 2005. *Das Schiff von Uluburun*. Bochum: Deutsches Bergbau-Museum.

## Authors' addresses

- Professor Noel H Gale, University of Oxford, Isotrace Laboratory, Nuffield College, New Road, Oxford OX1 1NF, UK (skouries3@gmail.com)
- Dr Zofia Anna Stos-Gale, University of Surrey, UniSdirect, Nodus Building, Guildford GU2 7XH, UK (S.Stos@surrey.ac.uk)

# One hundred years on: what do we know about tin and bronze production in southern Africa?

*Shadreck Chirikure, Simon Hall and Duncan Miller*

*ABSTRACT*   Pioneer research on pre-European tin and bronze production in southern Africa was first published in the early 20th century, followed by a gradual but unsystematic accumulation of data over the years. Yet, almost a century on, there still exists a host of basic questions about the chronology, smelting technology and consumption of tin and bronze. Significantly, the origins of indigenous tin production are still uncertain but it is becoming increasingly clear that from the early 2nd millennium AD, tin-bronze progressively assumed a position of importance at elite sites such as Mapungubwe, Great Zimbabwe and Khami. Bronze metal was incorporated into the local socio-economic and political systems and tin has a complex history of local and international trade. This paper gives a synopsis of the available information regarding the archaeology and technology of tin and bronze production in southern Africa and suggests profitable avenues for future research.

*Keywords:* southern Africa, tin, bronze, archaeology, Rooiberg, smelting, mining.

## Introduction

Southern Africa (Fig. 1) during the Iron Age[1] has been viewed as a cultural backwater that was isolated from the rest of the world. Consequently, it was wrongly believed that the region lacked technical and cultural innovations (M. Hall 1984, 1987). Archaeological research at numerous southern African Iron Age sites continues to show that these societies were indeed innovative. One such sphere of innovation was the emergence of tin metallurgy and the appearance of bronze in the metallurgical record in the 12th century AD almost a millennium after iron and copper were established. Several questions emerge from this, but perhaps the most important is whether this innovation was an entirely local development or whether indigenous knowledge was applied to satisfy local demand, created by imports of bronze. What makes this a critical question is that while bronze first appears in the 12th century, the earliest evidence for local tin production seems to lag behind, and dates to the early 16th century (453 ± 45 BP – GrN 5138 cal AD 1426–1633, 2 sigma) (S.L. Hall 1985; Mason 1986). Much more data are needed to outline this sequence securely. Equally, innovations do not arise in isolation from social or historical contexts. Understanding southern African tin and bronze metallurgy therefore requires a combination of technical, archaeological and historical research.

Irrespective of the origin of bronze metallurgy, it is clear that by the middle of the 2nd millennium cal AD, southern African communities had the competence to prospect for and produce tin, to trade it and to make bronze; they also incorporated bronze and possibly tin into local social and political systems (S.L. Hall 1981; Killick 1991; Mason 1986; Miller 2002; Prendergast 1979; Recknagel 1908). The appearance of bronze after iron in the archaeological record of

**Figure 1** Map of southern Africa showing the location of Rooiberg and the major centres of political and economic power that acted as recipients of the tin.

southern Africa indicates the importance of its visual appeal (Miller 2003). With iron, indigenous metallurgists in southern Africa already possessed a metal with obvious superior physical properties when compared to bronze. The ease with which bronze could be reused and cast into different objects, together with the cultural beliefs associated with the metal, promoted its use in making ceremonial and decorative objects (Herbert 1984).

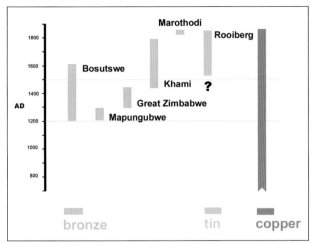

**Figure 2** Timeline showing the chronology of copper, tin and bronze in southern Africa during the Iron Age.

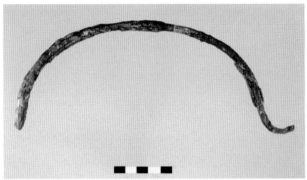

**Figure 3** Bronze 'bucket' handle with octagonal cross-section from Mapungubwe southern terrace with 6 wt% tin (Excavation Unit JS2a). Its function and area of origin are unknown, but it is possible that it was imported from Asia via the Swahili towns on the east African coast. Scale in cm.

**Figure 4** Map showing the Rooiberg area and some of the major archaeological sites in the region.

The chronology of tin and bronze provides the framework around which working hypotheses about the nature of this innovation can be developed. The earliest evidence of bronze is from the 12th and 13th century AD sites and polities with trade links with the southeast African coast and the Indian Ocean (Fig. 2). At Mapungubwe, the 13th-century capital of the Mapungubwe state, a bronze 'bucket' handle (Fig. 3) which may be Indian or Asian in origin, was found (Miller 2001: 97). It is possible that the first bronze artefacts were recycled from imported bronze; recent metallographic analyses of copper objects from other southern African archaeological sites show that many artefacts assumed to be made from copper are actually tin-bronzes. The occurrence of tin-bronze artefacts rose significantly within the 14th- and 15th-century Zimbabwe culture levels at Bosutswe, and at Great Zimbabwe itself (Fig. 2). This continued into the Khami period after AD 1500 and persisted into the early 19th century, as recorded at Tswana towns such as Marothodi (Fig. 4) (Friede 1975; Friede and Steel 1976; Grant *et al.* 1994; Hall *et al.* 2006; Miller 2002, 2003). The growing profile of tin-bronze well into the southern African Iron Age, demands a rethink of the view that most tin was produced for long-distance trade (Miller 2002, 2003).

Given the skeletal nature of the tin and bronze sequence in southern Africa, questions arise around the chronology of indigenous tin production and the status of the Rooiberg tin mines as either the sole supplier or one of several tin sources in southern Africa. Rooiberg is located in the southern Waterberg, Limpopo province (northern South Africa) and still remains the only known and unequivocal pre-colonial Iron Age tin mine (Fig. 4). Other potential sources such as the tin-bearing areas of the Bushveld Igneous Complex in the Waterberg at Zaaiplaats remain unproven because there is no archaeological evidence for pre-colonial exploitation (Killick 1991; Trevor 1912, 1930). Claims of pre-colonial tin mining near Rusape and Munyati in Zimbabwe are largely unconfirmed (Prendergast 1979). Furthermore, the tin-bearing pegmatites of Swaziland and Kamativi in Zimbabwe, which are the focus of modern mining, do not seem to have been exploited pre-colonially (Friede and Steel 1976). Consequently, most of the attention on pre-European tin production has focused upon the Rooiberg area (Baumann 1919a; Friede and Steel 1976; Grant 1994; Hall 1981; Mason 1986; White and Oxley-Oxland 1974). Although calculations of the amount of tin extracted are difficult, the Rooiberg production was not inconsiderable, and it has been suggested that close to 2000 tons of tin metal were extracted before the onset of modern mining (Baumann 1919b) (Fig. 5).

In contrast to the high visibility of pre-colonial tin mining at Rooiberg, the issue of tin and bronze production in southern Africa has largely remained a curious footnote and somewhat speculative. Major tin-smelting sites such as the Blaauwbank Donga and Smelterskop are a short distance from the core mine area. They have been continuously picked-over and excavated in an unsystematic manner, mainly to collect material for technological description, with little attention given to archaeological, historical and cultural context. Even in the early 20th century, Baumann (1919a: 127) lamented that 'the value of Smelterskop as a record of the past was not recognised and no attempt was made to properly investigate the smelting site'. Additionally, little is known about the extractive metal-

113

lurgy of tin in southern Africa despite encounters with the material for almost a century. This paper reviews the available data on tin production and bronze in southern Africa in order to highlight the gap in our knowledge and suggest potential avenues for future research.

## Was Rooiberg the only centre of pre-colonial tin production in southern Africa?

Tin ores are uncommon in southern Africa so it is not surprising that there are few reports of pre-colonial tin workings (Crocker *et al.* 1976; Grant 1994; Killick 1991; Trevor 1912). The Rooiberg area, situated approximately 60 km west of Bela Bela (formerly Warmbaths) in northern South Africa (Fig. 4), is assumed to be the source of most pre-European tin. Early geologists and miners who worked in the area at the start of modern mining operations were quick to recognise the pre-colonial workings and took a keen interest in them (Baumann 1919a; Kynaston and Mellor 1909; Recknagel 1908; Trevor 1912, 1913; Trevor *et al.* 1919; Wagner 1926). They described the geological nature of the tin mineralisation which is associated with the emplacement of the granites of the Bushveld Igneous Complex. Tin exists in the form of cassiterite associated with minerals such as tourmaline, quartzite, orthoclase, nickel sulphide, tungsten minerals and chalcopyrite (Cairncross and Dixon 1995; Crocker *et al.* 1976; Schwellnus 1935; Wagner 1924).

The Rooiberg cluster of sites, comprising mines, smelting areas and domestic homesteads, is extensive. There is a central mining core of about 70 acres that includes an extensive drainage basin called the Blaauwbank Donga, and a prominent hill called Smelterskop (Baumann 1919a; Miller and Hall in press), both of which still preserve plentiful evidence of tin smelting. The pioneer researchers at Rooiberg reported collapsed mine tunnels (Fig. 5), crumbled furnaces, vitrified tuyere tips, slag, and the remains of ore and 'tin feed/smelter concentrate' at Smelterskop (Fig. 6) and the Blaauwbank Donga (Baumann 1919a, 1919b; Trevor 1912). The collapsed shafts and stopes – and a seemingly haphazard maze of mined cassiterite pockets and tunnels – are testimony to the deliberate extraction of hard rock cassiterite in pre-European times (Baumann 1919b; Recknagel 1908). Early miners and geologists mistakenly thought that these were iron mines. The evidence indicates, however, that hard rock cassiterite was initially crushed in the immediate vicinity of the mines. Baumann (1919a) illustrated that grinding stones used for pounding ores found in one mine tunnel and similar stones were still abundant on Smelterskop. Tin feed and collapsed furnaces were also recorded in the Blaauwbank Donga (Trevor 1930; Wagner and Gordon 1929).

Early analysts recognised that both copper- and iron-smelting debris came from the same contexts as tin remains at Rooiberg. Without adequate dating, however, it is difficult to evaluate if all the metals were worked simultaneously. As suggested by Wagner (1926), it is possible that some of the iron- and copper-working debris was the product of tin smelting. Early discoveries of poorly provenanced tin ingots and tin and arsenical bronzes from the Blaauwbank Donga contrib-

**Figure 5** Colonial incline shaft at Rooiberg that broke through into pre-colonial tin mines.

**Figure 6** Photograph of Smelterskop. Most of the hilltop is covered with tin-smelting debris, currently believed to be in excess of 3 tons.

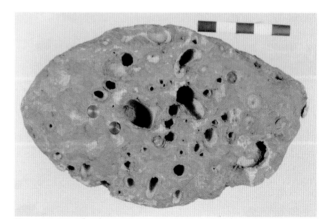

**Figure 7** A tin 'bun' ingot, no. 21/39/3, recovered from Rooiberg and currently stored in the University of Witwatersrand collection. It weighs 2100 g and contains 99.4 wt% tin with copper being a significant trace element at 0.008 ppm (Friede and Steel 1976: 464). Scale in cm.

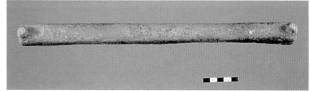

**Figure 8** A tin bar ingot recovered from Buffelshoek, about 30 km northwest of Rooiberg. It is possible that the ingot was produced in this area but using Rooiberg tin, adducing that both metal and ore were traded over fairly large distances. Scale in cm.

uted to heated debates over the local history of bronze metallurgy at Rooiberg (Trevor 1930; Wagner and Gordon 1929). One issue was whether there was intentional production of bronze, given that the amount (between 15 and 20 wt%) of arsenic in one 'ingot' rendered the metal unworkable by the available technology and could have been merely accidental. Another question is whether metallic tin production preceded bronze production. Early researchers did not have an archaeological context for these questions. Given the relative lack of tin artefacts from southern African Iron Age sites, it seems unlikely that the Rooiberg tin was exploited if the intention had not been to make bronze. This, however, is still an issue to be empirically addressed.

A substantial amount of work carried out in the first three decades of the 20th century focused on a collection of tin and 'bronze' ingots, and slags (Figs 7–10). Analysis focused predominantly on technical issues with little attention given to the

**Figure 9** Two tin rod ingots 14/47/1 and 14/47/2 collected near Polokwane (Pietersburg) and housed in the University of Witwatersrand collection. Each weighs about 500 g and contains 99 wt% tin with impurities in the ppm levels. Scale in cm.

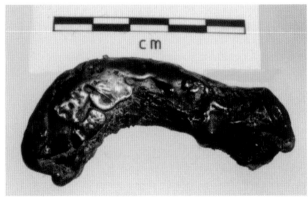

**Figure 10** An arsenical bronze 'ingot' no. 6 from the Blaauwbank Donga. Compositional analyses by Stanley (1929) showed that the ingot contained 81.2 wt% copper, 14.2 wt% arsenic, 3.26 wt% iron and 1.55 wt% tin.

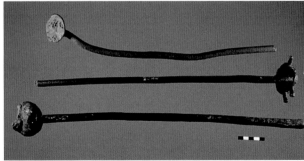

**Figure 11** Special type of ingots known as *lerale* consisting of a metal rod terminating in a club-like head. Weighing about 500 g and 450 mm long, the top ingot was made of tin while the others were made from copper. Scale in cm.

archaeological context. Consequently no effort was invested in recording the furnace types used for the smelting of different metals, some of which seem to have still been in a reasonable state of preservation at the start of the 20th century.

Chemical and metallurgical analyses of tin slags, tin ingots, tin-bronzes and arsenical bronze artefacts assumed to have their origins at Rooiberg continued after this initial phase of enquiry. The results of earlier analyses were checked, new and larger samples were analysed and more widely contextualised within a growing corpus of archaeological data. Slag analyses, for example, show that the tin content of most Rooiberg slags vary extensively – from nearly 50% tin oxide ($SnO_2$) to values ranging downwards to less than 1% (Miller and Hall in press; Wagner and Gordon 1929). The higher percentage of tin oxide shows that a significant amount of tin was trapped in the slag, suggesting that the tin-smelting process was inefficient at times (Friede and Steel 1976; Stanley 1929a, 1929b; Wagner and Gordon 1929). Consequently, tin-rich slags may have been recycled as smelter feed, although the high iron content noted in the slags may have made this impractical (Grant 1994; Killick 1991).

Turning to the metal, Friede and Steel (1976) reassessed indigenous tin and bronze production, analysing the chemical composition of bronzes and copper objects from the former Transvaal, speculating that all the tin was probably produced at Rooiberg. While some of the tin ingots found at Rooiberg were plano-convex in shape (for example specimens Rooiberg 21/39/1 and 21/39/3) (Fig. 7), there were other ingot shapes and forms which are also known from other parts of southern Africa (Killick 1991; Thompson 1949, 1954) such as the Messina *lerale* ingot (Fig. 11) or the bar ingot from Great Zimbabwe. The recurring problem of poor provenance afflicts many of these finds and the differences in ingot types may also be complicated by variability in ingot styles through time, both at the source of the tin and in the receiving communities. It may have been, however, that plano-convex ingots were a 'neutral' trade form that receivers reworked according to cultural preference.

With growing sophistication in analytical techniques, sourcing of tin ingots was pursued more rigorously. Grant (1990a), for example, carried out detailed archaeometallurgical analyses which involved the description, analysis and interpretation of all available tin and bronze ingots. Grant (1999) attempted to source tin and bronze artefacts from southern Africa in order to elucidate the history of their trade. He conducted trace element studies using neutron activation analysis (NAA) and suggested that tin ingots recovered from the Rooiberg area form a tight grouping on both lanthanum/europium (rare earths) and tantalum/scandium ratios, and are similar to tin ingots from Zimbabwe and other parts of northern South Africa. Grant concluded that most tin artefacts studied so far had been produced by smelting Rooiberg cassiterite. This is corroborated by other analyses of elemental composition, and Killick (1991) argues that the Messina tin *lerale* (Fig. 11) matched the Rooiberg source more closely than Zaaiplaats tin because the Zaaiplaats cassiterite is known to be associated with tungsten. Killick's 1991 analyses of the Messina *lerale* also showed that relatively high levels of so-called 'hardhead' (various tin-iron intermetallics) were tolerated in primary production and primary tin may have been refined through drossing or liquation to reduce iron content when required.

The distribution of tin ingots fans out from Rooiberg towards the north and northeast and, significantly, none are known to the south of Rooiberg (Fig. 2). This distribution is towards centres of significant political power that grew in the Shashe/Limpopo area and Zimbabwe from the 13th century. Grant *et al.* (1994) suggest that the spur for developing a tin production industry at Rooiberg probably came from the Indian Ocean trade with Zimbabwe culture capitals that increasingly became both consumers of tin and trade conduits for the metal. The association of tin with expansive class-based political systems and international trade networks directed at the east and southeast African coast may need to be tempered, however. Recent analyses by Miller (2002, 2003) show that there is a high incidence of tin-bronze artefacts from a number of southern African archaeological sites where no tin ingots were recovered and these are located to the north and south of Rooiberg. In the 2nd-millennium AD Iron Age, the demand for and use of tin in southern Africa did not come only from state-level systems. Preliminary analyses show that tin-bronze was more widely circulated among Iron Age communities, including Tswana towns in the early 19th century. It is clear that much of the ornamental metal from southern African Late Iron Age settlements that has been classified as copper needs more detailed analysis. According to Recknagel (1908), cassiterite was found in direct association with copper ore at Rooiberg and thus it is possible that tin-bronze was produced by smelting copper ore contaminated with cassiterite. Other than traded tin ingots, there are few if any tin artefacts from the southern African Iron Age and it appears that tin by itself had little utility for Iron Age farmers.

The existing chronological evidence for tin production at Rooiberg does not convince us that the process started much before the 15th or early 16th century (Fig. 12). This evidence is based on direct dating and the general chronology of the known local Iron Age sequence. Grant (1990b) dated charcoal fragments embedded in an arsenical copper 'ingot', and obtained a date of 860 ± 50 BP cal 1025 to 1260 (ETH 7770) but, as this 'ingot' may have been a metallurgical accident, his suggestion that arsenical copper production may have preceded tin production does not mean much. A Rooiberg plano-convex tin ingot was dated from charcoal inclusions to between cal AD 1436 and 1648 (ETH 5127) (Grant 1990b). Furthermore, the available chronology for hard rock mining indicates that this was underway by cal AD 1500. Radiocarbon dates and ceramic typologies for Smelterskop place the tin production debris there in the 18th century (1760 ± 30 BP, Pta 2850) (S.L. Hall 1981; Mason 1986), although the resolution of radiocarbon in this period is much reduced. Cassiterite and tin-smelting debris at Rooikrans to the south of Rooiberg have been dated to the mid-17th century. The similarity between Rooikrans pottery and that recovered from the settlement of Buffelshoek to the north suggests that the two sites may date to the same period.

The cultural affinities of most archaeological settlements associated with the Rooiberg mines, the Blaauwbank Donga and on and around Smelterskop, belong within the middle and later phases of the Moloko. This is the archaeological label for ancestral Sotho/Tswana-speakers, who dominated

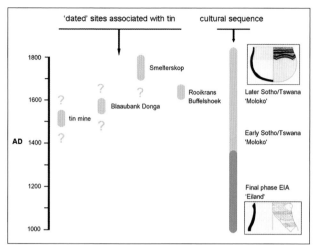

**Figure 12** Timeline showing the chronology of tin production at Rooiberg and from other sites in the Rooiberg region.

this landscape increasingly from AD 1500. A terminal phase of the Early Iron Age, referred to as Eiland (11th–14th century), preceded the Moloko and these people also occupied the Rooiberg and much of the former Transvaal (Hall 1981). Eiland pottery has been recovered from Mapungubwe (1220–90), providing evidence of interaction between these contemporary but geographically separate Iron Age communities. Whether this interaction involved the supply of tin by Eiland communities from Rooiberg is doubtful. There is nothing to indicate that Eiland people were involved in any way with the Rooiberg tin but this possibility still needs to be tested by fieldwork on Eiland sites in the Rooiberg region. At this stage, however, we agree with Grant *et al.* (1994) that the evidence as it currently exists only links the development of tin extraction from Rooiberg with Sotho/Tswana-speakers from the 15th and 16th century AD. The question of where the tin in the bronze, or the bronze itself, came from prior to this date still remains open. Was it locally produced and the tin sources have not yet been identified?

These developmental scenarios are complicated by more recent analyses of tin slags and tuyeres from Smelterskop (Miller and Hall in press). Electron microprobe results show a high level of titanium ($TiO_2$) and zirconium ($ZrO_2$) in the slags which match slags analysed by Wagner and Gordon (1929) and others earlier in the 20th century. The slags from Smelterskop ($ZrO_2$ almost 8 wt%) revealed a robust and affirmative correlation between zirconium and titanium, which implies a granitic detrital source for part of the furnace charge. The discovery of alluvial cassiterite in the smelter concentrate/feed by Wagner and Gordon (1929) amply supports this observation. Alternatively, granitic sand containing ilmenite, rutile and zircon may have been added to the charge as a flux because of its elevated silica content. The analytical evidence could be used to support the hypothesis that both hard rock underground cassiterite and alluvial deposits were exploited at Rooiberg.

Other Moloko settlements at some distance from the Rooiberg mines have also produced evidence of tin smelting (S.L. Hall 1981; Miller and Hall in press). Although in excess of 3 tons of tin smelting debris have been observed at Smelterskop alone, it is possible that significant amounts of ore were smelted elsewhere. This regional perspective requires more attention and may potentially cast some light on how the Rooiberg tin deposits were politically managed. Were the tin mines open to all to 'dip into' or were the mines under some form of centralised control? Furthermore, miners of tin may not always have been smelters and smiths. It is possible that issues of access and ownership varied through time, again depending on regional political processes and the intensity of demand for tin. Managing access did not presumably reside solely within the realm of political economy but also within a cultural framework. Continuity in mining or smelting, for example, may have been vested within a particular lineage whose ancestors as previous miners and smiths were critical to the continued success of production (see Maggs 1991: 71). Outsiders may not have ventured into ancestral realms that they had no prior connection to or empathy with. These are difficult questions for archaeology to address but the presence of tin-smelting debris at some distance from the mines hints at a more complex situation than simply a core area of immediate production at Rooiberg.

There are other hints of regional networks and connections. Analysis of mid-17th-century pottery from Buffelshoek (Fig. 4), some 30 km north of Rooiberg, shows that tin was a prominent trace element in some of the pots. Control analyses of pottery from two Early Iron Age phases at the base of the hill (at Buffelshoek) and pottery from a late 19th-century settlement on the same site show that these ceramics were made from local clays that contained no tin whatsoever (Hall and Grant 1995). It seems most likely that the mid-17th-century pots were made within the Rooiberg catchment. Whatever the reason as to why and how these pots moved from their point of origin to Buffelshoek, such evidence indicates regional networks that implicate the Rooiberg area. Recent elemental analyses of tin-bronze jewellery (Fig. 13) from Marothodi, some 100 km south of Rooiberg, have shown a potential and continued connection to the Rooiberg tin fields within late Moloko Tswana towns dating within the first quarter of the 19th century (Hall *et al.* 2006).

## Conclusions

Curiosity about tin in the southern African Iron Age has been rejuvenated by the routine analysis of southern African 2nd-millennium 'copper' artefacts which shows that a significant amount of Iron Age ornamental metal is in fact made from tin-bronze. That Rooiberg is the only known Iron Age tin mine in southern Africa, and possibly even further afield, calls for detailed chronological and technological studies at this mining complex in order to understand the development of tin production in southern Africa. As emphasised above, analyses that seek to elucidate the technology of tin production must be of well-dated archaeological material.

A growing body of data suggests that the use of tin and bronze in southern Africa dates from the 12th and 13th centuries at sites such as Mapungubwe but whether this early bronze was locally produced or imported requires an intensive dating programme around the tin mining and production sequence at Rooiberg itself. Is the AD 1500 date for the advent of tin production at Rooiberg correct and if so where was the

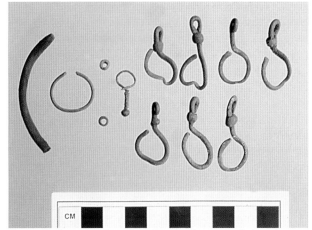

**Figure 13** Examples of metal earrings, beads, rings and a bracelet from Marothodi initially assumed to be made from copper, but analysis shows that a significant number are actually tin-bronze with an average of 6 wt% tin.

earlier bronze coming from? The discovery that alluvial tin was also worked at Rooiberg raises questions about the developmental sequence of tin extraction. Did alluvial extraction come first, followed by hard rock mining, or indeed, a mix of the two? Alternatively did ancestral Sotho/Tswana miners initially extract hard rock tin ores, with alluvial tin exploitation following after the hard rock deposits were exhausted or could not be exploited further with the available technology?

There is also an assumption that once the Rooiberg tin mines started to produce tin, exploitation was continuous and uninterrupted. The ebb and flow of southern African political power during the 2nd millennium AD should caution against this view. Related to this we must also ask whether technology was constant over time or whether various groups possessed different technological solutions to tin production.

The other important question, directly linked to tin technology, is the chronology of copper and iron smelting at Rooiberg. That iron and copper production in the southern African Iron Age considerably predates tin and bronze production is beyond doubt, but were these metals also worked simultaneously with tin at Rooiberg? Revisiting the published compositional data on Rooiberg tin slags, it is clear that most of the slags have elevated amounts of iron and tin. It is unlikely that both tin and iron were produced in the same furnace as suggested by Mason (1986). More likely, iron-rich ores were used as fluxes, leading to the production of these iron-rich slags. Cultural data are lacking regarding the working of tin and whether its spatial organisation was similar to that of copper or iron metals, which on cultural grounds were conceptually distinct. For example, was tin smelted outside villages in keeping with the procreative model in African smelting that required ritual seclusion for critical transformations?

In conclusion, despite close to a century of curiosity about the Rooiberg metallurgy and archaeology, fundamental questions remain unanswered, and in light of recent developments, several new ones have been raised. In the 1970s White and Oxley-Oxland (1974: 269–70), declared that 'very little is really known about ancient smelting techniques at Rooiberg ... and a great deal of archaeological work is necessary in the Rooiberg area to solve the puzzle of the ancient tin miners'. While the situation has improved since this statement, research is needed on both tin technology and its development, as well as on the cultural sequence within which tin production took place. The pre-colonial exploitation of Rooiberg tin by African Iron Age farmers still remains somewhat of a 'puzzle', but a better understanding of local and regional sequences means that we can further develop and test ideas about the contextual and regional demand for tin and its relationship to the rise of a wider bronze industry within the southern African Iron Age.

## Note

1. The Iron Age of Southern Africa began in the early 1st millennium AD giving way to the historical period from the 17th century onwards.

## References

Baumann, M. 1919a. Ancient tin mines of the Transvaal. *Journal of the Chemical, Metallurgical and Mining Society of South Africa* 19: 120–32.

Baumann, M. 1919b. Ancient tin mines of the Transvaal: reply to discussion. *Journal of the Chemical, Metallurgical and Mining Society of South Africa* 19: 32–4.

Cairncross, B. and Dixon, R. 1995. *Minerals of South Africa*. Johannesburg: Geological Society of South Africa.

Crocker, I.T. Coetzee, G.L. and Mehliss, M. 1976. Tin. In *Mineral Resources of the Republic of South Africa*, C.B. Coetzee (ed.), 5th edn. Pretoria: Council for Geosciences, 613–20.

Friede, H.M. 1975. Notes on the composition of pre-European copper and copper-alloy artefacts. *Journal of the South African Institute of Mining and Metallurgy* 75: 189.

Friede, H.M and Steel, R.H. 1976. Tin mining and smelting in the Transvaal during the Iron Age. *Journal of the South African Institute of Mining and Metallurgy* 76: 461–70.

Grant, M.R. 1990a. Trace elements in southern African artefact tins using NAA. In *Archaeometry '90*, E. Pernika and G. A. Wagner (eds). Basel: Birkhäuser Verlag, 165–72.

Grant, M.R. 1990b. A radiocarbon date on a tin artefact from Rooiberg. *South African Journal of Science* 86: 63.

Grant, M.R. 1994. Iron in ancient tin from Rooiberg, South Africa. *Journal of Archaeological Science* 21: 455–60.

Grant, M. R.1999. The sourcing of southern African tin artefacts. *Journal of Archaeological Science* 26: 1111–17.

Grant, M.R., Huffman, T.N. and Watterson, J.I.W. 1994. The role of copper smelting in the pre-colonial exploitation of the Rooiberg tin field. *South African Journal of Science* 90: 85–90.

Hall, M. 1984. Pots and politics: ceramic interpretations in southern Africa. *World Archaeology* 15: 262–73.

Hall, M. 1987. *The Changing Past: Farmers, Kings and Traders in Southern Africa, 200–1860*. Cape Town: David Philip.

Hall, S.L. 1981. *Iron Age Sequence and Settlement in the Rooiberg, Thabazimbi Area*. MA dissertation, University of the Witwatersrand.

Hall, S.L. 1985. Excavations at Rooikrans and Rhenosterkloof, Late Iron Age sites in the Rooiberg area of the Transvaal. *Annals of the Cape Provincial Museums (Human Sciences)* 1(5): 131–210.

Hall, S.L. and Grant, M. 1995. Indigenous ceramic production in the context of the colonial frontier in the Transvaal, South Africa. In *Proceedings of the 8th CIMTEC: The Ceramics Cultural Heritage*, P. Vincenzini (ed.). Faenza: Techna srl, 465–73.

Hall, S.L., Miller, D., Anderson, A. and Boeyens, J. 2006. An exploratory study of copper and iron production at Marothodi, an early 19th century Tswana town, Rustenburg district, South Africa. *Journal of African Archaeology* 4(1): 3–35.

Herbert, E. 1984. *The Red Gold of Africa: Copper in Pre-colonial History and Culture*. New York: University of Wisconsin Press.

Killick, D. 1991. A tin *lerale* from the Soutpansberg, northern Transvaal, South Africa. *South African Archaeological Bulletin* 46: 137–41.

Kynaston, H. and Mellor, E.T. 1909. *The Geology of the Waterberg Tin Fields*. South African Geological Survey Memoir 4, 90–98.

Maggs, T. O'C. 1991. My father's hammer never ceases its song day and night: the Zulu ferrous industry. *Natal Museum Journal of Humanities* 4: 65–87.

Mason, R.J. 1986. *Origins of Black People of Johannesburg and the Southern Western Central Transvaal AD 350–1880*. Occasional Paper 16, University of the Witwatersrand Archaeological Research Unit.

Miller, D. 2001. Metal assemblages from Greefswald areas, K2, Mapungubwe Hill and Mapungubwe southern terrace. *South African Archaeological Bulletin* 56: 83–103.

Miller, D. 2002. Smelter and smith: metal fabrication technology in the southern African Early and Late Iron Age. *Journal of Archaeological Science* 29: 1083–131.

Miller, D. 2003. Archaeological bronze processing in Botswana. *Proceedings of the Microscopy Society of Southern Africa* 33: 18.

Miller, D. and Hall, S.L. in press. Rooiberg revisited: the chronology and technology of tin and bronze production.

Prendergast, M.D. 1979. Cornucopia: phase 1 Zimbabwe stone buildings associated with an Iron Age tin mine? *Rhodesian Prehistory* 17: 11–16.

Recknagel, R. 1908. On some mineral deposits in the Rooiberg district. *Transactions of the Geological Society of South Africa* 11: 83–106.

Schwellnus, C.M. 1935. *The Nickel-Copper Occurrence in the Bushveld Igneous Complex West of the Pilandsbergen*. Geological Series Bulletin No. 5. Pretoria: Government Printer.

Stanley, G.H. 1929a. The composition of some prehistoric South African bronzes with notes on the methods of analyses. *South African Journal of Science* 26: 44–9.

Stanley, G.H. 1929b. Primitive metallurgy in South Africa: some products and their significance. *South African Journal of Science* 26: 732–48.

Thompson, L.C. 1949. Ingots of native manufacture. *Native Affairs Department Annual* 26: 14–16.

Thompson, L.C. 1954. A native-made tin ingot. *Native Affairs Department Annual* 31: 40–41.

Trevor, T.G. 1912. Some observations on ancient mine working in the Transvaal. *Journal of the Chemical, Metallurgical and Mining Society of South Africa* 12: 267–75.

Trevor, T.G. 1913. Some observations on ancient mine working in the Transvaal: a reply to the discussion. *Journal of the Chemical, Metallurgical and Mining Society of South Africa* 13: 148–9.

Trevor, T.G. 1930. Some observations on the relics of pre-European culture in Rhodesia and South Africa. *Journal of the Royal Anthropological Institute* 60: 389–99.

Trevor, T.G., Gray, C.J., Adam, H.R., Newton, S. and Waits, T.P. 1919. Ancient tin mines of the Transvaal: discussion. *Journal of the Chemical, Metallurgical and Mining Society of South Africa* 19: 282–91.

Wagner, P.A. 1924. *Magmatic Nickel Deposits of the Bushveld Complex in the Rustenburg District, Transvaal*. Geological Survey Memoir No. 21. Pretoria: Government Printer.

Wagner, P.A. 1926. Bronze from an ancient smelter in the Waterberg district, Transvaal. *South African Journal of Science* 23: 899–900.

Wagner, P.A. and Gordon, H.S. 1929. Further notes of ancient bronze smelters in the Waterberg district, Transvaal. *South African Journal of Science* 26: 563–74.

White, H. and Oxley-Oxland, G. St. J. 1974. Ancient metallurgical practices in the Rooiberg area. *Journal of the South African Institute of Mining and Metallurgy* 74: 269–70.

## Authors' addresses

- Shadreck Chirikure, Department of Archaeology, University of Cape Town, Rondebosch 7701, South Africa (shadreck.chirikure@uct.ac.za)
- Simon Hall, Department of Archaeology, University of Cape Town, Rondebosch 7701, South Africa (simon.hall@uct.ac.za)
- Duncan Miller (embo@telkomsa.net)

Brass and zinc

# Of brass and bronze in prehistoric Southwest Asia

*Christopher P. Thornton*

ABSTRACT    This paper presents a review of the numerous copper-zinc alloys (e.g. brass, gunmetal) that have been found in prehistoric contexts from the Aegean to India in the 3rd to the 1st millennium BC. Through a preliminary analysis of the available data, it is argued that there is a noticeable geographical and chronological correlation between early occurrences of copper-zinc alloys, tin-bronze, and rare examples of tin and tin-based metals. This association may have important implications not only for research into the great 'tin question' of Southwest Asia, but also for research into ancient technologies in general. It is here proposed that brass may have been confused with tin-bronze by local consumers ignorant of or ambivalent about the very different mechanical properties of these two alloys, and that the linguistic separation of these two metals in the 1st millennium BC may reflect larger changes in the sociocultural categorisation of materials.

*Keywords:* brass, tin-bronze, gunmetal, Near East, colour, symbolism, ethnocategories.

## Introduction

The alloys of copper and zinc, including brass (copper-zinc), gunmetal (copper-tin-zinc), and variants thereof, have never played a significant role in our understanding of Old World prehistory (Fig. 1). This is due in large part to their purported absence, or at best sporadic existence, in archaeological assemblages before the Greco-Roman period (Bayley 1998). Indeed, the origins of copper-zinc alloys have long been placed in Anatolia during the early 1st millennium BC (see Craddock 1978a; Forbes 1964: 268–9) – an assumption supported by and somewhat predicated on the discovery of early brass artefacts inside the Great Tumulus at the Phrygian capital of Gordion (Steinberg 1981). The fact that this tomb, labelled 'MM' or 'Midas Mound', was supposed to hold the remains of

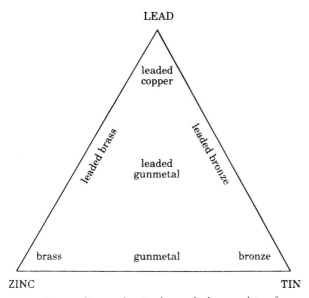

**Figure 1** Ternary diagram showing the standard nomenclature for copper-zinc-tin-lead alloys (from Bayley 1998).

the mythical king led Dorothy Kent Hill (1969: 61) to famously suggest that perhaps the story of Midas's 'golden touch' in fact referred to the earliest production of brass – i.e. turning copper into 'gold'.

This narrative for the development of copper-zinc alloys in Southwest Asia has gradually changed as more examples of these yellow- or 'golden'-coloured metals have been found, both in reviewing the available literature and through further analyses. As recent papers have shown, copper-zinc alloys occur sporadically in this region as early as the 3rd millennium BC, and continue to appear intermittently until the Greco-Roman period (Craddock and Eckstein 2003; Thornton and Ehlers 2003). Having established the existence of copper-zinc alloys in prehistory, we are left with a number of pressing questions including: where are the ancient textual references to the production of brass or gunmetal or to their distribution? Why do we discover only a few copper-zinc alloys in random sites separated by thousands of kilometres despite the innumerable prehistoric artefacts that are analysed every year? Should we care, or can we simply dismiss these sporadic occurrences of copper-zinc alloys as occasional acts of technological serendipity?

In this paper, an up-to-date (i.e. prior to 2006) compilation of prehistoric copper-zinc alloys in Southwest Asia is presented in six general chronologically and spatially defined groupings in an attempt to answer these questions. Although these groupings are based on limited data, they provide us with a new perspective on the where, when and how of early brass and gunmetal exploitation. Using these data, it is argued that there is a notable correlation between the occurrence of these rare metals and the appearance and utilisation of tin-bronze, a metal that (unlike brass) has figured prominently in discussions on the development of urban civilisations in western Asia (e.g. Franklin *et al.* 1978; Muhly 1973; Parzinger and Boroffka 2003; Penhallurick 1986; Pernicka 1998; Stech and Pigott 1986; Weeks 1999). Indeed, it is proposed that by studying ancient copper-zinc alloys, we may gain not only a richer understanding of the production and utilisation of different materials in prehistoric societies, but also discover a new avenue of research into the classic question on the origins of Near Eastern tin and tin-bronze first posed over 75 years ago by V. Gordon Childe (1928). For this reason, it is suggested that copper-zinc alloys are perhaps of greater importance than has long been assumed, and should begin to receive the scholarly attention that they are due.

## Of brass

Before delineating the six general groupings mentioned above, it is important to understand the basic processes through which copper-zinc alloys could have been produced in antiquity. As numerous scholars have argued, the production of copper-zinc alloys is more complicated than the creation of other copper-based alloys (Craddock 1998a; Pollard and Heron 1996: 196–204). This is due to the volatility of zinc above 906 °C, which is below the temperature at which zinc will reduce from its ores. Thus any attempt to smelt zinc ores causes zinc to sublimate (vaporise) (Forbes 1964: 263). Under extreme reducing conditions the gaseous zinc will condense along the inner wall of the furnace as zinc metal – a process known as distillation. If air is present – and most ancient furnaces were not designed to be completely airtight – then the zinc vapour condenses along the inner furnace lining in the form of zinc oxide (calamine). Given that distillation apparatuses capable of producing great quantities of metallic zinc were virtually unknown (or at least unused) before historical periods, the zinc had to be added either as a vapour to solid copper through a process known as cementation, or through a specially controlled mixed smelting of zinc and copper oxides. While the actual mechanism through which zinc enters the copper is essentially the same for the two processes, the outlook and the set of choices made by the metalworker is inherently different.

The crucial ingredient in both processes is zinc oxide, which can be found in its natural state as the ore smithsonite ($ZnCO_3$; aka calamine or zinc spar) or sublimated and oxidised by intentionally roasting zinc-sulphide ores such as sphalerite ($ZnS$; aka zincblende) or zinc-containing fahlerz.[1] Zinc oxide, or even droplets of zinc metal under extreme reducing conditions (cf. 'mock silver' or 'false silver') can also be created accidentally when zinc ores get mixed up with copper or iron ores in a smelt, or with argentiferous lead ores during the production of lead or silver (Craddock and Eckstein 2003: 217–18). In most cases, zinc oxide would be found by the metalworker as a white powder clinging to the inner lining of the furnace walls or in the flue (cf. Wertime 1980: 15). Given the rarity of naturally occurring zinc oxide ores in the Middle East and the ubiquity of mixed zinc-lead sulphide deposits, accidental production in a furnace seems highly probable as the source of ancient zinc oxide.[2] The presence of lead slags bearing significant quantities of zinc at sites such as 3rd-millennium BC Tepe Hissar in northeastern Iran (2.3–14.0 wt% zinc)[3] and 1st-millennium BC Balya in northwestern Anatolia (3.2–30.3 wt% zinc) would seem to confirm this association (Pernicka et al. 1984; Pigott et al. 1982).

After producing or obtaining zinc oxide, the next step is to create a copper-zinc alloy by separating the zinc from the oxide through sublimation while adding it to copper before it re-oxidises. For the purposes of this paper, an 'alloy' is defined as the *intentional* admixture of two or more elements in order to create a distinct material with certain desired physical properties. Thus, while under extreme reducing conditions the accidental smelting of mixed ores or high-zinc copper ores can produce copper metal with as much as 6–7 wt% zinc (see Pollard and Heron 1996), this is not an intentional alloy per se. Certainly this type of low zinc copper-base metal has some desirable physical differences from pure copper that may or may not have been noticed by the metalworker. Intentionality is not certain, however, especially if we are only confronted with isolated examples – for example the Early Cycladic dagger from Amorgos with 5.1 wt% zinc or the Middle Bronze Age axe from Beth-Shan with 6.5 wt% zinc mentioned by Craddock (1978a: 2). The addition of greater than ~8 wt% zinc produces a metal with at least one unmistakable characteristic – its golden colour – which, even if created fortuitously the first time, probably could have been reproduced if desired. Therefore, copper-zinc alloys will be defined for this paper somewhat arbitrarily as containing greater than ~8 wt% zinc.

Before the intentional production of zinc metal through distillation processes, which according to texts seems to have begun in India in the late 1st millennium BC (Craddock et al. 1998), there were two basic ways to create copper-zinc alloys. The first is the cementation method, in which zinc oxide is mixed with pieces of copper metal and charcoal in a tightly closed crucible, heated above 906 °C in order to vaporise the zinc and allow it to diffuse into the solid copper, before then introducing a final stage of raised temperature to melt the metal and thereby homogenise the mixture (Bayley 1998: 9–10). According to recent thinking (see Craddock and Eckstein 2003: 223–6), the temperature range was probably closer to 1000–1100 °C during Roman times, which would have produced in one step a molten copper-zinc alloy of 20–28 wt% zinc. Thilo Rehren (1999: 1085) has argued for a lower temperature range (c. 900–1000 °C) based on the material properties of early ceramic crucibles, while Jean-Marie Welter (2003) has suggested an upper limit of zinc uptake closer to 38–40 wt%. The amount of zinc in the resulting metal would also be significantly depleted if insufficient amounts of calamine were added to the mixture or if the base copper contained impurities such as tin or lead. In any case, subsequent melting of the copper-zinc alloy (or re-melting of scrap metal) would invariably cause the loss of about 10% of the zinc present in the alloy[4] (Caley 1964).

The other way to achieve a copper-zinc alloy using zinc oxides, and the one perhaps more likely to have produced the prehistoric examples discussed in this paper, is through a mixed-ore smelting process. This method has been demonstrated by many scholars including Sun and Han (1983–85: 268), who managed to create copper-zinc alloys with up to 34 wt% zinc by mixing chemically pure zinc oxide (ZnO) and cuprous oxide ($Cu_2O$) (1:1) in a graphite crucible. Of course, this was a modern smelting experiment and one not necessarily representative of ancient processes, but it should be noted that they also produced copper-zinc alloys (up to

18 wt% zinc) by smelting a mixture of naturally occurring malachite ($Cu_2(OH)_2CO_3$) and leaded smithsonite. The efficacy of the mixed-ore smelting method was also supported by the work of Rostoker and Dvorak (1991), who produced a copper-zinc alloy of 10.5 wt% zinc by smelting zinc oxide with malachite in an open crucible, although they note that the zinc uptake would have been higher with a closed crucible. Although we do not as yet have a means of distinguishing these two general methods of copper-zinc alloy production using only the artefacts themselves, the mixed-ore smelting method (which here includes the mixture of copper ores with manmade zinc oxide) seems more likely to have been utilised, given what we know about prehistoric metallurgical processes in Southwest Asia.

## Early copper-zinc alloys in Southwest Asia

Having delineated the possible means of copper-zinc alloy production in antiquity, it is time to turn to the evidence itself. As presented in Table 1, there are over 30 examples of prehistoric and protohistoric artefacts from Southwest Asia containing over 8 wt% zinc that have been reported in the literature. Also included are a number of artefacts containing 5–8 wt% zinc that may or may not be intentional alloys, but which are included here for the sake of reference. Notable finds east of the general area of Southwest Asia include the isolated and contested examples of copper-zinc alloys from the Yangshao and Longshan periods in central and eastern China (*c.* 4th–3rd millennia BC) (see An 2000; Han and Ko 2000; Mei and Li 2003: 112; Sun and Han 1983–85). In addition, there are two leaded gunmetal artefacts from Atranjikhera, an early/mid-1st millennium BC site in the Upper Ganga Basin of northern India (Gaur 1983), which precede the leaded brass vessels from Taxila in northern Pakistan (4th century BC)

and Begram in the Kabul valley of Afghanistan (2nd century BC) (Craddock *et al.* 1998: 27). While the Chinese examples remain suspect because of their extreme discordance with the development of Chinese metallurgy as we understand it, the finds from Atranjikhera and elsewhere on the Indian subcontinent coincide with (or slightly predate) the earliest textual references to brass and possibly even zinc metal from this region (Craddock *et al.* 1998; see also Rau 1974).

Contemporary with the Indian examples, there are also sporadic examples of copper-zinc alloys from the Hellenistic world as well as textual references to *oreichalkum* ('copper of the mountain'; cf. Latin *aurichalcum* or 'golden copper'),[5] beginning as early as the 7th century BC (see Craddock 1998b; Craddock *et al.* 1980). Also of note from Europe are the 3rd-century BC Etruscan statues analysed by Craddock (1978b) that contain 11.5 and 11.8 wt% zinc (and 0.68–3.0 wt% tin respectively), the 6th-century BC pin with 9.9 wt% zinc from a Phoenician tomb in Cadiz, Spain (Montero-Ruiz and Perea, this volume, p. 136) and the pre-Roman iron sword with the maker's stamps in brass (*c.* 15–20 wt% zinc) from Syon Reach outside London now dated to the 2nd century BC (Craddock *et al.* 2004).[6] These examples join the list of 21 prehistoric copper-zinc alloys compiled by Caley (1964: 3–8) in contesting the widely held belief that copper-zinc alloy production spread throughout Europe with the Roman conquests.

As stated above, the data from Southwest Asia can be grouped loosely into six spatially and chronologically definable clusters that predate the advent of the consistent utilisation of copper-zinc alloys in the late 1st millennium BC (Fig. 2). The first of the six groupings (Group 1) is located in the eastern Aegean region during the first half of the 3rd millennium BC and includes the eight copper-zinc alloys presented by Stos-Gale (1992) and Begemann *et al.* (1992) from the site of Thermi on the island of Lesbos. Here one or two examples of copper-zinc alloys (*c.* 5–17 wt% zinc) were found in almost every stratigraphic phase of the site, which is generally asso-

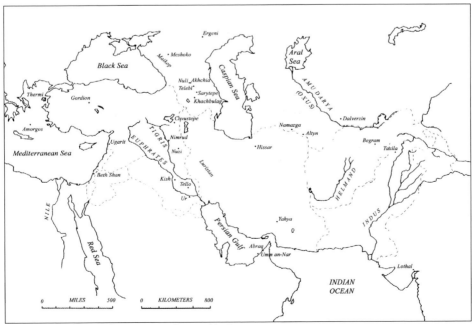

**Figure 2** Map showing the major sites discussed in the text.

**Table 1** Early copper-zinc alloys from Southwest Asia.

| Site (level) | Location | Type | Time | Method | Composition (wt%) | | | | Reference |
|---|---|---|---|---|---|---|---|---|---|
| | | | | | Sn | Zn | Pb | Other | |
| GROUP 1 | | | | | | | | | |
| Thermi (I-II) | Lesbos | pin | early 3rd mill. BC | XRF | | 6.8 | 3.9 | Sb: 4.22 | Stos-Gale 1992 |
| Thermi (I-II) | Lesbos | pin | early 3rd mill. BC | XRF | 0.3 | 6.2 | 3.4 | Sb: 0.9 | Stos-Gale 1992 |
| Thermi (II) | Lesbos | knife | early 3rd mill. BC | XRF/INAA | | 10.9 | | | Stos-Gale 1992 |
| Thermi (II-III) | Lesbos | pin | early 3rd mill. BC | XRF/INAA | | 6.6 | 9.2 | | Stos-Gale 1992 |
| Thermi (IIIa) | Lesbos | pin | early 3rd mill. BC | XRF/INAA | 4.2 | 5.1 | | Ag: 1.3 | Begemann et al. 1992 |
| Thermi (IIIb) | Lesbos | pin | mid-3rd mill. BC | XRF/INAA | | 8.5 | | As: 2.8 | Begemann et al. 1992 |
| Thermi (P.P.) | Lesbos | ornament | mid-3rd mill. BC | XRF/INAA | 2.2 | 10.3 | 0.7 | | Begemann et al. 1992 |
| Thermi (V) | Lesbos | pierced disk | mid-3rd mill. BC | XRF/INAA | 9.2 | 16.9 | | | Begemann et al. 1992 |
| GROUP 2 | | | | | | | | | |
| Kish (EDIII) | S. Iraq | toilet-knife #577 | mid/late 3rd mill. BC | XRF | | 14.0 | 0.6 | As: 2.3 | Hauptmann and Pernicka 2004 |
| Kish (EDIIIb-Akk) | S. Iraq | bowl #461 | mid/late 3rd mill. BC | XRF | 8.8 | 14.3 | | Fe: 3.6 | Hauptmann and Pernicka 2004 |
| Kish-Ingarra (?) | S. Iraq | bowl #456 | mid/late-3rd mill. BC? | XRF | 9.1 | 7.2 | | As: 1.3 | Hauptmann and Pernicka 2004 |
| Girsu/Tello (EDIII) | S. Iraq | blade rivet #849b | mid/late 3rd mill. BC | XRF | 1.0 | 6.0 | 10.7 | As: 1.7, Fe: 1.2 | Hauptmann and Pernicka 2004 |
| Ur (EDIIIb) | S. Iraq | bowl #1048 | mid/late 3rd mill. BC | XRF | 13.5 | 6.8 | | Fe: 1.0 | Hauptmann and Pernicka 2004 |
| Ur (EDIIIb-Akk) | S. Iraq | bowl #1057 | mid/late 3rd mill. BC | XRF | 12.2 | 5.8 | | | Hauptmann and Pernicka 2004 |
| Ur (EDIII-NeoSum) | S. Iraq | dagger #1474 | mid/late 3rd mill. BC | XRF | 5.0 | 7.1 | 2.0 | As: 3.1 | Hauptmann and Pernicka 2004 |
| Ur (EDIII?) | S. Iraq | helm #2067 | mid/late 3rd mill. BC | XRF | | 10.7 | 0.4 | Fe: 2.2; As: 0.9 | Hauptmann and Pernicka 2004 |
| GROUP 3 | | | | | | | | | |
| Ergeni | Kalmykia | knife | mid/late 3rd mill. BC | OES | | 11.0 | | | Egor'kov et al. 2004 |
| Ergeni | Kalmykia | hook | mid/late 3rd mill. BC | OES | | 8.0 | | | Egor'kov et al. 2004 |
| Telebi | E. Georgia | blade | late 3rd mill. BC? | OES? | | 6.6 | | As: 2.3 | Kavtaradze 1999 |
| Nuli | N. Georgia | discoid pin | late 3rd mill. BC? | OES? | | 5.7 | 0.4 | As: 4.5 | Kavtaradze 1999 |
| Akhchia (kurgan 3) | N. Georgia | twisted tube' | late 3rd mill. BC? | OES? | 5.8 | 5.2 | 1.3 | | Kavtaradze 1999 |
| Namazga | S Turkmenistan | needle | late 3rd mill. BC | OES | | 24.7 | | | Egor'kov 2001: 87 |
| Namazga | S Turkmenistan | seal | late 3rd mill. BC | OES | | 14.8 | | | Egor'kov 2001: 87 |
| Altyn depe (1966) | S Turkmenistan | blade | late 3rd mill. BC | OES | 6.6 | 16.0 | 12.0 | As: 2.2; Fe: 1.6 | Egor'kov 2001 |
| Altyn depe (1967) | S Turkmenistan | blade | late 3rd mill. BC | OES | 6.5 | <6.0 | 8.5 | Fe: 1.0 | Egor'kov 2001 |
| Altyn depe | S Turkmenistan | ring | late 3rd mill. BC | OES | 1.0 | 7.0 | | | Egor'kov 2001 |
| Dalverzin | S Uzbekistan | ring | early 2nd mill. BC | OES | | 25.0 | | Ni: 10 | Bogdanova-Berezovskaja 1962 |
| Dalverzin | S Uzbekistan | pin | early 2nd mill. BC | OES | 1-5 | 15-18 | | Fe: >5 | Bogdanova-Berezovskaja 1962 |
| GROUP 4 | | | | | | | | | |
| Umm an-Nar | UAE | dagger | late 3rd mill. BC | XRD | | 10.0 | | | Frifelt 1991 |
| Umm an-Nar | UAE | fragment | late 3rd mill. BC | AAS/XRD | | 8.6 | | | Frifelt 1991 |
| Tepe Yahya (IVA) | SE Iran | bracelet | early 2nd mill. BC | EMPA/ICP-MS | | 19.4 | 0.9 | | Thornton et al. 2002 |
| Tepe Yahya (IVA) | SE Iran | ribbon | early 2nd mill. BC | EMPA/ICP-MS | 0.3 | 17.0 | | | Thornton et al. 2002 |
| Tepe Yahya (IVA) | SE Iran | fragment | early 2nd mill. BC | EMPA/ICP-MS | 0.8 | 16.9 | 1.8 | | Thornton et al. 2002 |

| Site (level) | Location | Type | Time | Method | Composition (wt%) | | | | Reference |
|---|---|---|---|---|---|---|---|---|---|
| | | | | | Sn | Zn | Pb | Other | |
| **GROUP 5** | | | | | | | | | |
| Ugarit | NW Syria | ring | c. 1400 BC | OES/XRF/EMPA | ~3 | ~12 | ~1-2 | As: ~1 | Schaeffer-Forrer et al. 1982 |
| Ugarit | NW Syria | statuette | mid-2nd mill. BC | OES/XRF/EMPA | | ~12 | | | Schaeffer-Forrer et al. 1982 |
| Nuzi (II) | NE Iraq | ring | c. 1350 BC | INAA/ICP | 0.4 | 14.4 | 4.7 | | Bedore and Dixon 1998 |
| Nuzi (II) | NE Iraq | ring | c. 1350 BC | INAA/ICP | 6.3 | 12.2 | 3.4 | | Bedore and Dixon 1998 |
| **GROUP 6** | | | | | | | | | |
| Khachbulag | W Azerbaijan | bracelet | early 1st mill. BC | OES | 0.6 | 10.4 | | | Kashkai and Selimkhanov 1973 |
| Sarytepe | NW Azerbaijan | arrowhead | early 1st mill. BC | OES | | 9.4 | | | Kashkai and Selimkhanov 1973 |
| Sarytepe | NW Azerbaijan | figurine | early 1st mill. BC | OES | 10.2 | 8.0 | | As: 2.0 | Kashkai and Selimkhanov 1973 |
| Luristan Bronze | W Iran | pin head | early 1st mill. BC | EMPA-WDS | 10.6 | 8.4 | 1.9 | | Northover 1997 |
| Luristan Bronze | W Iran | brooch | early 1st mill. BC | EMPA-WDS | 2.6 | 8.2 | 1.1 | | Northover 1997 |
| Nimrud | N Iraq | bowl | 9th century BC | AAS | 8.4 | 6.4 | | | Hughes et al. 1988 |
| Nimrud | N Iraq | bowl | 9th century BC | AAS | 7.8 | 5.5 | | | Hughes et al. 1988 |
| Cavustepe | NE Anatolia | bracelet | 8th century BC | SEM-EDAX | | 11.0 | | S: 1.8 | Geckinli et al. 1986 |
| Gordion (MM) | cent Anatolia | fibulae | 8th century BC | OES | 10.0 | >10 | | Fe: 3.7 | Young 1981: 287 |
| Gordion (MM) | cent Anatolia | fibulae | 8th century BC | OES | 15.6 | >10 | | | Young 1981: 287 |
| Gordion (MM) | cent Anatolia | fibulae | 8th century BC | OES | 6.3 | >10 | | Fe: 3.8 | Young 1981: 287 |
| Gordion (MM) | cent Anatolia | fibulae | 8th century BC | XRF | 3.0 | 8.5 | 3.5 | | Young 1981: 290 |
| Gordion (MM) | cent Anatolia | bowl | 8th century BC | XRF | 4.0 | 12.0 | 2.0 | | Young 1981: 290 |
| **OTHER** | | | | | | | | | |
| Amorgos | Cyclades | dagger | mid-3rd mill. BC | OES? | av 1.8 | 5.1 | 0.5 | | Renfrew 1967 |
| Sanlihe | E China | awl | mid/late 3rd mill. BC | EMPA | | av 23.4 | av 2.8 | | Sun and Han 1983-85 |
| Lothal | W India | fragment | early 2nd mill. BC | unknown | | 6.0 | | Fe: 0.9 | Rao 1985 |
| Beth Shan | N Israel | axe | early 2nd mill. BC | unknown | 8.4 | 6.5 | 1.2 | | Oren 1971 |
| Atranjikhera | N India | unknown | early 1st mill. BC | unknown | 11.7 | 6.3 | 9.0 | Fe: 1.9 | Gaur 1983 |
| Atranjikhera | N India | pin | early/mid-1st mill. BC | unknown | 20.7 | 16.2 | 9.8 | Fe: 1.2 | Gaur 1983 |

XRF = X-ray fluorescence; INAA = instrumental neutron activation analysis; OES = optical emission spectroscopy; XRD = X-ray diffraction; AAS = atomic absorption spectroscopy; MS = mass spectroscopy; PA = performance analysis; ICP = inductively coupled plasma analysis; EMPA = electron microprobe analysis; WDS = wavelength dispersive X-ray spectroscopy; SEM = scanning electron microscopy; EDAX = energy-dispersive X-ray analysis ciated with the Troy I–early II periods or *c*. 3000–2500 BC (Pernicka *et al.* 2003). These artefacts, most of which were pins and other ornaments, often included other elements in significant quantities besides zinc, including arsenic (<2.8 wt%), tin (<9.2 wt%), lead (<9.2 wt%), and antimony (<4.2 wt%). These examples are quite isolated in that no other copper-zinc alloys are yet known from related sites in the region nor, indeed, from anywhere else at this date. Troy I–II period sites in the Troad and on the eastern Aegean islands of Lemnos, Lesbos, Chios, etc. however, are notable for the early (and often consistent) use of a wide array of metals including silver, gold, lead and tin, as well as copper alloyed with arsenic, zinc, tin, lead, silver, and antimony (Muhly 2002; Pernicka *et al.* 2003). It is also significant to note that the mines of Argenos on the northern shore of Lesbos contain deposits of copper oxides and sulphides as well as lead and zinc sulphides (Pernicka *et al.* 2003: 153).

The second group of copper-zinc alloys comes from the recent publication of the Frühe Metalle in Mesopotamien project at the University of Heidelberg, which presents the chemical analyses of nearly 3000 artefacts from southern Mesopotamia (Hauptmann and Pernicka 2004). Of these, only eight artefacts, dated to the mid/late 3rd millennium BC, contained greater than 8 wt% zinc, while an additional seven were found to contain between 6 and 8 wt% zinc. If we consider those artefacts analysed only on the surface patina by X-ray fluorescence (XRF) as potentially contaminated and focus specifically on drilled samples, however, then we are left with a helm, a toiletry article and a bowl with 10.7, 14.0 and 14.3 wt% zinc respectively, as well as five other artefacts containing 6–8 wt% zinc. It is worth noting that a number of these examples also contain significant levels of tin, arsenic and/or iron, but show a relative dearth of lead in all but one or two cases. Hopefully, future publications of the Heidelberg analyses of northern Mesopotamian artefacts and the long-awaited Mesopotamian Metals Project of the University of Pennsylvania will provide further prehistoric examples of copper-zinc alloys from this region.

The third grouping of reported copper-zinc alloys comes from the circum-Caspian region, notably from sites in the northern and southern Caucasus and southern Central Asia during the second half of the 3rd millennium BC. Unfortunately, most of the examples from the southern Caucasus and southern Central Asia can only be considered tentatively until further analyses have been conducted. For example, the two copper-zinc alloys from Namazga-depe in southern Turkmenistan, analysed in the 1950s and discussed by Egor'kov (2001: 87), were unstratified finds, as were the blade and the ring from Altyn-depe.[7] In addition, many of the spectrographic analyses performed in the 1960s have been shown to be questionable, such as the Altyn-depe blade that was reported to contain 16 wt% zinc when first analysed in 1966, but which upon re-analysis in 1967 was found to contain less than 6 wt% zinc (Egor'kov 2001). The copper-zinc alloys analysed by Bogdanova-Berezovskaja (1962) from late 3rd/early 2nd millennium BC Dal'verzin in Uzbekistan must also be regarded with a critical eye for these same reasons.

Fortunately, the nine copper-zinc alloys from kurgans at the site of Ergeni ('Yergueni') in northern Kalmykia (*c*. 2500–2200 BC) first reported by Gak (2004) are not as contentious,

because the metal artefacts were analysed in the past few years and the burials have been radiocarbon dated (Egor'kov *et al.* 2004). Although the zinc content in these artefacts is relatively low (1.3–5.6 wt% zinc), excluding the dagger (11 wt% zinc) and the hook (8 wt% zinc), it is undoubtedly significant that the other 31 contemporary copper artefacts analysed from this region contain considerable amounts of arsenic (<6.4 wt% arsenic, average: 3.0 wt% arsenic), an element which is missing from the nine artefacts mentioned above (average: 0.2 wt% arsenic) (Egor'kov *et al.* 2005). This may suggest that low zinc brass was being produced at, and imported from, an area not utilising arsenic-bearing copper. Alternatively, zinc was being added intentionally to pure copper (through cementation or mixed-ore smelting, as discussed above) by metalworkers (either locally or elsewhere) who perhaps recognised that zinc uptake is hindered by the presence of other alloying elements in the copper such as arsenic, lead and tin (Craddock *et al.* 1980: 60; Ponting 2002: 559–60). It is worth mentioning that Chernykh (1992: 66) reports a 3rd-millennium BC zinc metal ornament from the Maikop site of Meshoko also in the northern Caucasus, which (if authentic) may force us to reconsider the possible methods for early copper-zinc alloy production.

The fourth grouping of prehistoric copper-zinc alloys appears in the eastern Persian Gulf region in the late 3rd/early 2nd millennium BC in the cemetery of Umm an-Nar, United Arab Emirates (UAE), and at the village site of Tepe Yahya, Iran. At the former site, the copper-zinc alloys include a dagger (10 wt% zinc) and a fragment (8.6 wt% zinc) as well as six other fragments with 2.3–4.7 wt% zinc. All appear to be similar to the Ergeni examples in being relatively low zinc brasses that were made with fairly pure copper as the base metal (Frifelt 1991: 100). Given that 3rd-millennium BC raw copper and copper ingots from eastern Arabia generally have 60–600 ppm zinc (Weeks 2003: 85), the numbers from Umm an-Nar are significant and suggest (as at Ergeni) either importation of zinc-rich copper metal or the intentional production of low zinc brasses, perhaps through open-crucible smelting of zinc and copper oxides. Unfortunately, the few artefacts analysed from the Umm an-Nar cemetery were all fairly corroded and should perhaps be re-analysed, especially in light of the fact that the contemporary artefacts from the associated settlement contained less than 40 ppm of zinc (Hauptmann 1995).

The three copper-zinc ornament fragments from a domestic context at Tepe Yahya IVA (*c*. early 2nd millennium BC) have been reported in detail elsewhere and need not be repeated here (see Thornton and Ehlers 2003; Thornton *et al.* 2002). In general, these artefacts have higher zinc contents than at Umm an-Nar (and are less corroded) and show mixing with small amounts of lead and tin, perhaps suggestive of the recycling of scrap metal. Given the close cultural relations between southeastern Iran and the eastern Arabian Peninsula during the late 3rd millennium BC (see for example Lamberg-Karlovsky and Potts 2001; Méry 2000), the discovery of copper-zinc alloys in both regions is perhaps not surprising. Furthermore, the corroded fragment with ~6 wt% zinc found at the contemporary Late Harappan site of Lothal in Gujarat (Rao 1985: 660), which had significant trade relations with the eastern Persian Gulf region during this period (see Cleuziou and Tosi 2000; Frenez and Tosi 2005), may suggest a wider range for early copper-zinc alloys than previously thought.

The fifth grouping of prehistoric copper-zinc alloys occurs in two mid-2nd millennium BC sites in northern Mesopotamia. Schaeffer-Forrer *et al.* (1982) report two artefacts from Ugarit in western Syria that contain roughly 12 wt% zinc, including a ring with a Hittite-style stamp seal and an Egyptian-style zoomorphic statuette, both of which purportedly date to *c.* 1400 BC. The statuette, however, was not found in good context and can only be dated based on art-historical arguments. Contemporary to these pieces, Christine Ehlers analysed two rings from the destruction layer (*c.* 1350 BC) at Nuzi (Yorgan Tepe) in northeastern Iraq that proved to be leaded copper-tin-zinc alloys with 12.2–14.4 wt% zinc (see Bedore and Dixon 1998; Thornton and Ehlers 2003). The fact that these rings from Ugarit and Nuzi are the earliest substantiated copper-zinc alloys from northern Mesopotamia may suggest that more quotidian artefacts (instead of elite funerary goods) need to be analysed from earlier periods.

The final grouping occurs mainly in the region combining eastern Anatolia, northern Iraq, the southern Caucasus and western Iran in the early 1st millennium BC. There are substantially more examples from this grouping than from the previous five, but some are of questionable authenticity. For example, the horn of the unprovenanced Urartian tin-bronze bull head that was said to be 'a copper-tin-zinc alloy with relatively high concentration of Cr'[8] as reported by Hanfmann (1956: 207) has been found through recent XRF and inductively coupled plasma (ICP) analyses to contain merely traces of zinc.[9] Also unprovenanced (and therefore suspect) are the 'Luristan Bronze' pinhead and brooch analysed by Northover (1997) and found to contain just over 8 wt% zinc. On the other hand, these unprovenanced finds are roughly contemporary with excavated examples of copper-zinc alloys, including a twisted bracelet from the Urartian site of Cavustepe near Van in eastern Turkey containing 11 wt% zinc (Geckinli *et al.* 1986) and three excavated artefacts from Azerbaijan containing 8–10.4 wt% zinc (Kashkai and Selimkhanov 1973; see also Gasanova 2002; Schachner 2005). Also of note are the two 9th-century BC Assyrian bowls from Nimrud containing significant tin and what may be significant levels of zinc (5.5–6.4 wt%) (Hughes *et al.* 1988).

The most famous copper-zinc alloys from the first half of the 1st millennium BC come from just west of this core region at the aforementioned site of Gordion in central Anatolia. Here, three fibulae from the Great Tumulus[10] ('MM') were found to contain >10 wt% zinc as reported by Steinberg (1981: 286–9; see comment in Craddock and Eckstein 2003: 216). In addition, a fourth fibula and a bowl containing 8 and 12 wt% zinc, respectively, were also found in the Great Tumulus and analysed by W.J. Young in 1956 (in Young 1981: 289–90). Although the Greeks attributed the invention of copper-zinc alloys to the Phrygians, it is important to stress that Gordion was simply the last in a series of prehistoric and protohistoric Southwest Asian sites to consume, if not produce, these metals. Phrygia may even have imported these alloys or at least the technology from the core region to the east designated here in the sixth grouping.

## Of bronze

The growing list of reported prehistoric copper-zinc alloys presented above does not on its own provide any reasonable answers. While it seems quite likely that many of these copper-zinc alloys were being produced by a mixed-ore smelting method as opposed to cementation, this cannot be substantiated or refuted at present. Furthermore, where and when these rare alloys were being produced and by and for whom are all questions that remain to be answered. When juxtaposed with the relevant dataset for exploring the origins of tin and tin-bronze in the greater Near East, however, an intriguing pattern emerges that may hopefully provide some answers or, at least, some new ways of looking at the question.

As mentioned above, sites in the eastern Aegean witnessed an explosion of new metallic alloys during the Troy I–II periods (3000–2500 BC). All of these new metals, however, including the early copper-zinc alloys from Thermi, have been greatly upstaged in the literature by the early examples of tin-bronze known from the Troy I period at Thermi and recently discovered at the Troad site of Besik-Yassitepe (Begemann *et al.* 2003). Of perhaps even greater significance for the 'tin question' was the discovery of one of the earliest objects made of tin-base metal (also containing 22.7 wt% iron) from Thermi IV in the mid-3rd millennium BC (Begemann *et al.* 1992). It is also interesting to note that three of the four copper-zinc alloys reported from the later periods at Thermi (III–V) are actually made of gunmetal containing 2.2–9.9 wt% tin, which is a metal that remains prevalent throughout the entire sequence of early copper-zinc alloys.

Although uncertain about the origins of these early tin-bronzes and, perhaps by extension, the early copper-zinc alloys, Pernicka *et al.* (2003: 165–7) suggest an importation of tin metal from Central Asia based upon lead isotope data and the excavation of jade and nephrite axes at Troy. It seems equally likely that the Troad tin was coming from southeastern Europe, given the evidence for late 5th/early 4th-millennia BC tin-bronze artefacts and slag from this region (see Glumac and Todd 1991). The evidence for tin-bronze in this region continues into late 4th/early 3rd-millennia BC contexts such as the multiple artefacts from Sitagroi IV–V (Renfrew and Slater 2003) and the dagger from Velika Gruda, Montenegro (Primas 2002). It should be noted, however, that there are no reported copper-zinc alloys from prehistoric contexts in southeastern Europe, although perhaps future analyses will prove otherwise.

The Caucasus region provides another interesting, although less direct, association between brass and bronze. If we follow Kavtaradze (1999) in doubting the reported context of the two tin-bronzes found at Delisi in Georgia (*c.* early 4th millennium BC), then the earliest confirmed copper-tin alloys are the spiral ring with 10.2 wt% tin recently analysed by Laura Tedesco (2006)[11] from a tomb in Armenia dated to the late 4th millennium BC. There are also the tin-bronzes from graves at Velikent, Daghestan, reported by Kohl (2002) dating to the 3rd millennium BC (see also Chernykh 1992: 123–4). These dates coincide with the first reports of zinc appearing as a significant trace element (<2.5 wt%) in tin-bronzes from Kura-Araxes sites as well as the Meshoko zinc ornament discussed above (Chernykh 1992: 66). Tin-bronze becomes more widespread, however, during the second half of the 3rd millennium BC within the Sachkhere and Bedeni cultures of Georgia (Kavtaradze 1999), where purportedly numerous cases of arsenical copper and tin-bronze artefacts are said to contain up to ~5 wt% zinc (Chernykh 1992: 109).[12] It is particularly interesting to note that Chernykh (1992: 121) states that copper-tin-zinc 'alloys' (i.e. tin-bronzes with appreciable amounts of zinc) are specific to the Bedeni culture of eastern Georgia and are not found in the northern Caucasus. Although we must wait until these data are published in order to be entirely certain of the accuracy of these propositions, this statement does correspond well with the lack of tin in contemporary copper-zinc alloys from Ergeni mentioned above.

A third example of this interrelationship between copper-zinc alloys and the 'tin question' is to be found on both sides of the lower Persian Gulf region. As Weeks (1999, 2003) has shown, tin begins to appear as a minor but significant element (0.5–2.0 wt%) in eastern Arabian copper-based artefacts in the mid-3rd millennium BC, before blossoming into true tin-bronzes (>2 wt% tin) in the later 3rd millennium BC. This transition is contemporary with the appearance of the copper-zinc alloys at Umm an-Nar discussed above as well as with the earliest appearance of tin metal in the region in the form of a tin ring from a grave at Tell Abraq (Weeks 2003: 123). Intriguingly, this sequence of copper with minor amounts of tin transitioning to actual tin-bronzes and early brasses is paralleled almost exactly in southeastern Iran at Tepe Yahya.[13] There tin first appears as a minor element in the late 3rd millennium BC (Yahya IVB) before tin-bronzes, the three brasses mentioned above (two of which contain small but significant amounts of tin), and a lead-tin bangle appear in the early 2nd millennium BC (Yahya IVA) (Thornton et al. 2005).

Perhaps the most interesting examples that support an association between copper-zinc and copper-tin alloys in prehistory are the consistent copper-zinc-tin alloys that appear in these early periods. From Thermi to Gordion, roughly two-thirds of all the copper-zinc alloys listed in Table 1 and mentioned above contain appreciable amounts of tin, while roughly a half could qualify as 'tin-bronzes' in their own right (i.e. >1–2 wt% tin). This is reminiscent of the controlled mixing of copper, zinc and tin so prevalent during the Roman period (Bayley and Butcher 2004; Craddock 1978a), although markedly different in that prehistoric gunmetal appears to have been produced haphazardly with little regard for (or control of) the alloying process (Fig. 3). Indeed, it is not until the early 1st millennium BC (i.e. 'Group 6'), when almost all of the artefacts contain roughly 8–12 wt% zinc, that the amount of zinc seems to have been somewhat standardised, although the amount of tin in these same artefacts remains haphazard at best.

It may seem reasonable to suggest that prehistoric gunmetal was produced accidentally, perhaps as a result of the recycling of scrap metal for the manufacture of jewellery and other trinkets. If this is true, then why do we find only tin being alloyed with copper and zinc, and not more copper-arsenic-zinc or copper-antimony-zinc alloys? Indeed, the mixing of tin and zinc in copper is not an obvious combination from an archaeometallurgical standpoint, as tin and zinc are rarely found in the same geologic contexts and the presence of tin in copper will reduce the uptake of zinc twice as much as the presence of other alloying elements such as lead[14] (Craddock et al. 1980: 60; Ponting 2002: 560). How, then, are we to make sense of prehistoric gunmetal, which on the one hand seems to be haphazardly (perhaps serendipitously) produced, while on the other hand appears to have been intentionally associated with copper-tin alloys and tin metal?

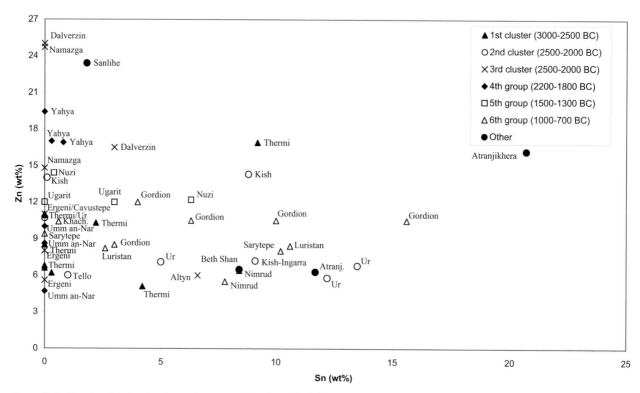

**Figure 3** Scatter-plot showing the zinc vs. tin compositions (in wt%) of the various copper-zinc alloys mentioned in the text. The lack of a definable correlation between the tin and zinc contents suggests an uncontrolled or unintentional production of these alloys

## Of aurichalcum: a case of technological ethnocategories?

One possible reason for the haphazard production of gun-metal is that brass and tin-bronze were both favoured in the production of items for display (e.g. pins, jewellery, blades) due to their golden colours and may have been confused in the past as they were in the medieval period of Europe and, indeed, even today. This statement is not as straightforward as it may seem. We distinguish copper-zinc and copper-tin alloys based on their chemical and physical properties, although we approach these materials from the privileged position of understanding what they are, how they can be produced, what their similarities and differences are, etc. While skilled pre-historic metalworkers would undoubtedly have realised the physical differences between these materials, we cannot be certain, without textual references such as those provided by the Greeks and Romans, that the ancient local consumers of these metals distinguished them quite so easily.

For example, in a recent paper it was argued, based on metallographic analyses of copper-based artefacts from five millennia of occupation at Tepe Yahya, that at least two of the three brass objects were created by a distinctly local style of metalworking that appears to remain unchanged at the site from the early 4th millennium BC until the 1st millennium BC (Thornton and Lamberg-Karlovsky 2004). This may suggest that these small brass ornaments were worked locally into their final forms, although the metal itself was probably imported given its rarity and the lack of metallurgical remains at the site (cf. similar comment by Begemann *et al.* (1992: 220) in regards to the Thermi examples). Indeed, we often assume only two major stages in the life-cycle of a metal artefact – production and consumption – but this ignores the possibility of intermediary metalworkers or of cottage-industry consumers who melted and crafted metals to fit their specific needs without necessarily understanding the technical properties of the materials with which they worked (cf. Pigott's (1980) discussion of the treatment of iron by Iranian bronzeworkers in the Early Iron Age). This is not to suggest that a master craftsman would not immediately notice the difference between tin-bronze and brass, just as he would undoubtedly notice the difference between a 2 wt% arsenical copper and a 7 wt% arsenical copper. Instead, it is meant to question whether the ancient consumers of the copper-zinc alloys at Yahya or Thermi or Nuzi categorised these metals as different from copper-tin alloys, or whether brass, bronze and gunmetal were treated as variants of the same genre of material that could be combined and exchanged at will.[15]

One thing that we can be fairly certain of is that the people *producing* copper-zinc alloys, who were most likely village-based or itinerant craftspeople located in regions peripheral to mainstream prehistoric societies, were aware that what they were making was not the same as tin-bronze. Disregarding the possibility of zinc metal distillation and ignoring for now the possibility of an early invention of the cementation method, the major distinguishing factor for ancient mixed-ore smelters would have been that zinc and tin ores are visually quite distinct and rarely ever found in the same geologic contexts. Indeed, zinc ores occur most frequently in lead and copper ore deposits, while the latter is often found in granites, in gold-bearing quartz veins and in streams that have eroded away the surrounding rock matrix but, like gold, have left nodules of the water-insoluble tin oxides (Charles 1978). This latter point has been used to explain the association of tin and tin-bronze with gold throughout the prehistoric Near East (Muhly 1973). In contrast it would appear that the only reason for copper-zinc alloys to be found together with tin-bronze is because they both possess a golden colour.

Although it may seem trite to announce that what brass and bronze had in common was their colour, the study of visual characteristics of artefacts can be a powerful means of understanding the past (see Jones and MacGregor 2002). As scholars such as Lechtman (1977, 1999) and Hosler (1994) have demonstrated, by focusing on the more qualitative aspects of metals, archaeometallurgy can go beyond purely technical studies and engage with ancient 'ethnocategories' or the ways in which people gave order and meaning to the world around them. The idea of ethnocategories derives from sociolinguistic and anthropological theories on the relationship between the structuring of language and the structuring of society and the natural world (e.g. the famous Boasian example of Arctic Inuits using multiple words to describe different types of snow). In the case of technology, the creation or adoption of a new material usually leads to a reconfiguration of such ethnocategories (e.g. the discovery of iron in western Asia led eventually to the creation of a 'blacksmith' as a specific social role distinct from other metalworkers), although there is often a lag between the initial introduction of that new material or object and the full understanding of its distinct qualities.[16] In the absence of texts, this type of conceptual delay is also observable archaeologically (e.g. early smelted iron maintaining the high value and prestige of meteoritic iron) and scientifically (e.g. early iron having been worked by bronze workers), and it should be the role of the archaeometallurgist to combine these lines of inquiry in order to study the patterns of social behaviour that led to the formation of new and distinct ethnocategories.

In the case of prehistoric Southwest Asia, it is here proposed that the earliest copper-zinc alloys were not necessarily considered as different from tin-bronze by anyone other than the smelter or by a skilled metalworker able to distinguish the differences in hardness and durability between the two materials. That is, within the greater sociocultural milieux of the 3rd and 2nd millennia BC, these new golden-coloured metals fell into the same ethnocategory as tin-bronze and other golden-coloured, copper-base alloys. In fact it was probably not until the 1st millennium BC that words signifying 'brass' (such as the Greek *oreichalkos*), as opposed to copper and tin-bronze (e.g. *xalkos, kuwkos*), first appeared in texts (Halleux 1973). This linguistic shift may coincide with the florescence of copper-zinc alloys in the sixth grouping discussed above. If that is the case, then the sixth grouping may represent a significant change in the way copper-base metals were categorised by ancient societies – from one based almost entirely on visual characteristics to one involving some level of technical (or 'chemical') knowledge.

## Conclusions

Although the number of prehistoric copper-zinc alloys from Southwest Asia has increased significantly since Paul Craddock (1978a, 1980) first began to research their origins, the answers to the questions posited above on the who, when, where and how of early brass production remain elusive. Indeed, it is seemingly impossible to connect such geographically, chronologically and culturally disparate sites as Thermi, Ergeni and Tepe Yahya in the hopes of inducing some larger pattern, although the association with tin and tin-bronze discussed above may provide an important clue for future study. What is needed is to target the copper-zinc artefacts as a distinct corpus for comparative chemical and metallographic analyses, and also to conduct more archaeo-metallurgical investigations of certain key areas – such as the southern Caucasus – which may contain evidence for early copper-zinc alloy production.

The dataset presented here of early examples of copper-zinc alloys in prehistory can no longer be ignored by all but a handful of scholars. These alloys, although rare, were intimately connected with the development of other important metals such as lead and silver, whence the production of zinc oxide perhaps originated, as well as tin-bronze, with which it may have been confused and inadvertently mixed due to their similar visual characteristics. This hypothesis is not meant to suggest that copper-zinc alloys could not have been intentionally produced by knowledgeable craftspeople in prehistory. Rather, it is meant to emphasise the multiple stages in the life-cycle of metal artefacts – from original production and subsequent trade, to secondary and tertiary manufacturing of consumable goods, to repair, reuse and/or recycling steps, and finally to discard or deposition – and the different episodes of cultural categorisation that these materials and objects must have undergone as they moved from one person or society to another.

Given the fairly extreme differences between the production and functioning of tin-bronze and brass, the strong archaeological association between the two metals (and, indeed, the haphazard mixing of these alloys in prehistory) deserves more research. While technical studies are important for our understanding of the manufacture of metal artefacts, we must temper these studies with a nuanced understanding of more cultural factors such as visual symbolism and how they may have affected and been affected by technological and social practices (cf. Ottaway 2001; Sofaer Derevenski and Sørensen 2002). This can only be achieved by a combination of chemical and metallographic analyses of excavated artefacts from good contexts, scientific experiments into the processes of manufacture and physical properties of copper-zinc alloys, and theoretical discussions on the role of metals and materials in ancient societies.

## Acknowledgements

The author wishes to thank the British Museum organisers of the *Metallurgy: A Touchstone for Cross-Cultural Interaction* conference and the Applied Mineralogy Group of the Mineralogical Society who funded his participation. A few colleagues contributed significantly to the numerous drafts of this paper including Justine Bayley, Ben Roberts and Vincent Pigott. In addition, a number of scholars lent their expertise, data and support to this research including: Susanne Ebbinghaus, Katherine Eremin and Henry Lie; Alexander Egor'kov, Evgenii Gak and Natalya Shishlina; Christine Ehlers; Giorgi Kavtaradze; Ignacio Montero-Ruiz; Ernst Pernicka; Thilo Rehren; Josef Riederer; Laura Tedesco; Lloyd Weeks; Michael Witzel and two anonymous reviewers. Finally, the author wishes to acknowledge the constant support and inspiration that he has received from Paul Craddock and his unparalleled contribution to research on early copper-zinc alloys.

## Notes

1. Although fahlerz ores such as tetrahedrite ($Cu_{12}Sb_4S_{13}$) and tennantite ($Cu_{12}As_4S_{13}$) often contain appreciable amounts of zinc (see Ixer and Pattrick 2003), very few of the early copper-zinc alloys discussed here contain significant levels of arsenic or antimony. Thus, fahlerz is an unlikely source material for these rare metals.

2. The absence of iron and manganese in the brasses from Tepe Yahya, Iran, may confirm that they were in fact made by manmade zinc oxide and not naturally occurring smithsonite (Thornton *et al.* 2002: 1459).

3. Note that the Hissar data were not corrected for oxides and are only semi-quantitative (Pigott, pers. comm.).

4. In Thornton and Ehlers (2003), this is reported erroneously as the loss of 10 *weight* percent zinc. In actuality, a 28 wt% zinc copper-zinc alloy that is re-melted will become a 25–26 wt% zinc alloy (a 10% loss) and not an 18 wt% zinc alloy.

5. 'Oreichalkum' is considered by Craddock (1998b) and others (e.g. Caley 1964) to refer to 'brass' as it did in the Roman period; contra 'oreichalkum' to mean arsenical copper (Eaton and McKerrell 1976; Forbes 1964; Heskel and Lamberg-Karlovsky 1980). For more on this topic, see Halleux 1973.

6. This artefact can be compared to a La Tene sword with brass inlay from Munich mentioned by Dannheimer (1975). Thanks to Josef Riederer for providing this reference.

7. Egor'kov, pers. comm., May 2005.

8. Henry Lie and Susanne Ebbinghaus (pers. comm., June 2005) of the Fogg Art and Sackler Museums at Harvard report that the horn was found to contain >10 wt% tin and 1–10 wt% zinc when it was analysed in the 1950s.

9. Katherine Eremin, pers. comm., October 2005.

10. The Great Tumulus is now securely dated with dendrochronology to *c.* 743–741 BC and is thus not likely to be the tomb of King Midas (see DeVries *et al.* 2003).

11. Laura Tedesco, pers. comm., April 2005.

12. Giorgi Kavtaradze, pers. comm., 2004.

13. In fact, this sequence is so similar that it may add fodder to the heated debate over the chronology of Yahya IVB–IVA in relation to the Tell Abraq sequence presented by Lamberg-Karlovsky and Potts (2001).

14. It is interesting to note that lead, although the element most commonly found with zinc in geologic contexts of Southwest Asia, is only prevalent in significant quantities (i.e. >1 wt% lead) in roughly one-third of the total corpus in Table 1. Whether the lead came with the zinc or as an independent alloying agent is a question that needs to be addressed in future research.

15. Much in the same way that English speakers today continue to refer to aluminum foil as 'tin foil' despite almost a century since actual tin foil went out of common use.

16. For example, as Renfrew (1978: 102) has noted, the automobile was first referred to as a 'horseless carriage' and only later was considered to be a distinct entity (i.e. the 'car').

# References

An Zhimin 2000. On early copper and bronze objects in ancient China. In Linduff *et al.* 2000, 195–9.

Bartelheim, M., Pernicka, E. and Krause, R. (eds) 2002. *Die Anfange der Metallurgie in der Alten Welt*. Rahden/Westf.: Verlag Marie Leidorf GmbH.

Bayley, J. 1998. The production of brass in antiquity with particular reference to Roman Britain. In Craddock 1998a, 7–26.

Bayley, J. and Butcher, S. 2004. *Roman Brooches in Britain: A Technological and Typological Study Based on the Richborough Collection*. London: Society of Antiquaries of London.

Bedore, C. and Dixon, C. 1998. New discoveries in an old collection. *Context* 13: 3–4.

Begemann, F., Schmitt-Strecker, S. and Pernicka, E. 1992. The metal finds from Thermi III–V: a chemical and lead-isotope study. *Studia Troica* 2: 219–40.

Begemann, F., Schmitt-Strecker, S. and Pernicka, E. 2003. On the composition and provenance of metal finds from Besiktepe (Troia). In Wagner *et al.* 2003, 173–201.

Bogdanova-Berezovskaja, I.V. 1962. Khimicheskij sostav metallicheskikh izdelij Fergany epokhi bronzy i zheleza. In *Drevnezemledel'cheskaja Kul'tura Fergany*, Y.A. Zadneprovskij (ed.). Leningrad: MIA, no. 118, 219–30.

Caley, E.R. 1964. *Orichalcum and Related Ancient Alloys*. New York: American Numismatic Society.

Charles, J.A. 1978. The development of the usage of tin and tin-bronze: some problems. In *The Search for Ancient Tin*, T.A. Wertime (ed.). Washington, DC: Smithsonian Institution, 25–32.

Chernykh, E.N. 1992. *Ancient Metallurgy in the USSR*. Cambridge: Cambridge University Press.

Childe, V.G. 1928. *New Light on the Most Ancient Near East*. New York: Frederick Praeger.

Cleuziou, S. and Tosi, M. 2000. Ra's al-Jinz and the prehistoric coastal cultures of the Ja'alan. *Journal of Oman Studies* 11: 19–73.

Craddock, P.T. 1978a. The composition of the copper alloys used by the Greeks, Etruscan and Roman civilizations: the origins and early use of brass. *Journal of Archaeological Science* 5: 1–16.

Craddock, P.T. 1978b. Europe's earliest brasses. *MASCA Journal* 1(1): 4–5.

Craddock, P.T. 1980. The first brass: some early claims reconsidered. *MASCA Journal* 1(5): 131–3.

Craddock, P.T. (ed.) 1998a. *2000 Years of Zinc and Brass*, 2nd edn. British Museum Occasional Paper 50. London.

Craddock, P.T. 1998b. Zinc in classical antiquity. In Craddock 1998a, 1–6.

Craddock, P.T. and Eckstein, K. 2003. Production of brass in antiquity by direct reduction. In Craddock and Lang 2003, 216–30.

Craddock, P.T. and Lang, J. (eds) 2003. *Mining and Metal Production through the Ages*. London: British Museum Press.

Craddock, P.T., Burnett, A. and Preston, K. 1980. Hellenistic copper-based coinage and the origins of brass. In *Scientific Studies in Numismatics*, W.A. Oddy (ed.). British Museum Occasional Paper 18. London, 53–64.

Craddock, P.T., Freestone, I.C., Gurjar, L.K. and Willies, L. 1998. Zinc in India. In Craddock 1998a, 27–72.

Craddock, P.T., Cowell, M. and Stead, I. 2004 Britain's first brass. *Antiquaries Journal* 84: 339–46.

Dannheimer, H. 1975. Zu zwei älteren keltischen Fundstücken aus der Münchner Schotterebene. *Archäologisches Korrespondenzblatt* 5: 59–67.

DeVries, K., Kuniholm, P.I., Sams, G.K. and Voigt, M.M. 2003. New dates for Iron Age Gordion. *Antiquity* 77(296) (http://www.antiquity.ac.uk/projgall/devries/devries.html).

Eaton, E.R. and McKerrell, H. 1976. Near Eastern alloying and some textual evidence for the early use of arsenical copper. *World Archaeology* 8(2): 169–91.

Egor'kov, A.N. 2001. Особенности состава металла Алтын-депе [Features of the metal composition of Altyn-depe]. In *Особенности производства поселения Алтын-депе в эпоху палеометалла [Features of Production of the Settlement of Altyn-depe in the Early Metal Age]*, V.M. Masson (ed.). St Petersburg: Institut istorii material'no kul'tury PAH, 85–103.

Egor'kov, A.N., Gak, E.I. and Shishlina, N.I. 2004. Могильник Ергени в Калмыкии: латунь в бронзовом веке [Ergeni cemetery in Kalmykia: brass in the Bronze Age]. In *Древний Кавказ: ретроспекция культур [Ancient Caucasia: Retrospective on Cultures]*, L.T. Yablonskij (ed.). Moscow: Institut Arkheologii RAN, 77–8.

Egor'kov, A.N., Gak, E.I. and Shishlina, N.I. 2005. Chemical composition of metal from Kalmykia in the Bronze Age. *Archaeological News* 12: 71–6.

Forbes, R.J. 1964. *Studies in Ancient Technology*, vol. 8 (2nd edn. 1971). Leiden: E.J. Brill.

Franklin, A.D., Olin, J.S. and Wertime, T.A. (eds) 1978. *The Search for Ancient Tin*. Washington, DC: Smithsonian Institution Press.

Frenez, D. and Tosi, M. 2005. The Lothal sealings: records from an Indus Civilization port town at the eastern end of the maritime trade routes across the Arabian Sea. In *Studi in Onore di Enrica Fiandra*, M. Perna (ed.). Contributi di Archeologia Egea e Vicinorientale 1. Paris: De Boccard.

Frifelt, K. 1991. *The Island of Umm An-Nar. Vol. 1: Third Millennium Graves*. Jutland Archaeological Society Publications XXVI: 1. Aarhus: Aarhus University Press.

Gak, E.I. 2004. Eastern Europe's earliest brasses. Paper presented at the 34th International Symposium on Archaeometry, Zaragoza, Spain, 3–7 May 2004.

Gasanova, A. 2002. *История Познания и Использования Цинка и Латуни (Епохи Бронзы-Средневековье) [History of the Knowledge and Utilization of Zinc and Brass (Middle Bronze Age)]*. Baku: Elm Publishers.

Gaur, R.C. 1983. *Excavations at Atranjikhera: Early Civilisations of the Upper Ganga Basin*. New Delhi: Aligarth Muslim University/Motil Banarsidass.

Geckinli, A.E., Bozkurt, N. and Basaran, S. 1986. Cavustepe metal buluntularinin metallografik analizi. *Aksay Unitesi Bilimssel Toplanti Bildirileri I*. Tubitak Pub. No. 648: 229–46.

Glumac, P.D. and Todd, J.A. 1991. Early metallurgy in South-East Europe: the evidence of production. In *Recent Trends in Archaeometallurgical Research*, P.D. Glumac (ed.). Philadelphia: MASCA Research Papers in Science and Archaeology 8, 8–19.

Halleux, R. 1973. L'orichalque et le laiton. *Antique Classique* 42: 64–81.

Han, R. and Ko, Ts. 2000. Tests on brass items found at Jiangzhai, phase I. In Linduff *et al.* 2000, 195–9.

Hanfmann, G.M.A. 1956. Four Urartian bulls' heads. *Anatolian Studies* 6: 203–13.

Hauptmann, A. 1995. Chemische Zusammensetzung von Metallobjekten aus der Siedlung von Umm an-Nar. In *The Island of Umm an-Nar. Vol. 2: The Third Millennium Settlement*, K. Frifelt (ed.). Jutland Archaeological Society Publications XXVI: 2. Aarhus: Aarhus University Press, 246–8.

Hauptmann, H. and Pernicka, E. (eds) 2004. *Die Metallindustrie Mesopotamiens von den Anfängen bis zum 2. Jahrtausend v. Chr.* Orient-Abteilung des Deutschen Archäologieschen Instituts, Orient-Archäologie Bd. 3. Rahden/Westf.: Verlag Marie Leidorf GmbH.

Heskel, D.L. and Lamberg-Karlovsky, C.C. 1980. An alternative sequence for the development of metallurgy: Tepe Yahya, Iran. In Wertime and Muhly 1980, 229–66.

Hill, D.K. 1969. Bronze working: sculpture and other objects. In *The Muses at Work*, C. Roebuck (ed.) Cambridge, MA: MIT Press, 60–95.

Hosler, D. 1994. *The Sounds and Colors of Power: The Sacred Metallurgical Technology of Ancient West Mexico*. Cambridge, MA: MIT Press.

Hughes, M.J., Lang, J.R.S., Leese, M.N. and Curtis, J.E. 1988. The evidence of scientific analysis: a case study of the Nimrud Bowls. In *Bronzeworking Centres of Western Asia*, J.E. Curtis (ed.). London: Kegan Paul, 311–16.

Ixer, R.A. and Pattrick, R.A.D. 2003. Copper-arsenic ores and Bronze Age mining and metallurgy with special reference to the British Isles. In Craddock and Lang 2003, 216–30.

Jones, A. and MacGregor, G. 2002. *Colouring the Past: The Significance of Colour in Archaeological Research*. Oxford: Berg.

Kashkai, M.A. and Selimkhanov, I.R. 1973. *Из Истории Древней Металлургии Кавказа* [*On the History of Early Caucasian Metallurgy*]. Baku: Akademija Nauk Azerbajdzhanskoj CCP.

Kavtaradze, G.L. 1999. The importance of metallurgical data for the formation of a Central Transcaucasian chronology. In *The Beginnings of Metallurgy*, A. Hauptmann, E. Pernicka, Th. Rehren and Ü Yalçin (eds). *Der Anschnitt* Beiheft 4. Bochum: Deutsches Bergbau-Museum, 67–101.

Kohl, P.L. 2002. Bronze production and utilization in Southeastern Daghestan, Russia: c. 3600–1900 BC. In Bartelheim *et al.* 2002, 161–84.

Lamberg-Karlovsky, C.C. and Potts, D.T. (eds) 2001. *Excavations at Tepe Yahya, Iran 1967–1975: The Third Millennium*. American School of Prehistoric Research Bulletin 45. Cambridge: Harvard University Press.

Lechtman, H. 1977. Style in technology: some early thoughts. In *Material Culture: Styles, Organization, and Dynamics of Technology*, H. Lechtman and R.S. Merrill (eds). St. Paul: West Publishing Co., 3–20.

Lechtman, H. 1999. Afterword. In *The Social Dynamics of Technology*, M.-A. Dobres and C.R. Hoffman (eds). Washington, DC: Smithsonian Institution Press, 223–32.

Linduff, K.M., Han, R. and Sun, Sh. (eds) 2000 *The Beginnings of Metallurgy in China*. New York: Edwin Mellen Press.

Mei, J. and Li, Y. 2003. Early copper technology in Xinjiang, China: the evidence so far. In Craddock and Lang 2003, 111–21.

Méry, S. 2000. *Les céramiques d'Oman et l'Asie moyenne: une archéologie des échanges à l'Age du Bronze*. Monographie du CRA, no 23. Paris: CNRS.

Muhly, J.D. 1973. *Copper and Tin: The Distribution of Mineral Resources and the Nature of the Metals Trade in the Bronze Age*. Hamden, CT: Archon Books.

Muhly, J. 2002. Early metallurgy in Greece and Cyprus. In *Anatolian Metal II*, Ü Yalçin (ed.). *Der Anschnitt* Beiheft 15. Bochum: Deutsches Bergbau-Museum, 77–82.

Northover, J.P. 1997. The analysis of early copper and copper alloys. In *The Art of Ancient Iran: Copper and Bronze*, H. Mahboubian (ed.). London: Philip Wilson, 325–42.

Oren, E.D. 1971. A Middle Bronze Age I warrior tomb at Beth-Shan. *Zeitschrift des Deutschen Palästina-Vereins* 87: 109–39.

Ottaway, B.S. 2001. Innovation, production, and specialization in early prehistoric copper metallurgy. *European Journal of Archaeology* 4(1): 87–112.

Parzinger, H. and Boroffka, N. 2003. *Das Zinn der Bronzezeit in Mittelasien I*. Archäologie in Iran und Turan, Bd. 5. Mainz am Rhein: Verlag Philp von Zaubern.

Penhallurick, R.D. 1986. *Tin in Antiquity*. London: Institute of Metals.

Pernicka, E. 1998. Die Ausbreitung der Zinnbronze im 3. Jahrtausend. In *Mensch und Umwelt in der Bronzezeit Europas*, E. Hänsel (ed.). Kiel: Oetker-Voges Verlag, 135–47.

Pernicka, E., Seeliger, T.C., Wagner, G.A. *et al.* 1984. Archaeo-metallurgische Untersuchungen in Nordwestanatolien. *Jahrbuch des Romisch-Germanischen Zentralmuseums Mainz* 31: 533–602.

Pernicka, E., Eibner, C., Oztunali, O. and Wagner, G. A. 2003. Early Bronze Age metallurgy in the north-east Aegean. In Wagner *et al.* 2003, 143–72.

Pigott, V.C. 1980. The Iron Age in Iran. In Wertime and Muhly 1980, 417–61.

Pigott, V.C., Howard, S.M. and Epstein, S.M. 1982. Pyrotechnology and culture change at Bronze Age Tepe Hissar (Iran). In *Early Pyrotechnology: The Evolution of the First Fire-Using Industries*, T.A. Wertime and S.F. Wertime (eds). Washington, DC: Smithsonian Institute Press, 215–36.

Pollard, A.M. and Heron, C. 1996. *Archaeological Chemistry*. Cambridge: Royal Society of Chemistry.

Ponting, M.J. 2002. Roman military copper-alloy artefacts from Israel: questions of organization and ethnicity. *Archaeometry* 44(4): 555–71.

Primas, M. 2002. Early tin-bronze in central and southern Europe. In Bartelheim *et al.* 2002, 303–14.

Rao, S.R. 1985. *Lothal: A Harappan Port Town (1955–1962)*, vol. 2. Memoirs of the Archaeological Survey of India 78. New Delhi: Archaeological Survey of India.

Rau, W. 1974. *Metalle und Metallgeräte im vedischen Indien*. Abhandlungen der Geistes- und Sozialwissenschaftlichen Klasse. Jahrg. 1973, Nr. 8. Wiesbaden: Akademie der Wissenschaft zu Mainz, F. Steiner.

Rehren, Th. 1999. Small size, large scale: Roman brass production in Germania Inferior. *Journal of Archaeological Science* 26: 1083–7.

Renfrew, C. 1967. Cycladic metallurgy in the Aegean Early Bronze Age. *American Journal of Archaeology* 71, 1–120.

Renfrew, C. 1978. The anatomy of innovation. In *Social Organisation and Settlement: Contributions from Anthropology, Archaeology, and Geography*, D. Green, C. Haselgrove and M. Spriggs (eds). BAR International Series (Supplementary) 47. Oxford: Oxford University Press, 89–117.

Renfrew, C. and Slater, E.A. 2003. Metal artifacts and metallurgy. In *Prehistoric Sitagroi: Excavations in Northeast Greece, 1968–1970. Vol. 2: The Final Report*, E.S. Elster and C. Renfrew (eds). Monumenta Archaeologica 20. Los Angeles: Cotsen Institute of Archaeology at UCLA, 301–24.

Rostoker, W. and Dvorak, J. 1991. Some experiments with co-smelting to copper alloys. *Archaeomaterials* 5: 5–20.

Schachner, A. 2005. Von Bronze zu Eisen: Die Metallurgie des 2. und frühen 1. Jahrtausends v. Chr. im östlichen Transkaukasus. In *Anatolian Metal III*, Ü Yalçin (ed.). *Der Anschnitt* Beiheft 18. Bochum: Deutsches Bergbau-Museum, 175–90.

Schaeffer-Forrer, C.F.A., Zwicker, U. and Nigge, K. 1982. Untersuchungen an metallischen Werkstoffen und Schlacken aus dem Bericht von Ugarit (Ras Shamra, Syrien). *Mikrochimica Acta* 1: 35–61.

Sofaer Derevenski, J. and Sørensen, M.-L. S. 2002. Becoming cultural: society and the incorporation of bronze. In *Metals and Society*, B.S. Ottaway (ed.). BAR International Series 1061. Oxford: Archaeopress, 117–22.

Stech, T. and Pigott, V.C. 1986. The metals trade in southwest Asia in the third millennium BC. *Iraq* 48: 39–64.

Steinberg, A. 1981. Analyses of selected Gordion bronzes. In Young 1981, 286–9.

Stos-Gale, Z. 1992. The origin of metal objects from the Early Bronze Age site of Thermi on the island of Lesbos. *Oxford Journal of Archaeology* 11(2): 155–77.

Sun Shuyun and Han Rubin 1983–85. A preliminary study of early Chinese copper and bronze artifacts, J.K. Murray (trans.). *Early China* 9–10: 261–89.

Tedesco, L.A. 2006. *Redefining Technology in Bronze Age Transcaucasia: Copper-alloy Metallurgy in Armenia in the Third to Second Millennium BC*. PhD dissertation, Department of Anthropology, New York University.

Thornton, C.P. and Ehlers, C. 2003. Early brass in the ancient Near East. *IAMS Journal* 23: 3–8.

Thornton, C.P. and Lamberg-Karlovsky, C.C. 2004. A new look at the prehistoric metallurgy of Southeastern Iran. *Iran* XLII: 61–76.

Thornton, C.P., Lamberg-Karlovsky, C.C., Liezers, M., and Young, S.M.M. 2002. On pins and needles: tracing the evolution of copper-base alloying at Tepe Yahya, Iran, via ICP–MS analysis of common-place items. *Journal of Archaeological Science* 29(12): 1451–60.

Thornton, C.P., Lamberg-Karlovsky, C.C., Liezers, M. and Young, S.M.M. 2005. Stech and Pigott revisited: new evidence concerning the origins of tin-bronze in light of chemical and metallographic analyses of the metal artefacts from Tepe Yahya, Iran. In *Proceedings of the Third International Symposium on Archaeometry, 22–26 April 2002, Amsterdam*, H. Kars and E. Burke (eds). Geoarchaeological and Bioarchaeological Studies 3. Amsterdam: Vrije Universiteit, 395–8.

Wagner, G.A., Pernicka, E. and Uerpmann, H.-P. (eds) 2003. *Troia and the Troad: Scientific Approaches.* Berlin/Heidelberg: Springer-Verlag.

Weeks, L. 1999. Lead isotope analyses from Tell Abraq, United Arab Emirates: new data regarding the 'tin problem' in western Asia. *Antiquity* 73: 49–64.

Weeks, L.R. 2003. *Early Metallurgy of the Persian Gulf.* Boston: Brill Academic Publishers, Inc.

Welter, J.-M. 2003. The zinc content of brass: a chronological indicator? *Techne* 18: 27–36.

Wertime, T.A. 1980. The pyrotechnologic background. In Wertime and Muhly 1980, 1–24.

Wertime, T.A. and Muhly, J. (eds) 1980. *The Coming of the Age of Iron.* New Haven, CT: Yale University Press.

Young, R.S. 1981. *Three Great Early Tumuli.* The Gordion Excavations Final Reports 1. Philadelphia: University Museum Monograph 43.

## Author's address

Christopher P. Thornton, University of Pennsylvania, Department of Anthropology, 3260 South St., Philadelphia, PA 19104, USA (cpt2@sas.upenn.edu)

# Brasses in the early metallurgy of the Iberian Peninsula

*Ignacio Montero-Ruiz and Alicia Perea*

*ABSTRACT*   In this paper we explore the use of brass in pre-Roman metallurgy in the Iberian Peninsula through the elemental analysis of metal objects. The earliest brass alloy appears in the 6th century BC in a colonial context and therefore could be imported, but other true brasses have been detected in the Iberic culture dating to the 4th and 3rd centuries BC. High zinc percentages (greater than 2%) are also found in metals of this area, which could be related to the re-melting of brasses, indicating a more regular presence of brass than that detected analytically to date.

*Keywords:* brass, alloy, XRF, analysis, Iberian Peninsula, pre-Roman, Iron Age, Iberic culture.

## Introduction

In the history of metallurgical technology, copper-zinc alloys were a later development than those that used copper with tin and lead. Although very ancient brasses have recently been documented in various parts of the world (Craddock 2003, and see Thornton in this volume, pp. 123–35), it was not until well into the 1st millennium BC that one can speak of a controlled and deliberate knowledge of alloying in the Mediterranean and European area. Brass clearly existed in the Hellenistic world and was known to the Etruscans (Craddock 1978; 1995: 294). Some brasses dating to before the Roman conquest have also been identified recently in Celtic Europe (Craddock 2003) and studies have been made of the historical evolution of the use of zinc alloys, both in Roman coins (Caley 1964; Dungworth 1996) and in items of adornment (Bayley 1990; Dungworth 1997). Within this panorama little is known about the situation in the Iberian Peninsula, a region in which research needs to take into account two important factors: its mineral wealth[1] and the development of trade with the eastern Mediterranean through the processes of Phoenician and Greek colonisation. In this study our aim is to review the available analytical data[2] and offer a brief synthesis of the pre-Roman history of zinc in the Iberian Peninsula.

## Pre-Roman brasses

It seems clear that, like other regions of the western Mediterranean, copper-zinc alloys were not generally produced in the Iberian Peninsula until the 1st century BC, within the Roman period. Of the more than 3000 Iron Age (7th–2nd century BC) copper-based objects analysed in the Iberian Peninsula Archaeometallurgy Project by energy-dispersive X-ray fluorescence (ED–XRF),[3] only a few contain zinc. The presence of brasses in the pre-Roman period can in some cases be explained by imports from the eastern Mediterranean because they have been recovered in colonial contexts. From the 2nd century BC onwards, however, the exceptions become more frequent until the turning point in the mid-1st century BC when the production of brass in the Roman world became more widespread thanks to the cementation method.

## Catalogue

The oldest known brass in Spain is a pin with a circular cross-section and rounded head that appeared in the Phoenician tomb at Casa del Obispo (Cadiz). This is a monumental tomb constructed with stone ashlars and set on a platform or podium of ashlars. Although part of it had been plundered, it was possible to document the presence of a wooden coffin and the inhumation of an individual whose remains were found in a very poor state of preservation. The grave goods recovered included a gold finger-ring and several sheets of gold concentrated in the region of the lower limbs (Perea *et al.* 2004). The burial architecture and the grave goods belong to a high-status man. The pin has a circular cross-section and spherical head, barely 4 cm long, although the point is missing (Fig. 1). Proton-induced X-ray emission (PIXE) analysis[4] gave a value of 9.9% zinc, with no significant impurities of other elements (see Table 1). The date proposed for the tomb is the 6th century BC. Given the context in which the pin appeared and the absence of brasses in the Iberian Peninsula at that date, we can assume that it had been imported from the eastern Mediterranean, where brass was known at that period.

The next object made from a copper-zinc alloy comes from the site of Torre de Doña Blanca, also in Cadiz. The excavation evidence indicates that the site was occupied from the 8th century BC to the end of the 3rd century BC, although it was occupied again during the medieval period (Ruiz-Mata 1994:

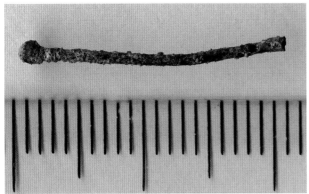

**Figure 1** Circular cross-section and rounded head pin from Casa del Obispo (Cadiz) and dated in the 6th century BC.

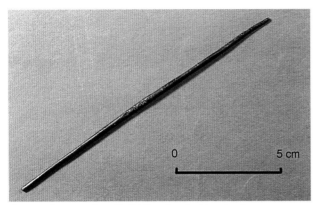

**Figure 2** Decorated needle (13.5 cm long) with a circular cross-section from Torre de Doña Blanca (Cadiz).

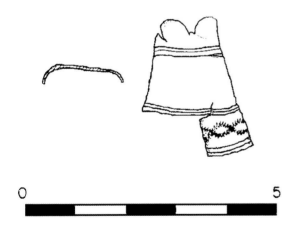

**Figure 3** Decorated sheet from the Iberic votive hoard of El Amarejo (Albacete).

4). The piece that interests us is a decorated needle 13.5 cm long with a circular cross-section (Fig. 2). It was recovered from pit C1, in a level beneath a medieval waste ditch. This level corresponds to an Ibero-Turdetan occupation dated to between the 4th and 3rd centuries BC.[5] The composition is 7.8% zinc and 2.4% lead, and the metallography indicates that after casting it was forged and finally annealed.

A fragment of decorated sheet, 2.5 cm long (Fig. 3) from the settlement of El Amarejo (Albacete) is of a similar age. This piece was recovered together with other metal, ceramic and glass finds from a rock-cut pit 3.5 m deep, interpreted as a votive deposit (Broncano 1989). This sheet brass fragment was recovered from layer 12 (between −1.65 and −1.80 m), about halfway down the pit, which was first used around the middle of the 4th century BC and last used when the settlement was abandoned between the end of the 3rd century and the beginning of the 2nd century BC. Of the 98 metal objects recovered from the deposit, there are two gold objects, 20 lead, 23 iron and 52 copper-based. Thirty-nine objects belonging to the latter group were analysed, of which only one is brass with a zinc content of 16.6%. We know of two other brasses from the same site, but their superficial position and their typology (bells) for which we have found no safe parallels in the Iberian world make us cautious about dating them, although their composition is very similar to the sheet of brass from the votive hoard, but with more tin and lead impurities.

Of slightly later date is the discoidal sheet from Castilmontán (Soria), dating to the 2nd century BC. It was recovered from the structure of a house (cut 4, room 3). Its composition contains higher levels of other elements than the earlier pieces; in addition to 17.6% zinc the alloy contains 1.2% tin and 5.2% lead.

Finally, a fragment of rod (AC.20.88.4(1)) with 9.8% zinc, 1.6% tin and 1.7% lead was recovered from the late Iberian phase (2nd–1st century BC) of the site at El Acequión (Albacete), a settlement with no signs of Roman occupation.[6]

The presence of more than 1% zinc is very occasionally found in the pre-Roman metallurgy of the Iberian Peninsula. Although they are not true brasses, we should mention these exceptions, since the presence of zinc could mean that brasses were melted down and reused, thus indicating the indirect presence of this alloy. The most convincing case in support of this hypothesis is an annular brooch in the Museum of Albacete which probably came from the necropolis of Hoya de Santa Ana. The typology of the piece indicates it is pre-Roman, although we cannot be more precise because it belongs to subtype 10, which was in use from the 5th to the 1st century BC (Sanz Gamo et al. 1992: 118). Its composition (see Table 2), with 4% zinc, could be explained as a base metal obtained from the reduction of copper-zinc minerals. We reject that

**Table 1** Pre-Roman objects made from zinc alloys (% in weight, nd = not detected).

| Analysis | Site | Description | Chronology | Fe | Ni | Cu | Zn | As | Ag | Sn | Sb | Pb | Bi |
|---|---|---|---|---|---|---|---|---|---|---|---|---|---|
| Namur | Casa del Obispo | Pin | 6th century BC | | | 90,1 | 9,9 | | | | | | |
| AA1428 | Torre Doña Blanca | Needle | 4th–3rd century BC | 0.20 | 0.62 | 88.2 | 7.8 | 0.52 | nd | 0.03 | 0.008 | 2.41 | nd |
| PA0708 | El Amarejo | Sheet | 4th–3rd century BC | 0.75 | 0.47 | 81.9 | 16.6 | nd | 0.002 | nd | 0.013 | nd | nd |
| PA2628 | Castilmontán | Sheet | 2nd century BC | 0.24 | 0.16 | 75.3 | 17.6 | nd | 0.043 | 1.24 | 0.080 | 5.24 | nd |
| PA1575 | El Acequión | Rod | 2nd–1st century BC | 0.20 | 0.23 | 85.8 | 9.83 | nd | 0.041 | 1.63 | 0.096 | 1.70 | nd |

**Table 2** Pre-Roman objects with high zinc impurities (% in weight, nd = not detected).

| Analysis | Site | Description | Chronology | Fe | Ni | Cu | Zn | As | Ag | Sn | Sb | Pb | Bi |
|---|---|---|---|---|---|---|---|---|---|---|---|---|---|
| PA2858 | Hoya de Santa Ana? | Annular brooch | 5th–1st century BC | 0.19 | 0.13 | 77.8 | 4.0 | nd | nd | 16.7 | 0.078 | 0.99 | nd |
| PA2887 | Puntal Horno Ciego | Finger-ring | 6th–5th century BC | 0.07 | 0.13 | 65.6 | 3.9 | nd | 0.016 | 10.3 | 0.021 | 19.9 | nd |
| 1J-144/10 | Cerro de los Santos | Finger-ring | 2nd–1st century BC | 0.33 | 0.24 | 58.6 | 5.1 | nd | 0.11 | 19.6 | 1.57 | 14.4 | nd |
| 1J-144/6 | Cerro de los Santos | Finger-ring | 2nd–1st century BC | 0.25 | 0.13 | 72.8 | 2.3 | nd | 0.20 | 15.9 | 0.96 | 9.64 | nd |

option, however, particularly taking into account that high zinc impurities are exceptional in the metallurgy of the Iberian region and in the metallurgy of the Iberian Peninsula as a whole since its earliest days.[7]

The Puntal del Horno Ciego (Villagordo del Cabriel, Valencia) is a sanctuary cave with a stratigraphic sequence dated in the Early Iberian period from the end of the 6th to the end of the 5th century BC. A great number of small caliciform pottery vessels were recovered together with six metal objects (Martí Bonafe 1990). Only a finger-ring (Fig 4: no. 7079) contains zinc at a high level (3.9%) in a leaded bronze alloy. The rest of the metal items, including another ring, have no traces of zinc in their composition.

Other Iberian pieces with contexts close to the Roman period that could have been produced by using melted-down metal are two finger-rings from the Iberian sanctuary of Cerro de los Santos (Albacete). Both finger-rings are leaded bronzes that contain 5.1% and 2.3% zinc respectively.

Another series of pieces made from zinc alloys dating to the 2nd and 1st centuries BC (Ibero-Roman period) that have been analysed do not have a precise context and the dates attributed to them could be doubtful. Similarly, the Royal Academy of History catalogue of materials has recently been published (Almagro *et al.* 2004) incorporating the composition of the metal objects, and some pieces typologically dating to this period contain varying percentages of tin, lead and zinc, the latter ranging from 5 to 10%.

## Conclusions

In conclusion we can say first of all that pre-Roman brasses were present in the Iberian Peninsula (Fig. 4) from at least the 6th century BC, although some of them may have been imports from the eastern Mediterranean; that between the 4th and 1st century BC some true brasses have been identified in the Iberian area and a tendency towards the presence of zinc in low proportions can be detected in other pieces that, while not constituting true brasses, could be classified as gunmetal or leaded gunmetal; and that this presence of alloyed zinc is not observed in the metallurgy of the previous period, when no metals with more than 1% zinc are recorded, and therefore the explanation of their natural formation through the use of copper-zinc minerals does not seem likely. If, on the other hand, they were produced by the recycling of metals, they would indicate a more regular presence of brass than that detected analytically to date. The observed concentration in the Albacete province (Fig. 5) reflects only the predominance of analytical work in this province (166 objects), while

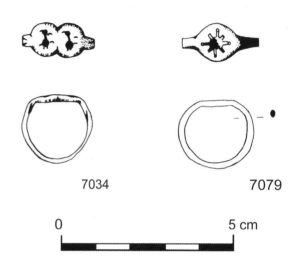

**Figure 4** Finger-rings from Puntal del Horno Ciego: no. 7079 contains 3.9% zinc.

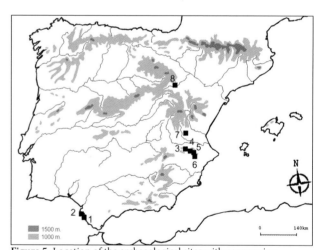

**Figure 5** Location of the archaeological sites with copper-zinc alloys mentioned in the text: 1. Cadiz; 2. Torre de Doña Blanca; 3. El Amarejo; 4. El Acequión; 5. Hoya de Santa Ana; 6. Cerro de los Santos; 7. Puntal del Horno Ciego; 8. Castilmontan.

the available information from Valencia (20), Murcia (10) or Alicante (4) provinces is scarce. None of the 189 objects analysed from Catalonia, however, dating to the Iberian period, contains zinc. The important pre-Roman mining of lead-zinc ores in the Murcia-Albacete district (Orejas and Montero-Ruiz 2001) could explain the early appearance of zinc in some metals from this area. From the mid-1st century BC onwards brasses with high percentages of zinc (greater than 15%), but never more than 30%, are generally used both in coins and ornaments all over the Iberian Peninsula.

## Notes

1. Sulphide (sphalerite) and non-sulphide zinc (calamine) occurrences in Spain are frequent. The most important were located in the north of Spain, but also in Andalusia and in Murcia. Of particular note are the mines from La Union (Cartagena, Murcia) exploited since the 2nd century BC (Antolinos 2001; Domergue 1987).

2. This paper is based mainly on the analytical results of the Iberian Peninsula Archaeometallurgy Project. This project began in 1982 and is still ongoing. Few other analyses of Iron Age metals have been published and none contain copper-zinc alloys.

3. Most of the analyses were carried out with a Kevex-7000 spectrometer working at 40 KeV and with an area of 80mm². Measuring time was fixed in 500 s. Post-2002 a new spectrometer was used: a portable METOREX X-MET 920MP using a $^{109}$Cd and $^{241}$Am gamma ray source and a Si (Li) detector. Measuring time was 300 s. Analytical procedures were similar with both sets of equipment: all the objects, except those which were highly mineralised, were mechanically polished to clean the surface. Calibration is based on the analysis of standards with known elemental compositions. Detection limits vary for each element: 10 ppm (silver and antimony), 100 ppm (arsenic, lead, tin, gold, iron) to 200 ppm (nickel) or 1000 ppm for zinc.

4. The analysis was carried out at LARN in Namur (Belgium) as part of the Joint Spanish/Belgian Project (200BE0002) Tartessian Precious Metal Work. We should like to thank Guy Demortier and all the laboratory staff for their help.

5. The village was abandoned in the last decade of the 3rd century BC (Ruiz-Mata 1994: 14).

6. We thank Dolores Fernández-Posse and Concepción Martín for this useful information.

7. Zinc higher than 1% is not found in any of 4500 objects from the Chalcolithic to Late Bronze Age analysed in the Iberian Peninsula Archaeometallurgy Project. The only exception is a rod with biconical terminals from the La Lloseta hoard in Mallorca (Delibes de Castro and Fernández-Miranda 1988: 36–40, 163) which contains 15% tin and 2.4% zinc. The hoard is dated to the Talayotic (9th–8th centuries BC).

## References

Almagro, M., Casado, D., Fontes, F., Mederos, A. and Torres, M. 2004. *Prehistoria. Antigüedades Españolas I.* Madrid: Real Academia de la Historia. Catálogo del Gabinete de Antigüedades.

Antolinos, J.A. 2001. Les mines de la Sierra de Cartagena. In *Atlas historique des zones minières d´Europe*, A. Orejas (dir.). Commission Européenne. Action COST G2 'Paysages anciens et structures rurales'. Luxemburg: Dossier II.

Bayley, J. 1990. The production of brass in antiquity with particular reference to Roman Britain. In *2000 Years of Zinc and Brass*, P.T. Craddock (ed.). British Museum Occasional Paper 50. London, 7–27.

Broncano, S. 1989. *El depósito votivo ibérico de El Amarejo. Bonete (Albacete).* Excavaciones Arqueológicas en España, 156. Madrid: Ministerio de Cultura.

Caley, E. R. 1964: *Orichalcum and Related Ancient Alloys.* New York: American Numismatic Society.

Craddock, P.T. 1978. The composition of the copper alloys used by the Greek, Etruscan and Roman civilizations. 3. The origins and early use of brass. *Journal of Archaeological Science* 5: 1–16.

Craddock, P.T. 1995. *Early Metal Mining and Production.* Edinburgh: Edinburgh University Press.

Craddock, P.T. 2003. Europe's earliest brasses? *Historical Metallurgy Society News* 54: 3–4.

Delibes de Castro, G. and Fernández-Miranda, M. 1988. *Armas y utensilios de bronce en la Prehistoria de las Islas Baleares.* Studia Archaeologica, 78. Universidad de Valladolid.

Domergue, C. 1987. *Catalogue des mines et des fonderies antiques de la Péninsule Ibérique.* Madrid: Casa de Velázquez Diffusion de Boccard.

Dungworth, D. 1996. Caley's 'zinc decline' reconsidered. *Numismatic Chronicle* 156: 228–34.

Dungworth, D. 1997. Roman copper alloys: analysis of artefacts from northern Britain. *Journal of Archaeological Science* 24: 901–10.

Martí Bonafe, M.A. 1990. Las cuevas del Puntal del Horno Ciego. Villargordo del Cabriel, Valencia. *Saguntum* 23: 141–82.

Orejas, A. and Montero-Ruiz, I. 2001. Colonizaciones, minería y metalurgia prerromanas en el levante y sur peninsulares. In *De la Mar y de la Tierra. Producciones y productos fenicio-púnicos. XV jornadas de arqueología fenicio-púnica (eivissa 2000).* Treballs del Museu Arqueològic d´Eivissa i Formentera 47: 121–59.

Perea, A., Montero, I., Cabrera, A., Feliu, M.J., Gayo, M.D., Gener, J.M. and Pajuelo, J.M. 2004. El ajuar de oro de la tumba fenicia del Obispo, Cádiz'. In *Tecnología del oro antiguo: Europa y América. Anejos de Archivo Español de Arqueología*, XXXII, A. Perea, I. Montero and O. García Vuleta (eds), 231–41.

Ruiz-Mata, D. 1994. El poblado fenicio del Castillo de Doña Blanca. Introducción al yacimiento. In *Castillo de Doña Blanca. Archaeo-environmental Investigations in the Bay of Cádiz, Spain (750–500 BC)*, E. Roselló and A. Morales (eds). BAR International Series 593. Oxford: Archaeopress, 1–19.

Sanz Gamo, R., López Precioso, J. and Soria, L. 1992. *Las fíbulas de la provincia de Albacete.* Albacete: Instituto de Estudios Albacetenses, 66.

## Author's address

Corresponding author: Ignacio Montero-Ruiz, Dpto. Prehistoria, Instituto de Historia (CSIC), C/Albasanz, 26-28, 28037 Madrid, Spain (imontero@ih.csic.es)

# The beginning of the use of brass in Europe with particular reference to the southeastern Alpine region

## J. Istenič and Ž. Šmit

ABSTRACT    Research on Roman brooches of the 1st century BC has shown that the beginning of the relatively regular use of brass in Europe should be dated to *c*. 60 BC, about 15 years before the first issue of brass coins in Europe in 46/45 BC. Brass was initially used for military equipment (which included brooches) and not for coins, as has been generally accepted until now.

*Keywords:* brass, brooches, PIXE, Roman, Italic, coin, Europe.

## Introduction

It is generally assumed that it was the Romans who spread the use of brass (a metal that they called *aurichalcum* or *orichalcum*) through Europe (Caley 1964: 31; Craddock 1978: 8–9). It seems likely that they encountered brass in Asia Minor, where deliberate production had probably commenced at the beginning of the 1st century BC. The earliest intentionally made and reliably dated items of brass in the region comprised a coin series, issued by Mithradates VI, the king of Pontus, in the period *c*. 75–65 BC. By about 50 BC, brass was frequently used for coinage in the Roman provinces of Asia, Bithynia-with-Pontus and Cilicia (Burnett *et al.* 1982). Brass was probably produced by a direct process reacting zinc ores and copper metal together in a closed crucible (Cowell *et al.* 2000: 670, 677) and by zinc oxide reacting with molten copper (Craddock and Eckstein 2003).

An important landmark in the use of brass in Europe was the coinage reform instigated by the Roman emperor Augustus at the mint of Rome in *c*. 23 BC. As part of this reform, brass was introduced as a new base metal (Burnett 1987: 54; Burnett *et al.* 1982). From the Augustan period onwards, and especially in the 1st century AD, brass was also widely used for military equipment, brooches and medical instruments (Bayley 1990: 13–21; Bayley and Butcher 1995; 2004: 209–10; Craddock and Lambert 1985; Jackson and Craddock 1995: 89–99; Ponting 2002; Riederer 2001: 225–35; 2002a: 109–20; 2002b: 286–90).

There has been little research on the use of brass in Europe before Augustus. The existence of an issue of brass coins made in 46–45 BC by Julius Caesar's *praefectus Clovius*, probably in Cisalpine Gaul (present-day northern Italy) has already been known for a century (Bahrfeldt 1905: 42; 1909: 78–84). There is also evidence of Caesar's brass coinage in the province of Macedonia, dated to 44 BC. In Grant's opinion (1969:

13–19, 87–90), it was Caesar himself who authorised this coinage; he would probably have profited significantly from the *orichalcum* issues because brass coins were considerably overvalued. Grant's view (1969: 7–11, 13–19) that brass had its origin in these coin issues was widely accepted (Burnett *et al.* 1982: 264, n. 11; Caley 1964: 92). Caley (1964: 31, 92) stressed that initially the Romans used *orichalcum* only for coinage. This view was not challenged because numismatics aside, there was no evidence for pre-Augustan brass in Europe until recently. In 1996, a monograph appeared on the copper-alloy objects from the American excavations at the Titelberg, Luxembourg. It included three fragments of brass dated to 100–50 BC (Hamilton 1996: 59–60, fig. 48). Unlike Hamilton, who interpreted them as evidence that the Gauls were using brass earlier than the Romans, we see no reason to doubt that these objects were Roman (see discussion below). Moreover, four years after that publication, a well-preserved Roman sword scabbard with brass fittings was published. It was found in Slovenia and was dated to the beginning of the second half of the 1st century BC at the very latest (Istenič 2000a, 2000b; Šmit and Pelicon 2000). From this it was apparent that the Romans did not use brass for coinage alone, even initially, and therefore research on pre-Augustan Roman finds other than coins might add to our knowledge of the emergence of the use of brass in Europe.

The territory of the southeastern Alpine region (present-day Slovenia) seemed to be an appropriate geographical framework for this research. Being an immediate neighbour of Italy to the northeast, it was the zone traversed by all the land routes from Italy to the Balkans and the Danube region and it had contacts with the Romans from the 2nd century BC onwards. In the period under discussion, i.e. the middle of the 1st century BC, the western part of this territory was already under Roman control, and Roman influence was gradually spreading eastwards and southwards (Istenič 2000b). There is also evidence for the presence of the Roman army in this period (Isteničٔ 2005a; forthcoming). This is important, as the army was a major user of brass in the Augustan period – and also presumably before that as the Ljubljanica sword scabbard and perhaps also the sword scabbard fitting from Titelberg (see below) suggest.

## Research strategy

We anticipated that in addition to the sword scabbard from the river Ljubljanica there would be other Roman pre-Augustan objects made of brass. In particular, brass might reasonably be expected in items of Roman military equipment. Among pre-Augustan weapons, which are rare finds generally, items made of iron predominate significantly. There is a class of artefact worn by soldiers as well as by civilians, however, which was not usually made of iron: brooches. Because they are numerous in the southeastern Alps, these objects seemed to be an appropriate subject for our research.

For the purposes of comparison we included in our research not only Roman brooches but also contemporary brooches characteristic of the indigenous peoples of the southeastern Alps and also of northern Italy. We did not expect that they would include brass items. In fact, it seemed unlikely that the non-Roman population of this wider region would have already mastered the production of brass in pre-Roman times.

They might have reused imported Roman brass objects by melting them down, however, either on their own or mixed with bronze. In either of these instances, the alloy from such processes would inevitably be reflected in the percentage of zinc in the alloy.

## Selection of objects for analysis

A total of 77 brooches (all found in Slovenia) belonging to seven different groups was included in this research. On archaeological grounds, a division between the presumed Roman and the non-Roman brooches could be made for most of the groups. The Italic and Roman brooches belonged to three groups: Almgren 65, Alesia and Jezerine I.

Almgren 65 group brooches (Fig. 1) were manufactured in Italy. They were produced in large quantities and widely distributed through trade, especially to Celtic *oppida* in central Europe. It is assumed that their production began between 90 and 70 BC. The latest examples, which already exhibited some of the characteristics of the *Flügelfibeln*, were probably post-Caesarean (Demetz 1999: 27–38). Thirteen brooches in this group were analysed.[1]

The brooches of the Alesia group take their name from the site of Alesia, the Gaulish stronghold besieged and captured by Caesar in 52 BC. This was an explicitly Roman class of brooch, the earliest type to have had a new mechanism: a hinge instead

**Figure 1** Brooches of the Almgren 65 group. The largest brooch (at the bottom) is made of brass. (Source: Archive of the National Museum of Slovenia, photograph by Tomaž Lauko.)

**Figure 2** Brooches of the Alesia group. All the brooches were made of brass except the upper three brooches (upper left corner), which were made of bronze, and the brooch fragment in the middle of the second row at the bottom. (Source: Archive of the National Museum of Slovenia, photograph by Tomaž Lauko.)

141

**Figure 3** Brooches of the Jezerine I group, both made of brass. (Source: Archive of the National Museum of Slovenia, photograph by Tomaž Lauko.)

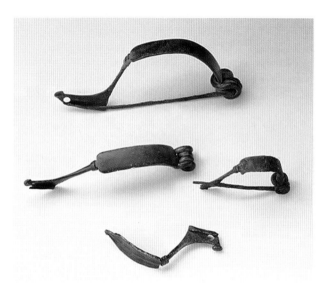

**Figure 4** A selection of brooches of the Jezerine II group made of brass (on the right), bronze (on the left) and gunmetal (top and bottom). (Source: Archive of the National Museum of Slovenia, photograph by Tomaž Lauko.)

of the spring which had characterised all previous brooches (Demetz 1999: 156–64). They were worn mostly by Roman soldiers, who are thought to have been the most important factor in their wide geographic distribution. The Alesia group brooches date roughly from the period of Caesar's Gallic wars (58/51 BC) to the period of the civil war following his death in 44 BC. Their production most probably ceased in the early years of the Augustan period (beginning in 27 BC) at the latest (Brouquier-Reddé and Deyber 2001: 295, 298, pl. 91/48; Istenič 2005a, 2005b; Ocharan Larrondo and Unzueta Portilla 2002: fig. 2/10). Eighteen brooches of this group were included in the research (Fig. 2).[2]

Jezerine I is a small group (20 brooches) distributed mainly in northern Italy and southern Gaul. Their production probably began in Italy around 50/40BC and perhaps continued into the Augustan period (Demetz 1999: 99–105, maps 29–31). Two brooches from this group[3] were submitted for analysis (Fig. 3).

The production of the large and widespread Jezerine II group of brooches probably started in northern Italy in about 40 BC, and continued in the Augustan period when they were presumably also produced in other regions such as the southeastern Alps and the Balkans. In fact, the brooches in this group were clearly concentrated in northern Italy and also on the eastern side of the Adriatic (Demetz 1999: 99–105). Twenty-five brooches of the Jezerine II group were analysed (Fig. 4).[4]

Three groups of non-Roman brooches of the 1st century BC were chosen for analysis: the southeastern Alpine group of so-called *Palmettenfibeln* (Fig. 5),[5] which were probably made locally in the southeastern Alps, the so-called *Schüsselfibeln* (Fig. 6),[6] probably by the non-Romanised population of northern Italy (Demetz 1999: 76–7, 64–72), and the Nauheim-type group of brooches (Fig. 7),[7] which were of Celtic origin. Of the

**Figure 5** *Palmettenfibeln*, all made of bronze. (Source: Archive of the National Museum of Slovenia, photograph by Tomaž Lauko.)

Figure 6 *Schüsselfibeln*, all made of bronze. (Source: Archive of the National Museum of Slovenia, photograph by Tomaž Lauko.)

Nauheim-type brooches analysed,[8] all except one belonged to a subgroup, Nauheim II. Brooches of this subgroup were found mostly in northeastern Italy and Slovenia, and were most probably produced in northeastern Italy during approximately the same period as the Almgren 65 brooches (Božič 1993: 141–3, 150–51; Salzani 1996: 51, pl. 26, C 8a).[9]

## Analytical methods

Two techniques were used to investigate the material from which the brooches were made: energy-dispersive X-ray fluorescence spectroscopy (XRF) and proton-induced X-ray emission spectrometry (PIXE). The other analytical methods at our disposal (e.g. inductively coupled plasma–atomic emission spectroscopy (ICP–AES)) were not appropriate, as most of the investigated objects were extremely thin, and for some of them hardly any metal core had survived. We expected, however, that the results obtained by the PIXE method would be accurate enough for the purposes of our research. In addition, PIXE spectrometry was also suitable for the analysis of the surface coatings (not discussed here).[10]

The presence or absence of zinc in the brooch alloys was determined by the application of XRF analysis to the unprepared surface of the objects.[11] The measurements involved a circular area, 11 mm in diameter. Thus the unprepared surface, i.e. the corrosion layer on the surface of the brooches, was investigated. For this reason, the results provide only an estimate of the approximate composition of the alloy. As our main interest was to detect brass or gunmetal brooches, the presence or absence of zinc was of primary importance.

Due to de-zincification we expected that the proportion of zinc in the corrosion layer on objects made of brass or gunmetal could be very small, therefore the unprepared surfaces of the brooches were also analysed by PIXE. The detection limit for zinc with PIXE was relatively high at < 0.5%, mainly due to interference of zinc K-alpha line with a strong, asymmetric K-beta line of copper.

The brooches for which the XRF or PIXE analyses showed the presence of zinc on the unprepared surface were submitted for further PIXE analyses.[12] The technique was applied to prepared areas of 2–3 mm², one or more on each brooch, from which the corrosion layer had been removed as thoroughly as possible, down to the metal core. It was obvious that all the corrosion products could not be removed from the selected area on the very poorly preserved brooch, RM Novo mesto inv. no. 1256. Observation of the prepared areas under a binocular

Figure 7 Nauheim group brooches, all made of bronze. (Source: Archive of the National Museum of Slovenia, photograph by Tomaž Lauko.)

microscope showed that, even with relatively well-preserved brooches, the complete removal of corrosion products from the surface area was a rather difficult task. This is also the main reason for the variation of concentrations obtained in different areas of the same brooch. The results of the measurements therefore provide approximate information on the composition of the brooches.

Very small areas were measured with a narrow beam (0.3 mm) in addition to measurements made with a beam of 2 mm diameter (as used normally). Unless otherwise stated, the values given in Table 1 were obtained with a 2 mm beam. In the calculation of concentrations given in Table 1, the sum of all metal constituents was normalised.

## Results

Among the brooches of groups that are regarded as non-Roman (the eight brooches of southeastern Alpine *Palmettenfibeln*, three *Schüsselfibeln* and eight Nauheim-type brooches), not a single example was found to have been made of a zinc-containing alloy. They were all made of bronze or leaded bronze, alloys with a long prehistoric ancestry (cf. Giumlia-Mair 1998; Jerin 2001; Trampuž Orel 1999: 415–17).

Zinc alloys – brass and gunmetal (Table 1) – appeared in the three groups of brooches that were regarded as Italic or Roman, that is the Almgren 65, Alesia and Jezerine I groups. They also appeared among the brooches of the Jezerine II

**Table 1** Elemental concentrations (in wt%) measured by PIXE on small areas from which the corrosion layer was removed as thoroughly as possible. Only brooches with zinc are included. Abbreviations: for NMS, IPCHS NG and RMNG see notes 1 and 3; RM = Regional Museum, n.b. = narrow beam. Asterisk (*) indicates area where considerable traces of corrosion products remained.

| Brooch | Group | Meas. | Cu | Zn | Sn | Pb | Fe | Ag | Ni | As | Bi | Notes |
|---|---|---|---|---|---|---|---|---|---|---|---|---|
| NMS, inv. no. R18485 | Almgren 65 | area 1 | 76.9 | 21.1 | 0.9 | 0.26 | 0.8 | | | | | |
| NMS, inv. no. R17393 | Alesia | area 1 | 78.2 | 19.9 | 0.4 | 0.36 | 1.2 | | | | | |
| " | | area 2 | 78.4 | 19.8 | 0.4 | 0.31 | 1.1 | 0.07 | | | | |
| NMS, inv. no. P19282 | " | area 1 | 76.4 | 21.3 | 0.6 | 0.49 | 1.0 | | 0.15 | | | |
| " | | area 2 | 76.0 | 21.0 | 0.8 | 0.53 | 1.7 | | 0.08 | | | |
| RMNG, inv. no. 7 | " | area 1 | 90.0 | 3.1 | 5.4 | 1.06 | 0.3 | | | 0.09 | | |
| NMS, inv. no. R17281 | " | area 1 | 80.7 | 16.8 | 1.2 | 0.26 | 1.1 | | | | | |
| " | | area 2 | 81.4 | 15.9 | 1.2 | 0.30 | 1.2 | | | | | |
| NMS, inv. no. R18974 | " | area 1 | 77.9 | 20.1 | 0.6 | 0.98 | 0.5 | | | | | |
| " | " | | 80.8 | 18.1 | | 0.92 | 0.26 | | | | | n.b. |
| RMNG, inv. no. 10 | " | area 1 | 79.2 | 15.4 | 2.8 | 0.86 | 1.6 | 0.12 | | | | |
| NMS, inv. no. R17319 | " | area 1 | 79.2 | 19.9 | 0.5 | 0.28 | <0.5 | | 0.10 | 0.01 | | |
| " | | area 2 | 79.0 | 18.8 | 0.5 | 0.26 | 1.3 | | 0.06 | | 0.09 | |
| NMS, ident. no. Zn198/49 | " | area 3 | 79.6 | 19.1 | 0.6 | 0.16 | 0.6 | | | | | |
| " | | area 5 | 77.1 | 20.6 | 0.7 | 0.33 | 1.4 | | | | | |
| " | | area 6 | 76.9 | 21.2 | 0.6 | 0.42 | 0.7 | 0.09 | 0.10 | | | |
| NMS, inv. no. R19078 | " | area 3 | 78.1 | 20.4 | 0.6 | 0.39 | 0.6 | | | | | |
| NMS, inv. no. P19946 | " | area 1 | 81.5 | 17.9 | 0.3 | 0.34 | <0.5 | | | | | |
| IPCHS NG, ident. no. K1874 | " | area 1 | 79.9 | 18.1 | 0.5 | 0.21 | 1.2 | 0.06 | | | | |
| " | " | | 78.1 | 21.7 | | | 0.17 | | | | | n.b. |
| NMS, inv. no. P12982 | " | area 1 | 81.2 | 17.0 | 0.7 | 0.26 | 0.7 | | | 0.07 | | |
| RMNG, inv. no. 8 | " | area 1 | 77.7 | 20.0 | 0.5 | 0.81 | 0.9 | | | | | |
| RM Novo mesto, inv. no. 1256 | " | area 1 | 78.7 | 16.1 | 0.8 | 0.67 | 3.6 | 0.1 | | | | * |
| NMS, inv. no. 24045 | " | area 3 | 78.2 | 19.5 | 0.9 | 0.36 | 1.1 | | | | | |
| MNG, inv. no. 24 | Jezerine I | area 1 | 80.2 | 16.7 | 0.7 | 0.87 | 1.5 | | | | | |
| NMS, inv. no. R19077 | " | area 1 | 80.1 | 18.2 | 0.5 | 0.22 | 1.0 | 0.07 | | | | |
| " | | area 2 | 78.4 | 19.3 | 0.5 | 0.20 | 1.5 | 0.07 | | | | |
| NMS, inv. no. R18758 | Jezerine II | area 1 | 78.7 | 19.9 | 0.7 | 0.26 | 0.5 | | | 0.05 | | |
| NMS, inv. no. P19838 | " | area 1 | 80.4 | 17.9 | 0.8 | 0.75 | <0.5 | | 0.20 | | | |
| NMS, inv. no. R18756 | " | area 1 | 78.1 | 21.2 | 0.5 | 0.23 | <0.5 | | | | | |
| IPCHS NG, ident. no. K1873 | " | area 2 | 80.0 | 16.5 | 1.0 | 1.35 | 1.0 | 0.07 | | 0.11 | | |
| NMS, inv. no. P19462 | " | area 1 | 77.7 | 21.1 | 0.7 | 0.23 | <0.5 | | 0.13 | 0.03 | | |
| NMS, inv. no. P19281 | " | area 1 | 74.4 | 21.7 | 0.6 | 1.45 | 1.7 | 0.09 | | 0.05 | | |
| NMS, inv. no. 1918 | " | area 1 | 77.9 | 19.5 | 0.6 | 0.48 | 1.5 | 0.08 | | | | |
| NMS, ident. no. Zn210/10 | " | area 1 | 84.3 | 3.2 | 12.3 | 0.22 | <0.5 | | | 0.07 | | |
| " | | area 2 | 86.1 | 3.2 | 10.5 | 0.18 | <0.5 | | | 0.07 | | |
| NMS, ident. no. Zn210/9 | " | area 3 | 83.1 | 5.3 | 9.8 | 0.50 | 1.1 | | | 0.17 | | |
| " | | area 4 | 83.2 | 5.0 | 9.9 | 0.54 | 1.3 | | | 0.11 | | |

group, which were most probably produced in the Roman (Romanised) communities in Italy as well as in the non-Roman (not yet Romanised) communities in the southeastern Alps and the Balkans.

- In the Almgren 65 group, which was the oldest, 12 brooches were made of bronze and one of brass. The brass brooch belonged typologically to the latest sub-group (Demetz 1999: 27–38).
- Fourteen brooches of the Alesia group (which generally date to approximately the period of Caesar's Gallic wars, 58/51 BC, and the following two decades) were made of brass, one was made of gunmetal and three were made of bronze (Fig. 2). These three bronze brooches all belonged to the same subgroup.[13]
- Of the Jezerine I brooches (an explicitly Roman type, probably produced from c. 40 BC onwards), both those analysed proved to be made of brass.
- Among the 25 brooches belonging to the large and widespread Jezerine II brooch group, 16 were made of bronze, seven of brass and two of gunmetal.

For the brass brooches, the analyses showed that the zinc content fluctuated between at least 15% and almost 22%, with most of the zinc values above 18%.

## Discussion

The Alesia group brooches constituted the oldest group of brooches for which the new alloy, brass, was widely used. The three bronze brooches all belonged to the same subgroup (Alesia Ic), and were most probably copies of the regular Alesia-type brooches made in small local workshops. We may, therefore, conclude that, except for the presumed copies, brass was used for the Alesia-type brooches. Concomitant with this type of brooch (produced most probably from about 60 BC onwards), the Romans had also introduced a new type of brooch mechanism, a hinge instead of a spring. Brooches of the earlier, Italic type, i.e. the Almgren 65 type, were made of bronze. The only example in brass can be explained as a product belonging relatively late in the series for which the use of a new metal was influenced by the more recent, but chronologically partially overlapping Alesia-type brooches.

The analyses of the two Jezerine I brooches indicate that, for this very small group, brass was also widely or exclusively used. The production of this Roman type of brooch had probably begun a decade or two later than that of the Alesia brooch group.

The use of pure brass, gunmetal and bronze for the Jezerine II group brooches both complements and supplements our archaeological knowledge of these brooches. It can be assumed that the brass brooches had originated in Italy, where they most probably came into production later than Alesia-type brooches, from c. 40 BC onwards. The high proportion of bronze brooches in the Jezerine II group supports the assumption that they were also produced in the southeastern Alps and the Balkans, where they were very numerous. In these regions, Jezerine II brooches were copied by non-

Roman communities in the period immediately preceding the Roman conquest. These peoples were able to imitate the brooch form but could not obtain brass. The few examples made of gunmetal may indicate that they were able to use the metal of the imported Roman brass brooches for their own brooch production. Future research will concentrate on the correlations between subgroups of the Jezerine II brooches and the material used for their manufacture.

From our research, it is evident that brass was not used for non-Roman brooches. This is not surprising. The only hint of any Celtic, pre-Roman use of brass derives from Elisabeth Hamilton's publication of the metal finds from the site of the Titelberg in Luxembourg (Hamilton 1996), which is briefly discussed here.

During the American excavations at Titelberg, three items were found which were made of pure brass with 18.2–21.9% zinc, and dated to before Caesar's invasion. These brass items were taken as proof that the Gauls (Celts) were using brass earlier than the Romans. Hamilton (1996: 59, 60, 79, fig. 48) speculated that the Gauls in Europe had come into the possession of brass through contacts with Gaulish tribes in Galatia in Asia Minor, or even that the knowledge of the cementation process had spread from Galatia to Gaul.

There are serious objections to this view. First, the brass object that was identified as 'stylistically German' (Hamilton 1996: fig. 48c) is, without doubt, a transverse fitment of a Roman sword scabbard (cf. Unz and Deschler-Erb 1997: pls 6–7, nos. 78–110). The other two fragments of brass objects cannot be culturally assigned so there is no reason to believe they were not Roman as well. The reliability of the dating of these three objects (to c. 100–50 BC) can also be questioned, as no detailed evidence of their stratigraphic context was given in the publication. In conclusion, we can claim that the three brass items from Titelberg do not provide evidence for the use of brass by the Gauls in the pre-Roman period.

It is unfortunate that very few of the brooches included in the present research derive from excavations or provide other means of precise dating (Istenič 2005a, 2005b). For this reason, most of them can be dated on typological criteria alone, which are relatively broad. For example, we cannot prove that any of the Alesia group brooches that were analysed belong to the early production period. But, in our opinion, it is highly probable that brass was used for this group of brooches from the beginning of their production, as they also represent something entirely new in brooch construction. Nevertheless, to be completely sure that this group of brooches was made of brass from its inception, a brooch found in a stratified context from the siege of 51 BC at Alesia should be analysed.

Unexpectedly, supporting evidence for our argument (that brass was used for brooches of the Alesia group from the beginning of their production), emerged recently, and quite fortuitously, from the investigation of objects which are not Roman but Celtic. Research on the coins of the Arverni (a Celtic tribe in central-southern Gaul) has shown that six examples, all found at Alesia, were made of brass which contained 10–15% zinc. They were of the same type as the contemporary gold staters (which they copied), and two of them bore the name of Vercingetorix, the leader of the Gallic revolt of 53/52. Most probably they were struck at Alesia during the Roman siege of 52 BC, and brass was used because of a shortage of gold. The source of the brass used for these coins was recycled Roman brass objects which comprised brooches (Nieto 2004). Thus, the coins of Arverni provide firm supporting evidence for the Roman use of brass during the period of Caesar's wars in Gaul.

## Conclusions

The results of this research on the pre-Augustan brooches, together with the research on the coins of the Arverni, have several wider implications. They confirm that it was the Romans who introduced the use of brass to Europe and indicate that the beginning of the relatively regular use of brass in Roman Europe should be dated to about 60 BC. Brass was initially used for military equipment, which included brooches. Slightly later, in 46/45 BC, brass was also used for a coin issue. It seems, however, that Caesar's death in 44 BC halted the initial use of brass for coins. It was not until over a quarter of a century later, in the first decade of the reign of the Emperor Augustus, that brass began to be used for coins in the mint of Rome and on a large scale.

## Acknowledgements

J. Istenič would like to thank Dr Dragan Božič for his specialist advice on the classification and dating of the brooches as well as for discussions concerning late La Tène and late Republican periods. She is also indebted to Dr Vivien Swan who edited the English.

## Notes

1. National Museum of Slovenia (NMS), inv. nos. R3419, R16720, R18464, R18485, R18555, R18565, R18687, P19286, P19704, P19962, P19465, R24088; Institute for the Protection of the Cultural Heritage of Slovenia, Nova Gorica district unit (IPCHS NG), ident. no. K 238. All the brooches except IPCHS NG, ident. no. K 238 can be seen in Figure 1.
2. In addition to the brooches listed in Table 1, the following brooches of the Alesia group were analysed: NMS inv. nos. R1464, R19080 and P19283. All the brooches except IPCHS NG, ident. no. K1874 are illustrated in Figure 2.
3. NMS inv. nos. R19077 and Regional Museum Nova Gorica (RMNG), inv. no. 24. Both brooches are illustrated in Figure 3.
4. NMS inv. nos. R1918, R11005, R16728, R18627, R18756, R18758, R19074, R19075, R19091, R24085, P19276, P19277, P19279, P19280, P19281, P19462, P19838, P19944, P19945, Zn 201/9, 10; IPCHS NG, inv. nos. K237, K1873, K2520, K2523. Brooches NMS inv. nos. R19091, Zn201/9, 10 and R1918 are illustrated in Figure 4.
5. Eight brooches of this group were submitted for analysis: NMS inv. nos. R17674, R18624, R18625, R18626, R18766, P19284, P19285 and P19955; all can be seen in Figure 5.
6. Three brooches of this group were included in the research: NMS inv. nos. R17668, P19587 and P19705; all are included in Figure 6.
7. Eight brooches of this group were included in the research: NMS inv. nos. R3760, R17675, R18595, R18755, R18827, R24086, P15234 and P19940; all are illustrated in Figure 7.
8. NMS inv. no. R18595.

9. The chronology and the precise dating of the brooches within the 1st century BC are still the subject of debate. The majority of the *Palmettenfibeln*, the *Schüsselfibeln* and the Nauheim-type brooches should, in general, be dated to approximately the first half of the 1st century BC (Dragan Božič, pers. comm.).
10. Tinning and also thin silver foil soldered to the substrate occur on the brooches of the Alesia group (Šmit *et al.* 2005a, 2005b: 32–3, fig. 5). Tinning appears also on few brooches of the Jezerine II group (not yet published).
11. The analyses were carried out at the NMS using a Model PEDUZO 01/Am/Sip-250 X-ray analyser manufactured at the Jožef Stefan Institute. The radiation source was Am-241 with an activity of 25 mCi. The X-ray detector was a Peltier-cooled Si PIN diode with a resolution of 150 eV at 5.9 keV. The diode window was made of beryllium 25 μm thick. The spectrometer energy region (preset during manufacture) extended between 3 and 30 keV.
12. The PIXE measurements, using a proton beam in the air, were made at the Tandetron Accelerator of the Jožef Stefan Institute. The protons were accelerated up to energy of 2.5 MeV, but their impact energy at the target was reduced to about 2.2 MeV. The proton beam hit the target at an angle of 22.5 degrees with respect to the surface. The X-ray detector was positioned at the same angle. The detection of induced X-rays was performed by a Si(Li) detector with an energy resolution of 160 eV at 5.9 keV. During the measurements, the detector was equipped with an aluminium absorber of 0.3 mm thickness. With such a thick absorber we improved the relative sensitivity for hard X-rays around tin, so the minimum detection limit in this region was 0.1%. Such a thick absorber also improved discrimination between the K X-ray lines of arsenic and the L X-ray lines of lead, as the filter increased the relative intensity of arsenic $K_\beta$ lines and lead $L_\beta$ lines in the spectra. The minimum detection limit for arsenic was 0.03%. A drawback of the absorber was the partial overlap of the iron K X-ray lines with the escape peaks of copper. The elemental concentrations were determined with a precision better than 5% for major elements and better than 10% for minor and trace elements. The uncertainties were greater for silver and tin due to low counting statistics (up to 20%), and for iron due to the subtraction of the copper escape peak; the resulting absolute uncertainty in iron concentration was 0.5–1%.
13. A detailed report on brooches of the Alesia group from Slovenia is given in Istenič 2005b.

## References

Bahrfeldt, M. 1905. Die Münzen der Flottenpräfekten des Marcus Antonius. *Numismatische Zeitschrift* 37: 9–56.

Bahrfeldt, M. 1909. Die letzten Kupferprägungen unter der römischen Republik. *Numismatische Zeitschrift* 42: 67–86.

Bayley, J. 1990. The production of brass in antiquity with particular reference to Roman Britain. In *2000 Years of Zinc and Brass*, P.T, Craddock (ed.). British Museum Occasional Paper 50. London, 7–27.

Bayley, J. and Butcher, S. 1995. The composition of Roman brooches found in Britain. In *Ancient Bronzes*, S.T.A.M. Mols, A.M. Gerhartl-Witteveen, H. Kars *et al.* (eds). Nijmegen: Provinciaal Museum G.M. Kam, 113–19.

Bayley, J. and Butcher, S. 2004. *Roman Brooches in Britain: A Technological and Typological Study based on the Richborough Collection*. London: Society of Antiquaries.

Božič, D. 1993. Slovenija in srednja Evropa v poznolatenskem obdobju (Slowenien und Mitteleuropa in der Spätlatènezeit). *Arheološki vestnik* 44: 137–52.

Brouquier-Reddé, V. and Deyber, A. 2001. Fourniment, harnachement, quincaillerie, objets divers. In *Alésia*, M. Reddé and S. von Schnurbein (eds). Paris: Diffusion de Boccard, 293–333.

Burnett, A. 1987. *Coinage in the Roman World*. London: Seaby.

Burnett, A.M., Craddock, P.T. and Preston, K. 1982. New light on the origins of orichalcum. In *Proceedings of the 9th International Congress of Numismatics I*, T. Hackens and R. Weiller (eds). Louvain-la-Neuve, Luxembourg: Association internationale des numismates professionels, 263–8.

Caley, E.R. 1964. *Orichalcum and Related Ancient Alloys*. New York: American Numismatic Society.

Cowell, M.R., Craddock, P.T, Pike, A.M. and Burnett, A.M. 2000. An analytical survey of Roman provincial copper-alloy coins and the continuity of brass manufacture in Asia Minor. In *XII. Internationaler Numismatischer Kongress Berlin 1997, Akten – Proceedings – Actes I*, B. Kluge and B. Weisser (eds). Berlin: Staatliche Museen Preussischer Kulturbesitz, Münzkabinett, 670–77.

Craddock, P.T. 1978. The composition of the copper alloys used by the Greek, Etruscan and Roman civilizations. 3. The origins and early use of brass. *Journal of Archaeological Science* 5: 1–16.

Craddock, P.T. and Eckstein, K. 2003. Production of brass in antiquity by direct reduction. In *Mining and Metal Production through the Ages*, P.T. Craddock and J. Lang (eds). London: British Museum Press, 216–30.

Craddock, P.T. and Lambert, J. 1985. The composition of the trappings. In A group of silvered-bronze horse-trappings from Xanten (Castra Vetera), I. Jenkins. *Britannia* 16: 141–64.

Demetz, S. 1999. *Fibeln der spätlatène- und frühen römischen Kaiserzeit in den Alpenländern*. Rahden/Westfalen: Marie Leidorf.

Giumlia-Mair, A. 1998. Studi metallurgici sui bronzi della necropoli di S. Lucia – Most na Soči. *Aquileia Nostra* 69: 29–136.

Grant, M. 1969. *From Imperium to Auctoritas: A Historical Study of Aes Coinage in the Roman Empire 49 BC–AD 14*, 2nd edn. Northampton: J. Dickens & Co.

Hamilton, E.G. 1996. *Technology and Social Change in Belgic Gaul: Copper Working at the Titelberg, Luxembourg, 125 BC–AD 300*. Philadelphia: Museum Applied Science Center for Archaeology.

Istenič, J. 2000a. A late Republican gladius from the river Ljubljanica (Slovenia). *Journal of Roman Military Equipment Studies* 11: 1–9.

Istenič, J. 2000b. A Roman late Republican gladius from the river Ljubljanica (Slovenia). *Arheološki vestnik* 51: 171–82.

Istenič, J. 2005a. Evidence for a very late Republican siege at Grad near Reka in western Slovenia. *Carnuntum Jahrbuch* 2005: 77–87.

Istenič, J. 2005b. Brooches of the Alesia group in Slovenia. *Arheološki vestnik* 56: 187–212.

Istenič, J. forthcoming. Caesarian militaria from the river Ljubljanica (Slovenia). *Bibracte* 10.

Jackson, R.P.J and Craddock, P.T. 1995. The Ribchester hoard: a descriptive and technical study. In *Sites and Sights of the Iron Age*, B. Raftery (ed.). Oxford: Oxbow Press, 75–102.

Jerin, B. 2001. *Čolničaste fibule v Sloveniji*. Degree thesis, Oddelek za arheologijo, Filozofska fakulteta, Univerza v Ljubljani.

Nieto, S. 2004. Monnaies arvernes (Vercingétorix, Cas) en orichalque. *Revue Numismatique* 160: 5–25.

Ocharan Larrondo, J.A. and Unzueta Portilla, M. 2002. Andagoste (Cuartango, Álava): un nuevo escenario de las guerras de conquista en el norte de Hispania. In *Arqueología militar Romana en Hispania*, Á. Morillo Cerdán (ed.). Madrid: Ediciones Polifemo, 311–25.

Ponting, M.J. 2002. Roman military copper-alloy artefacts from Israel: questions of organization and ethnicity. *Archaeometry* 44(4): 555–71.

Riederer, J. 2001. Die Berliner Datenbank von Metallanalysen kulturgeschichtlicher Objekte. III Römische Objekte. *Berliner Beiträge zur Archäometrie* 18: 139–259.

Riederer, J. 2002a. Die Metallanalyse der Funde aus Kupferlegierungen von Haltern. In *Die römischen Buntmetallfunde von Haltern*, M. Müller (ed.). Mainz: Philipp von Zabern, 109–45.

Riederer, J. 2002b. The use of standardised copper alloys in Roman metal technology. In *I bronzi antichi: Produzione e technologia*, A. Giumlia-Mair (ed.). Montagnac: Éditions Monique Mergoil, 284–91.

Salzani, L. (ed.) 1996. *La necropoli gallica e romana di S. Maria di Zevio (Verona)*. Mantova: Società Archeologica Padana.

Šmit, Ž. and Pelicon, P. 2000. Analysis of copper-alloy fitments on a Roman gladius from the river Ljubljanica. *Arheološki vestnik* 51: 183–7.

Šmit, Ž., Istenič, J., Gerdun, V. *et al.* 2005a. Archaeometric analysis of Alesia group brooches from sites in Slovenia. *Arheološki vestnik* 56: 213–33.

Šmit, Ž., Pelicon, P., Simčič, J. *et al.* 2005b. Metal analysis with PIXE: the case of Roman military equipment. *Nuclear Instruments and Methods in Physics Research, Section B* 239: 27–34.

Trampuž Orel, N. 1999. Archaeometallurgic investigations in Slovenia. *Arheološki vestnik* 50: 407–29.

Unz, C. and Deschler-Erb, E. 1997. *Katalog der Militaria aus Vindonissa*. Brugg: Gesellschaft Pro Vindonissa.

## Authors' addresses

- J. Istenič, National Museum of Slovenia, Prešernova 20, SI-1000 Ljubljana, Slovenia (janka.istenic@nms.si)
- Ž. Šmit, University of Ljubljana, Faculty of Mathematics and Physics, Jadranska 19, SI-1000 Ljubljana, Slovenia; Jožef Stefan Institute, Jamova 39, P.O.B. 3000, SI-1001 Ljubljana, Slovenia (ziga.smit@fmf.uni-lj.si)

# Roman brass and lead ingots from the western Mediterranean

*Gerd Weisgerber*

*with contributions by P.T. Craddock and N.D. Meeks, U. Baumer and J. Koller*

ABSTRACT   During the early 1980s the Deutsches Bergbau-Museum Bochum obtained a collection of brass and lead ingots from a sunken ship. The accompanying sherds from three amphorae suggest the cargo came from Roman Hispania and date the wreck to the late 1st century AD or a little later. The plano-convex oval brass ingots are the only ones of this type so far known from the Roman period. The resin in one of the vessels was identified as low-temperature carbonisation tar from pine wood.

*Keywords:* ingot, Roman, brass, lead, amphora, resin, analysis.

## Introduction

The Deutsches Bergbau-Museum Bochum (DBM) (German Mining Museum at Bochum) in Germany acquired a group of brass and lead ingots and a few pottery fragments around 1980. Since then they have been on exhibition. The objects are thought to have come from a shipwreck off the coast of eastern Corsica; most of the objects have marine incrustations and many shipwrecks have been reported around the island (Bebko 1971). Here also was the famous prehistoric settlement and Roman town of Aleria (Jehasse and Jehasse 1973, 1985). The purpose of this paper is to make these finds better known.

Besides amphorae, metal artefacts (mainly ingots) are the commonest finds from shipwrecks especially of the classical period in the Mediterranean.[1] More than 1500 Roman lead ingots have been found in the Mediterranean Sea or in surrounding countries, and more than 800 shipwrecks had been discovered by 1983 of which about 540 were found in the western Mediterranean. About 26% date from the 1st millennium AD (Meier 1990, 1994e: 768, 772).

The collection at Bochum consists of ten brass ingots and two lead ingots together with fragments of pottery from at least three large amphorae.

## Brass ingots

None of the brass ingots shows any trace of a stamp. All the ingots have a plano-convex profile and an oval surface (Fig. 1 and Table 1).[2] The term 'bun ingot' is deliberately avoided as misleading; the flat surface is the top.[3] The weights vary between *c.* 3 and 6.2 kg, with an average of more than 5 kg; only one is much smaller and lighter.

**Figure 1** General view of the brass ingots, presumed to have come from a Roman wreck off the east coast of Corsica.

**Table 1** Dimensions of the brass ingots.

| DBM Bochum inv. | Length (cm) | Width (cm) | Thickness (cm) | Weight (kg) |
|---|---|---|---|---|
| 080 503 183 000 | 22.4 | 11.2 | 5.0 | 5.125 |
| 080 503 184 000 | 24.5 | 12.0 | 4.4 | 5.150 |
| 080 503 185 000 | 22.4 | 12.9 | 5.3 | 6.130 |
| 080 503 186 000 | 24.3 | 13.0 | 4.9 | 6.210 |
| 080 503 187 000 | 23.4 | 11.9 | 4.5 | 5.020 |
| 080 503 188 000 | 22.2 | 10.8 | 4.1 | 3.880 |
| 080 503 189 000 | 21.2 | 10.9 | 3.3 | 3.160 |
| 080 503 190 000 | 18.2 | 10.0 | 4.3 | 3.190 |
| 080 503 191 000 | 18.5 | 5.1 (half) | 4.7 | 2.895 (half ingot) |
| 080 503 192 000 | 17.3 | 8.5 | 2.9 | 1.650 |

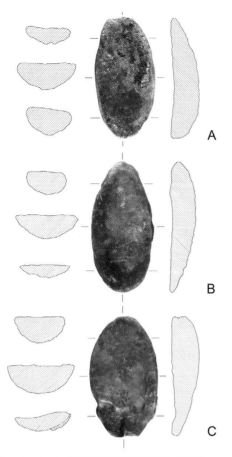

**Figure 2** Brass ingots. A: DBM 080 503 183; B:184; C: 185.

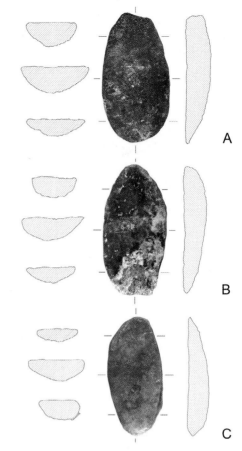

**Figure 3** Brass ingots. A: DBM 080 503 186; B:187; C: 188.

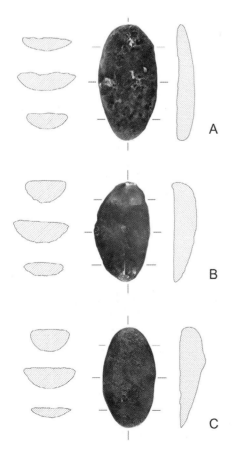

**Figure 4** Brass ingots. A: DBM 080 503 189; B: 190; C: 191.

Unusually for plano-convex copper-alloy ingots these are oval, not round. There are other differences: usually copper ingots are more or less flat, whereas the long sides of these brass ingots are steeply sloping. Moreover the longitudinal section of these ingots is shallow at one end and steep at the other. This reflects the shape naturally formed by hand-scooping a hollow in sandy ground. There is one other similar Roman brass ingot[4] from Claydon Pike in the upper Thames valley that has been analysed by Peter Northover as 20.6% zinc, 0.4% iron and 0.9% lead. It is not fully published, but is described as being a bun ingot (Bayley 1998: 11; Northover 1998). Another well-known Roman brass artefact from Britain is the 1st-century AD stamped sheet excavated from the Roman industrial complex at Sheepen, in Colchester. The sheet, 91 cm long, 15 cm wide and 0.5 cm thick, weighs 9.358 kg and has 27% zinc (Musty 1975 and Table 2). Pewter ingots such as the Roman Battersea ingots have an oval surface or a shape more or less round to oval. Hughes (1980: 47) suggested that the surface shapes were related to the different tin contents. Some (much larger) ancient Greek lead ingots were also of this simple form because they were cast into a depression in the ground.

### Roman brass

The Bochum ingots are a copper-zinc alloy with zinc contents typical of freshly made Roman brass (see appendix 1).

**Table 2** Analyses of three ingots from the Bochum group, together with a large sheet of Roman brass from Colchester for comparison.

| BM lab. no. | Bochum inv. no. | Cu | Sn | Pb | Ag | Fe | Ni | Zn | As | Sb | Bi | Mn | Cd |
|---|---|---|---|---|---|---|---|---|---|---|---|---|---|
| *Roman brass ingots:* | | | | | | | | | | | | | |
| 2878 | 503 191 | 70.0 | 0.6 | 2.50 | 0.090 | 0.39 | 0.050 | 26.6 | 0.40 | 0.80 | | 0.003 | 0.001 |
| 3286 | 503 183 A | 71.5 | | 3.30 | 0.068 | 0.45 | 0.039 | 24.5 | 0.09 | 0.12 | | 0.004 | |
| 3287 | 503 183 B | 71.5 | | 3.52 | 0.063 | 0.44 | 0.039 | 24.3 | 0.10 | 0.11 | | 0.004 | |
| 3288 | 503 183 C | 70.1 | | 3.73 | 0.064 | 0.42 | 0.037 | 24.7 | 0.11 | 0.11 | | 0.004 | |
| 3289 | 503 186 1 | 70.4 | 0.4 | 1.34 | 0.043 | 0.29 | 0.026 | 24.7 | 0.08 | 0.05 | | 0.005 | |
| 3290 | 503 186 2 | 71.2 | 0.4 | 1.45 | 0.045 | 0.28 | 0.025 | 25.3 | | 0.05 | | 0.005 | |
| 3291 | 503 186 3 | 70.5 | 0.5 | 1.64 | 0.048 | 0.29 | 0.024 | 25.3 | | 0.06 | | 0.006 | |
| *Brass sheet from Colchester:* | | | | | | | | | | | | | |
| 129 | Colchester | 72.1 | tr | 0.1 | 0.007 | 0.14 | 0.020 | 26.8 | 0.40 | 0.05 | | | |

This is not the place to repeat the debate concerning Pliny's aurichalcum and the derivation of the term from the Greek *oreichalkos* (metal of the mountain) (see Craddock 1995: 292–302; Tylecote 1987). It is well known that brass was introduced for Roman coinage from the mid-1st century BC (coinage of Clovius under Julius Caesar) (and see Istenič and Šmit in this volume, pp. 140–47). After Nero's short-lived reform, *as, semis* and *quadran*s of AD 63 were struck in orichalcum, constituting a single and comprehensive monetary system for the *aes* (MacDowall 1971: 86). During the 1st century AD, the content of zinc in the coins diminished slowly from 27.6% in 45 BC to 15.9% in AD 79 and to 7.8% in AD 161 whereas the contents of tin and lead increased (Caley 1964; Moesta 1983, 1993; Riederer 1974a,1974b; Rolandi and Scacciati 1956). In northern Britain the study of more than 1000 copper alloy artefacts demonstrated that brass and bronze scrap were often mixed (Dungworth 1997) and Bayley and Butcher (2004) have made an important study of the composition of brooches in Britain. Other Roman brass analyses have been published by Schramm (1939), Burnett *et al.* (1982), Erdrich (1995) and Hammer (2001). In central Europe, the Roman so-called 'Hemmoor buckets' of brass sold very well and also obviously served as the raw material for many small brass objects (den Boesterd and Hoekstra 1965; Werner 1970; Willers 1906).

Brass was produced by the cementation process in which calamine, a zinc-rich mineral, is heated together with copper under reducing conditions (Craddock 1998; Tylecote 1987: 144). Crucibles believed to be for brass production are known, with examples from Roman Colchester and Canterbury (Bayley 1984; Tylecote 1987: fig. 6.9), and Germany (Zwicker *et al.* 1985). The technique was also used in medieval times (Drescher 1987): remains of a foundry working during the 12th century AD at Bonn-Schwarzrheindorf included furnaces and more than 100 crucibles; the largest one could contain about 1.2 kg of metal. In some of them brass was produced, in others it was melted, all in close accordance with the descriptions given in Theophilus's *Presbyter* compiled in the 13th century (Drescher 1987: 223; Theophilus 1933: 122–5; 1963: 140–45).

The weight of the Bochum ingots, up to 5 kg, is far more than could have been produced in the Roman or medieval crucibles which are so far known. According to current understanding, the brass for one ingot had to be produced by cementation in several crucibles. This was then melted together and poured in one go into the mould, which was probably just a

shallow hollow in the ground. It must be said, however, that we have no physical evidence or written record of such large crucibles or furnaces. A transverse section through one of the Bochum ingots shows a uniform texture with little porosity and no sign that the metal had been poured from more than one crucible.

One important brass object has so far been largely neglected. In the Schweizerisches Landesmuseum at Zurich there is an inscription on an octagonal-sided metal sheet, 22.5 cm long and 10.7 cm wide from Augusta Raurica/Augst near Basle (Fig. 5). It once was fixed by nails to the base of a statue dedicated to the god of the sun. According to analysis, this metal sheet is of brass[5] as was the figurine or statue of the god. The inscription reads: 'DEO INVICTO TYPUM AUROCHALCINUM SOLIS' (This brass statue is dedicated to the invincible god Sol) (Furger and Riederer 1995) – golden brass for the golden sun. This agrees well with Pliny's 'aurichalcum which was leading with regard to quality and admiration' (*Natural History* 34: 2).

**Figure 5** Votive inscription for the god Sol on a brass plate from Augst, Switzerland (Schweizerisches Landesmuseum, inv. no. 3478, neg. 16879).

## Lead ingots

Generally there are three types of lead ingot: (1) flat planoconvex tongue- or cake-shaped; (2) rectangular blocks with semicircular (or parabolic) cross-section and (3) rectangular blocks with trapezoidal cross-section (Meier 1994e: 772). The Bochum lead ingots fit the last criteria (*tronc de pyramide à longue rectangulaire*) rather than *parallélépipède* (Domergue 1966: 42; 1987; Domergue *et al.* 1974, 1983).

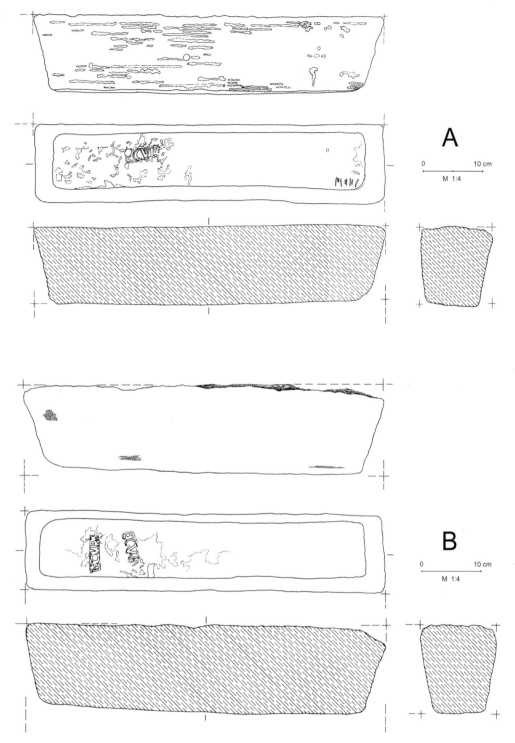

**Figure 6** Two lead ingots from the brass wreck: DBM 080 503 202 (A) and 203 (B).

The ingots are more than half a metre long and trapezoidal in cross-section.[6] The inscriptions (Fig. 7) were stamped in the mould and so appear on the underside of the ingot. No further inscriptions were stamped after casting though some small nicks were made at one edge of ingot no. 202 (Fig. 6A). These inscriptions are undeciphered and among the hundreds of known stamped Roman lead ingots they have no parallels. They cannot even be explained as a declaration of weight (see Parker 1974: 148).

On two places on the surface of one of the lead ingots there are impressions from coarse textile. This may be evidence of how the ingots were cast. There are ethnological parallels for using coarse textiles together with hot water in the moulds for casting bar ingots in Asia (Lewin and Hauptmann 1984: 92, fig. 9; Smith 1983: fig. 9, fig. F).

The weight of these lead ingots is remarkable. During Republican and early Imperial times the standard weight was 100 Roman pounds, equivalent to 32.7 kg (Domergue 1966; Eck 1994; Meier 1994e), but in practice it fluctuated between 32 and 42 kg with the extra weight being sometimes incised on the sides. In 1920, Besnier noted that most of the ingots from Hispania and Sardinia ranged from 30 to 35 kg (i.e. 95

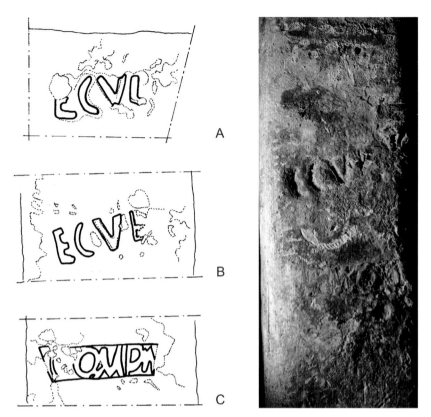

**Figure 7** Inscriptions on the two lead ingots. A: DBM 080 503 202; B and C 080 503 203.

to 105 Roman pounds, a Roman pound weighing 327.45 g). The lightest ingot from Spain weighed only about 11 kg, the heaviest 56.75 kg, but in Britain, according to Besnier, the lightest ingot weighed 22.7 kg, the heaviest 101.583 kg, with 19 ingots weighing between 70 and 88 kg. An ingot from Arbon in Switzerland weighs 145 kg, in near agreement with its inscription of 450 pounds (147.353 kg). The metal probably came from the Eifel in Germany (Meier 1990: 80) as does a lead ingot of the Legio XIX from Haltern in Westphalia in Germany with its weight of 64 kg (Schnurbein 1971), but there are no analyses to confirm this.

Most rectangular block ingots with trapezoidal cross-sections came from Britain, but they bear clear and informative inscriptions and have an average weight of 75–77.7 kg (Meier 1994e: 773). The oldest dated lead pig in Britain dates from the year AD 49 (Collingwood and Wright 1990: RIB 2404, 1–2); another old one comes from Flintshire and dates from the year AD 74, and some from Yorkshire appear in AD 81. Their export boomed during the 1st to 3rd centuries AD (Boon 1971: 469). There are parallels, however, to the greater weight of about 85 kg. A lead pig from Hexgrave Park near Mansfield in Nottinghamshire, now in the British Museum, weighs 83 kg (Meier 1992b: 17). On the lower side there is a large and long cast inscription with the typical Britannia EX ARG inscription.

Whittick (1961) states that 'Of the surviving full-size ingots of demonstrably Roman date from Britain, only three weigh less than 90 lb (40.82 kg) and two others less than 140 lb (63.50 kg), five weigh less than 150 lb (68.1 kg), nine less than 190 lb (86.18 kg), and to these may be added four more now incomplete which must from their dimensions have weighed

approximately 190 lb (86.18 kg) when intact ... the heaviest surviving ingot weighs about 606 lb (274.88 kg)'.[7] He pointed out that by comparison of small details of the ingots it is possible to demonstrate that several ingots have been cast in the same mould (Whittick 1961: 110f.), although no moulds are presently known. This absolute similarity is not the case for the two Bochum lead ingots under discussion. One of the ingots, as with almost every Roman lead pig, does not show any trace of the casting processes, no striations, no laminae and no intermediate solidification (compare Whittick 1961); the other, however, has some striations on the sides which could have resulted from pouring simultaneously from several crucibles (Fig. 6A). We have to suppose that most of the Roman lead-smelting furnaces or lead-melting installations

**Figure 8** Textile impression on lead ingot DBM 080 503 203.

were big enough to produce the mass of lead under discussion for a continuous casting.[8]

Heavy ingots were produced in Spain as well as in Britain (Besnier 1920: 240–41) and in Sardinia (Besnier 1920: 222, CIL X 8073, 2; Meier 1994e: 773). About 95 Spanish lead ingots were found in Fretum Gallicum between Corsica and Sardinia (Domergue 1984: 371). The most recent shipwreck from the western coast of Sardinia had the largest cargo containing no less than 1,300 lead ingots (Alessandrello *et al.* 1991; Fiorini 1991; Meier 1994e: 780; Salvi 1992, 1999; Salvi *et al.* 1992).

## Amphorae

No complete amphorae are included in the Bochum collection but there are a number of recognisable Roman types (Figs 9 and 10).[9] Amphorae are well-known Roman containers which have a wide distribution and are commonly found in shipwrecks. They were used for specific commodities and therefore are valuable indicators of trade (see for example Peacock 1974; Remesal-Rodrigez and Revilla-Calvo 1991; Rotroff 1994). The amphora fragment 1 (Fig. 9) corresponds to the Spanish amphorae Dressel 7/11 which are often found in Mediterranean shipwrecks (e.g. Colls *et al.* 1986: pl. 2). The wreck Cabrera 5 yielded a very similar selection of amphorae (Colls *et al.* 1986: fig. 2, 1–3) which indicates a date around the end of the 1st century or the early 2nd century AD. In contrast, the five different types of amphorae found in the Mahdia wreck off Tunisia date to around 100–80 BC (Hellenkemper-Salies *et al.* 1994) and are quite different from those described here, indicating that the wreck was not of Republican date. The Mahdia ingots came from the lead deposits of Carthagena (Begemann and Schmitt-Strecker 1994).

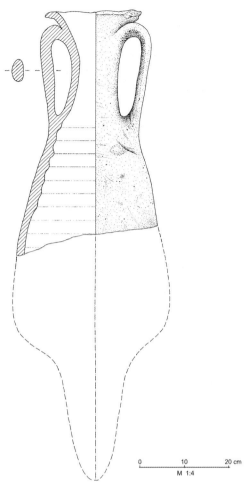

**Figure 9** Large fragment of an amphora from the brass wreck DBM 080 503 195.

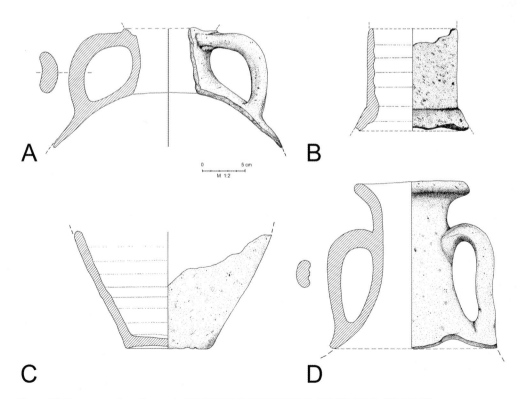

**Figure 10** Fragments of amphorae. A: 080 503 196; B: 080 503 197; C: 080 503 198; D: 080 503 201.

## Appendix 1: Composition of the brass ingots (by P.T. Craddock and N.D. Meeks, Dept. of Conservation, Documentation and Science, British Museum)

The six plano-convex ingots have been abraded by shifting sand but are otherwise in very good condition. Large samples of clean drillings were provided for analysis as well as a cut section of one of the ingots to check for inhomogeneity in the metal and especially for de-zincification. About 20 mg of drillings was taken from three sample positions in two of the ingots and from one from a third ingot and analysed by atomic absorption spectrometry (AAS) using a Cambridge SP9 spectrometer (Table 2). Full details of the analytical method are given in Hughes *et al.* (1976). In the table the precision of the results is ± 2% for the copper and zinc and about ± 20% for the remaining minor and trace elements. Most elements could be detected down to 0.005% in the metal; tin could be detected down to 0.1%.

Analyses were also performed on the section cut from ingot 080 503 191 in the scanning electron microscope. Seven area analyses were carried out from the centre to the edge of the ingot on relatively small areas at high magnification to avoid lead inclusions and porosity which may have contained corroded material. The analyses show very little surface depletion of zinc except for a small loss of perhaps 1% in the first 5 microns, indicating the metal is very homogeneous in composition. The structure shows the ingot is in an as-cast condition with the inclusions (principally lead) lying at the grain boundaries. The atomic absorption analysis of the three samples from each of two ingots shows great uniformity.

### *Discussion*

The analyses of all of the samples from the ingots show that their overall compositions are similar; that is approximately 25% zinc, some lead, and substantial traces of iron. Traces of tin were present in two of the three ingots analysed. The compositions are typical of brass made by the direct process adding zinc ore to molten copper in a crucible with charcoal (Craddock 1978; Craddock and Eckstein 2003). This is believed to have begun in Asia Minor, probably in Bithynia and Phrygia in the 2nd century BC and very rapidly spread through the Roman Empire; by the 1st century AD brass was in common usage. Very few brass ingots have previously been recognised apart from a large cast sheet of brass found at Colchester, England (Musty 1975). There is also little contemporary documentary or archaeological evidence of how the metal was made.

Werner (1970) and Haedecke (1973) showed that if either copper or a 40% zinc brass was heated with charcoal and zinc oxide in a sealed tube for several hours at 1000 °C then the product in both cases was a brass with 28% zinc. In the 16th century, Ercker (1951: 257) claimed that 29% was the maximum zinc content that could be obtained. William Champion claimed that 28% was the limit in the early 18th century at Bristol (Day 1973). Lead and tin in the copper suppress the absorption of zinc, and thus reasonably pure copper was always used in the process. The small quantities of lead

in the metal almost certainly come from the zinc ores, which are always associated with lead, and their physical separation is difficult. Thus with this lead content, the zinc contents of the Roman ingots suggest they are near the practical limit set by the direct process. Brass made by the direct process often has a higher iron content than is found in the contemporary bronzes (compare analyses of Roman copper and brass coinage for example – Craddock *et al.* 1980; Riederer 1974a, 1974b). This is due to the presence of iron in the zinc ore and its subsequent reduction and absorption by the copper in the crucible. During each re-melting the brasses could be expected to lose about 10% of their zinc (Caley 1964: 99), and thus one should regard these as ingots of prime metal coming straight from the crucibles.

The production of brass seems to have been an Imperial monopoly and much of the prime metal seems to have gone straight into coinage or metalwork for the military. Thus the compositions of early Imperial *sestertii* and *dupondii* (Caley 1964; Craddock *et al.* 1980; Riederer 1974a) and of military brasswork (Jackson and Craddock 1995) are very similar to the composition of the ingots. Contemporary civilian brasses tended to be diluted with scrap bronze or copper and typically contain well under 20% zinc, an alloy with about 10–15% being most common (Craddock 1978).

## Appendix 2: Composition and interpretation of the resinous material (by Ursula Baumer and Johann Koller, Doerner Institut, Bayerische Staatsgemäldesammlungen, Munich)

In the corner of a base fragment of a large Roman amphorae from the wreck (Fig. 10) a resinous material was preserved, apparently from the original contents of the container. Two samples of this resinous material were provided for analysis by gas chromatography (GC) and GC combined with mass spectrometry (GC–MS). Both samples were dissolved in a solution of 10% (w/v) oxalic acid in methanol and the resulting solutions were injected into the GC without prior derivatisation. The analyses revealed that the two samples were identical.

The gas chromatogram in Figure 11 shows that the resinous samples consist of acidic and neutral diterpenoid constituents. A diterpenoid resin acid (dehydroabietic acid (DA)) is the major component and its oxidised and methylated derivatives, methyloxodehydroabietate (MODA) and methyl dehydroabietate (MDA), are also present. The neutral components consist of retene (R), abietatriene (Ab), tetrahydroretene (T) and nor-abietatriene (N). These components are formed by decarboxylation and progressive aromatisation processes during smouldering of resin-rich pine wood (Baumer and Koller 2002; Beck *et al.* 1997), that is, during the production of pine tar.

Tars are viscous products which were used as caulking materials on historic ships. Since all volatile and liquid constituents of the tar have evaporated over time or have decayed through chemical reactions, a solid resinous product has been formed, which is known as pitch. Similar Mediterranean pitches used as caulking material have been identified on Hellenistic shipwrecks (Beck and Borromeo 1990), on the

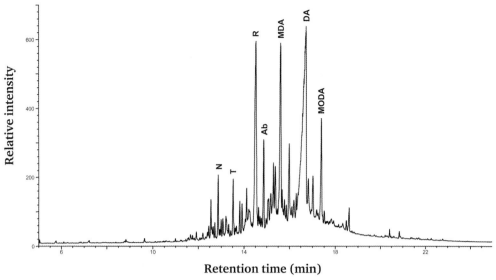

**Figure 11** Gas chromatogram of the acidified methanolic extract (10% w/v oxalic acid in methanol) of a sample of resinous material from the base of a large amphora, DBM 080 503 198 (Fig. 10C). (DA = dehydroabietic acid; MDA = methyl dehydroabietic acid; MODA = methyl 7-ketodehydroabietate; Ab = abieta-8,11,13-triene; R = retene; T = 1,2,3,4-tetrahydroretene; N = 18-norabieta-8,11,13-triene)

Mahdia shipwreck (Koller and Baumer 1994) and on a shipwreck found in the Strait of Bonifacio (Baumer and Koller 2002).

## Experimental

For GC analyses an Agilent (former Hewlett-Packard) GC model 6890 was used. The liquid samples were injected in the split/splitless injector ($T_{Inj}$ = 270 °C). The subsequent separations were performed on a 15 m capillary column (0.32 mm inner diameter, 0.1 μm film thickness, DB5-ht from J&W). A heating programme was used for the column oven:

$$T_{Start} = 55\ °C,\ \text{heating rate R} = 11\ °C/min,\ T_{Final} = 360\ °C$$

The resulting peaks were detected by a flame ionisation detector (FID, $T_{FID}$ = 360 °C). Helium 5.0 (purified) was used as carrier gas with a constant flow of 1.8 ml/min.

## Acknowledgements

For many discussions the author thanks Michael Prange and Andreas Hauptmann. Digitisation of the drawings was carried out by Wolfgang Weber and Jan Cierny, and of the photographs by Guillaume Ewande. Dr Rudolf Degen provided the photograph of the Augst votive plate. Claude Domergue helped with the ingot stamps and Paul Craddock aided with the editing of the paper. The author would like to thank them all here cordially.

## Notes

1. For example the wreck of the VOC ship the *Mauritius* which sank in 1609, contained about 20,000 ingots of zinc, each weighing between 1 and 13 kg (L'Hour 1989).

2. – DBM 080 503 183 (Fig. 2A). The upper surface is largely smooth but with irregularities, partly caused by corrosion. The underside is rough.
   – DBM 080 503 184 (Fig. 2B). The surface is corroded. The underside has rough sand impressions with irregularities. One edge turns sharply inwards.
   – DBM 080 503 185 (Fig. 2C). The surface is half smooth, half with waves, partly roughly corroded. The underside is rough, sand cast, with concretions.
   – DBM 080 503 186 (Fig. 3A). The surface is flat and corroded, the underside has sandy impressions and is rough, corroded, with concretions. Three drill holes from sampling.
   – DBM 080 503 187 (Fig. 3B). The surface is rough, corroded and partly covered with concretions; the underside is mostly smooth.
   – DBM 080 503 188 (Fig. 3C). The surface is smooth in general but some spots are slightly corroded, the edge slightly inward turning. The underside is smooth, some crystallisation is visible, covered partly by black concretions.
   – DBM 080 503 189 (Fig.4A). The surface has wavy structures with rough corrosion, and a little deeper corrosion. The underside is fairly smooth, with little corrosion and remains of large oval concretions.
   – DBM 080 503 190 (Fig. 4B). The surface is very smooth, mostly flat, with textile impressions. The thick end rises up, and there are two small rectangular depressions in the top surface. The underside is rough.
   – DBM 080 503 191 (ingot sectioned, see appendix 1) (Fig. 4C). The surface is fairly smooth, becoming increasingly rough towards the centre. The rim is abnormally rounded. The centre is structured with a small cone protruding down. The section is without porosity, but there are some black spots of corrosion.
   – DBM 080 503 192. This is the smallest ingot; the shape is irregular and the surface mostly covered by concretions of *Pomatoceros triqueter* (tubeworm). The edges are somewhat rounded or eroded. The underside is smooth and wavy.

3. An incorrect reconstruction of the production of similar but round ingots has recently been published in Czajlik (1996).

4. In 1891 a hoard of 57 'Roman' bar ingots was dredged from the Rhine near Mayence. They are up to 25 cm long, 1.5 cm wide and up to 1.1 cm thick, having a weight from 187 to 282.6 g. One of them has traces of an unreadable stamp. These bar ingots were

once considered to be Roman (Bachmann and Jockenhövel 1974: 139), but a comparison with other brass ingots of the same bar shape and the same length has shown that they are medieval (Eiwanger 1996). The production of this type of brass bar ingot was described by Theophilus (1933: 124; 1963).

5. I thank Dr Rudolf Degen, Zurich, for testing the material for me (letter of 2 April 1984).

6. – DBM 080 503 202 (Fig. 6A). Rectangular block with trapezoidal cross-sections. Length on top 58.0 cm, length underneath 51.7 cm, width on top 13.2 cm, width underneath 8.9 cm, thickness 12.9 cm, weight 84.8 kg. The top is heavily corroded in places (compare Becker and Kutzke 1994). There is a hole of 1 cm diameter and 6.8 cm length in one side of the ingot. On the bottom there is one stamp: ECVL (Fig. 7A).
 – DBM 080 503 203 (Fig. 6B). Rectangular block with trapezoidal cross-sections. Length on top 58.4 cm, length underneath 51.5 cm, width on top 13.6 cm, width underneath 9.1 cm, thickness 14.3 cm, weight 89.3 kg. On one side there is an imprint of a rough textile (Fig. 6B and Fig. 8). On the edge of the upper side there is a row of incisions near one corner. On the bottom there are two stamps: ECVL and an unclear collection of letters C QADN? or OMDN? or OAMDN? or C QM QP (Fig. 7B, C).

7. 1 lb = 0.4536 kg.

8. To give some idea of where large quantities of lead could be used, at Bath in England, a large basin has walls and bottom of lead sheets of 1 × 1.5 m, weighing in total 825 kg (Boon 1971: 470). Roman anchors regularly weighed more than 600 kg (Cochet 2000; Merlin 1912: 392).

9. – DBM 080 503 195 (Fig. 9). Large fragmentary upper part of an amphora. Preserved height 57 cm, diameter of the rim 19.5 cm, colour pale brown. This amphora corresponds to Dressel 11 in shape and handles (see typology in Callender 1965: fig. 1; Schallmeyer 1982: fig. 2). The vessel is of Hispanic origin. Normally they were used for the export of fish sauces (*garum*). The funnel-shaped mouth, large handle and slim body are typical (Hofheim Type 72); Claudian until the end of the 1st century AD (Gechter 1984: tables 6, 7; Gose 1950: 435). Good parallels come from Beltrán in the south of Spain (Histoire et archéologie, les dossiers, no. 65, août 1982, 5-85, type IIB; Beltran-Lloris 1970, 1978). Another well-dated parallel comes from Germania Superior at Kellereiplatz in Ladenburg, from a cellar (total height 1.30 m) (in *Archäologische Ausgrabungen in Baden-Württemberg* 1983: fig. 96). The building was abandoned during Trajan's fourth consulate, AD 101–102.
 – DBM 080 503 196 (Fig. 10A). Fragmentary part of a neck with one handle with marine incrustations (*Pomatoceros triqueter*?), fabric very light brown, max. height 17.6 cm. Since Augustan times these wide vessels were used for salted fish and sauces. Dressel 20, 23 shows a wide amphora for olive oil from Spain.
 – DBM 080 503 197 (Fig. 10B). Cylindrical fragment of the neck of a vessel, reddish clay with much sand tempering. Max. height 11.4 cm, max. diam. 10.7 cm. Clear grooves from wheel turning inside.
 – DBM 080 503 198 (Fig. 10C). Bottom of a large vessel with ring base, height 13.5 cm, diam. 10.2 cm. Inside is a resin (see appendix 2).
 – DBM 080 503 199. Five sherds, hand size, from at least three large vessels in red, yellow, light brown. Some of them with heavy incrustations (*Pomatoceros triqueter*?).
 – DBM 080 503 200. Twenty pottery sherds, some hand-size body sherds of amphorae, colours red, buff, pink and yellow, *Pomatoceros triqueter* incrustations. One large piece of the lower part of an amphora, diameter 28 cm, thickness 0.9–1.8 cm.
 – DBM 080 503 201 (Fig. 10D). Neck with two handles of an amphora. Preserved height 18.6 cm, diameter of the rim 13.4 cm, thickness of sherd 1.2 cm, pale brown. The handles suggest it probably could have been Dressel 10.

# References

Alessandrello, A., Cattadori, C., Fiorentini, G. *et al.* 1991. Measurements on radioactivity of ancient Roman lead to be used as shield in searches for rare events. *Nuclear Instruments and Methods in Physics Research* B 61(1): 106–17.

Bachmann, H.-G. and Jockenhövel, A. 1974. Zu den Stabbarren aus dem Rhein bei Mainz. *Archäologisches Korrespondenzblatt* 4: 139–46.

Baumer, U. and Koller, J. 2002. Verfahren der Holzverschwelung und die Verwendung ihrer Produkte – von der Antike bis zur Gegenwart [The process of wood charing and the use of its products – from ancient times to the present]. In *Plinius und die nordeuropäische Pechsiederei*, G. Heil and D. Todtenhaupt (eds). Berlin, 1–113.

Bayley, J. 1984. Roman brassmaking in Britain. *Journal of the Historical Metallurgy Society* 18(1): 42–3.

Bayley, J. 1998. The production of brass in antiquity with particular reference to Roman Britain. In Craddock 1998, 7–26.

Bayley, J. and Butcher, S. 2004. *Roman Brooches in Britain*. London: Society of Antiquaries.

Bebko, W. 1971. *Les épaves antiques du Sud de la Corse*. Bastia: Cahiers Corsica, 1–3.

Beck, C.W. and Borromeo, C. 1990. Ancient pine pitch: technological perspectives from a Hellenistic shipwreck. In *Organic Contents of Ancient Vessels: Materials Analysis and Archaeological Investigation*, W.R. Biers and P.E. McGovern (eds). Philadelphia: MASCA Research Papers in Science and Archaeology 7, 51–8.

Beck, C.W., Stout, E.C. and Jänne, P.A. 1997. The pyrotechnology of pine tar and pitch inferred from quantitative analyses by gas chromatography–mass spectrometry and carbon-13 nuclear magnetic resonance spectrometry. In *Proceedings of the First International Symposium on Wood Tar and Pitch*, W. Brzeziński and W. Piotrowski (eds). Warsaw: State Archaeological Museum, 181–90.

Becker, P. and Kutzke, H. 1994. Korrosion und Konservierung der Bleifunde. In Hellenkemper-Salies *et al.* 1994, 1077–80.

Begemann, F. and Schmitt-Strecker, S. 1994. Das Blei von Schiff und Ladung: Seine Isotopie und mögliche Herkunft. In Hellenkemper-Salies *et al.* 1994, 1073–6.

Beltran-Lloris, M. 1970. *Lás anforas romanas en España*. Zaragoza: Monografias arqueologicas 8.

Beltran-Lloris, M. 1978. *Ceramica Romana: Tipologia y clasificacion*. Zaragoza: Libros Portico.

Besnier, M. 1920. Le commerce du plomb à l'époque romaine d'après les lingots estampillés. *Revue archéologique* 5(12): 211–44.

Besnier, M. 1921. Le commerce du plomb à l'époque romaine d'après les lingots estampillés. *Revue archéologique* 5(13): 36–73.

den Boesterd, M.H.P. and Hoekstra, E. 1965. *Spectrochemical Analyses of the Bronze Vessels in the Rijksmuseum G.M. Kam at Nijmegen*. Leiden: Oudheidkundige mededelingen uit het Rijksmuseum van oudhelden te Leiden, No. 45.

Boon, G.C. 1971. Aperçu sur la production des métaux non ferreux dans la Bretagne romaine. *Apulum* 9: 453–503.

Burnett, A.M., Craddock, P.T. and Preston, K. 1982. New light on the origin of orichalcum. In *Proceedings of the 9th International Congress of Numismatics, Berne, September 1979*, T. Hackens and R. Weiller (eds). Louvain-la-Neuve: Association Internationale des Numismates Professionnels, Publication 6, 263–8.

Caley, E.R. 1964. *Orichalcum and Related Ancient Alloys*. New York: American Numismatic Society.

Callender, M.H. 1965. *Roman Amphorae with Index of Stamps*. London: Oxford University Press.

Cochet, A. 2000. Le Plomb en Gaule romaine. Techniques de fabrication et produits. *Monographies Instrumentum* 13: 1–30.

Collingwood, R.G. and Wright, R.P. 1990. *The Roman Inscriptions of Britain II: Instrumentum Domesticum: Fasciulei*. Gloucester: Alan Sutton, 38–40.

Colls, D., Domergue, C. and Guerrero Ayuso, V. 1986. Les lingots de plomb de l'épave romaine Cabrera 5 (Ile de Cabrera, Baléares). *Archaeonautica* 6: 31–80.

Craddock, P.T. 1978. Origins and early use of brass. *Journal of Archaeological Science* 5(1): 1–16.

Craddock, P.T. 1995. *Early Metal Mining and Production*. Edinburgh: Edinburgh University Press.

Craddock, P.T. (ed.) 1998. *2000 Years of Zinc and Brass*, 2nd edn. British Museum Occasional Paper 50. London: British Museum Press.

Craddock, P.T. and Eckstein, K. 2003. Production of brass in antiquity by direct reduction. In *Mining and Metal Production through the Ages*, P.T. Craddock and J. Lang (eds). London: British Museum Press, 216–30.

Craddock, P.T., Burnett A.M. and Preston, K. 1980. Hellenistic copper-base coinage and the origins of brass. In *Scientific Studies in Numismatics*, W.A. Oddy (ed.). British Museum Occasional Paper 18. London: British Museum Press, 53–64.

Czajlik, Z. 1996. Ein spätbronzezeitliches Halbfertigprodukt: Der Gußkuchen. Eine Untersuchung anhand von Funden aus Westungarn. *Archaeologia Austriaca* 80: 165–80.

Day, J. 1973. *Bristol Brass*. Newton Abbot: David and Charles.

Domergue, C. 1966. Les lingots de plomb romains du Musée archéologique de Carthagène et du Musée Naval de Madrid. *Archivo Español de Arqueologia* 39: 41–72.

Domergue, C. 1984. Mines d'or romaines du nord-ouest de l'espagne. Les 'Coronas': techniques d'exploitation ou habitats? In *Papers in Iberian Archaeology*, T.F. Blagg, R.F.J. Jones and S.J. Keay (eds). BAR International Series 193(II). Oxford: BAR, 370–95.

Domergue, C. 1987. Les lingots de plomb de l'épave romaine de Valle Ponti (Comacchio). *Epigraphica* 49: 109–68.

Domergue, C., Laubenheimer-Leenhardt, F. and Liou, B. 1974. Les lingots de plomb de L. Carulius Hispallus. *Revue archéologique de Narbonnaise* 7: 119–37.

Domergue, C., Mas Garcia, J. and Mas, J. 1983. Nuevos descubrimientos de lingotes de plombo romanos estampillados. *XVI Congreso nacional de arqueologia*, 905–16.

Drescher, H. 1987. Ergänzende Bemerkungen zum Gießereifund von Bonn-Schwarzrheindorf. In Janssen 1987, 201–35.

Dungworth, D. 1997. Roman copper alloys: analysis of artefacts from northern Britain. *Journal of Archaeological Science* 24: 901–10.

Eck, W. 1994. Die Bleibarren. In: Hellenkemper-Salies *et al.* 1994, 89–95.

Eiwanger, J. 1996. Barrenmessing – ein mittelalterliches Handelsgut zwischen Europa und Afrika. *Beiträge zur Allgemeinen und Vergleichenden Archäologie* 16: 215–26.

Ercker, L. 1951. *Treatise on Ores and Assaying*, A.G. Sisco and C.S. Smith (trans.). Chicago: University of Chicago Press.

Erdrich, M. 1995. Zur Herstellung von Hemmoorer Eimern. In *Proceedings of the 12th International Congress on Ancient Bronzes, Nijmegen 1992*, S.T.A.M. Mols *et al.* (eds). Nijmegen: Provincial Museum G.M. Kam, 33–8.

Fiorini, E. 1991. L'imtiego del piombo romano nelle richerce die eventi rari. *Il nuovo Saggiatore* 7(5): 29–39.

Furger, A.R. and Riederer, J. 1995. Aes und aurichalcum. *Jahresberichte aus Augst und Kaiseraugst* 16: 115–80.

Gechter, M. 1984. *Stammformen römischer Gefäßkeramik in Niedergermanien*. Bonn: Rheinland-Verlag.

Gose, E. 1950. *Gefäßtypen der römischen Keramik im Rheinland*. Keverlaer: Butzon & Bercker.

Haedecke, K. 1973. Gleichgewichtsverhältnisse bei der Messingherstellung nach dem Galmeiverfahren. *Erzmetall* 26: 229–33.

Hammer, P. 2001. Messing. In *Reallexikon der germanischen Altertumskunde* 1, H. Beck, D. Geuenich and H. Steuer (eds). Berlin: Springer.

Hellenkemper-Salies, G., Prittwitz und Gaffron, H.-H. and Bauchhenß, G. (eds) 1994. *Das Wrack. Der antike Schiffsfund von Mahdia*. Cologne: Rheinland Verlag.

Hughes, M.J. 1980. The analysis of Roman tin and pewter ingots. In *Aspects of Early Metallurgy*, W.A. Oddy (ed.). British Museum Occasional Paper 17. London: British Museum Press, 41–50.

Hughes, M.J., Cowell, M.R. and Craddock, P.T. 1976. Atomic absorption techniques in archaeology. *Archaeometry* 18: 19–36.

Jackson, R.P. and Craddock, P.T. 1995. The Ribchester hoard: a descriptive and technical study. In *Sites and Sights of the Iron Age*, B. Rafferty (ed.). Oxford: Oxbow, 75–102.

Janssen, W. 1987. Eine mittelalterliche Metallgießerei in Bonn-Schwarzrheindorf. *Beiträge zur Archäologie des Rheinlandes* 27: 135–235, tables 33–66.

Jehasse, J. and Jehasse, L. 1973. *La nécropole préromaine d'Aééria (1960–1968)*. Paris: Gallia, Supplement 25.

Jehasse, J. and Jehasse, L. 1985. Aleria et la métallurgie du fer. In *Il commercio etrusco arcaico. Atti dell' Incontro di Studio, 5–7 dicembre 1983*. Rome: Quaderni del centro di Studio per l'archeologia etrusca-italica, 95–101.

Koller J. and Baumer U. 1994. Organische Materialien von Bronzen und von der Kalfaterung des Mahdia-Schiffes. In Hellenkemper Salies *et al.* 1994, 1067–72.

Lewin, B. and Hauptmann, A. 1984. *Kodō-zuroku*. Illustrierte Abhandlung über die Verhüttung des Kupfers 1801. Bochum: Veröffentlichung aus dem Deutschen Bergbau-Museum 29.

L'Hour, M. 1989. Naufrage d'un navire 'experimental' de L'Oost-Indische Compagnie: Le Mauritius (1609). *Neptunia* 173: 8–25.

MacDowall, D.W. 1971. The economic context of the Roman Imperial countermark NCAPR. *Acta Numismatica* 1: 83–106.

Meier, S.W. 1990. *Bleibergbau in der Antike*. Zurich: Diss.

Meier, S.W. 1992a Bleigewinnung in der Antike. *Bergknappe* 61(3): 12–18.

Meier, S.W. 1992b. Bleigewinnung in der Antike. *Bergknappe* 62(4): 10–19.

Meier, S.W. 1994. Der Blei-Fernhandel in republikanischer Zeit. In Hellenkemper-Salies *et al.* 1994, 767–800.

Merlin, A. 1912. *Lingots et ancres trouvés en mer près de Mahdia (Tunisie)*. Paris: *Mélanges Cagnat* (Recueil de Mémoires concernant l'épigraphie et les antiquités romaines).

Moesta, H. 1983. *Erze und Metalle – ihre Kulturgeschichte im Experiment*. Berlin: Springer Verlag.

Moesta, H. 1993. Einige mikroskopische Beobachtungen am Römischen Münzmessing aus Kleinasien. *Blesa* 1: 461–4.

Musty, J. 1975. A brass sheet of the first century AD from Colchester (Camulodunum). *Antiquaries Journal* 55(2): 409–11.

Northover, P. 1998. Exotic alloys in antiquity. In *Metallurgica Antiqua*, Th. Rehren, A. Hauptmann and J.D. Muhly (eds). Bochum: Deutsches Bergbau-Museum, 113–21.

Parker, A.J. 1974. Lead ingots from a Roman ship at Ses Salines, Majorca. *International Journal of Nautical Archaeology* 3(1): 147–50.

Peacock, D.P.S. 1974. Amphorae and the Baetican fish industry. *Antiquaries Journal* 54: 323–44.

Remesal-Rodriguez, J. and Revilla-Calvo, V. 1991. Weinamphoren aus Hispania Citerior und Gallia Narbonensis in Deutschland und Holland. *Fundberichte aus Baden-Württemberg* 16: 389–439.

Riederer, J. 1974a. Metallanalysen römischer Sesterzen. *Numismatik und Geldgeschichte* 24: 73–98.

Riederer, J. 1974b. Roman needles. *Technikgeschichte* 4(2): 153–72.

Rolandi, G. and Scacciati, G. 1956. Ottone e zinco presso gli antichi. *L'industria mineraria* 7: 750–70.

Rotroff, S.I. 1994. The pottery. In Hellenkemper-Salies *et al.* 1994, 133–52.

Salvi, D. 1992. Le 'massae plumbae' di Mal di Ventre. *L'Africa romana* 9: 660–72.

Salvi, D. 1999. Lingotti, ancore ed altri reperti di etè romana nelle acque di Piscinas – Arbus (CA). *Pallas* 50: 75–88.

Salvi, D., Santoni, V., Fiorini, E., Pascolini, A., Mocchegiani Carpano, C. and Lattanzi, G. 1992. La Nave del piombo. *Archeologia Viva* 5: 56–66.

Schallmeyer, E. 1982. Wegmarken des antiken Welthandels – Römische Amphoren aus Baden-Württemberg. *Denkmalpflege in Baden-Württemberg* 3: 116–23.

Schnurbein, S.v. 1971. Ein Bleibarren der 19. Legion aus dem Hauptlager von Haltern. *Germania* 49: 132–6.

Schramm, A. 1939. '*oreichalkos*'. *Paulys* Realencyclopädie *der classischen Altertumswissenschaft*. Berlin.

Smith, C.S. (ed.) 1983. Kodō Zuroku – *Illustrated Book on the Smelting of Copper by Masuda Tsuma*. Norwalk, CT: Burndy Library.

Theophilus Presbyter 1933. *Technik des Kunsthandwerks im zehnten Jahrhundert*. Des Theophilus Presbyter Diversarium Artium Schedula. Berlin: W. Theobald.

Theophilus Presbyter 1963. *On Divers Arts*, J.G. Hawthorne and C.S. Smith (trans.). Chicago: University of Chicago Press.

Tylecote, R.F. 1987. *The Early History of Metallurgy in Europe*. London: Longman.

Werner, O. 1970. Über das Vorkommen von Zink und Messing in Altertum und im Mittelalter. *Erzmetall* 23: 259–69.

Whittick, C.G. 1961. The casting technique of Romano-British lead ingots. *Journal of Roman Studies* 51: 105–11.

Willers, H. 1906. *Die römische Messing-Industrie in Nieder-Germanien. Ihre Fabrikate und ihr Ausfuhrgebiet*. Bonn: Carl Georgi.

Zwicker, U., Greiner, H., Hofman, K.H. and Reitinger, M. 1985. Smelting, refining and alloying of copper and copper alloys in crucible furnaces during prehistoric up to Roman times. In *Furnaces and Smelting Technology in Antiquity*, P.T. Craddock and M.J. Hughes (eds). British Museum Occasional Paper 48. London: British Musuem Press, 103–16.

## Author's address

Gerd Weisgerber, Deutsches Bergbau-Museum, Am Bergbaumuseum 28, D-44791 Bochum, Germany (gerd.weisgerber@bergbaumuseum.de)

# Copper-based metal in the Inland Niger delta: metal and technology at the time of the Empire of Mali

*Laurence Garenne-Marot and Benoît Mille*

*ABSTRACT*   Elemental analyses have recently been conducted on two series of copper-based metal objects unearthed at the beginning of the 20th century in two large burial mounds in the Lakes region north of the Inland Niger delta (Mali): Koï Gourrey (Killi) and El Oualadji. These artefacts probably date from the heyday of the Islamic trans-Saharan trade (11th–13th century AD) when this region was part of the Mali Empire. The analyses reveal some surprises such as the exceptional occurrence of high zinc brass for two small animal figurines. Although the range of metals of these two burial mounds reveals ties with metal from the Islamic world, the techniques involved in working them were local and the choice of specific alloys – showing a strong emphasis on the colour of the metal – was culturally determined and obviously dictated by the 'values' attached to gold and brass.

*Keywords:* ancient Africa, Mali Empire, Sahara, trade, Islamic, metalwork, copper, brass, gold, ICP–AES, tumulus.

## Introduction

Elemental analyses have been recently conducted by the Centre de Recherche et de Restauration des Musées de France (C2RMF) on two sets of copper-based objects. The objects were unearthed at the very beginning of the 20th century along with other objects from two large burial mounds – tumuli – situated in the Lakes region north of the Inland Niger delta (modern state of Mali, West Africa) (see Fig. 1). Excavations were conducted by a lieutenant of the French colonial infantry, Louis Desplagnes, based at Goundam, who took great interest in the history of the region. The excavated ceramic and metal

objects were then shipped to the Musée de l'Homme in Paris where they were stored as 'the Desplagnes collections'.

## Archaeological context of the analysed objects: the tumuli of the Lakes region of the Inland Niger delta (Mali)

When Desplagnes conducted his excavations in 1901 and 1904, methods of archaeological research were not much different from those of treasure hunting and no precise account was

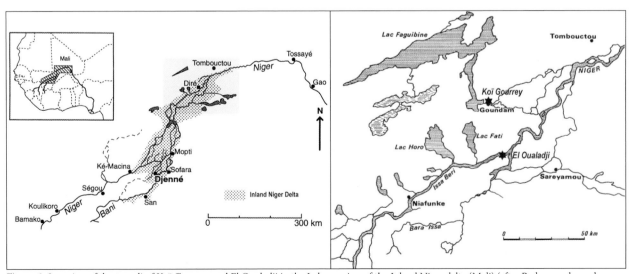

**Figure 1** Location of the tumuli of Koï Gourrey and El Oualadji in the Lakes region of the Inland Niger delta (Mali) (after Bedaux and van der Waals 1994; Lebeuf and Pâques 1970).

159

given of the stratigraphic location and spatial associations of the remains. Thus, the precise context of the analysed objects is usually difficult to ascertain. Some information, however, can be drawn from the excavation reports for each of the tumuli. The first to be excavated was the Koï Gourrey monument in the Killi district (Desplagnes 1903). A trench was run straight through the middle of the monument and the record of the finds follows the boring of the trench and, therefore, the progress of the excavation. First, two bodies with arms and iron ornaments were encountered; then numerous pots, complete but crushed; then, further inside the mound in what might be considered the central part of the tumulus, 20 to 25 entangled bodies with numerous copper-based ornaments; followed by another group of ceramics, this time glazed and well preserved; and finally, at the southern end of the monument, a single body with some iron and a necklace of bone and wood beads.

The copper-based metal objects were all found within this central part together with the entangled bodies. They are mostly ornaments such as bracelets, anklets and beads, with the exception of three very curious animal figurines – a crocodile, a bird and a lizard – around 10 cm in length or height, of a type previously unknown in the Inland Niger delta (Fig. 2).

Three years later Lieutenant Desplagnes excavated a second, larger monument, El Oualadji (the publication of which appeared posthumously some 45 years later; Desplagnes 1951). With the experience gained from the Koï Gourrey excavations, accounts of the El Oualadji dig were more precisely recorded, leading to a better understanding of the spatial organisation of the finds. What could be interpreted as a chamber made of palm trunks and straw mats, containing the remains of two bodies, was discovered within the central part of the earth mound (Fig. 3). The surface of the fire-hardened mound was in some places pitted by large ash-filled depressions –'sacrificial hearths' as Desplagnes called them – the larger of which, 5 m in diameter, contained burned horse bones.

The copper-based objects of El Oualadji can be sorted into two groups (see Figs 3 and 4). The first group consists of personal ornaments – bracelets, rings, beads – all found with one of the two bodies lying in the burial chamber (surprisingly and interestingly, the second body was accompanied solely by

iron objects). The second group, from a completely different location, was associated with the horse remains in the larger ash-filled depression at the surface of the mound. This group consists of what seem to be horse harnesses and ornamental trappings: bells, plaques and harness parts.

Accurate dating of the tumuli remains a problem. For the El Oualadji tumulus a charcoal sample was recovered in 1984 at an unexcavated location in the tumulus at a depth sufficient to date the 1904 finds. It gives a date of 900 ± 70 BP, which calibrates at 2σ to a calendar age from 1030 cal AD to 1220 cal AD (Fontes 1991: 263–5 and table 1; Orsay 2271). Based on Muslim traders' descriptions of life at the court of ancient Ghana, in the mid-11th century AD the Andalusian geographer al-Bakri wrote a detailed account of the burial of a king of Ghana under an earth tumulus (al-Bakri in Levtzion and Hopkins 2000: 80–81), which fits with the archaeological data. So, the construction of the El Oualadji monument can be placed with a fair degree of confidence during the 11th or 12th century AD.

For the Koï Gourrey tumulus things are more difficult. Due to the extent of Desplagnes' excavation trench, the monument has completely disappeared and therefore no postdating of the monument was possible. Oral local traditions, however, place the date of the monument construction before AD 1326 (Desplagnes 1903). All we can say, given the current state of research, is that both monuments belong to the heyday of the Islamic trans-Saharan trade and reflect the economic and social situation of this region of West Africa at the time of the Sahelian empires of ancient Ghana and Mali (geographical extent and time period of the empires are shown in Fig. 5).

## Scarcity of compositional analyses for archaeological copper-based objects in the Inland Niger delta

The need for compositional analyses of the Desplagnes copper-based objects was dictated by the scarcity of data for this region. Numerous copper-based objects have been found in the Inland Niger delta but very few have archaeological contexts. Although some were accidentally discovered, most

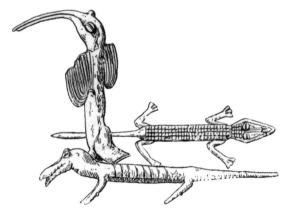

**Figure 2** The three animal figurines – a crocodile, a bird and a lizard – as drawn by Lieutenant Louis Desplagnes (Desplagnes 1903: 165, figs 35–37).

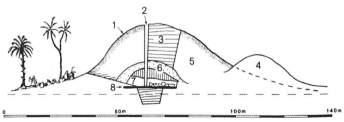

**Figure 3** Section through the El Oualadji tumulus (height 16 m; diameter 72 m). 1: Fired clay surface. 2: Shaft. 3: Excavated area. 4. Smaller mound. 5: Clay mound. 6: Domed roof of straw and wood c. 13 × 6.5 m and 2.85 m high. 7: Chamber. 8: Layer of red sand. (Drawing after Connah 1987: 108, fig. 5.5; after Desplagnes 1951: fig. 3).

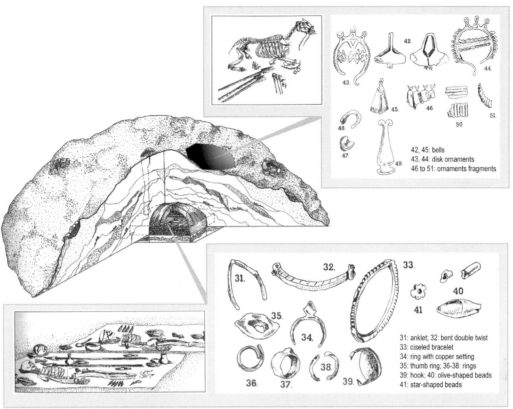

Figure 4 The tumulus of El Oualadji: location of the two groups of copper-based objects. (After (modified) Annita Barbara Schneider in McIntosh 1998: 225, fig. 8.1; the drawings of the objects are those published in Desplagnes 1951.)

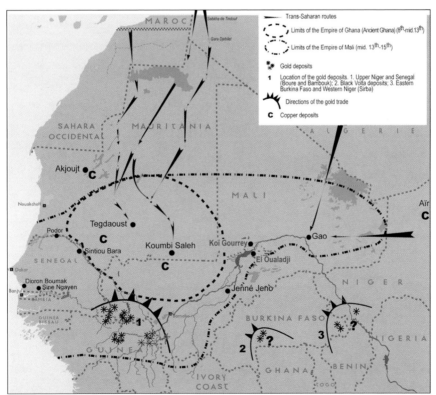

Figure 5 Map of the western part of Africa, showing: (1) The trans-Saharan routes for this part of Africa (the westernmost routes are precisely recorded thanks to the writings of al-Bakri (11th century) and al-Idrissi (12th century). All known routes are indicated and mapped without taking into account the fluctuations in the frequency and in the itineraries, which occurred between the 11th and the 14th century AD. (2) The estimated limits of ancient Ghana (8th to mid-13th century) and Mali (mid-13th to 15th century). (3) The gold deposits: among them, those of Eastern Burkina Faso and Western Niger geologically known but with no archaeological or textual evidence to document their exploitation at the time of the Sahelian empires of ancient Ghana and Mali. (4) The copper deposits bearing evidence of ancient workings.

were looted in the widespread plundering campaigns which have affected this region so much during the past 30 years.

Apart from one analysis from the Toguere Galia site (Bedaux *et al.* 1978: 147), our sole knowledge of copper alloys in an archaeological context comes from the excavations conducted at Jenné-Jeno by S.K. and R. McIntosh. These yielded an interesting but sparse corpus of nine analyses covering a span of ten centuries of occupation. According to McIntosh (1995: 386): 'The evidence from Jenné-Jeno indicates the possibility of an evolution from copper to bronze by AD 850 and the replacement with brass by AD 1000. Much further work in

the region is required to substantiate or refute this potential metallurgical sequence, and carefully controlled, stratigraphic excavation must lie at the heart of this research.'

## Metal composition of the Koï Gourrey and El Oualadji artefacts

Sixteen artefacts were sampled providing 18 results for elemental analysis by inductively coupled plasma–atomic emis-

**Table 1** ICP–AES results of the elemental analyses on the copper-based artefacts from the Desplagnes collections, musée de l'Homme. The base metal, copper, is not reported here. Key: nd = below detection limit; ° = below quantification limit.

*(a) Major and minor elements, weight %*

| Inv. no. | Site | Description | Results in wt% | | | | | | | | |
|---|---|---|---|---|---|---|---|---|---|---|---|
| | | | Zn | Sn | Pb | Ag | As | Fe | Ni | S | Sb |
| 03 8 2 | Koï Gourrey | Crocodile figurine | 27 ± 3 | nd < 0.0013 | 0.26 ± 0.03 | 0.012 ± 0.001 | 0.005 ± 0.0005 | 0.043 ± 0.004 | 0.017 ± 0.002 | 0.033 ± 0.003 | 0.013 ± 0.001 |
| 03 8 3 | Koï Gourrey | Bird figurine | 31 ± 3 | nd < 0.0012 | 0.43 ± 0.04 | 0.016 ± 0.002 | 0.007 ± 0.0007 | 0.045 ± 0.004 | 0.016 ± 0.002 | 0.024 ± 0.002 | 0.023 ± 0.002 |
| 03 8 103 | Koï Gourrey | Small bell | 5.2 ± 0.5 | nd < 0.0014 | 1 ± 0.1 | 0.055 ± 0.006 | 1.2 ± 0.1 | 0.54 ± 0.05 | 0.033 ± 0.003 | 0.24 ± 0.02 | 0.17 ± 0.02 |
| 03 8 99 | Koï Gourrey | Bracelet | 8.5 ± 0.9 | nd < 0.00091 | 0.8 ± 0.08 | 0.033 ± 0.003 | 0.27 ± 0.03 | 0.073 ± 0.007 | 0.059 ± 0.006 | 0.094 ± 0.009 | 0.061 ± 0.006 |
| 03 8 100 | Koï Gourrey | Bracelet | 14 ± 1 | 0.076 ± 0.008 | 1.3 ± 0.1 | 0.081 ± 0.008 | 0.57 ± 0.06 | 0.17 ± 0.02 | 0.034 ± 0.003 | 0.16 ± 0.02 | 0.095 ± 0.009 |
| 03 8 112 | Koï Gourrey | Bracelet | 1.6 ± 0.2 | 3.8 ± 0.4 | 6.7 ± 0.7 | 0.038 ± 0.004 | 0.22 ± 0.02 | 0.73 ± 0.07 | 0.074 ± 0.007 | 0.31 ± 0.03 | 0.097 ± 0.010 |
| 03 8 6a | Koï Gourrey | Half-bracelet | 0.46 ± 0.05 | nd < 0.00092 | 0.22 ± 0.02 | 0.22 ± 0.02 | 4.6 ± 0.5 | 0.063 ± 0.006 | 0.018 ± 0.002 | 0.22 ± 0.02 | 0.18 ± 0.02 |
| 03 86b | Koï Gourrey | Half-bracelet | 3.3 ± 0.3 | 0.018 ± 0.002 | 0.95 ± 0.10 | 0.19 ± 0.02 | 2.1 ± 0.2 | 0.33 ± 0.03 | 0.028 ± 0.003 | 0.13 ± 0.01 | 0.14 ± 0.01 |
| 03 8 67a | Koï Gourrey | Half-bracelet | 11 ± 1 | 4.8 ± 0.5 | 1.2 ± 0.1 | 0.072 ± 0.007 | 0.89 ± 0.09 | 0.18 ± 0.02 | 0.042 ± 0.004 | 0.082 ± 0.008 | 0.19 ± 0.02 |
| 03 8 67b | Koï Gourrey | Half-bracelet | 11 ± 1 | 4.9 ± 0.5 | 1.2 ± 0.1 | 0.074 ± 0.007 | 0.91 ± 0.09 | 0.18 ± 0.02 | 0.043 ± 0.004 | 0.087 ± 0.009 | 0.2 ± 0.02 |
| 03 8 117 | Koï Gourrey | Ring fragment | 7.7 ± 0.8 | nd < 0.00094 | 0.82 ± 0.08 | 0.033 ± 0.003 | 0.32 ± 0.03 | 0.32 ± 0.03 | 0.019 ± 0.002 | 0.11 ± 0.01 | 0.047 ± 0.005 |
| 06 30 135 | El Oualadji | Small bell | 2.4 ± 0.2 | 3.5 ± 0.4 | 3.4 ± 0.3 | 0.061 ± 0.006 | 0.27 ± 0.03 | 0.2 ± 0.02 | 0.055 ± 0.006 | 0.21 ± 0.02 | 0.45 ± 0.05 |
| 06 30 131 | El Oualadji | Bracelet | 0.036 ± 0.004 | nd < 0.00093 | 0.24 ± 0.02 | 0.006 ± 0.0006 | 0.25 ± 0.02 | 1.2 ± 0.1 | 0.072 ± 0.007 | 0.53 ± 0.05 | 0.025 ± 0.002 |
| 06 30 129 | El Oualadji | Bracelet | 5.4 ± 0.5 | 0.99 ± 0.10 | 0.95 ± 0.09 | 0.082 ± 0.008 | 0.49 ± 0.05 | 0.25 ± 0.03 | 0.041 ± 0.004 | 0.2 ± 0.02 | 0.39 ± 0.04 |
| 06 30 142 | El Oualadji | Bracelet | 1 ± 0.1 | 0.010 ± 0.0010 | 0.08 ± 0.008 | 0.005 ± 0.0005 | 0.1 ± 0.01 | 0.014 ± 0.001 | 0.002 ± 0.0002 | 0.004 ± 0.0004 | 0.006 ± 0.0006 |
| 06 30 132 | El Oualadji | Bracelet | 0.86 ± 0.09 | nd < 0.00010 | 1.3 ± 0.1 | 0.010 ± 0.0010 | 0.33 ± 0.03 | 0.069 ± 0.007 | 0.005 ± 0.0005 | 0.017 ± 0.002 | 0.041 ± 0.004 |
| 06 30 139 | El Oualadji | Ring | 3.4 ± 0.3 | 1.2 ± 0.1 | 1.2 ± 0.1 | 0.045 ± 0.005 | 0.85 ± 0.09 | 0.74 ± 0.07 | 0.054 ± 0.005 | 0.17 ± 0.02 | 0.17 ± 0.02 |
| 06 30 141 | El Oualadji | Ring fragment | 7.8 ± 0.8 | 0.066 ± 0.007 | 1.2 ± 0.1 | 0.052 ± 0.005 | 0.98 ± 0.10 | 0.64 ± 0.06 | 0.028 ± 0.003 | 0.16 ± 0.02 | 0.12 ± 0.01 |

sion spectrometry (ICP–AES). The results are shown in Table 1 (Garenne-Marot *et al.* 2003). The metal was collected by drilling a hole of 1 mm diameter. Corrosion products were avoided by discarding the patina layers encountered on the sampled surface and by carefully monitoring the drilling with a stereoscopic microscope. A Perkin Elmer Optima 3000 SC ICP-AES instrument was used for the analyses (the preparation of the sample and analytical protocol employed are fully described in Bourgarit and Mille 2003). In total 22 chemical elements were tested for, with the accuracy and precision of the results better than 10%.

Among the 16 artefacts analysed, no fewer than six different copper alloy types have been identified (Fig. 6): unalloyed copper, copper-arsenic, quaternary alloys containing more or less equal parts of lead, tin and zinc, ternary alloys with a high zinc content, and brass for the largest group (nine artefacts). Interestingly true bronze is lacking. The high zinc content of the crocodile (27 wt%) and the bird (31 wt%) is noteworthy, and will be discussed in detail. Lead is present at the percentage level in most of the objects, but does not seem to be correlated to the manufacturing technique: the only possible wrought pieces are the small

*(b) Trace elements, weight ppm*

| Inv. no. | Results in wt ppm | | | | | | | | | | | | |
|---|---|---|---|---|---|---|---|---|---|---|---|---|---|
| | Au | Ba | Bi | Cd | Co | Cr | In | Mg | Mn | P | Se | Te | Ti |
| 03 8 2 | 6.9 ± 0.7 | 3.3 ± 0.3 | nd < 9.9 | nd < 0.4 | 4.1 ± 0.4 | 33 ± 3 | 3.4° < 8.6 > 2.6 | 74 ± 7 | nd < 0.1 | 80 ± 10 | 8.3° < 19 > 5.7 | 39 ± 9 | 5.5 ± 0.5 |
| 03 8 3 | 5.4 ± 0.5 | 4 ± 0.4 | nd < 8.6 | nd < 0.3 | 4.1 ± 0.4 | 28 ± 3 | 8 ± 3 | 85 ± 9 | nd < 0.1 | 33° < 37 > 10 | 9.2° < 16.7 > 4.9 | 30 ± 20 | 4.5 ± 0.4 |
| 03 8 103 | 8.8 ± 0.9 | 4.7 ± 0.5 | 640 ± 60 | 1.2 ± 0.8 | 69 ± 7 | 33 ± 3 | 11 ± 3 | 110 ± 10 | 1.7 ± 0.2 | 43° < 44 > 13 | 39 ± 4 | 40 ± 20 | 5.5 ± 0.5 |
| 03 8 99 | 14 ± 1 | 0.7 ± 0.1 | 86 ± 9 | nd < 0.3 | 270 ± 30 | 22 ± 2 | nd < 1.9 | 31 ± 3 | nd < 0.1 | 42 ± 4 | 8.8° < 13.7 > 4 | 24 ± 10 | 3.7 ± 0.4 |
| 03 8 100 | 27 ± 3 | 2.9 ± 0.3 | 200 ± 20 | nd < 0.3 | 67 ± 7 | 23 ± 2 | 20 ± 3 | 61 ± 6 | nd < 0.1 | 31 ± 5 | 17 ± 3 | 50 ± 9 | 4.3 ±0.4 |
| 03 8 112 | 31 ± 3 | 12 ± 1 | 79° < 80 > 23 | nd < 0.8 | 120 ± 10 | 78 ± 8 | nd < 6.2 | 280 ± 30 | nd < 0.1 | 150 ± 20 | 86 ± 9 | 100 ± 10 | 14 ± 1 |
| 03 8 6a | 6.4 ± 0.6 | 2.8 ± 0.3 | 180 ± 20 | nd < 0.3 | 18 ± 2 | 22 ± 2 | 10 ± 2 | 60 ± 6 | nd < 0.1 | 28° < 31 > 9 | 70 ± 7 | 40 ± 10 | 4 ± 0.4 |
| 03 86b | 8.1 ± 0.8 | 2.9 ± 0.3 | 92 ± 9 | nd < 0.3 | 37 ± 4 | 23 ± 2 | 16 ± 2 | 45 ± 5 | nd < 0.1 | 30 ± 10 | 28 ± 3 | 27 ± 3 | 4 ± 0.4 |
| 03 8 67a | 8.1 ± 0.8 | 4.1 ± 0.4 | 100 ± 10 | nd < 0.3 | 26 ± 3 | 22 ± 2 | 14 ± 2 | 59 ± 6 | 1.3 ± 0.1 | 28° < 30 > 8 | 26 ± 3 | 41 ± 5 | 4.7 ± 0.5 |
| 03 8 67b | 8.3 ± 0.8 | 3.1 ± 0.3 | 100 ± 10 | nd < 0.3 | 26 ± 3 | 24 ± 2 | 13 ± 1 | 69 ± 7 | 1.5 ± 0.1 | 32 ± 3 | 27 ± 3 | 33 ± 8 | 4.1 ± 0.4 |
| 03 8 117 | 16 ± 2 | 1.4 ± 0.2 | 1100 ± 100 | nd < 0.3 | 31 ± 3 | 24 ± 2 | 16 ± 5 | 40 ± 4 | 2.8 ± 0.3 | 95 ± 10 | 14° < 15 > 4 | 43 ± 5 | 3.6 ± 0.4 |
| 06 30 135 | 20 ± 2 | 6.5 ± 0.6 | 87 ± 9 | nd < 0.3 | 99 ± 10 | 25 ± 3 | 18 ± 2 | 79 ± 8 | 1.5 ± 0.1 | 160 ± 20 | 52 ± 5 | 66 ± 7 | 5.4 ± 0.5 |
| 06 30 131 | 4 ± 0.4 | 2.2 ± 0.2 | 18° < 24 > 7 | 5.9 ± 0.6 | 150 ± 20 | 29 ± 3 | 18 ± 2 | 46 ± 5 | 0.8 ± 0.1 | 17° < 31 > 9 | 19 ± 2 | 36 ± 4 | 4 ± 0.4 |
| 06 30 129 | 21 ± 2 | 1.9 ± 0.2 | 170 ± 20 | nd < 0.3 | 70 ± 10 | 23 ± 2 | 18 ± 3 | 64 ± 6 | nd < 0.1 | 110 ± 10 | 24 ± 3 | 44 ± 4 | 4 ± 0.4 |
| 06 30 142 | 0.1 ± 0.01 | 0.6° < 2.5 > 0.7 | 3.3 ± 0.3 | 0.9 ± 0.1 | 0.9 ± 0.1 | nd < 0.6 | 1.5 ± 0.2 | 9.5 ± 0.9 | 0.1° < 0.5 > 0.1 | 7 ± 0.7 | 1° < 3.2 > 0.9 | 1.6 ± 0.5 | 0.0° < 2.1 |
| 06 30 132 | 0.6 ± 0.1 | 0.5 ± 0.05 | 8 ± 0.8 | nd < 0.1 | 7.9 ± 0.8 | 2.4 ± 0.2 | 1.6 ± 2 | 6.1 ± 0.6 | 0.2 ± 0.02 | 2.6° < 3.2 > 9 | 3.9 ± 0.4 | 2.7 ± 0.3 | 0.4 ± 0.04 |
| 06 30 139 | 20 ± 2 | 6.5 ± 0.7 | 400 ± 40 | nd < 0.6 | 110 ± 10 | 53 ± 5 | 7° < 13.9 > 4.2 | 180 ± 20 | 1.8 ± 0.2 | 80 ± 20 | 32 ± 0.6 | 61 ± 6 | 9.4 ± 0.9 |
| 06 30 141 | 8.1 ± 0.8 | 2.2 ± 0.2 | 530 ± 50 | nd < 0.3 | 49 ± 5 | 23 ± 2 | 18 ± 4 | 63 ± 6 | 1.2 ± 0.1 | 240 ± 20 | 25 ± 3 | 33 ± 3 | 4.2 ± 0.4 |

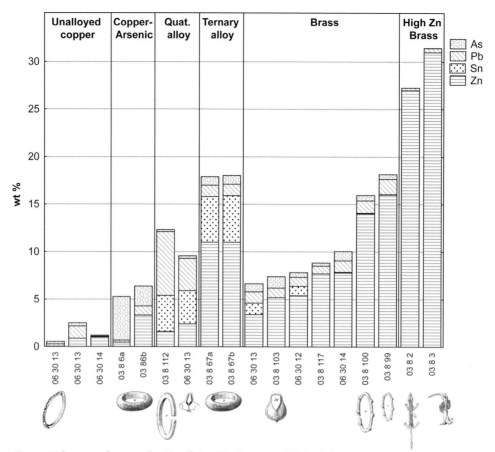

**Figure 6** The types of copper alloy identified at Koï Gourrey and El Oualadji.

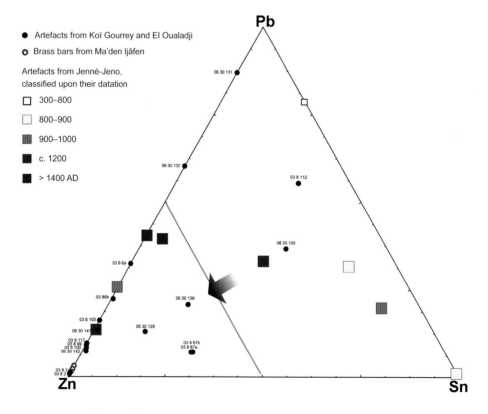

**Figure 7** Ternary diagram showing the relative amounts of zinc, tin and lead. The Koï Gourrey and El Oualadji artefacts are compared to the Jenné-Jeno sequence (nine analyses, McIntosh 1995) and to the three Ma'den Ijâfen brass bars (Werner and Willet 1975). Note that the relative amount of zinc increases through time for the Jenné-Jeno sequence, and that most of the Koï Gourrey and El Oualadji artefacts plot in the zinc-rich corner.

bells and their lead content is no lower than that of the cast pieces.

At first sight, the variety of alloy composition in burials such as Koï Gourrey or El Oualadji, where artefacts have been deposited at one time, is as great as in the Jenné-Jeno levels, which covers ten centuries of continuous occupation (McIntosh 1995). The assumption of linear evolution of alloy composition in Jenné-Jeno, as summarised in McIntosh's quotation reported above, is thus weakened and so it must be questioned whether a particular alloy corresponds to a specific period. Diversity of alloys depends not only on the evolution of their composition over time, but also reflects diversity in manufacturing techniques, in the function of the artefacts, in their provenance, with the additional problem of a possible long lifetime for these objects of which only the depositional date is known. Despite the small number of analyses conducted for the Jenné-Jeno sequence, the relative proportion of zinc is clearly increasing through time, as can be deduced from a Pb-Sn-Zn ternary diagram (see Fig. 7). This tendency is interesting for discussion of the date of the Koï Gourrey and El Oualadji artefacts. With their zinc predominance, Koï Gourrey and El Oualadji alloys are not inconsistent with the proposed chronological framework of the 11th to 13th century AD, a period where brass-based alloys also dominate at Jenné-Jeno.

## The Islamic trans-Saharan trade: the importance of copper-based metal

From the 11th to the 13th century, the heyday of the Islamic trans-Saharan trade, copper, both red (un-alloyed) and yellow (brass), as well as salt were exchanged for the gold of the Bilad al-Sudan (the 'land of the Blacks'). The 'importance of copper in the trans-Saharan trade runs as a leitmotiv through virtually all the written sources' wrote E. Herbert (1984: 113). The best proof of this trade is provided by the 2085 brass bars of the Ma'den Ijâfen – a caravan load buried in the Mauritanian desert dunes of the Ijâfen around the 12th century[1] and rediscovered by Theodore Monod in December 1964 (Monod 1969). Saharan salt was certainly an important commodity brought down by the northern caravans in exchange for the gold mined upstream on the Senegal and Niger rivers (see Fig. 5). But no salt remains in the archaeological record and archaeological gold is rarely found. Copper remains the key material with which to document the routes of this trade and any changes in the nature of the goods or in the routes over time. Compositional analyses of copper-based alloys might in specific cases provide valuable information for the reconstruction of the exchange networks in medieval western Africa.

The archaeological record shows that from at least the beginning of the 1st millennium BC, copper was mined and smelted south of the Sahara in places such as Akjoujt in Mauritania or within the Aïr massif in Niger (Lambert 1983) (see Fig. 5). Elemental analyses of copper-based artefacts from the sites of Tegdaoust and Koumbi Saleh confirm what had been suspected earlier by archaeologists who worked on these sites (see Vanacker 1983): at the time of the Islamic trans-Saharan trade (documented as early as the 8th century

AD), copper may still have been mined from the Mauritanian copper deposits and may have provided a great part of the material traded to the more southern countries where gold was mined (Garenne-Marot and Mille 2005). Some of the un-alloyed copper (red) was presumably produced locally, whereas brass (yellow) was imported. Colour certainly played a significant role in distinguishing the two metals, both in their commercial and social values. This distinction was probably made on an economic basis in the regions where copper was mined (Mauritania, West Mali, Niger), which were also the areas with close relations with north African Muslim traders (see Fig. 5) and where the Muslim metal value system was adopted. This was probably different in the regions of gold mining where non-alloyed copper and brass were both imported as exotic goods. The colour of the copper-based metal had its importance, however, as indicated by archaeometallurgical evidence. In the Tegdaoust workshop, copper-based metal ingots were manufactured to supply the trade with the gold-mining regions: imported brass was thus diluted with local copper but care was taken to preserve the golden tint of the metal by maintaining adequate zinc content (Garenne-Marot *et al.* 1994; Garenne-Marot and Mille 2005). Also, there is a clear distinction between red copper and yellow brass in the jewellery on the bodies recovered from two burial sites from Senegal – Sine Ngayen and Dioron Boumak – both dated around the same period as the tumuli, i.e. 12th–13th century AD. The ornaments were either of copper or brass but the two metals were never worn together. This attests to the existence of a strong social or symbolic meaning linked to the colour (Garenne-Marot 2004).

## The use of a high zinc brass for casting in the context of the Islamic trans-Saharan trade

The high zinc level detected in the crocodile and bird figurines from Koï Gourrey is unusual (the third figurine, missing from the collection, could not be analysed). The composition of the figurines has been plotted against 184 analyses reported in the literature for medieval Islamic artefacts from the 11th to the 14th century: 178 Fatimid period objects (Ponting 2003; Shalev and Freund 2002), and a selection of six Jordanian vessels dated from the 14th century (Al Saa'd 2000) (see Fig. 8). For Islamic metalwork, as first evidenced by Craddock *et al.* (1998), the choice of alloy is clearly correlated to the manufacturing techniques. Cast pieces are generally made with ternary or quaternary alloys, whereas hammered objects are principally made from a high zinc brass (from 15 to 25% Zn).[2] The Ma'den Ijâfen ingots (*c.* 20% zinc) prove that this kind of alloy reached the sub-Saharan area. At 27–31%, the two little figurines from Koï Gourrey exhibit even higher zinc content, yet they were cast. Examination and experiments have clearly shown the use of a lost-wax casting process (Robion 2000).

During medieval times, Islamic brass was prepared by the so-called cementation or calamine process, and it has been claimed since Werner's experiments that the upper threshold for the zinc content in such brass was 28%. Based on this argument (among others), the antiquity of a set of Ghanaian (Ashanti) brass lamps (25–34% zinc) has been questioned

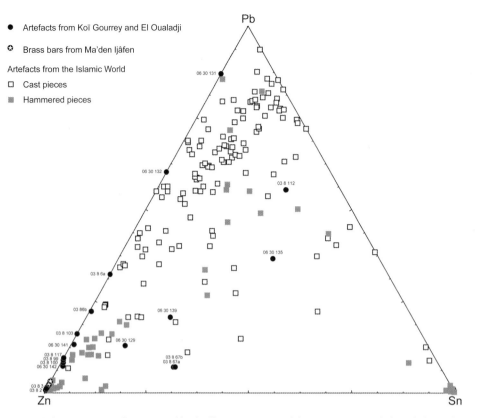

**Figure 8** Relative amounts of zinc, tin and lead. Alloy composition of the Koï Gourrey and El Oualadji artefacts versus objects from the medieval Islamic world (11th–14th century AD). The crocodile and bird figurines fit within the field of the Islamic high zinc brass hammered vessels.

(Craddock and Shaw 1984). Nevertheless, new experiments on the calamine process shift the threshold for the zinc content up to almost 40%, challenging the use of the zinc content as a chronological indicator (Welter 2003). Although the three animal figurines are typologically unparalleled, they were found in the same archaeological context as the other objects predating the mid-14th century AD, which we take to indicate the ancient date of the brass alloy of the figurines.

## Imitating gold when gold is in plentiful supply

With 27–31% zinc, the final colour of the animal figurines has a golden tint very close to that of true gold (Fig. 9).[3] The tumulus of Koï Gourrey dates to a peak time in the exploitation of the gold deposits of upper Senegal and Niger rivers (beginning around the end of the 13th century). This was also the time for *Mande* traders (Mali empire traders) to search for new gold supplies, going as far south as modern Ghana and the east Ivory Coast in their quest. Why then were these figurines made of an alloy that imitates gold and not simply made of gold?

The texts written by Arabic geographers and travellers at the time of ancient Mali give us some clues. They stress two things. First, the Muslim rulers of the Mali Empire of the 13th–14th century AD were tolerant of the pagan state of peoples in their empire. In some cases, moreover, they seem to have been more than just tolerant – they were eager to have

them maintained in this pagan state. These were populations living close to the gold mines and involved in gold mining.[4] Secondly, these populations were mining gold in order to obtain salt, but they were also more than willing to exchange gold for another metal, copper. To acquire this copper, the pagan natives accepted a totally unfair exchange rate: up to ⅔ of its weight in gold[5] whereas, at the same time, among the Mali elite, this same copper had little value.[6]

'From al-Husayn about 950 AD to the Benin Punitive Expedition in 1897, outsiders never ceased to wonder at the odd preference of Africans for cuprous metals.' This remark by

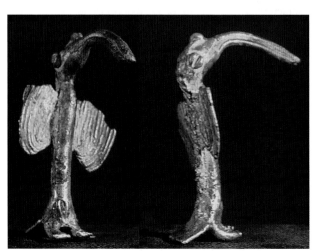

**Figure 9** Face and profile photographs of the bird figurine. (Photograph: Manuel Valentin, Département d'Afrique noire, musée de l'Homme, Paris.)

166

Herbert (1984: 296) should be qualified slightly when it refers to 'Africans' as if they were a whole and indivisible people. The phenomenon certainly did not affect the local elite who were in close contact with the Muslim traders but it fits perfectly with what prevailed among heathen populations living in the gold fields of the Mali Empire.

This 'odd' preference for a metal other than gold, highly prized in European and Mediterranean medieval and modern contexts, is not unique in the record of past societies. Examples such as those from South America are exemplified by the recent archaeological and the archaeometallurgical work on the early contact period cemetery of El Chorro de Maita, Cuba (Martinon-Torres *et al.* 2007). Ethnohistorical sources report that when the Europeans arrived on the island, they soon discovered that the metals they brought with them were perceived differently by the locals: brass, an unknown copper-based alloy, was more esteemed and sacred than pure gold. The Europeans bartered brass items – including cheap and dispensable lace-tags – for pure gold at astounding exchange rates (see discussion and references in Martinon-Torres *et al.* 2007: 8–10). The reasons for this preference for brass might be found in the complex relations the indigenous peoples of Cuba had with another alloy, a mixture of gold, copper and silver characterised by its 'sweet smell' (Martinon-Torres *et al.* 2007: 202; see also Bray 1993: 190).[7] Brass arriving with the Europeans was soon absorbed by the value system already in place.

The reasons for the preference for copper among the gold-mining populations of sub-Saharan Africa (or at least in the western part of Africa) are not easy to define. Not much can be deduced from Arabic sources, but some clues may be found in ethnographic surveys among present-day populations involved in gold mining in the very same regions. These populations evoke the supernatural nature of the gold: possession of the underground *jinè*. These *jinè* (genii) are responsible for any impediment, including fatalities among the gold miners, in extracting of the gold (see Guillard 1993; see also recent ethnographic work for other more eastern regions conducted in southwest Burkina Faso by Kadja Werthmann (2003)).

The fear of gold, or rather, of the deadly forces behind that gold, might then have ancient origins. One can postulate that this fear might have been turned to profit by the medieval black rulers. Xavier Guillard proposed that the geological characteristics of the Bambouk and Boure deposits (upstream Niger and Senegal rivers), both extremely dispersed and of low productivity, were one of the underlying reasons why the emperors of Mali were prevented from appropriating the gold fields and setting up a slavery system or at least some sort of coercive system. It would have been too difficult and too costly. We propose, based on Guillard's ideas (Guillard 1993) and clues drawn from the Arabic texts, that there was a more subtle form of subjection of the gold-mining populations by the Malian rulers. Some 'gathering' of gold was conducted through tribute, but the final control over the gold produced within the kingdom was exercised through economic constraints. In order to acquire vital salt and also copper for social and ritual activities, the 'pagan Blacks' were willing to mine gold and to exchange it for these highly valued goods at quite an unfair rate, at least by 'northern' standards.

Gold became a necessary medium with a high economic value but the hard and sometimes fatal conditions under which it was produced led to the perception of it as an 'evil' metal by the mining populations. On the other hand, copper was associated with salt, certainly a source of health and well-being. Brought by the same traders, it had the same exotic origins. Copper was endowed with innate qualities that gold lacked: it was a protective metal with apotropaic qualities.

These are some pointers to the reasons for the choice of high zinc brass. One can postulate that this particular alloy was used because it combined the external aspect of gold with the protective qualities ascribed to copper. We have previously drawn on the help of various arguments to stress the possible ritual character of these animal figurines, even through the peculiar casting device – the mould-cum-crucible process – that might have been used in their manufacture (Garenne-Marot *et al.* 2003). The tumuli of Koï Gourrey and El Oualadji are there to remind us that animism was still widespread among the elite.

## Conclusions

The range of copper-based metals from these two burial monuments shows evident ties with metal from the medieval Islamic world. Metal such as brass was imported from North Africa, but the techniques of working the metal were local. We see a long tradition of metalworking in the region: copper mining and working goes back to the beginning of the 1st millennium BC and iron metallurgical remains are widespread, not only in the Inland Niger delta and along the Senegal river valley but also in many places in western Africa where wood and water were both present.

When plotted against elemental analyses from other African sites such as Jenné-Jeno, zinc, as the principal alloying element, points to the period of the medieval trans-Saharan trade. When focusing on the figurines however, the choice of specific alloys seems to have been culturally determined following a local value system: they are made of high zinc brass, which, in Islamic contexts, is the alloy intended for hammered vessels but here the figurines were manufactured by lost-wax casting. The choice of alloy was not based solely on technical grounds but had more ideological ones, the mechanisms of which we have tried to decipher through the archaeological evidence, ethnographic data and the few Arabic texts available. In the regions where gold was mined and where miners were 'maintained' in their animist faith, copper-based metal was preferred over gold. Gold, due in large part to the conditions under which it was mined, was considered an 'evil' metal whereas copper was a metal with protective and curative qualities. The use of high zinc brass for these figurines may be associated with the interesting combination of the external aspect of gold with beliefs concerning the inner protective qualities attached to the copper material. In addition to the characterisation of the alloys, archaeometallurgy in these sub-Saharan regions sheds new light on the question of trans-Saharan exchanges and mechanisms of control over the West African gold resources and trade.

## Acknowledgements

We wish to express our thanks to Matthew Ponting, who very willingly and supportively provided us with unpublished information on the techniques of Fatimid metalwork.

## Notes

1. (I-2769): 785 ± 110 BP and (Dak-1): 860 ± 108 BP, which calibrate at 2σ to a calendar age from 1024 cal AD to 1329 cal AD for the first date and from 980 cal AD to 1303 cal AD for the second (using Stuiver and Reimer 1993 calibration chart). In fact the 'calendar dates' (1165 ± 110 AD and 1090 ± 108 AD) proposed at the Fort Lamy colloquium (Monod 1969: 310–11) and quoted frequently in the literature are uncalibrated; the proposed results are simply what emerges when the radiocarbon dates are subtracted from 1950.
2. Notice the peculiar group of high tin bronze hammered vessels; the corresponding manufacturing technique (hot forging, quenching) is described by Ponting 2003.
3. Confusion between gold and brass is abundantly documented in literature. For example, recently, US soldiers seized 4100 'gold bars' worth $700 million to $1 billion in Iraq (*Washington Post*, 17 June 2003). These gold-coloured bars appear to be melted-down shell casings made of brass containing 34% zinc (*New York Times*, 2 August 2003).
4. See in particular Al-'Umari's account of what the emperor of Mali (the Mansa ('emperor') Musa) told Egyptian officials in Cairo on his way to Mecca in 1324, that he did not wish to extend his authority over the uncouth infidels of the land of gold, for 'the kings have learnt by experience that as soon as one of them conquers one of the gold towns and Islam spreads and the muezzin calls to prayer there the gold there begins to decrease and then disappears, while it increases in the neighbouring heathen countries' (in Levtzion and Hopkins 2000: 262, ref. manuscript 26b). See also the account of Ibn al-Dawadan who wrote his work only 7–10 years after the pilgrimage of Mansa Musa and his visit to Cairo and who collected first-hand information on the Mansa's words (in Levtzion and Hopkins 2000: 250).
5. Quoted also in Al-'Umari, about the exploitation of a copper mine that Mansa Musa has within his kingdom: 'We send [the copper] to the land of the pagan Sudan and sell it for two-thirds of its weight in gold, so that we sell 100 mithqals of this copper for 66 2/3 mithqals of gold' (in Levtzion and Hopkins 2000: 272, ref. manuscript 36a).
6. See discussion in Garenne-Marot and Mille 2005.
7. Bray cites Fra Bartolomeo de las Casas who notes that in the Antilles gold-copper (sometimes with silver) was preferred to pure gold 'for the smell which they perceive in it, or for some virtue which they believed there was in it'.

## References

Al Saa'd, Z. 2000. Technology and provenance of a collection of Islamic copper-based objects as found by chemical and lead isotope analysis. *Archaeometry* 42(2): 385–97.

Bedaux, R.M.A. and van der Waals, J.D. (eds) 1994. *Djenné. Une ville millénaire au Mali*. Leiden: Rijksmuseum voor Volkenkunde.

Bedaux, R., Constandse-Westermann, L., Hacquebord, L., Lange, A.G. and van der Waals, J.D. 1978. Recherches archéologiques dans le Delta intérieur du Niger (Mali). *Palaeohistoria* XX, Acta et Communicationes Instituti Bio-Archaeologici Universitatis Groninganae, Harlem: Fibula-Van Dishoeck, 91–220.

Bourgarit, D. and Mille, B. 2003. The elemental analysis of ancient copper-based artefacts by inductively coupled plasma–atomic emission spectrometry (ICP–AES): an optimized methodology reveals some secrets of the Vix Crater. *Measurement Science and Technology* 14: 1538–55.

Bray, W. 1993. Techniques of gilding and surface enrichment in pre-Hispanic American metallurgy. In *Metal Plating and Patination: Cultural, Technical and Historical Developments*, S. La Niece and P.T. Craddock (eds). London: Butterworth, 182–92.

Connah, G. 1987. *African Civilizations*. Cambridge: Cambridge University Press.

Craddock, P.T. and Shaw, T. 1984. Ghanaian and Coptic brass lamps. *Antiquity* 223: 126–8.

Craddock, P.T., La Niece, S. and Hook, D. 1998. Brass in the medieval Islamic world. In *2000 Years of Zinc and Brass*, P.T. Craddock (ed.), 2nd edn. British Museum Occasional Paper 50. London, 73–114.

Desplagnes, Lieut. L. 1903. Étude sur les tumuli du Killi. Dans la région de Goundam. *L'Anthropologie* 14: 151–72.

Desplagnes, Lieut. L. 1951. Fouilles du tumulus d'El Oualedji (Soudan). *Bulletin IFAN* XIII(4): 1159–73.

Fontes, P.-B. 1991. Sites archéologiques de la région des Lacs au Mali: éléments chronologiques. In *Recherches archéologiques au Mali. Les sites protohistoriques de la Zone lacustre*, M. Raimbault and K. Sanogo (eds). Paris: ACCT-Karthala, 258–71.

Garenne-Marot, L. 2004. In-depth study of copper-based artefacts: what can be hidden behind the patina? In *Archéométrie. Actes du XIVe Congrès UISPP*, Université de Liège, Belgique, 2–8 septembre 2001, édité par le Secrétariat du Congrès. BAR International Series 1270. Oxford: Archaeopress, 17–25.

Garenne-Marot, L., Wayman, M.L. and Pigott, V.C. 1994. Early copper and brass in Senegal. In *Society, Culture and Technology in Africa*, T. Childs (ed.). Supplement to vol. 11 of MASCA Research Papers in Science and Archaeology, Philadelphia, 45–62.

Garenne-Marot, L., Robion, C. and Mille, B. 2003. Cuivre, alliages de cuivre et histoire de l'empire du Mali. À propos de trois figurines animales d'un tumulus du Delta intérieur du Niger (Mali). In Le Métal, *Technê*, 18, C2RMF-CNRS-UMR 171: 74–85.

Garenne-Marot, L. and Mille, B. 2005. Les fils à double tête en alliage à base de cuivre de Koumbi Saleh: valeur du métal, transactions et monnayage de cuivre dans l'empire de Ghana. *Afrique: Archéologie & Arts* 3: 81–100.

Guillard, X. 1993. La production ancienne de l'or au Soudan occidental. In *Outils et ateliers d'orfèvres des temps anciens*, C. Éluère (ed.). Antiquités Nationales, Mémoire no 2, Musée des Antiquités Nationales, Saint-Germain-en-Laye, 277–88.

Herbert, E. 1984. *Red Gold of Africa: Copper in Pre-Colonial History and Culture*. Madison, WI: University of Wisconsin Press.

Lambert, N. 1983. Nouvelles contribution à l'étude du Chalcolithique de Mauritanie. In *Métallurgies Africaines, nouvelles contributions*, N. Échard (ed.). Mémoires de la Société des Africanistes 9: 63–87.

Lebeuf, A.M.D. and Pâques, V. 1970. *Archéologie malienne. Collections Desplagnes*, Catalogues du musée de L'Homme, Série C, Afrique noire, Museum national d'Histoire naturelle, supplement to vol. 10, *Objets et Mondes*. Paris: Revue du musée de l'Homme.

Levtzion, N. and Hopkins, J.F.P. (eds) 2000. *Corpus of Early Arabic Sources for West African History*, translated by J.F.P. Hopkins and edited and annotated by N. Levtzion and J.F.P. Hopkins. Princeton, NJ: Markus Wiener Publishers.

Martinón-Torres, M., Valcárcel Rojas, R., Cooper, J. and Rehren, Th. 2007. Metals, microanalysis and meaning: a study of metal objects excavated from the indigenous cemetery of El Chorro de Maita, Cuba. *Journal of Archaeological Science* 34(2): 194–204.

McIntosh, R.J. 1998. *The Peoples of the Middle Niger: The Island of Gold*. Oxford: Blackwell.

McIntosh, S.K. 1995. *Excavations at Jenné-Jeno, Hambarketolo and Kaniana: The 1981 Season*. University of California, Monographs in Anthropology. Berkeley, CA: University of California Press.

Monod, T. 1969. *Le Ma'den Ijâfen : une étape caravanière ancienne dans la Majâbat al-Koubrâ*. Actes du premier colloque international d'archéologie africaine, Fort Lamy, 286–320.

Ponting, M.J. 2003. From Damascus to Denia: scientific analysis of three groups of Fatimid period metalwork. *Historical Metallurgy* 37(2): 85–105.

Robion, C. 2000. *Étude des objets en alliage cuivreux de la collection Desplagnes*. Mémoire de DEA, Université Paris I, Panthéon-Sorbonne.

Shalev, S. and Freund, M. 2002. The archaeology of Islamic metals and the anthropology of traditional metal casting in Cairo today. In *Metals and Society: Papers from a Session held at the European Association of Archaeologists Sixth Annual Meeting in Lisbon 2000*. BAR International Series 1061. Oxford: Archaeopress, 83–97.

Stuiver, M. and Reimer, P.J. 1993. Extended 14C database and revised CALIB 3.0 14C age calibration program. *Radiocarbon* 35(1): 215–30.

Vanacker, C. 1983. Cuivre et métallurgie du cuivre à Tegdaoust (Mauritanie Orientale). In *Métallurgies Africaines, nouvelles contributions*, N. Echard (ed.). Mémoires de la Société des Africanistes 9: 89–107.

Welter, J.-M. 2003. The zinc content of brass: a chronological indicator? In Le Métal, *Technê*, 18, C2RMF-CNRS-UMR 171: 27–36.

Werner, O. and Willett, F. 1975. The composition of brasses from Ife and Benin. *Archaeometry* 17(2): 141–56.

Werthmann, K. 2003. Cowries, gold and 'bitter money': gold-mining and notions of ill-gotten wealth in Burkina Faso. *Paideuma: Mitteilungen zur Kulturkunde* 49: 105–24.

## Authors' addresses

- Laurence Garenne-Marot, Équipe 'Afrique', UMR 7041 ArScAn, Maison de l'Archéologie et de l'Ethnologie, 21 allée de l'université, 92 023 Nanterre cedex, France (lgmarot@wanadoo.fr)
- Benoît Mille, Centre de Recherche et de Restauration des Musées de France (C2RMF), UMR 171 CNRS-Culture, 14 quai François-Mitterrand, 75 001 Paris, France (benoit.mille@culture.gouv.fr)

# Preliminary multidisciplinary study of the Miaobeihou zinc-smelting ruins at Yangliusi village, Fengdu county, Chongqing

*Liu Haiwang, Chen Jianli, Li Yanxiang, Bao Wenbo, Wu Xiaohong, Han Rubin, Sun Shuyun and Yuan Dongshan*

*Abstract*   The Miaobeihou zinc-smelting site located at Yangliusi village in Fengdu county, Chongqing, was excavated between 2002 and 2004 in order to determine its date and the smelting techniques used. Slag, retort fragments and other materials were collected and analysed by chemical and metallurgical methods and dated by AMS-$^{14}$C radiocarbon. The results indicate that smithsonite and coal were used to produce zinc metal during the Ming dynasty. The zinc-smelting process found to have been carried out on this site closely matches the account given in the *Tian Gong Kai Wu*. The results of this excavation mark a major advance in our knowledge of the origins of zinc smelting in China: more field investigations, archaeological excavation and laboratory analysis work will be needed to take this further.

*Keywords:* zinc, smelting, China, metallurgy, archaeology, excavation.

## Introduction

Despite the occasional very early occurrences of accidentally produced brass in China, deliberate metallic zinc production did not occur until the Jiajing period (AD 1552–1566) of the Ming dynasty, when zinc began to be produced on a large scale and used in the manufacture of brass. This late occurrence was due to the difficulty of the smelting process. The evidence comes from field investigations in Guizhou province (Zhou Weirong and Dai Zhiqiang 2002), although other researchers believe that zinc smelting began earlier than this date (Liu Guangding 1991; Xu Li 1986, 1998).

The *Tian Gong Kai Wu*, written by Song Yingxing at the end of Ming Dynasty (E-tu-Zen Sun and Shiou-Chuan Sun 1966) recorded zinc-smelting techniques (Fig. 1). Since the 1920s Chinese scholars have carried out systematic, text-critical research into the origins, development and technology of zinc smelting in ancient China (Liu Guangding 1991; Mei Jianjun 1990; Xu Li, 1986, 1998; Zhang Hongzhao 1955; Zhao Kuanghua 1984, 1996; Zhou Weirong and Dai Zhiqiang 2002, and see Zhou Weirong in this volume, pp. 179–86). Others outside China have also researched this subject. Over the past 80 years, research into the origins of the zinc-smelting process has not shown great progress because of the lack of archaeological evidence. In the mid-1990s, Paul Craddock and Zhou Weirong (China Numismatic Museum) carried out field investigations into traditional zinc-smelting technology in southwest China to research its origin and development (Craddock and Zhou Weirong 1998, 2003).

In the 1980s, before the construction of the big dam of the Three Gorges of the Yangtze river, several smelting sites had

**Figure 1** Drawing of the zinc-smelting process recorded in *Tian Gong Kai Wu*.

been found on the southern bank of the Yangtze in Fengdu county, but no further investigation was carried out until 2002–04 when a site was excavated in Miaobeihou by the Henan Provincial Archaeology Institute (see Fig. 2). Many samples of furnaces, ore, slag and other smelting remains were unearthed, some of which were collected for analysis in our

**Figure 2** Map showing the location of the Miaobeihou zinc-smelting site.

laboratories. The results indicated that it was a zinc-smelting site. At the end of 2004, a research group with members from four institutions was established. More than 20 zinc-smelting sites were found in Fengdu county in the last year of field investigations and about 6000 square metres were excavated, of which the Miaobeihou site was just a small part. This paper reports on these new archaeological findings and the zinc-smelting technology seen at the Miaobeihou site.

## Fieldwork

Located in southwest China in the southeast Sichuan basin, the Miaobeihou zinc-smelting site at Yangliusi village, Fengdu county is part of Chongqing, a municipality directly under the control of central government. It shares borders with the provinces of Hubei, Hunan, Guizhou, Sichuan and Shaanxi. The Miaobeihou site was located on the southern bank of the Yangtze river (Fig. 3). Almost all of the zinc-smelting sites found in Fengdu county were located on the lower terraces of the river at an altitude of about 150–170 m, convenient for transporting the ore and fuel to the site and to take away the zinc produced. A number of very important features were found during the excavation of the site including a furnace and unused smelting retorts (Figs 4 and 5). The retorts vary in size, with an average height of about 27 cm. Figure 6 shows a used retort excavated from the site. From this figure it can be seen that the pocket (see also Fig. 7) and the gas passage are preserved, and furthermore that some zinc metal, zinc oxide and zinc white still adhere to the pocket. A great deal of zinc ore together with some iron tools (Fig. 8), fragments

of zinc ingots (Fig. 9), coal and other remains (Fig. 10) were also found.

## Scientific examination

The debris of metal production was examined, in particular to discover the date and manufacturing technology, including the ores, fuel sources, smelting and processing techniques, the composition of the metal and its mechanical qualities. In order to determine smelting technologies, slag, retort fragments and other remains were collected from the site and analysed using metallographic microscopy, scanning electron microscopy with energy-dispersive spectrometry (SEM–EDS), X-ray diffraction (XRD) and inductively coupled plasma–atomic emission spectrometry (ICP–AES). The dates were provided using AMS-$^{14}$C radiocarbon methods.

### Microstructural and chemical analyses

Compositional analysis of the alloys, slag and ores is significant to our understanding of the smelting techniques and the sources of ores. A scanning electron microscope (SEM) was used to observe the microstructures and compositions of polished sections of samples. An SEM with energy-dispersive spectrometer (EDS) was used to carry out the non-sampling quantitative analysis. This research was undertaken using a Japan Electron JEOM-850 SEM and a Philips PV9550 EDS. The excitation voltage was 20 kV. Since light elements such as carbon and oxygen whose atomic numbers are less than

171

**Figure 3** View of the terrain of the Miaobeihou site on the banks of the Yangtze river.

**Figure 4** The excavated foundation of No. 1 furnace.

**Figure 5** Unused retorts excavated from the site.

**Figure 6** Used retort excavated from the site.

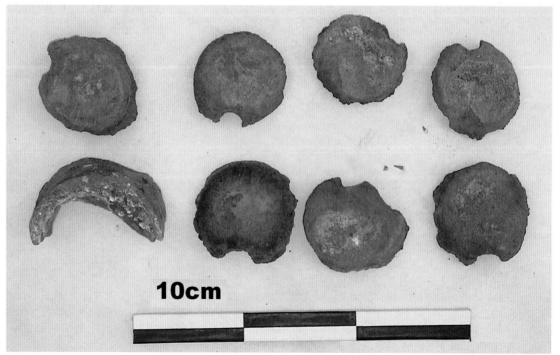

**Figure 7** The pockets for collecting the zinc at the top of the retorts.

**Figure 8** An iron tool found at the site. Very similar iron tools were recorded by Xu Li (1998: pl. 7), still in use in the mid-20th century.

**Figure 9** Part of a zinc ingot excavated from the Miaobeihou site.

**Figure 10** A piece of by-product, mainly oxidised zinc (which will have converted to carbonate).

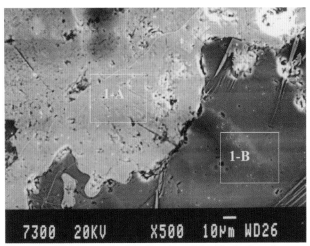

**Figure 11** SEM image of polished section of slag No. 1 adhered to retort wall, high zinc content in white area 1A (see Table 1 for analyses).

**Figure 12** SEM image of slag No. 2 adhered to retort wall, high zinc content in white area 1C.

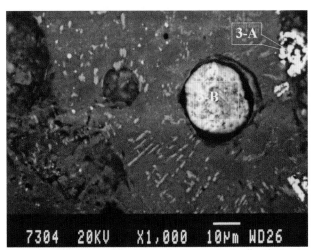

**Figure 13** SEM image of slag No. 3 adhered to retort wall with zinc-rich area 3B.

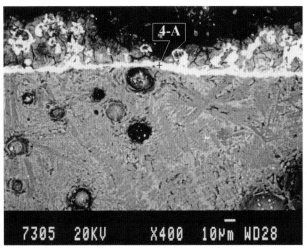

**Figure 14** SEM image of the matrix of slag No. 4, high zinc content at position 4A.

173

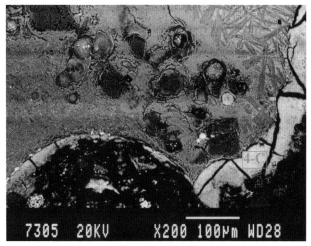

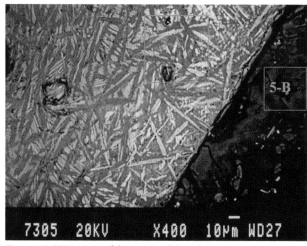

**Figure 15** SEM image of the matrix of slag No. 4 with high zinc content at 4C.

**Figure 16** SEM image of the matrix of slag No. 5.

11 could not be detected, only a qualitative analysis of the corroded objects or occluded trace elements could be given. Oxidised components could not be determined. To determine the average components, surface scanning was used with multifaceted scanning on different parts of the samples to discover the precise composition of each sample. Based on previous analyses, the lower confidence limits for this instrument may be established at 0.3 wt%; values below this limit can be taken as indicative only. The analysis results of eight samples are shown in Table 1 and the SEM images are shown in Figures 11–17. Samples 1–8 are slag and sample no. 12 is a piece of

by-product (shown in Fig. 10) which is a compound of zinc. In order to further elucidate their structure, other techniques were used such as metallography and XRD. Several were analysed with a Ricoh MD2/JADE 5 XRD apparatus carried out at the National Museum of Japanese History.

A total of eight samples of slag, by-product, zinc oxide, zinc ore and a zinc ingot was analysed quantitatively using ICP–AES. The instrument used was a Prodigy model (made by Leeman-Labs Inc., USA) at Peking University. The spectrometer had a wavelength coverage of 170–1010 nm and a resolution of 0.001 nm. The following analytical parameters

**Table 1** Normalised composition analysed by SEM–EDS (wt%).

| Sample no. | Analysis area | Mg | Al | Si | S | K | Ca | Ti | Mn | Fe | Cu | Zn | Figure no. |
|---|---|---|---|---|---|---|---|---|---|---|---|---|---|
| 1 | Average | 1.2 | 14.1 | 50.5 | 0.1 | 5.3 | 6.2 | 1.8 | 0.2 | 9.3 | 0.1 | 7.8 | 11 and 12 |
|  | 1-A | 1.5 | 2.7 | 17.5 | 0.1 | 0.2 | 0.1 | 0.1 | 0.2 | 13.3 | 0.1 | 64.1 |  |
|  | 1-B | tr | 0.3 | 1.4 | 0.1 | tr | 0.5 | 0.2 | tr | 95.2 | tr | 2.3 |  |
|  | 1-C | tr | tr | tr | 0.1 | 0.1 | 0.1 | tr | tr | 13.5 | 0.1 | 86.1 |  |
|  | 1-D | 0.2 | 7.6 | 54.7 | 0.2 | 5.3 | 6.5 | 0.4 | tr | 9.6 | tr | 15.6 |  |
| 2 |  | 1.2 | 2.7 | 25.6 | 0.2 | 0.7 | 0.9 | 0.3 | 0.4 | 11.7 | 0.3 | 56.2 |  |
|  |  | 0.3 | 10.2 | 48.6 | 0.4 | 5.0 | 5.1 | 1.1 | 0.7 | 13.1 | 0.1 | 15.5 |  |
|  |  | tr | 0.2 | 1.2 | 0.4 | tr | 0.4 | 0.1 | 0.3 | 94.0 | 0.6 | 2.6 |  |
| 3 | Average | 1.4 | 17.0 | 61.6 | tr | 5.7 | 1.9 | 1.4 | tr | 9.5 | tr | 1.7 | 13 |
|  | 3-A | 1.5 | 9.9 | 10.8 | 0.1 | 1.2 | 0.4 | 30.3 | 0.4 | 43.7 | 0.2 | 1.5 |  |
|  | 3-B | 0.2 | 2.0 | 4.2 | 0.8 | 0.2 | tr | 0.1 | 0.2 | 1.6 | 1.0 | 89.7 |  |
| 4 | Average | 0.8 | 22.1 | 64.4 | 0.2 | 5.2 | 0.5 | 1.2 | tr | 4.7 | 0.1 | 0.3 | 14 and 15 |
|  | 4-A | 2.3 | 19.5 | 18.5 | 0.1 | 1.3 | 1.0 | 0.7 | 0.2 | 17.0 | 1.4 | 37.9 |  |
|  | 4-B | 2.1 | 16.5 | 54.3 | tr | 5.4 | 8.9 | 1.5 | tr | 10.5 | tr | 0.9 |  |
|  | 4-C | 0.2 | 2.9 | 37.9 | 0.2 | 0.2 | 1.2 | 0.1 | tr | 1.0 | 0.4 | 56.1 |  |
| 5 | Average | 1.7 | 20.4 | 52.8 | 0.5 | 4.8 | 6.2 | 1.4 | 0.2 | 11.3 | 0.1 | 0.3 | 16 |
|  | 5-A | 4.9 | 13.2 | 41.8 | tr | 4.0 | 2.6 | 1.7 | 0.2 | 18.6 | 0.1 | 12.8 |  |
|  | 5-B | 0.4 | 25.2 | 47.3 | tr | 0.2 | 4.7 | 0.1 | 0.1 | 0.7 | tr | 21.4 |  |
|  | 5-C | 6.5 | 38.8 | 17.1 | tr | 0.8 | 3.1 | 1.4 | 0.2 | 26.6 | tr | 5.5 |  |
|  | 5-D | tr | tr | tr | 33.1 | 0.1 | 0.2 | 0.2 | tr | 65.2 | 0.6 | 0.1 |  |
| 6 | Average | 1.5 | 12.1 | 51.7 | tr | 1.8 | 7.1 | 0.9 | 0.1 | 16.1 | 0.1 | 8.7 | 17 |
|  | 6-C | 2.5 | 15.8 | 31.4 | 0.1 | 2.9 | 8.6 | 0.9 | 0.8 | 15.8 | 0.5 | 20.6 |  |
|  | 6-D | 1.6 | 10.3 | 25.6 | 0.1 | 2.8 | 16.0 | 1.0 | 0.7 | 18.4 | tr | 23.4 |  |
| 8 | Average | 0.2 | 2.3 | 26.8 | 0.4 | 0.9 | 1.1 | 0.1 | tr | 13.2 | 0.3 | 54.3 |  |
| 12 | Average | tr | 0.2 | 1.3 | 0.1 | 0.1 | 0.1 | 0.2 | tr | tr | tr | 98.0 |  |

Average = averages of three or five measurements of different areas in cross-section at low magnification
tr = trace

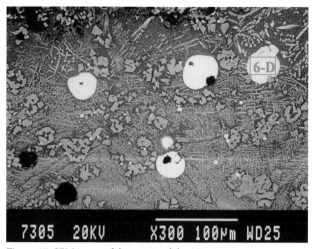

**Figure 17** SEM image of the matrix of slag No. 6.

were used: coolant flow rate 14 L/min; atomisation air pressure 25 psi; high-frequency power 1.1 kW; rotation speed of peristaltic pump 1.2 ml/min and an integration time of 30 s repeated three times. The analytical results obtained for zinc (Zn), lead (Pb), silver (Ag), cadmium (Cd), arsenic (As) and antimony (Sb) are shown in Table 2.

From the results given in Tables 1 and 2 it can be concluded that the process taking place at the site was distillation for zinc smelting. Of the metals used in the past, zinc was the most difficult to smelt because it volatilises at about the same temperature (around 1000 °C) that is needed to smelt the zinc ore. As a result it would form as a vapour in an ordinary furnace and immediately become re-oxidised and hence lost. Sample No. 12 (Fig. 10), which was collected from a slag pit, is an example of this phenomenon: the main component was smithsonite ($ZnCO_3$) with a little zinc oxide (ZnO).

### Radiocarbon dating

Seven charcoal samples were taken from the various levels of the site and from the bottom of some of the retorts and furnaces for AMS-[14]C dating. After acid-alkali-acid (AAA) treatment, charcoal samples were combusted by a Vario Elemental Analyser (Yuan Sixun *et al.* 2000). Pure carbon dioxide ($CO_2$) was collected and reduced to graphite on iron (Fe) powder by hydrogen gas in a vacuum system. Measurements of radiocarbon dates were performed by tandem accelerator mass spectrometry (AMS) at Peking University. All radiocarbon dates were converted to calibrated dates by OxCal v3.10 with the intcal104 calibration curve (Bronk Ramsey 2005). The results are shown in Table 3. Several retort and furnace wall samples were collected for dating by thermoluminescence (TL) and optically stimulated luminescence (OSL), the results of which will be discussed elsewhere, but it can be concluded that the Miaobeihou zinc-smelting site could date back about 400 to 500 years, somewhat earlier than the record in the *Tian Gong Kai Wu* published in 1637.

### Discussion

Based on the archaeological excavations and the chemical analysis of the components of excavated retort, ore and slag samples, we conclude that the smithsonite was smelted with the local coal. The results are discussed below.

### Raw materials used for zinc smelting

Many pieces of smithsonite ore were found during our excavations. There is evidence from geological investigation and

**Table 2** Composition analysed by ICP–AES.

| Sample no. | Sample | Zn (%) | Pb (%) | Ag (ppm) | Cd (ppm) | As (ppm) | Sb (ppm) |
|---|---|---|---|---|---|---|---|
| 1 | Slag adhered to retort wall | 6.24 | 0.40 | 22 | | 1838 | 30 |
| 2 | Slag adhered to retort wall | 8.16 | 0.42 | 10 | | 2608 | 183 |
| 7 | Slag | 0.58 | 0.39 | 1 | 4 | 883 | 131 |
| 12 | A piece of by-product | 57.3 | 0.38 | 10 | 412 | 257 | 145 |
| 13 | Slag adhered to retort wall | 4.03 | 0.38 | 14 | | 1604 | 78 |
| 14 | Zinc oxide adhered to pocket | 34.6 | 0.40 | 46 | 433 | 138 | 139 |
| 15 | Zinc ore | 16.1 | 0.39 | 9 | 527 | | 9 |
| 16 | Zinc ingot | 99.2 | 0.72 | 21 | 125 | | |

**Table 3** Radiocarbon dates of the Miaobeihou site.

| Lab code | 14C age (BP, 1σ) | Calendar age AD (68.2% probability) |
|---|---|---|
| BA04196 | 400 ± 40 | 1440–1520 (57.6%) or 1600–1620 (10.6%) |
| BA04199 | 385 ± 40 | 1445–1520 (51.8%) or 1590–1620 (16.4%) |
| BA04200 | 345 ± 40 | 1480–1525 (24.4%) or 1555–1635 (43.8%) |
| BA04201 | 325 ± 40 | 1510–1605 (53.7%) or 1615–1640 (14.5%) |
| BA04203 | 385 ± 40 | 1445–1520 (51.8%) or 1590–1620 (16.4%) |
| BA04204 | 330 ± 40 | 1490–1530 (19.6%) or 1535–1605 (37.1%) or 1615–1635 (11.5%) |
| BA04206 | 330 ± 40 | 1490–1530 (19.6%) or 1535–1605 (37.1%) or 1615–1635 (11.5%) |

field explorations that there are many large-scale smithsonite mines in Fengdu county and neighbouring areas. One mine is only about 50 km from the Miaobeihou site and the Longhe (Dragon river) could have been used for transporting the zinc ores. Coal was also found on the site and is still mined nearby today. The location of the site is thus very convenient for zinc production.

### Structures of the furnace and retort

Furnace and retort structures are important to the study of zinc-smelting technology. The retorts are very similar to those depicted in the *Tian Gong Kai Wu* (Song Yingxing 1637): the majority are of earthenware with sand temper, but some are of earthenware with silt as filler. The average height of a retort is about 27 cm, the internal diameter of its mouth is 8 cm and the external diameter is 11 cm. The maximum diameter of the ventral part is 15 cm and the base is about 9 cm (Figs 5 and 6). The manufacture of the retorts involved a series of processes such as kneading the clay, modelling, forming and firing over a slow fire. The zinc ore was reduced in the retort with the coal acting as the reducing agent. The zinc vapours ascended to the low temperature region through the gas passage and condensed as molten zinc on the iron lid before dripping down into the pocket. The gas hole was essential to the control of air pressure and the discharge of waste gases during the process (Fig. 18).

It is instructive to compare the excavated No. 1 furnace foundation with the illustration in the *Tian Gong Kai Wu*. The excavation indicated that to build the furnace, a pit was dug about 55 cm deep with a 3–5 cm thick layer of coal powder and earth at the bottom to provide damp-proofing. The hearth was built on this. The interior of furnace 1 is about 305 cm long and 235 cm wide. The mouth and air vents were also found. This rectangular form of furnace found in the Miaobeihou site is similar to those depicted in the *Tian Gong Kai Wu* (the apparent triangular shape depicted is in reality no more than an attempt at perspective). A furnace with rectangular foundations was also excavated at the Jiudaoguai site not far away from the Miaobeihou site as shown in Figure 19. The Jiudaoguai furnace is about 550 cm in length and 100 cm in width. The structure of this furnace is the same as the *yan* (horse trough) furnaces described by Xu Li (1986) that operated in the mid-20th century in Guizhou province.

### Smelting process

It is believed that the coal, generally broken and mixed with the smithsonite ore, was the main fuel and reducing agent for the smelting process at the site, and that the charcoal, which was found in small quantities confined to the bottom of the retorts, had been used only to ignite the charge.

Inferences concerning the smelting procedure can be made from the artefacts found on the site and also from recent and present-day smelting practice (Craddock and Zhou Weirong 2003; Xu Li 1986, 1998). After the smithsonite ore and coal were crushed and sifted, they were well mixed in the correct proportions. The ratio between ore and coal was correlated with the quality of the ore: the higher the ore quality, the larger the amount of coal. The mixture was packed into the retorts and a layer of fine-grained furnace ash with a little slurry was put on top of it. This was then carefully pressed down with a stone or iron hammer until the charge was packed tightly and a depression was created on the top. An iron plate was inserted to one side of the ashes on the top to keep a passage open through which the zinc vapour would pass to the top of the retort where it condensed, as illustrated in Figures 6 and 18. Clay was coiled around the rim of the retort to build up a pocket about 10 cm tall. Finally an iron lid was placed on top of the pocket and sealed with more mud, leaving only a small

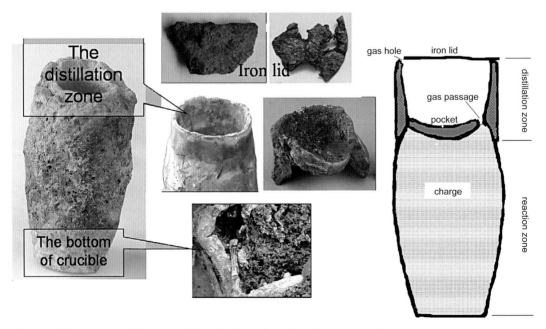

**Figure 18** Used retorts and lids excavated from the Miaobeihou site with explanatory diagram.

**Figure 19** A furnace with a rectangular foundation excavated at the Jiudaoguai site.

hole to act as a gas passage during the smelting process. The zinc condensed on the underside of the lid and dripped down into the collecting pocket.

The next step was to arrange the retorts in the hearths of the furnace. Based on the dimensions of the excavated hearths and retorts, there could have been about 60–80 retorts on a platform. Between the hearth and the top of the walls of retorts there would have been gaps which were filled with suitably sized slag lumps and mud. This sealing was important to keep the temperature of the pockets above the retorts at a lower level than that of the retort during the smelting process and also to physically stabilise the retort. During the distilling procedure the temperature of that part of the retort where the reaction took place would have been about 1200 °C but the temperature in the upper pocket of the retort would have been about 800 °C, which is essential for the rapid condensation of zinc vapour. It is estimated that it would have taken about 20 hours for the materials in the retorts to react completely after the lid was placed on the pocket. In practice it would generally have relied on the experience of the furnace master to decide when the reaction was complete. This was done by observing the flame during the night-time and the smoke during the daytime as it was emitted by the retorts through the small hole in the iron lid. After cooling the retorts for about 5–6 hours, the zinc ingots could be taken out of the pockets. Several fragments of zinc ingot, excavated in the 1970s at this site and now in the Institute of Cultural Relics of Fengdu county, were analysed by atomic absorption spectrometry (AAS) by the Chongqing Iron and Steel Company and found to contain about 95–98% zinc. A piece of zinc ingot was analysed by ICP–AES for this paper and found to contain 99.21% zinc – a high purity metal.

### Date of the Miaobeihou site

The dating of the seven charcoal samples excavated from this site broadly agrees with the estimate by Zhou Weirong concerning the origin of zinc production in China (see Zhou Weirong, this volume, pp. 179–86). It is very unlikely, however, that Miaobeihou is the earliest site of zinc production in China because of its relatively developed technique and the large scale of the industry in this area. It must surely have taken a long time for the smelting technique to develop to the large scale and high technical level seen at this site.

### Conclusions

The Miaobeihou zinc-smelting site is the first Chinese early zinc-smelting installation to be excavated. By using SEM–EDS, XRD and radiocarbon dating, the artefacts from the Miaobeihou zinc-smelting site were investigated. The findings show that the techniques used at the site were similar to those recorded in the *Tian Gong Kai Wu* and thus can represent the zinc-smelting technology at least of the Ming period. This site is only one of a group of similar sites in the locality and as zinc-smelting techniques are unlikely to have become so highly developed during a short period, it is probable that the origins of zinc smelting in this area are much earlier. More field investigations, archaeological excavation and laboratory analysis will be needed to take this further.

### Acknowledgements

The research was supported by the National Natural Science Foundation of China (10405003), the State Administration of Cultural Relics of China (20050102) and the Japanese Society for the Promotion of Science. We would specially like to thank Professor Tsun Ko of the University of Science and Technology Beijing, Professor Tiemei Chen, Professor Liping Zhou, Dr Jianfeng Cui, Miss Yan Pan and Mr Shijun Gao of Peking University, Professor Mineo Imamura, Dr Tsutomu Saito, Dr Minoru Sakamoto and Dr Emi Koseto of the National Museum of Japanese History, for their useful supervision, advice and help in the research. Many thanks also to the Cultural Relics Administration Council of the Three Gorges Project for providing samples and useful materials.

# References

Bronk Ramsey C. 2005. www.rlaha.ox.ac.uk/oxcal/oxcal.htm

Craddock, P. T. and Zhou Weirong, 1998. The survival of traditional zinc production in China. In *The Fourth International Conference on the Beginnings of the Use of Metals and Alloys (BUMA-IV), Shimane, Japan, 1998,* 85

Craddock, P.T. and Zhou Weirong, 2003. Traditional zinc production in modern China: survival and evolution. In *Mining and Metal Production through the Ages,* P.T. Craddock and J. Lang (eds). London: British Museum Press, 267–92.

E-tu-Zen Sun and Shiou-Chuan Sun (trans. and eds) 1966. *T'ien-kung K'ai-Wu: Chinese Technology in the 17th Century, by Sung Ying-Hsing.* Philadelphia: Pennsylvania State University Press.

Liu Guangding 1991. The history of zinc in China: re-investigation on the existence of Wo-chien in the Five dynasty. *Chinese Studies (Guo Xue Yan Jiu)* 9(2): 213–21 [in Chinese].

Mei Jianjun 1990. Traditional zinc smelting technology in modern China. *China Historical Materials of Science and Technology (Zhong Guo Ke Ji Shi Liao)* 11(2): 22–6 [in Chinese].

Song Yingxing 1637 *Tian Gong Kai Wu. Vol. 14: Wu Jin* [*Five Metals*] [in Chinese].

Xu Li 1986. Traditional zinc-smelting technology at Magu, Hezhang county. *Studies in the History of Natural Sciences* 5(4): 361–9 [in Chinese].

Xu Li 1998. Traditional zinc-smelting technology in the Guma district of Hezhang county of Guizhou province. In *2000 Years of Zinc and Brass,* 2nd edn, P.T. Craddock (ed.). British Museum Occasional Paper 50. London, 115–32.

Yuan Sixun, Wu Xiaohong, Gao Shijun *et al.* 2000 The CO2 preparation system for AMS dating at Peking University. *Nuclear Instruments and Methods in Physics Research B* 172: 458.

Zhang Hongzhao 1955. The origin of zinc use in China. *Science (Ke Xue)* 8(3): 1923 [in Chinese].

Zhao Kuanghua 1984. A re-study of the origin of zinc use in China. *Zhong Guo Ke Ji Shi Liao (China Historical Materials of Science and Technology)* 3: 15–23 [in Chinese].

Zhao Kuanghua 1996. An analysis of the chemical composition of Northern Song coins and a tentative study for the tinned coins. *Studies in the History of Natural Sciences* 5(3): 229–46 [in Chinese].

Zhou Weirong and Dai Zhiqiang 2002. *Qian Bi Shi He Ye Jin Shi Lun Ji* (*Papers on Numismatics and the History of Metallurgy*). Beijing: Zhonghua Book Company [in Chinese].

# Authors' addresses

- Corresponding author: Chen Jianli, School of Archaeology and Museology, Peking University, China, 100871 (jianli_chen@pku.edu.cn)
- Liu Haiwang (Henan Provincial Archaeology Institute)
- Bao Wenbo, Wu Xiaohong (Peking University)
- Li Yanxiang, Han Rubin and Sun Shuyun (University of Science and Technology Beijing)
- Yuan Dongshan (Chongqing Archaeology Institute)

# The origin and invention of zinc-smelting technology in China

## Zhou Weirong

*ABSTRACT*   The technology of metallic zinc production by distillation in China originated in the late 16th or early 17th century during the Wanli period (1573–1620)[1] of the Ming dynasty, developing from Chinese cementation brass-smelting processes. The arguments presented in this paper are as follows:

- The earliest historical records, official and unofficial, relating to zinc production date to the late Wanli period and thereafter.
- There is no documentary evidence for the production of brass by mixing copper and zinc metals before AD 1620.
- The analytical results (especially of the trace element cadmium) of a very large sample of Chinese brass coins and other brass artefacts suggest that brass was made by mixing copper and zinc metals from the early 17th century.
- The zinc-smelting remains found in Fengdu, Chongqing, near the Yangtze river compare well with the description of zinc smelting in the *Tian Gong Kai Wu*, written in the 17th century. This date is consistent with the date for the remains as determined by Peking University (Liu Haiwang *et al.* in this volume, pp. 170–78).

*Keywords:* China, zinc, smelting, distillation, cadmium, analysis.

## Introduction

The history of zinc smelting in early China is a controversial issue. Most Chinese scholars believe that China began to use and smelt metallic zinc before the Ming dynasty (AD 1368). There are three opinions regarding the origin of zinc production in early China: the first is that it originated in the Han dynasty (206 BC–AD 220), the second is that it originated in the Five dynasties (AD 907–960) and the third is that it originated in the Song dynasty (AD 960–1279).

The common viewpoint in China is that zinc in early Chinese literature was called 倭铅, pronounced *woyuan* in pinyin. So, for more than half a century the study of the history of the use of zinc has been a search for the term *woyuan* in early Chinese texts. It will be argued here, however, that *woyuan*, written 倭铅 was originally the term signifying lead, and that the term for zinc, confusingly also pronounced *woyuan* in pinyin, was written 窝铅.

## Evidence that the term for zinc was *woyuan*, 窝铅

When researching the materials used to cast coins in the Ming and Qing dynasties it was found that zinc at this period was called both *woyuan*, 窝铅 and *woyuan*, 倭铅. For example, in a memorial to the emperor Chong Zhen concerning the casting of coins produced by Hou Xun (head of the Ministry of Revenue), there are such sentences as 'Copper ore originating from stone will become red copper through three smelting processes. Then, melting the copper with *woyuan*, 窝铅,

it will become brass. According to the current standard, for 100 catties of brass, 57 catties of red copper and 43 catties of *woyuan*, 窝铅 are used.'[2] In the local gazetteers of the southwest of China the term 窝铅 is also frequently found.[3]

Field investigations of Chinese traditional zinc-smelting techniques in the southwest of China (Craddock and Weirong Zhou 1998, 2003) clarified the related terms: 窝铅 pronounced *woyuan* combines 铅, pronounced *yuan*, zinc, and 窝, pronounced *wo*, pocket or nest. This is a reference to the smelting process where the zinc is collected in the condensation pocket at the top of the retort during the distillation. The pocket is called 'the nest of *yuan*' and the round lumps of rough zinc are called *woyuan*, 窝铅 (Figs 1 and 2).

Although the *Tian Gong Kai Wu* written by Song Yingxing, published in 1637, clearly recorded *woyuan*, 倭铅 as zinc, it does not appear in formal literature and documents of the Ming dynasty,[4] or in the many kinds of memorials to the Ming emperors. Hence it would seem that although the term for zinc in early China was occasionally written as *woyuan*, 倭铅, the true name was *woyuan*, 窝铅.

*Woyuan*, 倭铅 certainly meant zinc in the late 16th and early 17th century AD, based on the relevant description in the *Tian Gong Kai Wu*. There is a section in the metals volume of this text, which not only explains the properties of *woyuan*, 倭铅, but also states that it comes from smithsonite and includes pictures of the smelting process. However, another important source, the *Compendium of Materia Medica* written by Li Shizeng (1518–1593) and published earlier than the *Tian Gong Kai Wu* places *woyuan*, 倭铅 in the Lead Section. It states: '倭铅 can help smelting other metals.'[5] Here 倭铅 is difficult to explain as zinc but instead should be translated as lead. This is

179

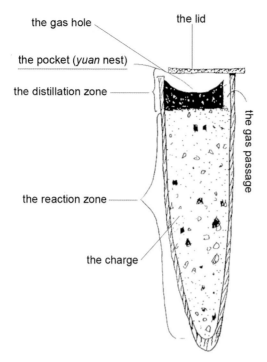

**Figure 1** Drawing of a section through a recent zinc-smelting retort showing the '*yuan* nest' pocket for collecting the zinc.

**Figure 2** Lumps of zinc (*woyuan*) from Yunnan smelted in 1994.

linked to the comment made elsewhere in the *Tian Gong Kai Wu* that lead can be used to smelt tin. It states that 'smelting tin usually uses a big furnace … the tin ore is relatively difficult to melt. If so, some pieces of lead are put into the furnace. It will become easy to do.'[6] It is well known that lead was used to help to smelt silver (by cupellation) in early China. There are also many works on alchemy which support the view that 倭铅 is lead. For example, Sanyuan Dadan Mifan Zhenzhi (三元大丹秘范真旨) wrote, 'the ore of 倭铅 looks like pebbles, grains or pellets. The local people break them into small pieces, then fire them with carbon. The metal, remaining at the bottom, is called 倭铅.'[7] There is another source for the origin of the term 倭铅 in *The Miscellanies of South of the Five Ridges* (written by Wu Zhenfang in the Qing dynasty): 'White lead is produced in the Chu area, where its price is 3 taels of silver for 100 catties. It is transported to Guangdong by dealers, the price is 6 taels of silver. But from 100 catties of white lead 16 taels of silver can be extracted and black lead is left behind, also locally

called *woyuan* 倭铅. And originally no *woyuan* 倭铅 yields.' According to Wu, *woyuan*, 倭铅 was the lead from which silver had been extracted by cupellation. So, it was a kind of 'black lead' or 'poor lead'. This may be the origin of the name *woyuan*, 倭铅. It is said that in provinces such as Fujian and Zhejiang in Southeast China the word *wo*, 倭 means poor or inferior (Zhang Kuanghua 1984).

## 倭铅 – lead, tin or zinc?

It is a modern interpretation that 倭铅 meant only lead, but in early times the meaning was more complex. Before the Ming dynasty the words for tin and lead were frequently interchangeable. For example, the white powder lead hydroxycarbonate (cerussite, $Pb(OH)_2CO_3$) was called tin powder, and lead was often called black tin. After the Ming dynasty the confusion of lead and tin was infrequent, but instead lead and zinc were often confused. In most books zinc was recorded as *yuan* or *yan*, rarely as *woyuan* (either 窝铅 or 倭铅). This has continued until today in the southwest of Guizhou where zinc and lead are still recorded by the same word written 铅. However, where *putonghua* (the official pronunciation) is accepted, 铅 (lead) is pronounced *qian* and 铅 (zinc) is pronounced *yuan*.[8] On the other hand, many uneducated people still pronounce both as *yuan*.

From the confusion of these names and the fact that zinc and lead have similar properties (apart from the ease with which zinc evaporates), it would appear that some alchemists and metallurgists did use the term *woyuan*, 倭铅 to refer to zinc.

## The origins of zinc smelting in China

The date when zinc smelting began in China has been a very important issue for Chinese and also some Western scholars for a long time. In the 1920s, Zhang (Hongzhao), the first Chinese scholar to study the history of zinc use in China, put forward the 'Han theory' (the H-theory). He believed that in China, zinc smelting began in the Han dynasty (206 BC–AD 220) (Zhang Hongzhao 1955).[9] Almost at the same time Zeng (Yuanrong) professed that he had found evidence that metallic zinc had been used in the Five dynasties (907–960) (the F-theory – Wang Jin *et al.* 1955). Today there are many scholars who believe that China began to smelt zinc in the Song dynasty (960–1279) (the S-theory), or at least in the initial stage of the Ming dynasty (1368–1644) (the M-theory).

Originally there were few scholars who accepted Zhang's H-theory (and see Zhou Weirong (1991, 1993) disproving Zhang's basic evidence). The F-theory was based on '*Woyuan*, 倭铅 can be used to prompt some metals to be extracted', which was quoted in the *Compendium of Materia Medica* from *Bao Cang Chang Wei Lun* written by Xuanyuan Shu, who lived in the Five dynasties. Chinese scholars of the history of Chinese science and technology once universally accepted this viewpoint. In the 1980s, however, it was claimed that the all-important sentence was added by the scholar Li Shizeng

(1518–1593) in compiling the *Compendium of Materia Medica* (Zhao Kuanhua 1996). Subsequently, belief in the F-theory sharply decreased until Xu Li (1986, 1998) showed that there was apparently a record of zinc smelting dating to the Five dynasties in the local gazetteers. A few years later Liu Guangding (1991) reinstated the sentence in the Five dynasties compendium, the *Bao Cang Chang Wei Lun*. Thus Zeng's F-theory was apparently re-confirmed. However it follows from the discussion of terminology above that, whether or not the sentence '倭铅 can be used to prompt some metals to be extracted' was in the original *Bao Cang Chang Wei Lun*, it is irrelevant to the discussion of the history of zinc in China because 倭铅 refers to lead.

The following three citations by Xu Li from local gazetteers must also be examined:

1. According to the Dading Prefectural Gazetteer compiled by Zou Hanxun and Fu Ruhuai in the Daoguang period, the Weining Subprefectural Gazetteer says:
   At Yinchanggou [Silver Works Canal] in Tianqiao [Heavenly Bridge] black lead and white lead are produced; there are both long and high furnaces. [The works] are located on the border between Yunnan and Guizhou provinces. A tradition passed down by the elders has it that the works have been in operation since the Tianfu reign period of Emperor Gaozu of the Five dynasties period.

2. The same local gazetteer also has the following entry:
   Mt. Dabao is 105 *li* [*c.* 50 km] to the east of the city [Weining], by Yinchanggou. There is a silver and lead works called Tianqiaochang [Heavenly Bridge works]; its original name was Lianhuachang [Lotus Blossom works]. The extraction began in the Tianfu reign period of the Han dynasty of the Five dynasties period. By the old mine shafts there is an inscription on the side of the cliff to record this fact. The mineshafts are the oldest in Yunnan and Guizhou provinces.

3. In the 1924 Weining County Gazetteer compiled by Miao Boran and Wang Zuyi there is a similar entry:
   Mt. Dabao lies 105 *li* to the east of the city. Tianqiao, formerly Yinchanggou, is rich in lead mines. From *lian* very good silver can be extracted. The works called Yangjiao, Lianhua, Fulai, Tianyuan, Zhazi, Woyan and Sibao are adjacent to it, the furthest away at a distance of 50 *li* [*c.* 25 km], the closest at a distance of only 10–15 *li* [*c.* 5–7.5 km]. The extraction began in the Tianfu reign period of the Han dynasty of the Five dynasties period; above the old mineshafts there is an inscription on the side of the cliff recording this fact. It is the most ancient mine in the Southwest.

It can be argued that the three quotations were for the same thing; the second one continues 'Here extraction [for silver] was continuously made. The output peaked in the Jiaqing reign period (1796–1820), and every day ten thousand horses were loaded at the mines.' In this context *lian* is lead, not zinc (Zhou Weirong 1991, 1993), and what it described is cupellation. The extraction which originated from the Tianfu reign period

(947) of the later Han dynasty (947–950) of the Five dynasties (907–960) was for silver, rather than zinc. Production began at the Yinchanggou (Silver Works canal) at the time when the gazetteers were compiled in 947 (the Tianfu reign period), and both 'black lead and white lead' were produced in 'long and high furnaces'.

## The S-theory or M-theory

The S-theory and M-theory both depend on 倭源白水铅, pronounced wo*yuan baishuiyuan* (as recorded in *A Collection of Xuande Incense-burners*, written by Lu Zhen in 1428) being translated as zinc. This source says that wo*yuan baishuiyuan* was used in great quantity at the initial stage of the Ming dynasty. The earliest zinc-smelting technology would probably have only been suited to small-scale production because of the difficulty of the procedure, and so it is unlikely that it developed very quickly into an industrial process important enough to be recorded in the literature. Therefore, scholars of the history of chemistry and metallurgy have searched the Song dynasty for evidence of zinc smelting in China. Based on the credible historical literature, however, it was demonstrated that wo*yuan baishuiyuan*, 倭源白水铅 was not used to cast the Xuande incense burners and moreover that there was no evidence that wo*yuan baishuiyuan* was zinc (Zhou Weirong 1990). So, what did wo*yuan baishuiyuan* mean?

Based on the research into the terminology of *woyuan*, 倭铅 (see above), it is now clear that 倭源 is the same as *woyuan*, 倭铅 and that 白水铅, *baishuiyuan* is lead (*yuan*). In early China, it was normal practice to add some modifying words before a metal's name, for example lead, 铅 was recorded as black lead, 黑铅; tin was recorded as white tin, 白锡 and water tin, 水锡 etc. In particular the alchemists not only abused the modifying words but often used enigmatic words. For example, sometimes lead, 铅 was recorded as water metal, 水中金, *shui zhong jin* (a kind of metal from water). Hence, so-called wo*yuan baishuiyuan* is actually a combination of two names for lead, probably concocted by the alchemists.

Over ten years ago, it was proposed that the Chinese began to smelt zinc between the Jiajing (1522–1566) and Wanli (1573–1619) reign periods of the Ming dynasty (Zhou Weirong 1993, 1996a, 1996b). In recent years a series of researches – including on-the-spot investigations – was made of Chinese traditional technologies in the southwest of China. These included the excavation of the early zinc-smelting remains found in Fengdu, Chongqing, the study of the substantial local gazetteers and the analysis of many early brass coins. From this research it can be deduced that China did not begin to smelt zinc before the Wanli period of the Ming dynasty for the following reasons:

- The correct name for zinc in early China was *woyuan*, 窝铅. This name has not been found earlier than the Wanli period in the relevant historical literature and documents.
- In China, after the Ming dynasty, the main use for copper, lead, tin and zinc was to cast coins. All records relevant to the materials for coinage only mention

copper and tin, not zinc. For example, Hao Jing, head of the Department of Materials, the Ministry of Works, in the 26th year of the Wanli period, stated the regulation of casting coins as follows: 'One catty of copper and tin can be used to cast 130 coins.' The instructions for the proportion needed for the manufacture of coins is 100 catties of copper and three or four catties of tin, but after the Tianqi period (1621–1627), all records about coinage mention copper and 'poor lead' (zinc) (Zhou Weirong and Fan Xiangxi 1993).

- Song Yingxing in his famous work, the *Tian Gong Kai Wu* wrote: 'For the manufacture of brass, people used to put ten catties of copper and six catties of smithsonite into a clay vessel to smelt. But recently people have come to realise that smithsonite has a heavy smoke loss, so they substitute smithsonite with "poor lead" (zinc).' But the earlier *Compendium of Materia Medica* written by Li Shizeng (1518–1593) states: 'smithsonite looks like sheep's brains, at present all brass comes from this ore smelted with copper'. This book was published in 1596 (the 24th year of Wanli). It follows that, up to the early years of the Wanli reign period, metallic zinc was still not used for making coins.

  The above can be argued to be in line with the statement under the 'poor lead' entry of the *Tian Gong Kai Wu* that 'poor lead (zinc) cannot be found in ancient books; only recently can we find it'. Song was born in the 15th year of Wanli (1587) and died in the early years of the Kangxi reign period (1662–1722). He wrote this book between the Tianqi and Chongzhen periods, so assuming that 'recently' was during Song's lifetime this would indicate that China began to smelt zinc no earlier than the Wanli period of the Ming dynasty.

- In 2002 an early zinc-smelting site was found in Fengdu, Chongqing, near the Yangtze river (Chen Jianli *et al.* 2007; Liu Haiwang *et al.*, this volume, pp. 170–78). It is the earliest yet found in China. The retorts, lids and other relics found there (Fig. 3) fit the description of zinc smelting in the *Tian Gong Kai Wu*, written by Song Yingxing in the 17th century. Its date, determined by AMS-$^{14}$C in Peking University, spans the 15th to the 17th century AD, i.e. the late Ming dynasty.

**Figure 3** The 17th-century zinc smelting retorts unearthed in Fengdu.

## No need, no invention: the evolution of zinc smelting

It could be argued that any development comes from an actual need. Occasional instances of brass are found very early in China: based on the archaeological finds, artefacts of brass have been found at Late Neolithic sites. For example a brass flake and a brass tube were unearthed at a Neolithic site belonging to the Yangshao culture (4100–3600 BC) in Jiangzhai, Shaanxi province and two pieces of brass cone bars were unearthed at a Neolithic site belonging to the Longshan culture (2300–1800 BC) in Jiaoxian, Shandong province. The brass-smelting techniques of prehistoric China were not handed down, however, and did not develop. In the following dynasties, the Xia, Shang and Zhou or even the Han, there were only bronzes; no trace of brass smelting has yet been found. After the Eastern Han dynasty, over the period from the Wei and Jin dynasties to the Five dynasties (220–960) brass (*toushi*, 鍮石) came from the countries of the Western Regions along with Buddhism. The first indigenous brass-smelting practice that can be confirmed at present commenced in the 10th century (the end of the Five dynasties and the beginning of the Northern Song dynasty). However, even then only some alchemists knew how to make brass and brass smelting was not developed as a metallurgical technology. After the middle period of the Ming dynasty (16th century) the emergence and growth of brass coinage issues in China gave a strong impetus to the development of brass smelting and founding techniques which finally led to a method of producing metallic zinc (Zhou Weirong 1996a, 1996b, 2000, 2001).

## Production of cementation brass in China

In the Five dynasties period, the rise of the Xixia state to the west of China obstructed trade so imports of brass (*toushi*) from the West were curtailed and the Chinese had to produce brass for themselves. According to the extant historical sources, the alchemists found a method for making *toushi* from red copper and smithsonite after years of experiments with metals and ores. The earliest record is in a book entitled *Ri Hua Zi Dian Geng Fa* (*Methods of Metal Manufacture*), an alchemic work written in the last years of the Five dynasties and the early years of the Song dynasty.[10] It reads:

> One catty of pure copper and one catty of smithsonite from Taiyuan are mixed and poured into an iron box. Calcining the box for two days and nights, then taking the mixtures into a reducing stove, further firing them for six hours, at the end we can get some beautiful pellets.

Here, a typical cementation brass procedure is described. Its chemical reaction is as follows:

$$Cu + Zn\,CO_3 \xrightarrow{\text{calcination}} Cu + ZnO \xrightarrow{\text{reduction}} Cu\text{-}Zn \text{ (brass)}$$

During this process smithsonite plays a very important role and so alchemists in the Song-Yuan period called it 'stove master'. A short time later, Cui Fang, an alchemist, wrote a simple and clear description of this method in his work *Wai Dan Ben Cao*, 外丹本草: 'Three catties of copper and one catty of smithsonite react upon each other, one and a half catties of *toushi* produced.'[11]

By the Northern Song dynasty (early 11th century), this method of *toushi* manufacture became more generally known. The Economy Section (vol. 34) of the *Song Hui Yao Ji Gao*, 宋会要辑稿 (an official history of the Song dynasty), recorded the following story:

> In the third year of Jingde (AD 1006), a soldier named Zhao Rongfa said that he could change copper into *toushi* by a medicine. But the emperor did not allow him to do so, because at that time there was no copper among the common people and people usually had to melt coins to make *toushi*.

The 11th volume of *Rong Zhai San Bi*, 容斋三笔 mentioned that in the period of Dazhong Xiangfu (1008–1016) the government established a bureau to make *toushi* for the construction of the capital. The 71st volume of the *Xu Zi Zhi Tong Jian Chang Bian*, 续资治通鉴长 also says, 'In the second year of Dazhong Xiangfu (AD 1009) people in some places currently melted copper to make *toushi* and a lot of coins had been destroyed, so the government issued an order to prohibit it.'

For a relatively long time after this there were no further developments in brass production, possibly because of government prohibition. In China, after the Han dynasty, governments usually forbade civilians to possess copper and, from the Sui-Tang period onwards, also prohibited them from using *toushi* (brass). Since civilians did not have copper and were also prohibited from using it, brass-smelting techniques could not be developed by the people.

After the Yuan dynasty (1276–1368) the foreign term *toushi* was gradually abandoned and *huangtong* became the preferred term for brass. For example, *Ge Wu Cu Tan*, 格物粗谈 (*A Brief Discussion on Nature*), written in the Yuan dynasty, in some places called it *toushi* and in others huangtong.[12] But, by the Ming dynasty, *toushi* had become a historical term and brass was only called *huangtong* (Zhou Weirong 2001).

It must be pointed out that in China before the Yuan dynasty, *huangtong* meant bronze. For example, *Shen Yi Jing*, 神异经 (a book on legends written by Dongfang Shuo in the Han dynasty), recorded that 'there is a palace that is made of *huangtong* walls in the northwest'. Furthermore, in the Song dynasty, *huangtong* usually indicated the red copper produced from yellow copper ores such as chalcopyrite in order to distinguish it from copper produced from chalcanthite, hydrated copper sulphate (Zhou Weirong and Fan Xiangxi 1993). So, early Chinese books should be read with care in order to be correctly understood.

During this period, although *toushi* or *huangtong* was no longer used as a kind of official dress mark as it had been in the Tang dynasty, it was still regarded as a precious material, coming after gold and silver, and often used to make noble articles for royal families and temples. For example, in the Northern Song dynasty, *toushi* was even used to make three official seals of the imperial government (Tuo Tuo *et al.* 1985) and the decorations for imperial construction projects (see above). In the Ming dynasty, brass (now called *huangtong*) was

**Figure 4** A Xuande incense burner, 宣德炉 made of brass with detail of the stamp.

**Figure 5** The Wudang Mountain Metal House, 武当山金殿. Much brass was used in its construction.

used to make imperial sacrificial utensils, such as the Xuande incense burners, 宣德炉 (Fig. 4) (Zhou Weirong 1990) and temple metal houses, such as the Wudang Mountain Metal House, 武当山金殿 (Fig. 5).[13]

## The development of cementation brass

The middle period of the Ming dynasty marked a change in the course of the use of brass in China. It began to be used to make coins and this greatly accelerated the development of brass production. According to textual research (Zhou Weirong and Fan Xiangxi 1993) brass was used to make coinage issues commencing in the period of the emperor Jiajing (1522–1566). *Du Shi Fang Yu Ji Yao*, 读史方舆纪要, a geographical work written by Gu Zuyu in 1667, recorded 'The Ling'an county in Yunnan province yields smithsonite. Its mine had been closed. When Yunnan started to cast coins in the Jiajing years, the mine was reopened in order to put smithsonite into copper.' *Ming Hui Dian*, 明会典 an official book of the Ming dynasty, recorded that 'The directions for coinage issues in the Jiajing years are: for six million Tongbao coins, 47,272 catties of brass and 4,728 catties of tin are to be used.' These are the earliest documents concerning the use of brass to make coins. The analytical data of historical coins confirm that the occurrence of coins with high zinc content (10–20%) did indeed begin in the Jiajing Tongbao period (Zhou Weirong 2004).

Evidence suggests that when brass coinage was introduced it was thought that *huangtong* was a kind of good copper rather than a new alloy material. The first reason is that the official coinage regulations prescribed that the proportions of the materials, *huangtong* and tin, for coinage issues were 10:1, and the analytical data of coins of the Jiajing Tongbao perid show that the tin content is indeed about 8–9% in the Jiajing Tongbao period (Zhou Weirong 2004) (Fig. 6). It is obvious that they did not know the properties of brass. In fact, brass, unlike copper, has excellent casting properties. For brass casting, tin is unnecessary in the alloy and such a high tin content is actually harmful. Moreover, in the Ming dynasty although tin was in great demand it was in short supply and often had to be imported from the countries of Southeast Asia. The second reason is based on the use and manufacture of *huangtong* at that time. Many historical documents such as the *Xu Wen Xian Tong Kao*, 续文献通考 (*Continuation of a Comprehensive Study of the History of Civilization*, completed in 1586 and published in 1603) and the *Gong Bu Chang Ku Xu Zhi*, 工部厂库须知 (a handbook of the Ministry of Works in the Wanli period, written by He Shijing in 1616), among others, recorded that *huangtong* was produced at copper-extracting sites rather than at coin-casting sites and it was made from copper with the 'medicine' (smithsonite), although by then metallic zinc was already in production in China (Zhou Weirong 1991, 1993, 2001).

Once *huangtong* began to be used to produce coins, progress was made on its technical development. First, the zinc content of the coins was increased. In early China, the zinc content of early brass was generally between 10 and 20%. In the Jiajing period the zinc content of brass coins was also usually about 10–20%, but by the Wanli period the zinc content quickly reached more than 30% (Zhou Weirong 2004). Secondly, the tin content of the coins decreased. The official coinage regulations of the Jiajing period stated that the proportion of *huangtong* and tin for coinage issues was 10:1, but by the Wanli period the proportion was decreased to 15:1. Furthermore, the analyses show that less than 20% of the coins of the Wanli period contain tin.[14] Volume 7 of the *Gong Bu Chang Ku Xu Zhi* recorded the official coinage regulations of the Wanli years: 'For ten thousand coins 90 catties of *huangtong*, five catties and 11.2 taels of tin were usually used; but now tin was no more used.' This book was finished by He Shijing in the 43rd year of Wanli. So, by the late Wanli period tin was no longer used in the coinage alloy.

The rise of the zinc content and the reduction of tin content in the alloy were the two important factors in the improvement of brass smelting and casting techniques in early China. Hence, it can be said that brass production progressed through the Wanli period and by the late Wanli period it had reached maturity.

## The production of speltering brass

The earliest written records of speltering brass are from the Tianqi years onwards. As quoted above, Song Yingxing, in the *Tian Gong Kai Wu*, published in the tenth year of Chongzhen (1637) explains that although brass had been made by smelting smithsonite with copper in a clay vessel, recently smithsonite had been substituted with 'poor lead' (zinc) to avoid the heavy smoke loss of the cementation process. It follows that this change from cementation brass to speltering brass took place before 1637 and there is further historical evidence in support of the first year of the Tianqi period (1621). First, we found that before the first year of the Tianqi period all documents about brass manufacture recorded that brass was made by the cementation processes, and for coin casting, brass and tin were used. For example, the *Compendium of Materia Medica*, 本草纲目 published in 1596 (the 24th year of Wanli), as quoted above, clearly states that brass was made by smelting copper with smithsonite, which 'looks like sheep's brain'. *Guo Chao Jing Ji Lu*, 国朝经济录, a book on the economy of the Ming dynasty, recorded that 'in the early years of the Wanli period the emperor allowed national mints to buy brass with money borrowed from the government or turn red copper into brass with smithsonite to cast coins'. After this date, however, almost all related documents record 'using copper and "poor lead" (zinc)' for brass casting. For example, the Coinage Section of *Xu Wen Xian Tong Kao*, 续文献通考 recorded that in September of the third year of the Tianqi period (1624) the national historical officer Fengxiang You stated 'for the manufacture of coinage issues, before the alloy was made up of 70% copper and 30% poor lead but now it is made of 50% copper and 50% poor lead'.[15]

Elemental analysis has determined that there are internal differences between cementation brass and speltering brass. Chinese cementation brass usually has a very low cadmium content but speltering brass generally contains much higher levels. Over two hundred brass coins and several dozen brass articles were analysed by atomic absorption spectrometry (AAS). The analyses showed that before 1621 the cadmium concentration of the coins was generally below 10 ppm; after

**Figure 6** Coins of the Jiajing and Wanli *tongbao* made of cementation brass.

**Figure 7** A coin of the Tianqi Tongbao made of speltering brass.

this date it almost always exceeded 10 ppm. It was very variable but often much higher – up to 100–200 ppm of cadmium (Cowell *et al.* 1993; Zhou Weirong 2004; Zhou Weirong and Fan Xiangxi 1994).

Both zinc and cadmium are in the IIB group of the Periodic Table, and not surprisingly cadmium is commonly associated with zinc ore. Before the Tianqi period, brass was produced by the cementation method and because of the heavy smoke loss in the cementation process, little cadmium was retained in the alloy. After this date brass was made from copper and metallic zinc and because the zinc was made by a condensation process the cadmium was retained. Thus the cadmium content of brass provides an effective method of identifying the introduction of speltering brass, which closely followed the invention of the zinc distillation process in China.

## The invention of Chinese traditional zinc-smelting technology

On the basis of field investigations by the British Museum team and the studies of the relevant historical literature, it is believed that Chinese traditional zinc-smelting technology by distillation developed from the cementation method of brass production.

Zinc distillation technology in China emerged after the cementation method began to be used to produce brass in large quantities. In the Ming dynasty, after the Jiajing reign period, where it was recorded: 'add smithsonite into copper to make brass for casting coins',[16] the coinage alloy was gradually changed from bronze to brass. In the later years of the Wanli period the changeover was complete and a great deal of *si-huo* brass (a kind of high zinc content brass) was used to cast coins, consuming large quantities of brass each year (Zhao Kuanghua *et al.* 1988).

The distillation method for producing metallic zinc emerged because of the defects of the cementation method. The *Tian Gong Kai Wu* described it as follows: 'For the manufacture of brass, people first put self-blast coal into the furnace, then add ten catties of copper and six catties of smithsonite into a clay vessel to smelt.' Self-blast coal is a kind of coal briquette formed by mixing powdered coal with mud. During the smelting no bellows were needed, but the furnace burned red day and night.

The early distillation zinc-smelting method was very like the cementation brass method. After investigating the most primitive extant zinc-smelting techniques in the Magu district of Hezhang county, Xu Li (1998) wrote: 'In the traditional

distillation-smelting of zinc, coal directly from the mine is not used. Instead briquettes of various sizes and shapes are formed by mixing powdered coal with yellow mud, furnace ashes and water. These briquettes are used not only as fuel, but also for propping up the pointed retorts, making them stand stable and upright in the furnace.' Furthermore, the fire did not have bellows but 'fired red day and night' (Xu Li 1986, 1998). Thus both processes not only have the same clay vessel (the retort) and coal briquettes but also the same form of blast and fire. During field investigations in the southwest of China in 1994, the condensation in the distillation retorts was discussed with senior furnace masters at many of the sites visited. It was known that as long as the upper parts of the clay vessels were sufficiently above the fire zone then some zinc particles could be deposited, even without pockets, because the temperature of the upper parts is usually lower than 907 °C (boiling point of zinc), especially if the charge is pushed down in the retort to a suitable depth. According to the description stated above in the *Tian Gong Kai Wu*, this process is the same as the cementation brass method used in the Ming dynasty.

## Conclusions

On the basis of the researches above:

- Zinc in early China was originally called *woyuan*, 窝铅 because it came from the pocket (the *wo*) of the distillation retort. Another term *woyuan*, 倭铅 was derived from the word for lead because of the confusion between the names of lead and zinc in early China. The original meaning of 倭铅 was lead not zinc.
- The production of zinc for making brass coinage began in the Wanli period of the Ming dynasty (16th century), and developed from the existing cementation process. It followed naturally from the development of the Chinese brass industry. Although this technology emerged relatively late, it was unlike zinc production technology used elsewhere and thus there is no reason to believe that it came from outside China.

## Appendix

- Han dynasty (206 BC–AD 220)
- Five dynasties (907–960)
- Song dynasty (960–1279)
- Ming dynasty (1368–1644)
  - Jiajing period (1522–1566)
  - Longqing period (1567–1672)
  - Wanli period (1573–1620)
  - Tianqi period (1621–1627)
  - Chongzhen period (1628–1644)
- Qing Dynasty (1616–1911)
  - Kangxi period (1662–1722)
  - Yongzheng period (1723–1735)
  - Qianlong period (1736–1795)
  - Jiaqing reign period (1796–1820)

## Notes

1. See the appendix for chronology of Chinese dynasties and periods mentioned in the text.
2. Sun Chengze, *Chun Ming Meng Yu Lu*, 春明梦余录 a recorded book about the capital of the Ming Dynasty, vol. 38.
3. Miao Boran and Wang Zuyi, *Weining County Annals*, 1924.
4. This literature includes *Ming Shi* (history of the Ming dynasty), *Ming Shi Lu* (a chronological record about the statements and actions of the Ming emperors), *Ming Hui Dian* (an official book of the Ming dynasty), *Xu Wen Xian Tong Kao* (continuation of a comprehensive study of the history of civilisation, completed in 1586, published in 1603).
5. Song Yingxing, *Tian Gong Kai Wu. Vol. 14: Wu Jin (Five Metals)*.
6. *Ibid.*
7. *Wai Jin Dan*, 外金丹, a book on alchemy in the Ming dynasty, vol. 4.
8. Sometimes it is also pronounced *yan*.
9. Originally published in 1923.
10. See No. 594 of the Taoist Canon.
11. Li Shizhen 1991: 664. NB: there appears to be some mistake in the quantities quoted here by Li Shizhen.
12. *Ge Wu Cu Tan*, vol. 2. In: *A First Collection of a Series of Books* (丛书集成初编) (Beijing, Zhonghua Book Company, 1983), pp. 27 and 37.
13. As published by a joint compiling group for the China Defence Industry Press in 1980, *Coloured Casting Alloy and its Smelting*, pp. 4–5.
14. In fact, most Wanli *tongbao* coins only contain trace tin; see Zhou Weirong 2004.
15. See the fifth section of coins of *Xu Wen Xian Tong Ka*, 文献通考, *Continuation of a Comprehensive Study of the History of Civilization* (China).
16. *Gu Zuyu: Du Shi Fang Yu Ji Yao*, 读史方舆纪要 a geographical book, written in 1667, vol. 115.

## References

Cowell, M.R., Cribb, J., Bowman, S.G.E. and Shashoua, Y. 1993. The Chinese cash: composition and production. In *Metallurgy in Numismatics*, vol. 3, M.M. Archibald and M.R. Cowell (eds). London: Royal Numismatic Society, 185–98.

Craddock, P. T. and Zhou Weirong 1998. The survival of traditional zinc production in China. In *The Fourth International Conference on the Beginnings of the Use of Metals and Alloys (BUMA-IV), Shimane, Japan, 1998*, 85.

Craddock, P.T. and Zhou Weirong 2003. Traditional zinc production in modern China: survival and evolution. In *Mining and Metal Production through the Ages*, P.T. Craddock and J. Lang (eds). London: British Museum Press, 267–92.

Li Shizhen 1991. *Compendium of Materia Medica* (本草纲目), vol. 9 (smithsonite). Shanghai: Ancient Books Publishing House, 664 [in Chinese].

Liu Guangding 1991. The history of zinc in China: re-investigation on the existence of *Wo-chien* in the Five dynasties. *Chinese Studies* 9(2): 213–21 [in Chinese].

Tuo Tuo *et al.* 1985. The Dress Section of *Song Shi* (*History of the Song Dynasty* 宋史), vol. 154. Beijing: Zhonghua Book Company, 3582 [in Chinese].

Wang Jin *et al.* (eds) 1955. *Metal Chemistry and Alchemy in Ancient China*. Beijing: China Scientific Books and Instruments Company, 92 [in Chinese].

Xu Li 1986. Traditional zinc-smelting technology in the Magu district of Hezhang county. *Studies in the History of Natural Sciences* 5(4): 361–9 [in Chinese].

Xu Li 1998. Traditional zinc-smelting technology in the Guma district of Hezhang county of Guizhou province. In *2000 Years of Zinc and Brass*, 2nd edn, P.T. Craddock (ed.). British Museum Occasional Paper 50. London, 115–32.

Zhang Hongzhao 1955. The origin of zinc use in China. In Wang Jin *et al.* 1955, 21–8 [in Chinese].

Zhao Kuanghua 1984. A re-study of the origin of zinc use in China. *China Historical Materials of Science and Technology (Zhong Guo Ke Ji Shi Liao)* 3: 15–23 [in Chinese].

Zhao Kuanghua 1996. An analysis of the chemical composition of Northern Song coins and a tentative study for the tinned coins. *Studies in the History of Natural Sciences* 5(3): 229–46 [in Chinese].

Zhao Kuanghua, Zhou Weirong, Guo Baozhang, Liu Junqi and Xue Jie 1988. A study of the composition of Ming dynasty coins and the related problems. *Studies in the History of Natural Sciences* 7(1): 54–65 [in Chinese].

Zhou Weirong 1990. A study of zinc in Xuande incense burners. *Studies in the History of Natural Sciences* 9(2): 161–4 [in Chinese].

Zhou Weirong 1991. A new inquiry into the history of the use of zinc in China. *Studies in the History of Natural Sciences* 3: 259–66 [in Chinese].

Zhou Weirong 1993. A new study on the history of the use of zinc in China. *Bulletin of the Metals Museum* 19: 49–53.

Zhou Weirong 1996a. A new study on the history of zinc smelting in China. *Chinese Studies* 14(1): 117–25 [in Chinese].

Zhou Weirong 1996b. Chinese traditional zinc-smelting technology and the history of zinc production in China. *Bulletin of the Metals Museum* 25: 36–47.

Zhou Weirong 2000. A new transliterational study of *toushi* (tutty, 鍮石). *Bulletin of the Metals Museum* 32: 65–72.

Zhou Weirong 2001. The emergence and development of brass-smelting techniques in China. *Bulletin of the Metals Museum* 34: 87–98.

Zhou Weirong 2004. *Chinese Coin: Alloy Composition and Metallurgical Research*. Beijing: Zhonghua Book Company [in Chinese].

Zhou Weirong and Fan Xiangxi 1993. A study on the development of brass for coinage in China. *Bulletin of the Metals Museum* 20: 35–45.

Zhou Weirong and Fan Xiangxi 1994. Application of zinc and cadmium for the dating and authenticating of metal relics in ancient China. *Bulletin of the Metals Museum* 22: 16–21.

## Author's address

Zhou Weirong, China Numismatic Museum, No. 22, Xijiaominxiang St., Beijing 100031, China (chinumis@public2.bta.net.cn; chinumis@yahoo.com.cn)

# Iron and steel

# Slags and the city: early iron production at Tell Hammeh, Jordan, and Tel Beth-Shemesh, Israel

*Harald Alexander Veldhuijzen and Thilo Rehren*

*ABSTRACT*  The 'coming of the age of iron' (Wertime and Muhly 1980) around 1200 BC is an event of major historical importance. Bringing widespread changes to societies, it gave rise to the subsequent period being called the Iron Age. The Near East is the supposed origin of iron metallurgy, but finds of early production (pre-500 BC) are extremely scarce. So how did iron production start and how did it evolve? How was it embedded in society? Recently excavated early production finds (iron smelting at Tell Hammeh, Jordan; iron smithing at Tel Beth-Shemesh, Israel) present an exceptional opportunity to start answering such questions.

Uniquely studying smelting and smithing together, the *chaîne opératoire* of both technologies is reconstructed through science-based analyses. This paper treats the difference in material assemblage, layout, location and archaeometry of the two sites in order to discuss their sociocultural and economic frameworks. Emphasis is further placed on the role played by technical ceramics (tuyeres, furnace wall) in this early iron production. Hammeh and Beth-Shemesh indicate hitherto unknown cross-cultural relations through their particular tuyere design. And finally, both together provide a clear picture of what constitutes a smelting and a smithing operation respectively, allowing a reassessment of earlier, often disputed, claims for iron production.

*Keywords:* iron, metallurgy, Iron Age, Jordan, Israel, smelting, smithing, microstructure, metallography.

## Introduction

This paper discusses the recent finds of early iron-smelting operations at Tell Hammeh in Jordan (930/910 ± 40 cal BC), and secondary smithing of iron at Tel Beth-Shemesh, Israel (*c.* 900 cal BC). These virtually contemporary operations provide a unique opportunity to study the metallurgical and sociocultural characteristics of some of the earliest remains of iron production known thus far. They further allow a unique comparison of both the technology and the organisation of two consecutive processing stages of iron metallurgy: smelting and secondary smithing. Following a brief review and assessment of previously known production sites in the Near East, this paper discusses the characteristics of the Tell Hammeh smelting process and attempts to reconstruct how this activity was embedded in the local society. This is followed by, and compared to, the characteristics of the Tel Beth-Shemesh smithing activities.

## Historical background

The origin and early history of iron use are widely debated topics. To date, however, it has proved problematic to reconstruct the innovation and spread of this new technology. This is certainly aggravated by the scarceness of iron artefacts and especially the absence of iron production evidence (see discussions in Curtis *et al.* 1979; Pleiner 2000: 7–22; Waldbaum 1978, 1989, 1999; Wertime and Muhly 1980). Whereas (possibly) smelted iron artefacts do appear in quite early contexts in the Near East, evidence for their actual smelting and smithing prior to the middle of the 1st millennium BC has hardly been attested with the exception of a few sites with evidence for secondary smithing (Craddock 1995: 259; Pleiner 2000: 7–8; Waldbaum 1978: 65). There are some sites in the Near East where claims have been made in the 1970s and 80s for direct evidence for iron smelting; these, however, are now disputed. The following gives a necessarily very brief summary and assessment of the most prominent of these claims.

### Claims for the early smelting of iron

An early claim for iron smelting was made in the 1970s, and concerns a group of sites in the Black Sea coast region (ancient Colchis) in modern-day Georgia (Khakhutaishvili 1976, 2001, 2005). The activities here are often dated somewhere between 1100 BC and 700 BC (Pleiner 2000: 36–7, 58). No detailed archaeometallurgical information is at present available, however, to prove or disprove either these early dates or the exact nature of the metallurgical activity practised at these sites, that is to say, whether they are concerned with iron or copper smelting.

A second claim for early iron-smelting activity concerns Tel Yin<sup>c</sup>am in northern Israel. The excavators claim that furnace-like structures, surrounded by some iron slag and ochre ore, belong to an experimental stage of iron smelting dating to

**Figure 1** Map of the southern Levant, indicating the location of Tell Hammeh, Tel Beth-Shemesh, Mugharet al Warda, several of the iron production related sites discussed, as well as several of the major sites in the Jordan valley.

the 13th century BC (Liebowitz 1981: 82–4; 1983; Liebowitz and Folk 1980, 1984). Beno Rothenberg, however, invited by the excavators to the excavation, sampled and subsequently analysed some of the presumed slag and ore and concluded that the identification of an early instance of iron smelting was unfounded. His argument here is that the presumed slag contains less than 5 wt% iron oxide, i.e. far too little for any pre-modern iron smelting slag (Rothenberg 1983: 69–70). More recently, Vincent Pigott suggested (in our view) a much more likely interpretation of the finds at Tel Yin°am: as a possible production site for red ochre pigments by heating ochreous bog ores taken from a local swamp (Pigott 2003).

A third claim for early iron production, also dating to the 13th century BC, was made in the mid-1980s for the site of Kamid el-Loz in Lebanon. The authors describe how a few minute fragments of iron metal were found in a workshop area near some fragments of slag and lumps of hematite ore, the last furthermore in the vicinity of a furnace structure. They then speculate how this means that the iron was produced here (Frisch *et al.* 1985: 77–8). The authors further mention that the tiny fragments of slag (0.5–2 cm³) from this area are

green, grey-green and black in colour, and bubbly in appearance, but also that, in the field, they classified all material that was neither hematite nor finished product as 'slag'.

It is difficult to follow the reasoning and evaluate the authors' interpretation of the actual material, as all data are fragmented over various locations in the publication. Nevertheless, examination of the presented evidence raises strong doubts about an interpretation of any of the material as belonging to iron metallurgy. From an archaeological perspective, the excavators' own assertion that the ore, slags and especially the iron metal are all found in secondary contexts (Frisch *et al.* 1985: 77, 96) makes a proposed relation between them doubtful. The chemical and microscopical analysis of the few slag samples shows that all contain significant amounts of copper sulphide (Frisch *et al.* 1985: 107, 134–46, 161, 178–80, especially tables 149ff in the appendix). The authors propose that their discoveries represent an early attempt at iron smelting based on the existing Bronze Age copper metallurgy.

However, although the co-occurrence of iron and slag originally prompted the idea of iron smelting at Kamid el-Loz, they then went on to speculate about the iron metal resulting from a virtually slag-free process. In our opinion, the Kamid el-Loz material shows no evidence for iron metallurgy besides the presence of a few Late Bronze Age iron artefacts, and even tentative interpretation of experimental iron smelting there is certainly not substantiated by the available archaeological or archaeometric data.

In conclusion, the Lebanese and possibly the Georgian sites are most likely related to copper metallurgy and the Israeli site to pigment production. None produced any quantity of iron-smelting slag or other unequivocal evidence for iron metallurgy. Only from the mid-1st millennium BC onwards does reasonably well-documented evidence exist for iron smelting in the eastern Mediterranean and Middle East, primarily from Greece and Cyprus (Hjärthner-Holdar and Risberg 2003; Muhly *et al.* 1982; Pleiner 2000, and the references therein). These cases do provide a set of early iron technological data with which new finds can be compared.

## Claims for secondary smithing of iron

A small number of occurrences of secondary smithing of iron are reported from the Iron Age II (1000–586 BC) in the Near East. Most of these date around 700 BC or later and concern just a few pieces of slag to a few handfuls of slag without any sort of production context. Most lack a proper analysis of the material or closer technical assessment. One of the earlier reported instances is the find of a small number of slags in a room on Tell Afis in Syria, dated to *c.* 750 BC, but these lack both a production context and a clear determination of the processing stage (Ingo and Scoppio 1992; Ingo *et al.* 1992a; 1992b: 285–6).

Recently more cases of secondary smithing slag have been reported, such as a few samples of varying nature including some slags, but without a clear production context, from Tell Siukh Fawqani, Syria, dating to the 7th century BC (Luciani *et al.* 2003). Slags were excavated in 2004 at Tell Ahmar (Til Barsip), Syria[1] and at Tel Hamid in Israel.[2] Both date to the 7th century BC, and where the first is almost certainly related to smithing, the slag and tuyeres from the latter are as yet unassigned.

More secondary smithing slag was found at Tell es-Saʿidiyeh, Jordan, relatively close to Tell Hammeh (Mascelloni 2004), and Khirbet Mudayna, south of ʿAmman (Daviau and Steiner 2000), again both dating to the 7th century BC. Iron metallurgy is also reported from sites such as Tel Dan, Hazor, Megiddo and Tel Masos, all in Israel, but no data on these finds are available in the public domain. Several kilograms of Persian-period (*c.* 586–332 BC) secondary smithing slags were found at Tel Dor (Shai 1999).[3]

Clear evidence for secondary smithing was attested at Tel esh-Shariʾa (Tel Sera) in the northern Negev, Israel, where an Assyrian smithy (late 7th century BC) was found in the citadel. Here a hearth structure with two tuyeres and four pieces of magnetic slag were found, as well as hammerscale, an iron spike and a completely corroded piece of iron metal (Rothenberg and Tylecote 1991).

In summary, most published evidence for early iron smithing dates to around 700 BC or later, and little systematic archaeometric or contextual data on these finds are available

in the literature. It is in the light of both the paucity of actual production finds and the difficult nature of the early claims described above, that the finds at Hammeh and Beth-Shemesh gain importance. Both sites were excavated and studied in parallel, with a specific metallurgical focus and developing and using dedicated metallurgical excavation techniques. The metallurgical finds from both sites were subjected to the same range of widely used analytical techniques to facilitate comparison. The data provide for the first time a clear picture of what, in the early 1st millennium BC, constitutes a smelting site and what constitutes a smithing site.

An important feature of Hammeh and Beth-Shemesh is their dating. Both sites are active at around *c.* 900 BC, which makes them very early in the regional history of iron metallurgy, and places these activities right at the moment in time when iron is becoming the prime utilitarian metal in the Near East.

## Slags but no city: smelting at Tell Hammeh

### Chronology and stratigraphy

Tell Hammeh was excavated by a team from the University of Leiden, the Netherlands and Yarmouk University, Jordan, in three seasons in 1996, 1997 and 2000. The last season (directed by Harald Alexander Veldhuijzen), was specifically aimed at the iron production remains. During the fieldwork, dedicated methods of excavation were applied that were later expanded for the excavation of Beth-Shemesh (see below). Close to one tonne of debris from very early iron smelting and primary smithing (bloom consolidation) operations were found, comprising various types of slags, tuyeres, charcoal, molten technical ceramics and possible furnace structures. Radiocarbon analysis of two short-lived olive wood charcoal samples from the production phase (*Olea europaea*)[4] provides a date at 930/910 cal BC (± 40 years; 1σ ranges of 1000–900 and 940–850 cal BC; accelerated mass spectrometry (AMS) analysis with $^{13}C–^{12}C$ correction). With due caution (van Strydonck *et al.* 1999), these dates place Hammeh as the earliest known find of iron smelting in the Near East (Pleiner 2000; Waldbaum 1999). Taking the dating as a *terminus post quem* together with the archaeological evidence for purely non-metallurgical activity at Hammeh from *c.* 750 BC onwards, production at Hammeh may cover a period of 100 to 150 years (van der Steen 1997, 2001, 2004; Veldhuijzen 2005a, 2005b; Veldhuijzen and Rehren 2006; Veldhuijzen and van der Steen 1999, 2000).

The site itself is a relatively small tell in the central Jordan valley located where the Zarqa river valley opens into the Jordan valley, close to several larger tells (e.g. Tell Deir ʿAlla, Tell es-Saʿidiyeh). It has access to the natural resources desirable in metal production: water, outcrops of marly clays (see Veldhuijzen 2005b: 297) (see Fig. 1), and above all the only iron ore deposit of the wider region at Mugh004et al-Warda (Abu-Ajamieh *et al.* 1988; Bender 1968: 149–51; Pigott 1983; Pigott *et al.* 1982; van den Boom and Lahloub 1962).

Several periods are attested at Hammeh. From bedrock upward, remains of Chalcolithic (*c.* 4500–3000 BC) and Early

Bronze Age (*c.* 3000–2000 BC) occupation were found, followed by more substantial layers of Late Bronze Age (*c.* 1600–1150 BC) material. Hammeh appears continuously settled through the Late Bronze Age and Iron Age I (*c.* 1150–1000 BC), up to the start of iron production in the Early Iron Age II (see van der Steen 2004). At that point in time, domestic structures, at least in the excavated areas, cease to exist, and are covered immediately, i.e. without an observable period of abandonment of the site, by a stratigraphically well-defined phase of iron production. This phase has a complex internal layering, probably reflecting seasonal activity over an extended period of time. More extensive excavation of the levels below the iron-production phase is necessary to establish the exact nature and date of this transition (Veldhuijzen 2005a).

Very soon or immediately after iron production ceased, habitation of the site resumed. This later Iron Age II phase seems to form the last extensive occupation of Tell Hammeh. Based on examination of the extensive pottery finds from this post-smelting phase, it can be assumed that the iron-production activities must have ended no later then 750 BC.[5] No settlement structures contemporary to the iron-smelting phase are presently known from Tell Hammeh.

## Scale of production and dating

A substantial part of the original production area was removed from the eastern side of the tell by modern bulldozer activity (see Fig. 2). The iron production-related stratigraphy can be traced along large stretches of the resulting vertical bulldozer-cut face, and radiates at least 20–25 m westward into the tell from there (see Fig. 2). At the cut, the deposit is clearly present both to the north and to the south of square A/B7, but is often disturbed there by later (occupation) phases. A high concentration of iron production-related material was excavated in squares A/B7 to A/D7, which seems to indicate that here (or more to the east, in the lost part of the tell) may have been the epicentre of smelting activity. From the bulldozer cut westwards, as observed in trench 1 (see Fig. 2), the production debris slowly tapers off, but nevertheless does appear in both A/D5 and B/A5.

Based on the elevation levels of the remaining tell, together with the westward stretch of the production layer, it is estimated that half or more of the original production area was removed. No more than 5–10% of that projected production area has been excavated to date. So far, Hammeh has yielded

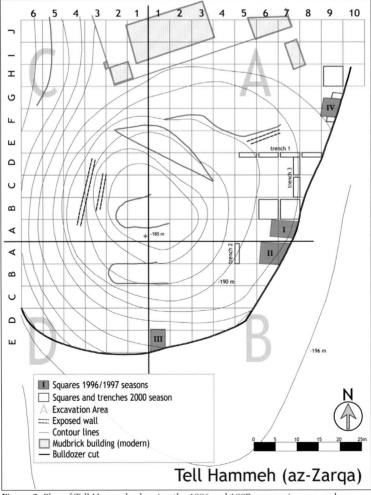

**Figure 2** Plan of Tell Hammeh, showing the 1996 and 1997 squares in grey, and the 2000 squares in red (plan based on the survey of the tell by Muwafaq Bataineh, Yarmouk University).

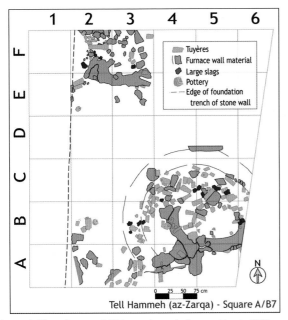

**Figure 3** Plan of square A/B7 near the end of excavation in 2000, with the 75 × 75 cm excavation grid in blue. Slag was found in large quantities throughout the square, only pieces of significant size are indicated here. 'Empty' grids were either excavated before the drawing was made (e.g. F4), or have not been fully excavated (e.g. row D).

roughly 700 kg of slags, present throughout the entire production layer, and more than 350 individual tuyere fragments, mainly from square A/B7 (see Fig. 3).

In order to assess the scale of operations, the excavated slag quantities were extrapolated to the original production area. The excavated slag quantities and available ore analyses were then used to estimate the yield of iron metal per unit of slag. Iterative mass balance calculations of the Hammeh tap slag and the local Warda ore (Veldhuijzen and Rehren 2006) show that 100 kg of average Mugharet al-Warda ore produces 57.5 kg of slag and *c.* 47 kg of iron metal, but requires an addition of *c.* 19 kg clay material as a flux. This means that the excavated 700 kg of slag corresponds to *c.* 570 kg of iron metal. Assuming that 5–10% of the original total production area has

been excavated thus far, this translates to a total production of metal between *c.* 5.7 to 11.5 tons. Such quantities show that the smelting activities at Hammeh were a well-established and substantial operation, as opposed to an early attempt or experimentation with a new technology. Spread over an assumed production period of about 100 years, this equals a production of 50–100 kg iron per annum, requiring something in the order of 100–200 kg of ore, or just a few donkey loads. This scale is more in keeping with a seasonal activity than a full-time specialisation, and would require only a single brief period of ore collection and smelting per year.

From the fact that the metalworkers at Hammeh were able to perform and sustain such an operation over a considerable period of time, one can conclude that they must have had access to the resources necessary in the process, particularly the ore. It seems unlikely that such access was possible without a larger socio-economic structure in which the iron production was embedded, but the political situation for this region at this period is not well enough understood to allow us to say more about the nature of this structure.

### Slags and tuyeres at Hammeh

The predominant material at Hammeh is slag. Five different types can be distinguished. Placed in a logical order, these are: furnace slags (incompletely reduced ore; *c.* 30% of the total), furnace bottom slag (<1%), tap slags (*c.* 60%), 'slags' rich in technical ceramic (*c.* 1%), and primary smithing slags (a few %). It is interesting to compare the chemical composition of these slags to the composition of the local ore, the local clay, the olive wood charcoal, and the technical ceramic of the tuyeres and local clay (see Tables 1 and 2). At the time of analysis, we had access to only two ore samples. Current research on approximately 20 specimens of the Warda ore by Yosha al-Amri at the German Mining Museum in Bochum, however, confirms that our samples reflect the range of composition of that ore as it is found today.[6]

The Hammeh slags range from a compositional similarity to the Mugharet al Warda ore in the furnace slags to a similar-

**Table 1** Comparison between the average major element compositions of the Warda ore, the various Hammeh slag types, the local clay and Hammeh tuyeres, and the Hammeh charcoal. In addition, the major elements of the Tel Beth-Shemesh secondary smithing slags are shown. Values are expressed in wt%, and normalised to 100% (except the charcoal which is given as analysed). Analysis by (P)ED-XRF (slag_fun calibration method, April 2005 (Veldhuijzen 2005a, 124–44; 2003)). The accuracy of the method, tested against certified reference materials (BCS-CRM 381, USGS BHVO-2, BCS-CRM 301-1, CCRMP SL-1) is better than 2.5% relative for oxides above 1 wt%. Precision was monitored through three repeat analyses of each sample and was found to be less than 1.5% relative for all oxides above 1 wt%. Although the (P)ED-XRF equipment identifies elements above magnesium at concentrations as low as 10 ppm, we report data only to the level of 0.01 wt% (100ppm).

| Material type (no. of samples averaged) | $SiO_2$ | $Al_2O_3$ | FeO | $TiO_2$ | MnO | CaO | MgO | $K_2O$ | $P_2O_5$ |
|---|---|---|---|---|---|---|---|---|---|
| Mugharet al Warda hematite ore (2) | 4.86 | 0.17 | 89.9 | 0.05 | 0.05 | 4.36 | 0.17 | 0.01 | 0.16 |
| Furnace slags (10) | 16.5 | 3.76 | 70.8 | 0.24 | 0.72 | 4.25 | 0.77 | 1.01 | 1.3 |
| Furnace bottom slag (1) | 17.6 | 2.18 | 69.6 | 0.13 | 0.06 | 8.10 | 0.99 | 1.06 | 0.21 |
| Tap slags (15) | 25.2 | 5.44 | 52.5 | 0.34 | 1.12 | 10.9 | 1.80 | 1.12 | 1.20 |
| Primary smithing slags (1) | 33.3 | 5.35 | 40.6 | 0.39 | 0.97 | 14.4 | 1.90 | 1.67 | 1.00 |
| Ceramic-rich 'slags' (10) | 46.4 | 5.98 | 21.3 | 0.48 | 0.69 | 19.6 | 2.48 | 1.70 | 0.74 |
| Lisan clay + Hammeh tuyere ceramic (3) | 57.2 | 13.2 | 4.78 | 0.96 | 0.07 | 17.5 | 2.96 | 2.84 | 0.22 |
| Hammeh (*Olea europea*) charcoal (1) | 0.40 | n.d. | 0.10 | n.d. | 0.00 | 8.40 | 2.10 | 2.50 | 0.10 |
| Tel Beth-Shemesh secondary smithing slags (5) | 24.8 | 3.94 | 50.0 | 0.27 | 0.08 | 16.6 | 1.44 | 1.85 | 0.85 |

**Table 2** Comparison between the average trace element compositions of the Warda ore, the various Hammeh slag types, the local clay and Hammeh tuyeres, and the Hammeh charcoal. In addition, the trace elements of the Tel Beth-Shemesh secondary smithing slags are shown. Analysis by (P)ED-XRF (slag_fun calibration method, April 2005 (Veldhuijzen 2005a, 124–44; 2003)). Values are expressed in parts per million (ppm), except sulphur, which is expressed in wt%; 'bdl' indicates when results are 'below detection limit'

| Material type (no. of samples averaged) | S | Ni | Cu | Zn | Rb | Sr | Y | Zr | Ba | Pb |
|---|---|---|---|---|---|---|---|---|---|---|
| Mugharet al Warda hematite ore (2) | 0.28 | bdl | 25 | 65 | bdl | bdl | 25 | 70 | bdl | bdl |
| Furnace slags (10) | 0.62 | bdl | 45 | 45 | bdl | 340 | 60 | 120 | 45 | bdl |
| Furnace bottom slag (1) | 0.06 | bdl | 20 | 15 | bdl | 250 | bdl | 110 | bdl | bdl |
| Tap slags (15) | 0.39 | bdl | 30 | 60 | bdl | 640 | 80 | 200 | 100 | bdl |
| Primary smithing slags (1) | 0.52 | 30 | 25 | 35 | bdl | 690 | 70 | 270 | 140 | bdl |
| Ceramic-rich 'slags' (10) | 0.21 | 10 | 25 | 35 | 20 | 670 | 60 | 330 | 170 | bdl |
| Lisan clay + Hammeh tuyere ceramic (3) | 0.33 | 45 | 25 | 90 | 60 | 270 | 30 | 410 | 170 | bdl |
| Hammeh (Olea europea) charcoal (1) | 0.10 | 10 | 20 | 10 | bdl | 790 | bdl | bdl | bdl | bdl |
| Tel Beth-Shemesh secondary smithing slags (5) | 0.10 | 25 | 60 | 30 | 10 | 410 | 20 | 210 | 1290 | bdl |

ity with the technical ceramics/local clay in the ceramic-rich 'slags' (see Fig. 4). The furnace slag is compositionally relatively similar to the ore, except for an increased silica and alumina content; the ceramic-rich 'slag' is very similar to the local clay except for an increased iron oxide content (see Table 1).

This apparent relation to two source materials is further seen in the prime indicator of a smelting process, the tap slags. The Hammeh tap slags are chemically similar to typical bloomery slags (e.g. Bachmann 1982; Kronz 1998), but show a distinct iron-oxide-poor and lime-rich composition. Mass balance calculations of the Warda ore and the Hammeh tap slags confirm that simple removal of iron oxide as iron metal (to simulate the smelting process) from the Warda ore by itself cannot result in the formation of Hammeh slags (Veldhuijzen and Rehren 2006). This means that either the smelters used a different (blend of) ore or other materials must contribute to the process. As no other iron ore exists within a very wide radius around Hammeh, the second option seems more plausible. The most likely additional material is the ceramic-rich 'slags'. These share all the macroscopic characteristics of

the tap slags (black/grey glassy with flowing patterns), but are chemically virtually identical to the technical ceramics, their only chemical link to the smelting process found in their elevated iron oxide content, indicating a mixing of molten ceramic material with smelting slag.

The suspected influence of the technical ceramics was tested by further mass balance calculations, iteratively adding technical ceramic to and removing iron oxide from the Warda ore until the calculated hypothetical slag closely matched the actual Hammeh (tap) slags (Veldhuijzen 2005b; Veldhuijzen and Rehren 2006).

The best match was found for an addition of c. 19 kg ceramic material per 100 kg of ore, resulting in the production of c. 47 kg metal and 57.5 kg slag. This substantial addition of ceramic material is necessary to obtain a low melting slag, even though it dilutes the ore and reduces the amount of iron metal extracted from it. Thus, the relatively low iron oxide content and high lime content of the Hammeh slag are a result of the particular nature of the ore body. This requires a silica-rich addition to facilitate slag formation within an Iron Age

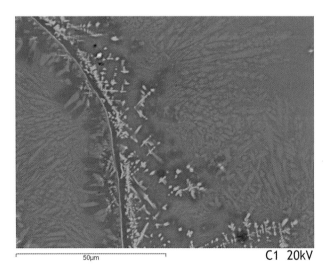

C1 20kV

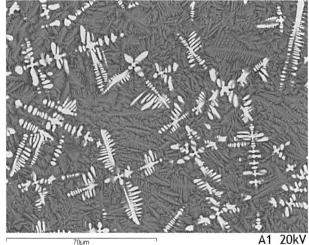

A1 20kV

**Figure 4** Backscattered electron micrograph of tap slag C1, showing a tapping band between two flows of slag. The edges of these flows consist predominantly of a glassy matrix (grey) with small formations of fayalitic laths (lighter grey) and some precipitation of thin dendrites of iron oxide (light grey) where the two flows touch.

**Figure 5** Backscattered electron micrograph of tap slag A1, showing a glassy matrix (dark grey), with feathery devitrification (mid-grey) and wüstite dendrites (light grey).

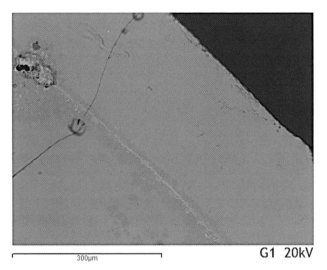

**Figure 6** Backscattered electron micrograph of 'ceramic-rich' slag G1, showing an amorphous mass of glass and a thin tapping band of iron oxide (possibly wüstite) flanked by darker conglomerates of glass that is richer in $Al_2O_3$ and $K_2O$ than the matrix.

that clearly exceeded individual consumption, but fell short of constituting a large-scale regional industrial centre. It is certainly conceivable that other smelting operations were active alongside Hammeh in the Mugharet al Warda area around the same time.

The fact that all charcoal excavated consists of short-lived olive wood, together with the high number of stratigraphic layers within the slag deposit, the absence of contemporaneous habitation of the site, the highly standardised shape of the tuyeres, and the apparent access to resources, strongly suggest a locally coordinated, seasonal activity that is tied in with other activities such as the harvesting and pruning of the olive trees. It takes place close to the necessary resources rather than the eventual consumer.

technology. It is not a sign of a particularly advanced smelting technology, similar to the much later blast furnace technology with limestone fluxing. From the high number and heavily vitrified state of the tuyeres at Hammeh (see Fig. 7), it seems clear that these, together with the furnace wall, provided the necessary ceramic material (Veldhuijzen 2005b).

In summary, the material found at Tell Hammeh and its archaeological context create a strong impression of a well-established and dedicated iron-smelting operation that forms part of a wider web of activities. The use of sacrificial tuyeres indicates, in our opinion, a clear understanding of the process by the smelters, using and adapting to the particularities of the locally available materials. Iron was produced in quantities

## Slags and the city: smithing at Tel Beth-Shemesh

### Location and excavation history

Tel Beth-Shemesh is located in the northeastern Shephelah, Israel, approximately 20 km west of Jerusalem and *c.* 75 km southwest of Tell Hammeh (see Fig. 1). It lies at what once formed the border area between the Philistine territory of the lower Shephelah and coastal plain, and the Judean hill country.

Three major expeditions have conducted excavations of the tell. The first, in 1911 and 1912, was directed by Palestine Exploration Fund archaeologist Duncan MacKenzie (MacKenzie 1914), the second by Elihu Grant (Grant and Wright 1939), from Haverford College, Haverford, Pennsylvania, in the late 1920s and early 1930s. Both exposed large areas of the site, digging deep trenches reaching down to bedrock. Shlomo Bunimovitz and Zvi Lederman from Tel

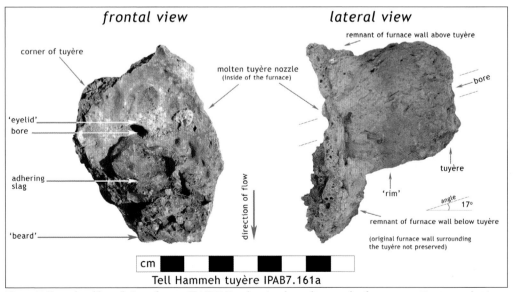

**Figure 7** Frontal and lateral view of tuyere IPAB7.161a, showing the molten nozzle of a square section tuyere that is partially fused with the technical ceramic of the furnace wall. Remnants of unvitrified furnace wall are also preserved, both below and above the tuyere. An 'eyelid', formed by molten technical ceramic flowing down and pushed up and forward by the airflow, is visible above the bore. A large 'beard' of molten ceramic material with adhering slag can also be seen. In the lateral view, some 3 cm behind the molten nozzle, can be seen the 'rim', a greyish band indicating the extent to which reducing conditions in the furnace affected the ceramic, simultaneously suggesting a downward angle for the tuyeres.

Aviv University initiated a third series of excavations in the early 1990s (Bunimovitz and Lederman 2003).

## Discovery and excavation of the smithy

In 2001, work started in Area E (see Fig. 8), a narrow area in the southwestern part of the site that had remained unexcavated between a 1912 MacKenzie trench and a 1930 Grant trench. In both large public buildings had been excavated. Area E contained several phases of industrial and commercial activity. After several phases of these activities were excavated, evidence for metallurgical activity began to appear in square E/T48, at which point the directors approached the present authors. Excavation of the smithy took place in July 2003 and June 2006 using specific and pre-developed excavation techniques.

The assemblage of metallurgical debris excavated at Beth-Shemesh consists of technical ceramics, metal artefacts, a single type of morphologically homogeneous slag: concavo-convex 'smithing hearth bottom' (SHB) slag (or PCB: 'plano-convex bottom' slag), and very fine magnetic material (i.e. hammerscale) (on smithing technology, see Serneels and Perret 2003).

Intriguingly, the Beth-Shemesh tuyeres are virtually identical to the Hammeh ones in all macroscopic aspects, from size, colour, feel and temper to shape. Less abundant here (*c.* 30 specimens) than at Hammeh, they are all square and approximately 5 × 5 cm in section, with a bore of *c.* 10 mm in diameter (see Fig. 9; see also Fig. 7). Radiocarbon analyses of three burned olive pits from the smithy resulted in a date of *c.* 900 cal BC (± 45 years; AMS analysis with $^{13}C$–$^{12}C$ correction).

From the nature of the material recovered in 2001, and especially the fact that only one type of slag, i.e. SHBs, was present, subsequent excavation of the workshop in 2003 and 2006 began on the assumption that the workshop at Beth-Shemesh represented a secondary smithing operation, as opposed to iron smelting and/or primary (bloom-)smithing. To confirm or deny this assumption, dedicated excavation techniques were developed, expanding the normal archaeological stratigraphical approach.

## Dedicated metallurgical excavation techniques

The development of these metallurgical excavation techniques drew on the experience at Hammeh as well as English Heritage guidelines (Bayley *et al.* 2001). With these methods, we sought to find and record the minute magnetic material associated with an iron-related metallurgical workshop, which is not recovered using standard archaeological excavation methods. This magnetic material, i.e. hammerscale and slag prills, is an important indicator of the type of metallurgy practised (Bayley *et al.* 2001: 14). It furthermore assists in determining otherwise invisible use of space and location of activities within the metallurgical workshop, e.g. the location of a hearth or anvil, by plotting its distribution within the workshop. A grid system of 25 × 25 cm was laid out over square E/T48. In the northern part of E/T48 the grids were

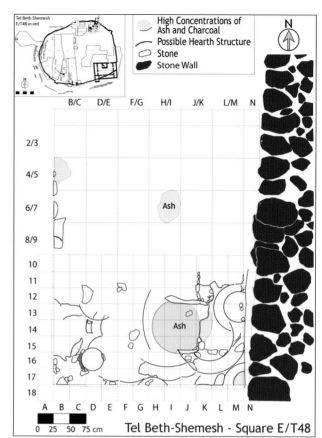

Figure 8 Plan of square E/T48, with the excavation grid system in grey. Inset: plan of Tel Beth-Shemesh showing the location of E/T48 (based on the survey of the tell by Zvi Lederman).

Figure 9 One of the (smaller) Beth-Shemesh tuyeres, which was excavated during the 2001 season. The tuyere clearly shows the narrow bore and square section, as well as the molten nozzle. This particular tuyere has a smaller cross-section (3.5 × 3.5 cm) than the other Beth-Shemesh tuyeres, which on average measure 5 × 5 cm.

bundled to 50 × 50 cm as this part was already excavated further (see Fig. 10).

The surface that was previously exposed in 2001 was used to test and refine the magnet-dragging techniques. Then, the soil from each unit (arbitrary vertical spits of 5 cm) within each grid (horizontal location, e.g. H15) was put in a separate bucket with a label, and subsequently spread out on a plastic sheet. A magnet held in a small sealable sample bag

**Figure 10** The excavation of the 25 × 25 cm grid in square E/T48 at Tel Beth-Shemesh. Soil from each grid is recovered separately. The inset shows the dragging of a magnet above the soil surface of a previously excavated grid to recover magnetic material.

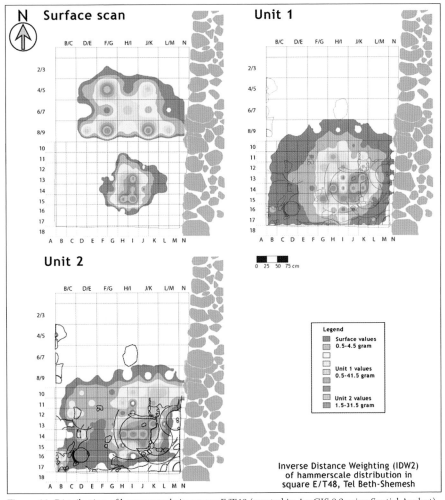

**Figure 11** Distribution of hammerscale in square E/T48 (created in ArcGIS 8.3 using Spatial Analyst). Please note that the plotting spills over into the empty grid row 8/9 for both units 1 and 2.

was then dragged just over or lightly touching the soil for a minute and a half. The soil was then jumbled up by hand, and a second round of dragging, again for a minute and a half, was performed.

Several structures were discovered in the second unit (see Fig. 10). The exact function or sequence of these intersecting structures is as yet unknown, and is one focus of the (ongoing) 2006 excavations. Although they show no vitrification or similar indicators for metallurgical use, several contain ash, and all contain hammerscale. Two ridges (in I12-J12, and in J15, see Fig. 11) contain a large ash lens between them.

### Use of space in the Beth-Shemesh smithy

The magnetic debris at Beth-Shemesh consists predominantly of hammerscale. Whereas hammerscales can also occur in the context of primary smithing, larger quantities of scales are usually associated with the more prolonged and extensively oxidising circumstances of secondary smithing (Jones 2001: 14).

To determine whether a spatial pattern could be discerned in the hammerscale scattering, the weights were plotted, per excavated unit, on the grid plan of E/T48 (see Fig. 11). The highest concentration of hammerscale coincides with the large ashy area observed archaeologically in all units excavated (see Fig. 1; compare Fig. 10). High concentrations can be expected inside a smithing hearth and in the vicinity of the anvil. When such structures or objects are no longer present or are difficult to identify, patterns within the hammerscale distribution may help the archaeological reconstruction of a workshop (Jones 2001: 14). At Beth-Shemesh, the observed correlation between high concentrations of hammerscale with those of ash in the same location suggests that this concentration reflects the (final) hearth structure rather than the location of an anvil.

### Slags and tuyeres at Beth-Shemesh

Smithing slags are often very similar to bloomery smelting slags in general chemical (see Tables 1 and 2) and microstructural composition (Pleiner 2000: 255). The morphology, size and uniformity of the Beth-Shemesh slag all point towards secondary smithing (see Figs 12 and 13). In total, 65 complete cakes and more than 150 fragments were found, all belonging to a single type: mostly round, concavo-convex shapes with a diameter of up to 10 cm, with rust adhering to the top and soil to the bottom.

These morphological features form a classic example of a 'smithing hearth bottom' (SHB) or 'plano-convex bottom' (PCB), where each specimen probably represents a single smithing operation (Serneels and Perret 2003: 473). Five samples were selected for further analysis. In their overall composition (see Tables 1 and 2), these slags compare quite well to other smithing slags discussed in the literature (e.g. Kronz 1998: 225), except that their lime content is considerably higher (c. 16 wt% compared to a more regular content between c. 0.5 and 5 wt% elsewhere). Their average iron oxide content is also lower than might be considered regular (almost 50 wt% compared to between 50 and 80 wt%). In contrast, the

Tel Beth Shemesh Smithing Hearth Bottom E/T48 SF

**Figure 12** Section of smithing hearth bottom slag E/T48 SF. The SHB clearly shows a porous slag core of concavo-convex shape, with rust adhering to the top and soil embedded at the bottom.

TBS 2001 SHB E/T48 2834.23

**Figure 13** Top view of SHB slag E/T48 2824.23, showing a depression at the centre, indicating where the airflow from the tuyere hit the forming slag. The approximately straight side at the bottom of the picture may be the place where the SHB was in contact with the hearth wall, indicating that the tuyere was located on this side of the SHB (see Serneels and Perret 2003).

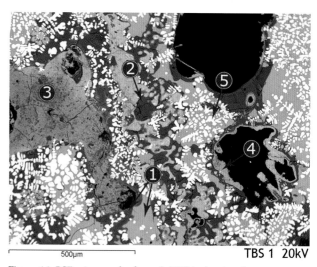

500µm                     TBS 1 20kV

**Figure 14** BSE micrograph of sample TBS 1, showing the very heterogeneous microstructure of this sample. It consists of a heterogeneou glassy matrix (mid grey and dark grey; 1), incorporating partly fused and thermally cracked quartz grains (dark grey; 2), a large inclusion of iron-rich ceramic material (left centre, with cracks; 3), iron hydroxide (around holes; 4), and wüstite (FeO) dendrites (whitish; 5).

Beth-Shemesh slags are quite similar in composition to both the Hammeh smelting (tap) and primary smithing slags, suggesting the use of similar materials and methods. The main discrepancies are in the lower manganese and sulphur levels in the Beth-Shemesh slags, as expected for secondary smithing slags, and the much higher barium concentrations; the latter may simply reflect locally different trace elements in the clays used to build the smithing hearth(s) and tuyeres. Work is ongoing to characterise these in comparison to the finds from Tell Hammeh. The relatively large quantities of free iron oxide in the form of wüstite in some areas of the SHBs (see Fig. 14) are probably due to the incorporation of hammerscale during the smithing operation.

## Summary

At Beth-Shemesh, a picture emerges of a smithing workshop that is operated regularly and at considerable (local) scale within the confines of one of the larger cities in the Iron Age Levant, catering to the needs and wants of that settlement. Plotting of the magnetic remains has created a picture of the spatial layout of the workshop, with the likely position of the hearth (i.e. the last hearth in use). The quantity of SHBs, where each specimen represents a separate smithing operation, cycle or workday (Serneels and Perret 2003: 472), indicates that the smithing work at Beth-Shemesh must have been a regular operation, as opposed to an experimental or just an occasional one. The choice of location of the Beth-Shemesh smithy, near substantial public buildings, may reflect the status of this relatively new metal as well as being determined by a desired closeness to the consumer, rather than by considerations such as the availability of raw materials.

Assuming that each SHB represents a single workday of smithing, the Beth-Shemesh smithy was most probably a workshop for the creation and/or repair of iron (and perhaps bronze) artefacts for strictly local consumption, as opposed to a centre of secondary production for the wider area.

## Conclusions

With the integrated excavation and analysis of both an early smelting and a smithing site, the sites discussed in this paper provide a template for distinction between the primary and secondary stages of iron-production technology, both during fieldwork and in subsequent analysis of the material, as well as for the re-evaluation of 'older' evidence and in the interpretation of future finds. Both sites are clearly distinct in terms of workshop layout, the nature and variety of material found, as well as the choice of location. The identification as either smelting or smithing slag is mostly based on their characteristic shape and archaeological context. The chemical and phase analyses support these interpretations, but would not on their own permit unambiguous assignation.

It is often suggested that the political fragmentation of the Iron Age I and II in the Levant stimulated local industries to exploit locally available raw materials (see Stech-Wheeler *et*

*al.* 1981; van der Steen 2004). This is exactly the picture that emerges from Hammeh, where local people use and adapt their technology to local materials. In our opinion, it is very likely that Hammeh represents a seasonal smelting operation, taking place close to the necessary resources and in synchronisation with the equally seasonal pruning of olive trees to obtain fuel for the furnace. The lack of contemporary habitation on the tell raises interesting questions about ownership, organisation and control of space, know-how and resources in the region, which are beyond the scope of this paper.

The smithing at Beth-Shemesh shows a quite different choice of location as it takes place within the confines of a large city, which corresponds with other finds identified as smithing in the region. Here the choice of location probably reflects a desire to be close to the consumer, rather than being close to resources.

The identical design characteristics of the Beth-Shemesh and Hammeh tuyeres are probably indicative of cross-cultural contacts, shared technological characteristics, or even a possible socio-ethnic link between the people conducting two consecutive processes at two different sites. The square cross-section of the tuyeres is certainly not a technological requirement for their use. This shape may therefore represent a *technological choice*, i.e. a choice not guided by technological constraints, but one made by the person performing the technological activity based on, for example, social or cultural considerations, perceived requirements or local traditions. The largely uniform size and shape of all the tuyeres further suggests organisation of production and standardisation (see discussions in, among others: Costin 2001; Costin and Wright 1998; Dobres and Robb 2000; Killick 2004; Lechtman 1977, 1988, 1999; Lechtman and Steinberg 1979; Lemonnier 1986, 1989; Pfaffenberger 1988, 1989, 1992).

In the light of ethno-archaeological observation from sub-Saharan Africa (e.g. Chirikure and Rehren 2004; MacKenzie 1975), it is tempting to speculate about smelters (seasonally) smelting on a site near the required resources (Hammeh), and then travelling around the surrounding area, smithing their product near the consumers, i.e. in settlement contexts such as Beth-Shemesh and perhaps Tell Deir ᶜAlla or Tell es-Saᶜidiyeh, where this travel is reflected in identical tuyere design. Neither the Beth-Shemesh smithing nor the Hammeh smelting shows any sign of innovation or development of the technology practised. Both sites represent well-established technological processes of a considerable scale and with indications of standardisation. This clearly suggests that iron production and working were known and practised in the region prior to the start of iron smelting at Hammeh around 930 BC, and the smithing at Beth-Shemesh in *c.* 900–850 BC.

## Notes

1. John Russell and Elizabeth Hendricks, pers. comm.
2. Samuel Wolff, pers. comm.
3. Adi Behar, pers. comm.
4. Eleni Asouti, pers. comm.
5. Gerrit van der Kooij, pers. comm.
6. Yosha al-Amri, pers. comm.

# References

Abu-Ajamieh, M.M., Bender, F.K. and Eicher, R.N. 1988. *Natural Resources in Jordan, Inventory-Evaluation-Development Program*. Amman: Natural Resources Authority.

Bachmann, H.-G. 1982. *The Identification of Slags from Archaeological Sites*. Institute of Archaeology Occasional Publication 6. London: Institute of Archaeology.

Bayley, J., Dungworth, D. and Paynter, S. 2001. *Centre for Archaeology Guidelines: Archaeometallurgy*. London: English Heritage.

Bender, F.K. 1968. *Geologie Von Jordanien (Beiträge Zur Regionalen Geologie Der Erde, Band 7)*. Berlin: Gebrüder Borntraeger Verlagsbuchhandlung.

van den Boom, G. and Lahloub, M. 1962. *The Iron-Ore Deposit 'Warda' in the Southern Ajlun District*. Unpublished report, Federal Institute of Geoscience and Natural Resources, Hannover (Unveröffentlichte Bericht Deutsche Geologische Mission Jordanien, Archiv Bundesanstalt für Bodenforschung)/National Resources Authority, Amman.

Bunimovitz, S. and Lederman, Z. 2003. Tel Beth Shemesh, 2001–2003. *Israel Exploration Journal* 53: 233–7.

Chirikure, S. and Rehren, Th. 2004. Ores, furnaces, slags, and prehistoric societies: aspects of iron working in the Nyanga agricultural complex, AD 1300–1900. *African Archaeological Review* 21: 135–52.

Costin, C.L. 2001. Craft production systems. In *Archaeology at the Millennium: A Sourcebook*, G.M. Feinman and T.D. Price (eds). New York: Kluwer Academic/Plenum Publishers.

Costin, C.L. and Wright, R.P. 1998. *Craft and Social Identity*. Arlington, VA: American Anthropological Association.

Craddock, P.T. 1995. *Early Metal Mining and Production*. Edinburgh, Edinburgh University Press.

Curtis, J.E., Stech-Wheeler, T., Muhly, J.D. and Maddin, R. 1979. Neo-Assyrian ironworking technology. *Proceedings of the American Philosophical Society* 123: 369–90.

Daviau, P.M.M. and Steiner, M. 2000. A Moabite sanctuary at Khirbat Al-Mudayna. *Bulletin of the American Schools of Oriental Research* 320: 1–21.

Dobres, M.-A. and Robb, J.E. 2000. *Agency in Archaeology*. London: Routledge.

Frisch, B., Mansfeld, G. and Thiele, W.-R. 1985. *Kamid El-Loz 6, Die Werkstätten Der Spätbronzezeitlichen Paläste*. Bonn: Dr. Rudolf Habelt.

Grant, E. and Wright, G.E. 1939. *Ain Shems Excavations (Palestine): Part V (Text)*. Haverford, PA: Haverford College.

Hjärthner-Holdar, E. and Risberg, C. 2003. The introduction of iron in Sweden and Greece. In *Prehistoric and Medieval Direct Iron Smelting in Scandinavia and Europe: Aspects of Technology and Society. Proceedings of the Sandbjerg Conference 16th to 20th September 1999*, L.C. Nørbach (ed.). Aarhus: Aarhus University Press.

Ingo, G.M. and Scoppio, L. 1992. Small-area Xps and Xaes study of early iron metallurgy slags. *Surface and Interface Analysis* 18: 551–4.

Ingo, G.M., Scoppio, L., Bruno, R. and Bultrini, G. 1992a. Microchemical investigation of early iron metallurgy slags. *Mikrochimica Acta* 109: 269–80.

Ingo, G.M., Scoppio, L., Mazzoni, S., Mattogno, G. and Scandurra, A. 1992b. Application of surface and bulk analytical techniques for the study of iron metallurgy slags at Tell Afis (N-W Syria). In *Material Issues in Art and Archaeology III*, P.B. Vandiver, J.R. Druzik, G.S. Wheeler and I.C. Freestone (eds). Pittsburgh: Materials Research Society, 285–90.

Khakhutaishvili, N. 1976. A contribution of the Kartvelian tribes to the mastery of metallurgy in the ancient Near East. In *Wirtschaft und Gesellschaft in Vorderasien*, J. Harmatta and G. Komoroczy (eds). Budapest: Akademiai Kido.

Khakhutaishvili, N. 2001. Alte Eisenproduktion an Der Östlichen Schwarzmeerküste. In *Georgien. Schätze Aus Dem Land Des Goldenen Vlies*, I. Gambaschidze, A. Hauptmann, R. Slotta and Ü. Yalçin (eds). Bochum: Deutsches Bergbau-Museum.

Khakhutaishvili, N. 2005. Development of iron metallurgy in west Transcaucasia (historical Kolkheti). *Metalla* 12: 80–87.

Killick, D.J. 2004. Social constructionist approaches to the study of technology. *World Archaeology* 36: 571–8.

Kronz, A. 1998. *Phasenbeziehungen Und Kristallisationsmechanismen in Fayalitischen Schmelzsystemen. Untersuchungen an Eisen- Und Buntmetallschlacken*. Bielefeld: Friedland.

Lechtman, H. 1977. Style in technology: some early thoughts. In *Material Culture: Styles, Organization, and Dynamics*, H. Lechtman and R.S. Merrill (eds). St Paul: American Ethnological Society/West Publishing, 3–20.

Lechtman, H. 1988. Traditions and style in Central Andean metalworking. In *The Beginning of the Use of Metals and Alloys. Papers from the Second International Conference on the Beginning of the Use of Metals and Alloys, Zhengzhou, China, 21–26 October 1986*, R. Maddin (ed.). Cambridge, MA: MIT, 344–78.

Lechtman, H. 1999. Afterword. In *The Social Dynamics of Technology: Practice, Politics, and World Views*, M.-A. Dobres and C.R. Hoffman (eds). Washington, DC: Smithsonian Institution Press, 223–32.

Lechtman, H. and Steinberg, A. 1979. The history of technology: an anthropological point of view. In *The History and Philosophy of Science*, G. Bugliarello and D.B. Doner (eds). Urbana, IL: University of Illinois Press, 135–62.

Lemonnier, P. 1986. The study of material culture today: toward an anthropology of technical systems. *Journal of Anthropological Archaeology* 5: 147–86.

Lemonnier, P. 1989. Towards an anthropology of technology. *Man* 24(3): 526–7.

Lemonnier, P. 1992. *Elements for an Anthropology of Technology*. University of Michigan Anthropological Paper 88. Ann Arbor, MI: University of Michigan.

Lemonnier, P. and Pfaffenberger, B. 1989. Towards an anthropology of technology. *Man* 24: 526–7.

Liebowitz, H. 1981. Excavations at Tel Yin'am: the 1976 and 1977 seasons, preliminary report. *Bulletin of the American Schools of Oriental Research* 243: 79–94.

Liebowitz, H. 1983. Reply to Beno Rothenberg. *Bulletin of the American Schools of Oriental Research* 252: 71–2.

Liebowitz, H. and Folk, R.L. 1980. Archaeological geology of Tel Yin'am, Galilee, Israel. *Journal of Field Archaeology* 7: 23–42.

Liebowitz, H. and Folk, R.L. 1984. The dawn of iron smelting in Palestine: the Late Bronze Age smelter at Tel Yin'am, preliminary report. *Journal of Field Archaeology* 11: 265–80.

Luciani, M., Zaghis, F., Salviulo, G., Calliari, I., and Ramous, E. 2003. Iron Age metallurgy: a preliminary study of slags from Tell Shiukh Fawqani (northern Syria). In *Archaeometallurgy in Europe (24–26 September 2003)*. Milan: Associazione Italiana di Metallurgia.

MacKenzie, D. 1914. *Excavations at Ain Shems (Beth-Shemesh)*. London: Harrison and Sons.

MacKenzie, J.M. 1975. Pre-colonial industry: the Njanja and the iron trade. *Native Affairs Department Annual* 11: 200–220.

Mascelloni, M.L. 2004. *Testing the Evidence for Local Metalworking: Metals, Slag and Vitrified Materials from Tell Es-Sa'idiyeh, Jordan*. Unpublished MA thesis, Institute of Archaeology, University College London.

Muhly, J.D., Maddin, R. and Karageorghis, V. (eds) 1982. *Early Metallurgy in Cyprus, 4000–500 BC. Proceedings of the International Archaeological Symposium, Larnaca, Cyprus, 1–6 June 1981*. Nicosia: Pierides Foundation and Department of Antiquities.

Pfaffenberger, B. 1988. Fetished objects and humanised nature: towards an anthropology of technology. *Man* 23(2): 236–52.

Pfaffenberger, B. 1989. Reply to Pierre Lemonnier 'Towards an anthropology of technology'. *Man* 24(3): 527.

Pfaffenberger, B. 1992. Social anthropology of technology. *Annual Review of Anthropology* 21: 491–516.

Pigott, V.C. 1983. The innovation of iron (as steel) in Palestine. *American Journal of Archaeology* 87: 252.

Pigott, V.C. 2003. Iron and pyrotechnology at 13th century – Late Bronze Age – Tel Yin'am (Israel): a reinterpretation. In *Man and Mining*, T. Stöllner, G. Körlin, G. Steffens and J. Cierny (eds). *Der Anschnitt* Beiheft 16. Bochum: Deutsches Bergbau-Museum, 365–75.

Pigott, V.C., McGovern, P.E. and Notis, M.R. 1982. The earliest steel from Transjordan. *Journal of the Museum Applied Science Center for Archaeology* 2: 35–9.

Pleiner, R. 2000. *Iron in Archaeology: The European Bloomery Smelters.* Praha: Archeologický Ústav Av Cr.

Rothenberg, B. 1983. Corrections on Timna and Tel Yin^cam in the Bulletin. *Bulletin of the American Schools of Oriental Research* 252: 69–70.

Rothenberg, B. and Tylecote, R.F. 1991. A unique Assyrian iron smithy in the northern Negev (Israel). *IAMS Journal* 17: 11–14.

Serneels, V. and Perret, S. 2003. Quantification of smithing activities based on the investigation of slag and other material remains. In *Proceedings of the International Conference Archaeometallurgy in Europe.* Milan: Associazione Italiana di Metallurgia, 469–79.

Shai, N. 1999. דור - בתל ברזל חרשי עבודת (The work of a smith at Tel-Dor). הטבע ומדעי ארכאולוגיה (*Archaeological and Natural Sciences*), 7: 38–49.

Stech-Wheeler, T., Muhly, J.D., Maxwell-Hyslop, K.R., and Maddin, R. 1981. Iron at Taanach and early iron metallurgy in the eastern Mediterranean. *American Journal of Archaeology* 85: 245–68.

van der Steen, E.J. 1997. Excavations at Tell El-Hammeh. *Occident and Orient* 3: 12–14.

van der Steen, E.J. 2001. Excavations at Tell El-Hammeh. *Studies in the History and Archaeology of Jordan* 7: 229–32.

van der Steen, E.J. 2004. *Tribes and Territories in Transition. The Central East Jordan Valley in the Late Bronze and Early Iron Ages: A Study of the Sources.* Leuven: Peeters.

van Strydonck, M., Nelson, D.E., Crombé, P., Bronk Ramsey, C., Scott, E.M., van der Plicht, J. and Hedges, R.E.M. 1999. What's in a ^14C date? In *^14C et Archéologie. Actes du 3^e Congrès International (Lyon 6–10 Avril 1998).* Mémoires de la Société Préhistorique Française, XXVI, 1999 et Supplément 1999 de la Revue d'archéométrie, 433–40.

Veldhuijzen, H.A. 2003. 'Slag_fun' – a new tool for archaeometal-lurgy: development of an analytical (P)Ed-Xrf method for iron-rich materials. *PIA (Papers from the Institute of Archaeology)* 14: 102–18.

Veldhuijzen, H.A. 2005a. *Early Iron Production in the Levant: Smelting and Smithing at Early 1st Millennium BC Tell Hammeh, Jordan, and Tel Beth-Shemesh, Israel.* Unpublished PhD thesis, Institute of Archaeology, University College London.

Veldhuijzen, H.A. 2005b. Technical ceramics in early iron smelting: the role of ceramics in the early first millennium BC iron production at Tell Hammeh (Az-Zarqa), Jordan. In *Understanding People through their Pottery: Proceedings of the 7th European Meeting on Ancient Ceramics (Emac '03)*, I. Prudêncio, I. Dias and J.C. Waerenborgh (eds). Lisbon: Instituto Português de Arqueologia (IPA), 294–302.

Veldhuijzen, H.A. and Rehren, Th. 2006. Iron smelting slag formation at Tell Hammeh (Az-Zarqa), Jordan. In *Proceedings of the 34th International Symposium on Archaeometry, 3–7 May 2004, Zaragoza, Spain*, J. Pérez-Arantegui (ed). Zaragoza: Institución 'Fernando el Católico', 245–50.

Veldhuijzen, H.A. and van der Steen, E.J. 1999. Iron production center found in the Jordan valley. *Near Eastern Archaeology* 62: 195–9.

Veldhuijzen, H.A. and van der Steen, E.J. 2000. Early iron smelting (Tell-Hammeh, Jordan). *Archaeology* 53: 21.

Waldbaum, J.C. 1978. *From Bronze to Iron: The Transition from the Bronze Age to the Iron Age in the Eastern Mediterranean.* Goteborg: Paul Astroms Forlag.

Waldbaum, J.C. 1989. Copper, iron, tin, wood, the start of the Iron Age in the eastern Mediterranean. *Archeomaterials* 3: 111–22.

Waldbaum, J.C. 1999. The coming of iron in the eastern Mediterranean: thirty years of archaeological and technological research. In *The Archaeometallurgy of the Asian Old World*, V.C. Pigott (ed.). Pennsylvania: MASCA Research Papers in Science and Archaeology 16: 27–57.

Wertime, T.A. and Muhly, J.D. 1980. *The Coming of the Age of Iron.* New Haven and London: Yale University Press.

## Authors' addresses

- Corresponding author: Harald Alexander Veldhuijzen, Institute of Archaeology, 31–34 Gordon Square, London WC1H 0PY, UK (h.veldhuijzen@ucl.ac.uk)
- Thilo Rehren, Institute of Archaeology, 31–34 Gordon Square, London WC1H 0PY, UK (th.rehren@ucl.ac.uk)

# Innovations in bloomery smelting in Iron Age and Romano-British England

*Sarah Paynter*

*ABSTRACT*    This paper discusses bloomery smelting in England in the Iron Age and Romano-British periods. Initially non-tapping furnaces are used, with tapping furnaces known from the Late Iron Age, and becoming widespread in the Romano-British period. Different variations of both furnace types are evident. The capabilities and operation of non-tapping and tapping furnaces are compared on the basis of the composition of the slag and its morphology. Differences in slag morphology result from the way the slag was removed from the furnace, rather than changes in slag composition or properties. Many aspects of smelting technology change little including temperature, redox conditions, and often the raw materials used, the main exception being the exploitation of siderite ore from the 1st century AD in the Weald. It is suggested that tapping furnaces offered advantages when demand for iron was great, combining ease of reuse with the desired capacity.

*Keywords:* iron, slag, smelting, England, Iron Age, Romano-British, SEM, EDS, analysis.

## Introduction

Although the period from about the 7th century BC to the mid-1st century AD in England is referred to as the Iron Age, comparatively little is known about the primary production of iron during this time. This paper focuses particularly on the different types of furnaces used and their capabilities. A previous related paper identified regional variations in slag composition, which were attributed to the raw materials used (Paynter 2006).

## Furnace types during the Iron Age and Romano-British periods

Substantial bloomery furnace remains, particularly ones with surviving superstructure, are rare from the Iron Age and Romano-British periods in England and reconstructing their likely form and mode of operation is very difficult. Variations in furnace structure and operation are only detected when the furnace remains are unusually well preserved and/or expertly interpreted (for example Cleere and Crossley 1985; Crew 1998a; Jackson and Ambrose 1978) and it is rare that industrial remains are precisely dated. Nevertheless, different typologies of bloomery smelting furnace have been distinguished (Cleere 1972; Tylecote 1990) on the basis of archaeological evidence (slag and furnace remains) as well as on the results of ethnographical and experimental work (Cleere 1971; Tylecote *et al.* 1971). Further studies have placed these types into broad chronological and cultural frameworks (Cleere and Crossley 1985; Tylecote 1990).

The main criterion used by Cleere (1972) to classify furnaces was how the slag waste was removed from the active part of the furnace. (Although there are many additional criteria that can be used to subdivide furnace groups, such as the type of superstructure and the use of natural or forced draught, these are not discussed here.) Non-tapping furnaces were characterised by a sunken hearth or pit below ground level where the slag collected. Tapping furnaces had a ground-level hearth and the slag flowed out while hot through an aperture in the furnace wall. Differences are apparent in the morphology of the slag associated with non-tapping and tapping furnaces.

In continental Europe, non-tapping furnaces of different periods have been extensively researched, but much of the evidence from England is less well understood and is not yet fully published. Non-tapping furnaces are known from a number of English sites of early medieval date (Tylecote 1990). An increasing number of non-tapping bloomery furnaces and/or slag assemblages, however, have also been identified from Iron Age and early Romano-British contexts. The quantities of slag are often fairly small although at some sites tonnes of slag have been recovered (Clogg 1999; Crew 1998b; McDonnell 1988).

Tapping furnaces and slag assemblages have been more thoroughly studied. Evidence of tapping furnaces is common in England from Romano-British sites, and Late Iron Age examples have also been found (Tylecote 1990). Slag assemblages from tapped furnaces are often large, with vast slag heaps reported at many sites (Cleere and Crossley 1985; Crew 1998a).

## Aims

Previous authors have dispelled the misconception that non-tapping furnaces of an early date were necessarily bowl types with little superstructure and very limited capacity (Cleere 1972; Clough 1985). In the absence of evidence to the contrary, however, there is often a tendency to assume that these furnaces were inefficient with a small capacity relative to the tapping furnaces that were widely adopted in the Romano-British period.

This paper aims to compare the ways in which furnaces were constructed and operated during the Iron Age and Romano-British periods, and to demonstrate the potential of slag assemblages for providing information that is otherwise difficult to obtain. The factors considered include the raw materials and smelting conditions used, the proportion of iron extracted from the ore, the way in which the slag collected and was removed from the furnace, the capacity of the furnaces and the changing cultural and economic context of the industry. These issues have been explored using compositional data to estimate the liquidus temperature and viscosity of the slag. The form and microstructure of samples of smelting slag produced over this period, from non-tapping furnaces in particular, have also been examined.

For the purposes of this overview, the furnaces and slag assemblages discussed have been categorised simply as tapping or non-tapping. It is clear, however, that there were many variations of these two basic furnace types over the time period considered. The picture will undoubtedly change as more archaeological evidence accumulates, but this is a starting point for further research.

## Sites, materials and methods

The locations of the sites in this study are shown in Figure 1 and site details are summarised in Table 1. Some examples are from recent work by the author and some from other researchers' work on comparable material, including slag from many of the same sites discussed by Paynter (2006). Slag samples were selected that were visually representative of the assemblage from each site, with clear features diagnostic of smelting waste (smithing slag was not included in the study). Smelting slag is heterogeneous and variable so a large number of samples are required to obtain representative information. Only a limited number of samples from each site could be examined with the resources available, however, and similarly much of the literature data relate to a restricted number of samples. Therefore the information provided here is intended only as a guide to what might be typical of each site. For analytical details and slag compositional data see Paynter (2006) and Tables 1 and 2.

### The morphology of Iron Age and Romano-British bloomery slag

The different morphological types of slag produced by tapping and non-tapping furnaces can be classified visually. Tap slag, the main diagnostic type from tapping furnaces, is easily recognised. The major diagnostic slag types from non-tapping furnaces of Iron Age and Romano-British date, which are varied and more often misidentified, are described briefly below. Although the slag is divided into distinct categories, a

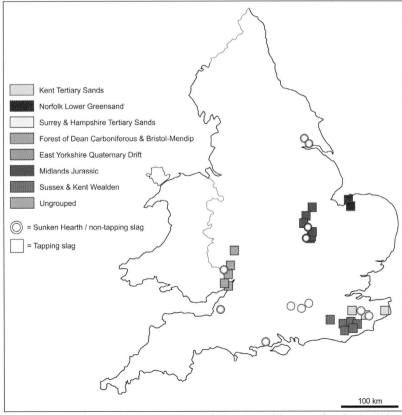

**Figure 1** A map of England and Wales showing the locations of the sites referred to in Table 1.

**Table 1** The average normalised slag composition for the sites discussed in this study from Paynter (2006) and with some new data (number of analyses in brackets, bd = below detectable limit, nm = not measured). The location of each site, the approximate date of the smelting activity (IA = Iron Age, R = Roman, M = medieval, L = Late, E = Early) and the type of slag assemblage (SH = sunken hearth/non-tapping, T = tapping) are also noted. See Figure 1 for locations.

| Site (no. of analyses) | Location | NGR | Date | Type | Source | $Na_2O$ | $MgO$ | $Al_2O_3$ | $SiO_2$ | $P_2O_5$ | $SO_3$ | $K_2O$ | $CaO$ | $TiO_2$ | $MnO$ | $FeO$ |
|---|---|---|---|---|---|---|---|---|---|---|---|---|---|---|---|---|
| Leda Cottages (6) | Kent | TQ965475 | LIA/RB | SH | Paynter 2006 | 0.4 | 0.4 | 5.4 | 17.7 | 1.6 | bd | 0.8 | 1.8 | 0.1 | 0.5 | 71.2 |
| Hawkinge (9) | | TR219398 | EIA | SH | Paynter 2006 | 0.6 | 0.4 | 5.9 | 24.3 | 2.0 | 0.2 | 0.8 | 1.2 | 0.3 | 0.3 | 64.2 |
| Brisley (11) | | TQ 99 40 | IA/RB | T | Paynter 2006 | 0.2 | 0.4 | 5.6 | 21.5 | 1.5 | bd | 0.8 | 1.5 | 0.2 | 0.6 | 67.6 |
| Westhawk (18) | | TR002399 | RB | T | Paynter 2006 | 0.2 | 0.4 | 5.8 | 23.3 | 1.6 | bd | 0.9 | 2.2 | 0.2 | 0.3 | 65.0 |
| Headcorn (5) | | TQ832432 | RB | T | New data | 0.6 | 0.5 | 8.7 | 21.5 | 2.2 | 0.2 | 2.3 | 2.7 | 0.2 | 0.8 | 60.3 |
| Snettisham (8) | Norfolk | TF 68 34 | RB | T | Chirikure and Paynter 2002 | 0.1 | 0.2 | 2.0 | 24.1 | 1.4 | 0.2 | 0.3 | 1.3 | nm | 0.5 | 69.8 |
| Ashwicken (1) | | TF684178 | RB | T | Tylecote 1962 | nm | 1.4 | 3.3 | 21.8 | 1.8 | nm | 0.0 | 0.4 | nm | 0.5 | 70.8 |
| Brooklands | Surrey | TQ068632 | EIA | SH | Hanworth and Tomalin 1977 | bd | 0.2 | 2.5 | 20.7 | 1.8 | nm | 0.6 | 0.7 | 0.2 | 0.0 | 73.3 |
| Thorpe Lea (9) | | TQ017697 | IA/RB | SH | Starley 1998 | 0.5 | 0.1 | 2.2 | 17.5 | 3.3 | 0.2 | 0.6 | 1.7 | 0.0 | 1.0 | 72.8 |
| Heckfield (6) | Hampshire | SU 72 60 | EIA | SH | Dungworth forthcoming | bd | 0.3 | 3.3 | 23.6 | 4.0 | nm | 0.7 | 1.0 | 0.2 | 0.5 | 66.5 |
| Stowe Hill (9) | Gloucestershire | SO570067 | EM? | SH | Paynter 2006 | 0.2 | 1.7 | 5.0 | 22.9 | 0.2 | bd | 1.8 | 4.9 | 0.2 | 0.1 | 62.7 |
| Woolaston (13) | | ST597987 | RB | T | Fulford and Allen 1992 | nm | 1.8 | 4.6 | 23.5 | 0.2 | nm | 1.8 | 2.2 | 0.3 | 0.2 | 65.5 |
| Pill House (3) | | ST 57 95 | RB | T | Allen 1988 | 0.1 | 1.7 | 3.5 | 23.3 | 0.3 | nm | 1.4 | 2.4 | 0.2 | 0.2 | 66.9 |
| Horse Pill (2) | | ST 58 98 | | T | Allen 1988 | 0.1 | 1.1 | 1.9 | 16.1 | 0.5 | nm | 0.5 | 1.1 | 0.1 | 0.2 | 78.5 |
| Ley Pill (2) | | SO 62 00 | | T | Allen 1988 | 0.1 | 1.5 | 3.0 | 21.0 | 0.3 | nm | 1.3 | 2.2 | 0.1 | 0.2 | 70.3 |
| Awre (6) | | SO 71 08 | | T | Allen 1988 | 0.1 | 0.6 | 3.8 | 24.4 | 0.5 | nm | 0.6 | 1.1 | 0.2 | 0.1 | 68.7 |
| Severn House Farm (1) | | ST 64 98 | | T | Allen 1988 | 0.2 | 0.8 | 7.1 | 28.8 | 0.4 | nm | 1.0 | 2.1 | 0.4 | 0.0 | 59.2 |
| Hills Flats (3) | | ST 62 97 | | T | Allen 1988 | 0.1 | 0.9 | 3.1 | 23.2 | 0.4 | nm | 0.7 | 1.4 | 0.2 | 0.2 | 69.8 |
| Oldbury Flats (2) | | ST 61 95 | | T | Allen 1988 | 0.2 | 1.2 | 3.0 | 22.9 | 0.4 | nm | 0.6 | 1.7 | 0.1 | 0.2 | 69.6 |
| Oldbury Pill (1) | | ST 60 93 | | T | Allen 1988 | 0.1 | 0.9 | 2.6 | 19.0 | 0.2 | nm | 0.5 | 1.2 | 0.1 | 0.2 | 75.3 |
| Worcester, The Butts (9) | Worcestershire | SO847551 | RB | T | Blakelock 2005 | 0.4 | 1.0 | 6.0 | 26.5 | 0.3 | 0.1 | 1.3 | 1.6 | 0.2 | 0.2 | 62.4 |
| Worcester, Deansway (21) | | SO849548 | RB | T | McDonnell and Swiss 2004 | 1.0 | 1.8 | 5.9 | 30.2 | 0.1 | nm | 1.6 | 2.2 | 0.3 | 0.2 | 57.0 |
| Ariconium (2) | Herefordshire | SO 64 24 | RB | T | Thomas and Young 1999 | 0.5 | 1.2 | 4.4 | 21.8 | 0.4 | nm | 1.5 | 2.7 | 0.2 | 0.2 | 67.1 |
| Chelm's Combe (16) | Somerset | ST 47 54 | IA | SH | New data | 0.8 | 0.6 | 3.5 | 24.6 | 0.2 | 0.1 | 1.2 | 3.5 | 0.2 | 0.1 | 65.1 |
| North Cave (15) | E. Yorkshire | SE 89 33 | IA/RB | SH | McDonnell 1988 | 0.2 | 0.1 | 2.9 | 26.4 | 0.9 | 0.2 | 0.5 | 1.4 | 0.1 | 0.7 | 66.6 |
| Welham Bridge (15) | | SE793341 | IA | SH | Clogg 1999 | bd | bd | 3.1 | 25.4 | 1.1 | nm | 0.6 | 1.4 | 1.5 | 0.9 | 66.6 |
| Wakerley (4) | Northants | SP941983 | IA/RB | SH | Fells 1983 | nm | 0.4 | 6.2 | 19.7 | 0.8 | nm | 0.6 | 2.3 | 0.5 | 0.6 | 69.0 |
| Great Oakley (2) | | SP881866 | EIA | SH | Fells 1983 | nm | 0.3 | 4.2 | 22.8 | 0.5 | nm | 0.4 | 1.0 | 0.3 | 0.5 | 69.9 |
| Bulwick (1) | | SP929939 | RB | T | Fells 1983 | nm | 0.3 | 6.6 | 25.4 | 0.1 | nm | 0.5 | 1.1 | 0.5 | 1.1 | 64.5 |
| Brigstock (1) | | SP963859 | RB | T | Fells 1983 | nm | 0.4 | 7.1 | 16.6 | 2.4 | nm | 0.5 | 2.7 | 0.3 | 0.3 | 69.7 |
| Blind Eye Quarry (2) | | TF002148 | RB | T | Fells 1983 | nm | 0.2 | 4.2 | 23.9 | 0.3 | nm | 0.0 | 0.9 | 0.4 | 0.6 | 69.5 |
| Laxton (5) | | SP966972 | RB | T | Crew & Young pers. comm. | bd | 0.3 | 6.6 | 28.4 | 0.2 | nm | 0.3 | 3.3 | 0.5 | 0.8 | 59.6 |
| Ridlington (4) | Rutland | SK835024 | LIA/RB | T | Schrüfer-Kolb 2004 | 0.6 | 0.6 | 6.3 | 22.3 | 1.1 | nm | 1.3 | 1.3 | 0.4 | 0.4 | 65.8 |
| Whitwell (4) | | SK926082 | RB | T | | 1.3 | 1.2 | 5.8 | 21.2 | 1.0 | nm | 0.9 | 1.6 | 0.5 | 2.3 | 64.3 |
| Great Cansiron (3) | West Sussex | TQ448383 | RB | T | Paynter 2006 | bd | 2.3 | 9.4 | 24.0 | 0.8 | 0.4 | 0.9 | 6.2 | 0.4 | 2.4 | 53.3 |
| Little Furnace W'd (23) | | TQ591243 | IA/RB | T | Paynter 2006 | 0.2 | 1.0 | 7.7 | 30.3 | 0.6 | 0.2 | 1.2 | 6.0 | 0.5 | 1.3 | 51.1 |
| Clappers Wood (5) | East Sussex | TQ594168 | RB | T | Paynter 2006 | 0.3 | 1.9 | 6.8 | 32.0 | 0.9 | 0.3 | 1.2 | 19.3 | 0.5 | 3.2 | 33.7 |
| Chitcombe (5) | | TQ813211 | RB | T | Paynter 2006 | 0.4 | 1.9 | 11.0 | 34.2 | 0.7 | 0.4 | 1.5 | 10.1 | 0.5 | 2.0 | 37.3 |
| Winser Farm (9) | Kent | TQ855308 | RB? | T | New data | 0.3 | 2.0 | 6.1 | 27.2 | 0.7 | 0.4 | 0.8 | 15.3 | 0.4 | 2.3 | 44.5 |
| Little Farningham Farm (8) | | TQ801352 | RB | T | New data | 0.4 | 3.0 | 7.1 | 26.2 | 0.6 | 0.4 | 0.9 | 7.7 | 0.4 | 2.3 | 50.9 |
| Longham Lakes (7) | Dorset | SZ061975 | EIA? | SH | | 0.3 | 0.5 | 6.1 | 30.5 | 0.2 | 0.1 | 1.5 | 1.9 | 0.3 | 1.0 | 57.5 |

range of forms can be produced in a single smelt and so large fragments often have characteristics of several categories.

The most diagnostic slags from non-tapping furnaces are the large masses often referred to as furnace bottoms or (as here) slag cakes (Cleere 1972; Clough 1985; Tylecote 1990). Slag cakes are often oval in plan, with one or more surfaces that were in contact with the sides and bottom of the cut feature in which the slag accumulated and solidified. The size of these roughly plano-convex accumulations varies, although those considered here have similar dimensions, for example Longham Lakes (8.4 kg, 320 mm at widest point) (Fig. 2) and

Leda Cottages (up to 13 kg, 320 mm at widest point) (Oxford Wessex Archaeology Joint Venture 2004) (Fig. 3). The majority of the slag cakes from Welham Bridge reported by Clogg (1999) were around 360 mm across weighing about 13 kg, although many larger examples, the biggest 74 kg, were also recovered. The dimensions of slag cakes from North Cave reported by McDonnell (1988) were on average 450 mm at the widest point. Angular, blocky fragments of smelting slag in non-tapping assemblages (also referred to as smelting slag or dense iron silicate slag in literature) frequently share many characteristics of the slag cakes described above, including

**Table 2** EDS compositional data (wt%, normalised) for smelting slags analysed in this study (bd = below detectable limit).

| Site & sample | Na$_2$O | MgO | Al$_2$O$_3$ | SiO$_2$ | P$_2$O$_5$ | SO$_3$ | K$_2$O | CaO | TiO$_2$ | MnO | FeO |
|---|---|---|---|---|---|---|---|---|---|---|---|
| Chelm's Combe 1 (slag cake) | 0.85 | 0.60 | 4.33 | 25.47 | bd | bd | 0.89 | 3.63 | 0.15 | bd | 63.77 |
| | 0.23 | 0.55 | 2.91 | 18.18 | bd | bd | 0.48 | 2.15 | 0.13 | bd | 75.09 |
| | 0.30 | 0.60 | 3.66 | 24.67 | bd | bd | 0.64 | 3.27 | 0.23 | bd | 66.45 |
| | 0.84 | 0.45 | 3.54 | 22.60 | bd | bd | 0.76 | 2.58 | 0.18 | bd | 68.92 |
| | 0.48 | 0.62 | 4.68 | 25.68 | 0.68 | 0.25 | 1.02 | 3.64 | 0.23 | bd | 62.67 |
| | 0.47 | 0.58 | 2.76 | 25.56 | bd | bd | 0.79 | 2.91 | 0.10 | 0.10 | 66.59 |
| | 0.41 | 0.45 | 3.84 | 24.99 | 0.24 | bd | 0.81 | 3.20 | 0.12 | bd | 65.76 |
| Chelm's Combe 2 (slag cake) | 2.38 | 0.59 | 3.82 | 26.28 | 0.28 | 0.30 | 1.88 | 4.14 | 0.23 | 0.13 | 59.99 |
| | 0.56 | 0.72 | 3.34 | 25.10 | bd | bd | 1.44 | 3.86 | 0.22 | bd | 64.44 |
| | 2.25 | 0.51 | 3.40 | 25.33 | 0.31 | 0.22 | 1.57 | 3.94 | 0.18 | 0.11 | 62.18 |
| | 0.52 | 0.78 | 3.20 | 25.47 | bd | bd | 1.40 | 3.68 | 0.15 | bd | 64.55 |
| | 1.02 | 0.74 | 3.82 | 25.45 | 0.26 | 0.20 | 1.73 | 4.06 | 0.15 | bd | 62.52 |
| Chelm's Combe 3 (slag cake) | 0.45 | 0.54 | 3.28 | 24.40 | bd | bd | 1.43 | 3.75 | 0.16 | bd | 65.75 |
| | 0.45 | 0.65 | 3.52 | 23.84 | bd | bd | 1.59 | 3.81 | 0.23 | 0.12 | 65.56 |
| | 0.48 | 0.78 | 3.35 | 26.10 | bd | bd | 1.47 | 3.82 | 0.19 | 0.10 | 63.41 |
| | 0.59 | 0.67 | 3.15 | 25.23 | bd | bd | 1.38 | 3.77 | 0.17 | bd | 64.71 |
| Great Winster 1 (tap) | 0.15 | 1.98 | 6.41 | 29.30 | 0.77 | 0.38 | 0.77 | 16.28 | 0.40 | 2.24 | 41.32 |
| | 0.25 | 2.00 | 6.93 | 29.89 | 0.87 | 0.39 | 0.89 | 16.47 | 0.45 | 2.19 | 39.67 |
| | 0.33 | 2.12 | 5.98 | 26.20 | 0.70 | 0.35 | 0.70 | 14.02 | 0.41 | 2.24 | 46.96 |
| Great Winster 2 (tap) | 0.28 | 1.86 | 5.84 | 26.18 | 0.72 | 0.34 | 0.72 | 14.63 | 0.42 | 2.21 | 46.80 |
| | 0.36 | 1.84 | 6.09 | 26.22 | 0.78 | 0.34 | 0.74 | 15.07 | 0.49 | 2.19 | 45.88 |
| | 0.33 | 2.03 | 5.96 | 27.33 | 0.71 | 0.33 | 0.75 | 15.33 | 0.42 | 2.34 | 44.47 |
| Great Winster 3 (tap) | 0.36 | 1.96 | 5.93 | 26.43 | 0.69 | 0.35 | 0.76 | 15.45 | 0.42 | 2.29 | 45.37 |
| | 0.34 | 1.99 | 6.05 | 26.67 | 0.68 | 0.28 | 0.75 | 15.66 | 0.42 | 2.36 | 44.80 |
| | 0.32 | 1.98 | 5.95 | 26.95 | 0.65 | 0.41 | 0.78 | 15.21 | 0.42 | 2.31 | 45.01 |
| Little Farningham (Cranbrook) 1 (tap) | 0.42 | 2.69 | 7.16 | 26.63 | 0.56 | 0.45 | 1.00 | 8.68 | 0.39 | 1.12 | 50.89 |
| | 0.43 | 3.32 | 6.14 | 25.92 | 0.45 | 0.34 | 0.82 | 7.45 | 0.34 | 2.25 | 52.54 |
| | 0.48 | 3.00 | 6.68 | 26.27 | 0.54 | 0.35 | 0.89 | 8.07 | 0.38 | 2.20 | 51.12 |
| | 0.45 | 3.16 | 6.22 | 26.06 | 0.53 | 0.35 | 0.83 | 7.81 | 0.34 | 2.27 | 51.98 |
| Little Farningham 2 (tap) | 0.43 | 2.72 | 8.07 | 25.54 | 0.75 | 0.39 | 0.95 | 7.03 | 0.63 | 2.51 | 50.99 |
| | 0.43 | 3.42 | 7.23 | 26.75 | 0.72 | 0.44 | 0.94 | 6.90 | 0.46 | 2.67 | 50.06 |
| | 0.49 | 2.69 | 8.40 | 26.24 | 0.82 | 0.54 | 1.22 | 8.70 | 0.54 | 2.62 | 47.75 |
| | 0.31 | 3.03 | 6.80 | 26.11 | 0.73 | 0.45 | 0.93 | 6.77 | 0.47 | 2.60 | 51.79 |
| Headcorn (Rolvenden Layne) 1 (tap) | 0.50 | 0.59 | 10.36 | 20.62 | 1.99 | bd | 2.27 | 2.42 | 0.19 | 0.82 | 60.10 |
| | 0.34 | 0.54 | 7.50 | 19.69 | 1.76 | bd | 1.76 | 2.43 | 0.22 | 0.86 | 64.70 |
| | 1.07 | 0.46 | 7.79 | 20.79 | 1.91 | 0.21 | 2.27 | 2.49 | 0.23 | 0.82 | 61.94 |
| | 0.50 | 0.46 | 9.21 | 23.81 | 3.30 | bd | 2.93 | 3.74 | 0.20 | 0.79 | 54.91 |
| | 0.42 | 0.52 | 8.52 | 22.82 | 1.91 | 0.21 | 2.34 | 2.54 | 0.19 | 0.84 | 59.70 |
| Longham Lakes 1 (slag cake) | 0.56 | 0.47 | 7.38 | 32.82 | bd | bd | 2.57 | 1.87 | 0.26 | 1.02 | 52.79 |
| | 0.34 | 0.67 | 4.78 | 30.10 | bd | bd | 1.05 | 1.87 | 0.40 | 1.19 | 59.46 |
| | 0.41 | 0.52 | 7.46 | 31.59 | bd | bd | 2.66 | 1.99 | 0.27 | 0.94 | 53.93 |
| | bd | 0.66 | 4.88 | 30.91 | 0.20 | bd | 1.02 | 1.94 | 0.22 | 1.17 | 58.76 |
| | bd | 0.54 | 5.22 | 30.79 | 0.20 | bd | 1.61 | 1.76 | 0.29 | 1.09 | 58.36 |
| Longham Lakes 2 (slag adhered to lining) | 0.29 | 0.45 | 8.18 | 30.53 | 0.24 | bd | 1.79 | 2.21 | 0.32 | 0.84 | 55.04 |
| | bd | 0.51 | 4.54 | 26.92 | 0.29 | 0.25 | bd | 1.65 | 0.28 | 0.94 | 64.46 |

**Figure 2** A section through the slag cake from Longham Lakes showing a well-consolidated lower layer and a porous, heterogeneous top layer.

impressions of organic material. These fragments appear to have broken from larger masses.

Many slag cakes and fragments of smelting slag contain impressions of organic matter. In the examples from Leda Cottages and Stowe Hill these impressions included clusters of fibrous, straw-like material at the extremities of some pieces (Fig. 4). In the Leda Cottages slag there were also numerous large, squarish fragments derived from trunk or branch wood from mature trees (Fig. 3). Well-preserved pieces were identified as oak, but it was not possible to tell whether charcoal or wood was present originally.[1] Impressions of large wood or charcoal fragments have also been described in slag from elsewhere (Clogg 1999; Hanworth and Tomalin 1977; McDonnell 1988; Paynter 2002; Tylecote 1990), sometimes arranged in a particular way (Schrüfer-Kolb 2004; Starley 1998). Often the orientation of the slag could be reconstructed from the direction of runs on the surface, for example the Leda Cottages slag cakes had formed in a feature with a steeply inclined side which channelled the slag in one direction (Fig. 3).

Flowed slag with a ropey surface texture, resembling tap slag or slag prills, is often present in assemblages associated with non-tapping furnaces. It generally constitutes a small proportion of the total and often has atypical features, for example with flows on both the top and bottom surfaces, or with the flows flattened, or with evidence of disturbance while still hot (Crew 1998b; McDonnell 1988; Tylecote 1990). The fragments sometimes have better defined individual runs compared to the broad flows in tap slag from tapping furnaces (see Fig. 5).

### The composition of Iron Age and Romano-British bloomery slag

Paynter (2006) compared the compositions of smelting slag from sites across England, grouped by geological region. The samples included slag with varied morphologies from non-tapping (sunken hearth) and tapping furnaces of the Iron Age and Romano-British periods. Few major compositional differences were found, regardless of when or where the slag was produced (the most notable exception being the Sussex and Kent Weald group discussed below). This suggested general continuity in the type of raw materials and smelting conditions, such as temperature and redox conditions (Tylecote *et al.* 1971) used over these periods, even though the morphology of the slag varied greatly.

**Figure 3** A slag cake from Leda Cottages. The angular cavities in the slag are impressions from large, squared pieces of wood or charcoal. The slag collected in a pit with one steeply inclined side, causing the slag to flow in one direction (to the left as shown in this figure).

**Figure 4** Impressions of straw-like, organic fibres (indicated by arrows), about 2 mm in diameter, in a slag cake from Leda Cottages.

**Figure 5** A block of multiple slag flows in an assemblage of slag from non-tapping furnaces at Leda Cottages. The slag has been disturbed while hot and some fired clay from the furnace structure has been incorporated.

However, there were consistent slight differences between the slag samples from each geological region in terms of the concentrations of elements such as phosphorus, lime, magnesia, alumina and titania. These compositional trends were attributed to the raw materials available in each area, influenced mainly by the type of ore smelted and to a lesser extent the materials used to construct the furnace. Rich limonite-type ores appeared to have been smelted at many of the sites included in the study. The slags from the Sussex and Kent Weald, dating from the Late Iron Age onwards, contained substantially less iron oxide and more alumina and lime than slags from elsewhere, which was attributed to the use of local siderite ore.

## Properties of Iron Age and Romano-British bloomery slag

The physical properties of the slag influence the resulting slag morphology. Therefore the properties of slag from non-tapping and tapping furnaces were compared to explore the conditions used in the furnace, and why and how different morphological types had formed.

Compositional data can be used to estimate the liquidus temperature of slag, which gives an indication of the temperatures achieved in the furnace. The same data can also be used to estimate slag viscosity at different temperatures. Morton and Wingrove (1969) used the $FeO$-$CaAl_2Si_2O_8$-$SiO_2$ phase diagram to estimate the liquidus temperature of Roman bloomery tap slags as about 1150 °C. With this approach the slag composition is reduced in complexity so that it can be plotted on a phase diagram for a simpler system, but this can introduce errors of the order of several hundred degrees (Freestone 1988).

Here a model by Nathan and Van Kirk (1978) has been used to estimate slag liquidus temperatures, as described by Freestone (1982, 1988) and Craddock et al. (1985). Using compositional data from Table 1, the estimated olivine liquidus temperatures for slag samples from both non-tapping and tapping assemblages were found to be broadly similar, spanning 1050–1160 °C. The tapped slags from the Sussex and Kent Weald occupied the higher part of this range as they contained higher levels of magnesia, lime and alumina. This model has limitations, for example neither manganese nor phosphorus are included, nor is the phase wüstite (the primary phase at equilibrium in a few very iron-rich samples) (Schairer 1942). Nonetheless, Freestone (1982) found that the model was quite accurate for experimental liquids with compositions similar to the Wealden slags, giving estimates only 50 °C lower than measured values.

The viscosity of slag at temperatures of 1200 and 1250 °C, roughly 100 °C above the estimated liquidus temperatures, was estimated using a model by Bottinga and Weill (1972). Craddock et al. (1985) demonstrated the accuracy of the model to be within a factor of 2. The model applies to silicate liquids containing more than 35 mol% silica however, which excludes most slag samples in this study except those from the Sussex and Kent Weald (Table 1). At 1200 °C the estimated viscosity for these slags was low, at 4–16 poise, and was 1–6 poise at 1250 °C. Most of the remaining slags from other regional groups discussed in this paper and by Paynter (2006) contained 25–35 mol% silica and lower concentrations of alumina, and so should have lower viscosities than the Wealden slags at the same temperature. Accordingly the few samples from other regional groups containing in excess of 35 mol% silica had estimated viscosities of around 3–6 poise at 1200 °C.

These data suggest similar temperatures, exceeding 1100–1200 °C, were probably attained in both non-tapping and tapping furnaces using a forced draught, and that at 1200–1250 °C the viscosity of the slag produced in both furnace types was similarly low. Therefore the morphological differences between the slag from non-tapping and tapping furnaces must be due predominantly to changes in the way the slag collected and was removed.

## Microstructure of Iron Age and Romano-British bloomery slag

Slag composition influences the phases formed on cooling. Olivine group phases (generally fayalite), wüstite and a glassy matrix were observed in most samples, but there were variations in the phases present in slag from different regional groups because of slight regional differences in slag composition (see Paynter 2006). In contrast, samples with similar compositions, from a single assemblage or from the same regional group, tended to contain the same phases regardless of the slag morphology.

Many of the samples from non-tapping assemblages, such as the slag cakes and fragments of smelting slag, had coarse microstructures consistent with slow cooling (Fig. 6). Porous and poorly consolidated areas of slag were particularly heterogeneous and contained more numerous charcoal fragments, often surrounded by an irregular perimeter of small iron-rich crystals from reduction of the slag (Fig. 7) (Tholander 1987). Some particles of partially reduced ore, also surrounded by iron-rich crystals, and small metallic iron particles were present in these unconsolidated regions but for the most part the slag and iron had separated cleanly. Samples of flowed slag

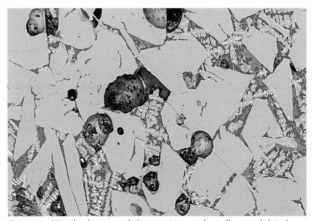

**Figure 6** SEM backscattered electron image of a well-consolidated region of slag cake from Longham Lakes (original magnification 280×, horizontal field of view = 0.95 mm). The microstructure includes large crystals of fayalite (mid-grey), wüstite dendrites (white, branching crystals) sheathed by fayalite (mid-grey layer), leucite (dark grey) and pyroxenes (augite and titanian augite) (small mid-grey crystals) in a glassy matrix.

tended to have more homogenous microstructures, which were finer towards the surface of the samples where cooling was faster, with dendritic wüstite and olivine laths.

Most tap slag samples were characterised by fine microstructures typical of more rapid cooling, particularly towards their upper surfaces. Some of the tap slag samples from the Sussex and Kent Weald contained an exceptionally large glassy component (Fig. 8) with very few crystalline phases present as a result of their atypical composition combined with fast cooling.

## Discussion

### *The formation of different morphological slag types*

The varied but distinctive morphology of the slag from non-tapping furnaces results from the method used to remove the slag from the active part of the furnace. The slag gradually accumulated in a sunken hearth or pit beneath the furnace superstructure. In some instances there is evidence that this pit was initially filled with organic material, such as large squared pieces of oak (it has not been possible to determine whether wood or charcoal) and straw, to support the weight of the charge above. As smelting progressed, the organic material gradually burnt out as slag was produced and filled the cavity. Due to the temperature differential between the hotter, active parts of the furnace and the cavity where the slag accumulated (for example see Mikkelson 1997), much of the slag was cooling slowly as it collected, with an increasing viscosity and surface tension. As a result, the impressions of organic matter are clearly preserved; the slag occasionally fails to wet other surfaces or to fully consolidate, and the profile of individual slag runs is often more distinct.

In contrast the slag in tapping furnaces accumulated in the hearth and was tapped in large flows that cooled and solidified rapidly once outside the furnace. For successful tapping, the furnace hearth would have to be maintained at sufficiently high temperatures to keep the slag in a fluid state. In addition to the air blast, other aspects of tapping furnace construction are likely to have contributed to this capability: the majority were small (an internal diameter of about 0.2–0.3 m), with thick insulating walls and a shallow hearth (Tylecote 1990). (Larger diameter tapping furnaces are also known, which had multiple air inlets to achieve the same effect (Crew 1998a)).

### *Characteristics of non-tapping furnaces*

It is unlikely that any of the slag in this study was produced by bowl-type furnaces. Many of the slag cakes recovered are large and intact, representing the majority of the waste from a smelt. Some kind of furnace superstructure would have been necessary to contain sufficient ore and charcoal to form a metal bloom plus slag cakes of this size. Additional evidence for a furnace superstructure is the relatively clean separation of the slag from the metal and the way the slag has accumulated beneath the bloom. This has only been recreated using furnaces with a superstructure, as discussed by Cleere (1972),

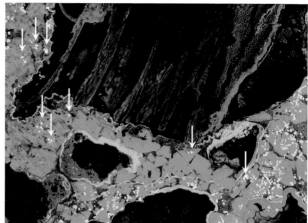

**Figure 7** SEM backscattered electron image of a poorly consolidated region of slag cake from Chelm's Combe (original magnification 170×, horizontal field of view = 1.57 mm) (voids are black). The phases present include wüstite dendrites (light grey), olivine group crystals (mid-grey fayalite bordered by darker grey kirschsteinite), and finer crystals throughout the surrounding glass matrix. The top half of the image is dominated by a piece of charcoal (grey with cell structure) with a perimeter of iron-rich crystals (light grey to white). Small particles of metallic iron (white) are arrowed.

**Figure 8** Black glassy tapped slag with a characteristic conchoidal fracture from Clappers Wood, a site in the Sussex Weald where siderite ore was smelted.

Clough (1985) and Tylecote (1990). Fragments of furnace superstructure are also found among some slag assemblages.

Although often only the basal parts of non-tapping furnaces survive, different forms are nonetheless apparent (Crew 1998b; Jackson and Ambrose 1978; Tylecote 1970). In the example thought to be the earliest in this study (Longham Lakes), the hearth was shallower and the slag remained hot for long enough to become well consolidated and compact before slowly cooling. In other cases (for example Leda Cottages) the slag flowed gradually into a deeper and laterally extended pit, further removed from the heat, particularly at the extremities. The profile of the slag cakes indicates that one side of this pit was steeply inclined to channel the slag in one direction (see Fig. 3). In most cases, the only surviving furnace remains are pits with some evidence of heating. These are often quite large (for example over 1 m diameter at Leda Cottages), although the internal diameters of the furnaces themselves were probably much smaller, as indicated by the diameter of the slag cakes. Some furnaces appear to have been used only once while others have repeatedly been repaired and reconstructed.

## *Quantities of iron produced by different furnace types*

The composition of the slag shows that the smelting conditions, including temperature and redox conditions, used in non-tapping and tapping furnaces were broadly similar. Therefore the different furnace types would probably be able to extract a similar proportion of iron from a given amount of the same ore. The amount of ore smelted in each batch, and thus the size of iron bloom produced, may however have differed.

Some literature data are available on the size of blooms and billets over this period but the degree of refining and completeness is often unclear and quoted weights are likely to have been affected by corrosion. It has been proposed that currency bars, many of which date to the 2nd and 1st century BC and weigh from 0.15 kg to 2 kg, may be the product of a single smelt, reduced in mass by subsequent smithing (Crew 1994). These weights agree well with an estimate of bloom size extrapolated from the weight of the slag cakes described previously in this paper. This estimate assumes losses at each stage of iron production in the order of 40 wt% (of the available iron metal) from ore to bloom, 50 wt% from bloom to billet, and 30 wt% from billet to bar, all based on experimental work (for example Crew 1991; Tylecote *et al.* 1971). Using these figures and compositional data for slag and ore from a number of non-tapping sites, a smelt resulting in a 13 kg slag cake would produce enough iron for a 1.8 kg billet or a 1.3 kg bar.

However, some slag cakes are much larger than 13 kg (Clogg 1999) indicating that iron workers probably were able to produce bigger blooms with non-tapping furnaces, although it appears that they often chose not to. The capacity of a non-tapping furnace would be limited by the amount of slag produced per smelt as the slag would eventually obstruct the operation of the furnace. Non-tapping furnaces could be made to accommodate more slag in order to form a bigger bloom, but it would also require more effort to remove the resulting slag mass, and more raw materials and time to repair the furnace in order to reuse it or to rebuild it elsewhere. Perhaps a compromise was reached between the capacity of the furnace and the ease of reusing it.

Refined billets of Roman date and lumps of iron used in beams and anvils, are mostly in excess of 1 kg, with bigger examples in the region of 5–12 kg (Dungworth 2001; Pleiner 2000; Starley 2002; Tylecote 1990). This suggests that larger amounts of iron tended to be produced per smelt in the Romano-British period when slag-tapping furnaces domi-nated. Tapping furnaces facilitated the production of larger blooms, since the slag was cleared at intervals during smelt-ing so that the bloom could continue to grow. Consequently the furnaces could be reused more easily, even when large blooms were produced. Tapping furnaces are also thought to be better for smelting less rich ores (Salter 1989), such as the Wealden siderite, as more slag was probably produced per unit of iron, which would have an adverse impact on the capacity of a non-tapping furnace.

The incentive to produce a large amount of iron per smelt must also be considered. It is evident from material culture and the scale of the remains of the industry that the demand for, and uses of, iron changed greatly over the period considered here (Salter 1989). Although iron production at some sites using non-tapping furnaces was on quite a large scale, this is dwarfed by the vast scale of production at a large number of Romano-British sites using tapping furnaces. In addition, in the Romano-British period, an increasing number of uses, such as architectural beams, required large masses of iron.

## Conclusions

This preliminary investigation of bloomery smelting in England in the Iron Age and Romano-British period has found elements of continuity as well as innovation. There are obvious differences in the morphology of slag waste from non-tapping bloomery furnaces as opposed to tapping ones, due to the way the slag was removed. Other aspects of the technology are broadly comparable, however, such as the smelting conditions and, in many cases, the types of raw materials used.

Non-tapping furnaces dominated in the earlier part of the time range considered here, but there were a number of changes by the Late Iron Age. Significant developments include the earliest indications for smelting siderite ore (notably in the Sussex and Kent Weald region) and the use of tapping furnaces. Tapping furnaces were then widely adopted in the Romano-British period. As demand for iron grew to previously unknown levels, these furnaces may have been preferred because they combined ease of reuse with a large capacity, regardless of the grade of ore smelted. In many areas, however, iron workers relied on the same raw materials that had been used previously.

The archaeological remains of different forms of non-tapping furnace have been found and variations are evident from the slag examined in this study: some furnaces were used once and others repeatedly; the dimensions and shape of the basal remains differ; and different types of organic matter, including large pieces of wood/charcoal and straw, were used to fill the cavity where slag collected. Accurately dated and identified evidence from many more sites is needed to draw meaningful regional and chronological comparisons.

## Acknowledgements

The author is very grateful to the conference organisers and to Leigh Allen and Paul Booth (Oxford Archaeology), Oxford Wessex Archaeology, Southern Archaeological Services, David Dungworth, Justine Bayley and Gill Campbell (English Heritage), Lynne Keys, Thilo Rehren (UCL) and Ian Freestone (Cardiff University) for slag samples, advice and improvements to this paper.

## Note

1. Gill Campbell, pers. comm.

# References

Allen, J.R.L. 1988. Chemical compositional patterns in Romano-British bloomery slags from the wetlands of the Severn valley. *Historical Metallurgy* 22(2): 81–6.

Blakelock, E. 2005. *The Metalworking Remains from 14–20 The Butts, Worcester*. Centre for Archaeology Report 18/2005. London: English Heritage.

Bottinga, Y. and Weill, D.F. 1972. The viscosity of magmatic silicate liquids: a model for calculation. *American Journal of Science* 272: 438–75.

Chirikure, S. and Paynter, S. 2002. *A Metallurgical Investigation of Metalworking Remains from Snettisham, Norfolk*. Centre for Archaeology Report 50/2002. London: English Heritage.

Cleere, H. 1971. Iron making in a Roman furnace. *Britannia* 2: 203–17.

Cleere, H. 1972. The classification of early iron-smelting furnaces. *Antiquaries Journal* 52: 8–23.

Cleere, H. and Crossley, D. 1985. *The Iron Industry of the Weald*. Leicester: Leicester University Press.

Clogg, P. 1999. The Welham Bridge slag. In *Rural Settlement and Industry: Studies in the Iron Age and Roman Archaeology of Lowland East Yorkshire*, P. Halkon and M. Millet (eds). Yorkshire Archaeological Report No. 4. Leeds: Yorkshire Archaeological Society, 81–95.

Clough, R.E. 1985. The iron industry in the Iron Age and Romano-British period. In Craddock and Hughes 1985, 179–87.

Craddock, P.T. and Hughes, M.J. (eds) 1985. *Furnaces and Smelting Technology in Antiquity*. British Museum Occasional Paper 48. London.

Craddock, P.T., Freestone, I.C., Gale N.H., Meeks, N.D., Rothenberg, B. and Tite, M.S. 1985. The investigation of a small heap of silver smelting debris from Rio Tinto, Huelva, Spain. In Craddock and Hughes 1985, 199–214.

Crew, P. 1991. The experimental production of prehistoric bar iron. *Historical Metallurgy* 25(1): 21–36.

Crew, P. 1994. Currency bars in Great-Britain: typology and function. In *La Sidérurgie Ancienne de L'Est de la France dans son Contexte Européen, Colloque de Besançon*, M. Mangin (ed.). Annales Littéraires de l'Université de Besançon 536. Série Archéologie 40. Paris, 345–51.

Crew, P. 1998a. Laxton revisited: a first report on the 1998 excavations. *Historical Metallurgy* 32(2): 49–53.

Crew, P. 1998b. Excavations at Crawcwellt West, Merioneth 1990–1998: a late prehistoric upland iron-working settlement. *Archaeology in Wales* 38: 22–35.

Dungworth, D. 2001. *Metalworking Debris from Elms Farm, Heybridge, Essex*. Centre for Archaeology Report 69/2001. London: English Heritage.

Dungworth D. forthcoming. *The Metalworking Waste from Heckfield, Surrey*. Research Department Report. London: English Heritage.

Fells, S. 1983. *The Structure and Constitution of Archaeological Ferrous Process Slags*. Unpublished PhD thesis, Aston University, Birmingham.

Freestone, I.C. 1982. Applications and potential of electron probe micro-analysis in technological and provenance investigations of ancient ceramics. *Archaeometry* 24(2): 99–116.

Freestone, I.C. 1988. Melting points and viscosities of ancient slags: a contribution to the discussion. *Historical Metallurgy* 22(1): 49–51.

Fulford, M.G. and Allen, J.R.L. 1992. Iron-making at the Chesters villa, Woolaston, Gloucestershire: survey and excavation, 1987–91. *Britannia* 23: 159–216.

Hanworth, R. and Tomalin, D.J. 1977. *Brooklands, Weybridge: The Excavation of an Iron Age and Medieval Site 1964–5 and 1970–1*. Research Volume of the Surrey Archaeological Society No.4. Guildford: Surrey Archaeological Society.

Jackson, D.A. and Ambrose, T.M. 1978. Excavations at Wakerley, Northants, 1972–75. *Britannia* 9: 115–242.

McDonnell, J.G. 1988. *The Ironworking Slags from North Cave, North Humberside*. Ancient Monuments Laboratory Report 91/1988. London: English Heritage.

McDonnell, G. and Swiss, A. 2004. Ironworking residues. In *Excavations at Deansway, Worcester 1988–1989. Romano-British Small Town to Late Medieval City*, C.H. Dalwood and R. Edwards (eds). CBA Research Report 139. York: CBA, 148–57.

Mikkelsen, P.H. 1997. Straw in slag-pit furnaces. In *Early Iron Production: Archaeology, Technology and Experiments*, L.Chr. Nørbach (ed.). Technical Report No. 3. Lejre: Historical-Archaeological Experimental Centre, 63–6.

Morton, G.R. and Wingrove, J. 1969. Constitution of bloomery slags. Part 1: Roman. *Journal of the Iron and Steel Institute* December: 1556–64.

Nathan, H.D. and Van Kirk, C.K. 1978. A model of magmatic crystallization. *Journal of Petrology* 19(1): 66–94.

Oxford Wessex Archaeology Joint Venture. 2004. *The Late Iron Age and Roman Settlement at Leda Cottages, Westwell, Kent*. CTRL Integrated Site Report Series. Oxford: Oxford Archaeology.

Paynter, S. 2002. *Metalworking Waste from Canterbury Road, Hawkinge, Kent*. Centre for Archaeology Report 34/2002. London: English Heritage.

Paynter, S. 2006. Regional variations in bloomery smelting slag of the Iron Age and Romano-British periods. *Archaeometry* 48(2): 271–92.

Pleiner, R. 2000. *Iron in Archaeology: The European Bloomery Smelters*. Praha: Archeologický Ústav Avčr.

Salter, C. 1989. The scientific investigation of the iron industry in Iron Age Britain. In *Scientific Analysis in Archaeology*, J. Henderson, (ed.). Oxford: Oxbow Press, 250–73.

Schairer, J.F. 1942. The system CaO-FeO-Al$_2$O$_3$-SiO$_2$. 1: Results of quenching experiments on five joins. *Journal of the American Ceramic Society* 25(10): 241–73.

Schrüfer-Kolb, I. 2004. *Roman Iron Production in Britain: Technological and Socio-economic Landscape Development along the Jurassic Ridge*. BAR British Series 380. Oxford: Archaeopress.

Starley, D. 1998. *Analysis of Metalworking Debris from Thorpe Lea Nurseries, near Egham, Surrey 1990–1994*. Ancient Monuments Laboratory Report 1/98. London: English Heritage.

Starley, D. 2002. Metallurgical study of an iron beam from Catterick bypass (site 433). In *Roman Catterick and its Hinterland: Excavations and Research, 1958–1997, Part II*, P.R. Wilson (ed.). CBA Research Report 129. London: English Heritage, 166–72.

Tholander, E. 1987. *Experimental Studies on Early Iron-making*. Unpublished dissertation, Royal Institute of Technology, Stockholm.

Thomas, G.R. and Young, T.P. 1999. The determination of bloomery furnace mass balance and efficiency. In *Geoarchaeology: Exploration, Environments, Resources*, A.M. Pollard (ed.). Geological Society Special Publications 165. London: Geological Society, 155–64.

Tylecote, R.F. 1962. Roman shaft furnaces in Norfolk. *Journal of the Iron and Steel Institute* 200: 19–23.

Tylecote, R.F. 1970. A pre-Roman iron-smelting site at Levisham, North Yorkshire. *Historical Metallurgy* 4(2): 79.

Tylecote, R.F. 1990. *The Prehistory of Metallurgy in the British Isles*. London: Institute of Metals.

Tylecote, R.F., Austin, J.N. and Wraith, A.E. 1971. The mechanism of the bloomery process in shaft furnaces. *Journal of the Iron and Steel Institute* 209: 342–63.

## Author's address

Sarah Paynter, Fort Cumberland, Fort Cumberland Road, Eastney, Portsmouth, PO4 9LD, UK (sarah.paynter@english-heritage.org.uk)

# Decisions set in slag: the human factor in African iron smelting

*Thilo Rehren, Michael Charlton, Shadreck Chirikure, Jane Humphris,*
*Akin Ige and Harald Alexander Veldhuijzen*

*ABSTRACT*   Slags are the most abundant and best-preserved product of traditional iron smelting and are thus a staple of archaeometallurgical research in this area. A wealth of technical information has been gleaned from these studies, identifying the bloomery process as the universal method of pre-industrial iron production across the Old World. Despite covering such a vast expanse of land and spanning more than two millennia, there is little fundamental variability in the resulting products – bloomery iron and fayalitic slag.

This is at odds with the numerous ethnohistorical studies of traditional iron smelting, particularly in Africa, that have documented a bewildering array of practices, both social and technical. This is most spectacularly obvious from the range of furnace designs, from a mere hole in the ground to elaborately decorated and substantial structures kept in semi-permanent use. The social status of smelters within their societies, or associated ritualised practices, are other examples of wide-ranging diversity in iron smelting. Such diversity is not restricted to Africa, but is matched by a similarly wide range of archaeologically documented furnace designs across prehistoric western Asia and Europe.

This paper attempts to chart some common ground among the extremes of technical engineering studies, ethnographic documentation and sociopolitical contextualisation in African iron smelting. First by exploring the inherent factors providing the envelope or constraints of technical possibilities, and then identifying degrees of freedom within this envelope which offer room for, or rather require, human decision-taking. Some of these decisions have left their traces in the slag, and teasing out these variables may eventually offer insights into social and cultural practices through technical studies.

*Keywords:* bloomery smelting, Africa, ternary diagram, slag composition, iron.

## Introduction

The study of traditional iron-smelting practices stems from several important roots. The two most predominant are the earth science and engineering approach – investigating the chemical and mineralogical compositions of iron, slag and other physical residues (Bachmann 1982; Miller and Killick 2004; Morton and Wingrove 1969, 1972; Tylecote 1962), and the ethnohistorical and anthropological approach – examining the cultural meaning and relationships between iron-makers, their products, behaviour and the whole of their environment (Haaland 2004; Herbert 1993; Schmidt 1978, 1997). For a number of reasons, most of the former research is centred on European archaeological evidence, while the latter has its stronghold in 20th-century African contexts, sometimes relying on re-enactments and oral histories as main sources. This separation suggests a dichotomy which is purely artificial and not inherent in the object of study. Unfortunately, the data obtained through these two approaches are often fundamentally incompatible although they clearly concern the very same activities. Even where they overlap, most visibly in the determination of iron yields, they are typically deduced through comparison of slag and ore composition on the one hand, and through direct observation on the other (Childs and Schmidt 1985; Schmidt 1997). Direct observation in most cases reports the quantities of ore and charcoal used, and iron metal produced; reconstructed yield data are typically based on assumed ore qualities and slag quantities as the only archaeologically measurable parameters, the two parameters least often reported in experimental or documented smelts. Both approaches are very much complementary, and allow very different aspects of the same operation to be documented, interpreted and possibly even understood. A survey of the literature, however, reveals an astonishing level of an 'either/or' approach, as if they were mutually exclusive (see Miller and van der Merwe 1994 for detailed surveys). Only more recently, approaches which view technologies as simultaneously material and sociocultural have somewhat reoriented technological studies of African iron smelting (Childs 1994; Killick 2004a).

It is not the intention of this paper to offer a complete and balanced view of the literature (see Miller and Maggs 1998), and we are consciously ignoring for the moment several very valuable contributions in this field. Most of the technical literature, however, is wholly descriptive or restricted in its interpretative ambition to provide technical explanations of specific features of the slag (estimated furnace temperature, phase content or slag morphology). Variability in process execution or raw material supply, both within short and long timescales, are only rarely investigated and reasons for it typically not explored (see Cline 1937). This latter is true for much of the ethnographic literature as well, often taking the one or two recorded smelts as representative of unchanged, or even unchangeable, past practice. Thus, the effects of cultural change and adaptation to varying environmental and economical constraints, and the potential for evolution are frequently overlooked or denied altogether, further contributing to a rather static image of early iron-smelting practice (Chirikure 2005; Chirikure and Rehren 2006). At the same time, both ethnographic and archaeological evidence offers a bewildering array of differing practices and furnace designs. These cannot possibly be the result of a multitude of independent inventions, but have to be seen as direct and strong evidence for large scale and significant evolution and development. Indeed, diffusion of iron technology from just a few points of discovery or entry into Africa has been high on the agenda for much of the last century (e.g. Kense 1983; van der Merwe and Avery 1982: 150–52; and see Killick 2004b; Quenchon 2002). Surprisingly, this seems not to have resulted in much consideration of the likely adaptation of processes and practices to the changing ecological, economic and political conditions under which they were performed. The slags, in stark contrast, are often found to be frustratingly similar to each other, and seemingly offer little representation of the variability in furnace designs (Miller and Killick 2004).

Having thus argued the case for widespread, even inevitable and necessary, change in African iron-smelting practice over space and time it is now appropriate to look for traces of such evolution in the archaeological record. Even if the archaeological record seems less informative regarding sociocultural patterns than ethnohistorical studies, it offers a much

deeper time-resolved view (Chirikure 2005). Studies of change and development clearly depend on a strong timeline, and it is our aim to suggest a few approaches for teasing out these historically accumulated configurations from the slags of past iron smelting. These methods allow us to discuss evolution as an observable phenomenon as well as something which has human causes (Charlton 2007).

## Slag homogeneity and past decisions

A screening of iron slag studies reveals a strong, systematic and highly repetitive pattern in slag composition (Bachmann 1982; Buchwald 2005; Killick and Gordon 1988; Miller *et al.* 2001; Morton and Wingrove 1969; Tylecote 1975). Most smelting slags plot around the fayalitic region of the FeO-$SiO_2$-$Al_2O_3$ phase diagram, typically drawn out parallel to the FeO-$SiO_2$ side linking two areas of particularly low melting temperature (Fig. 1). These temperature minima are here labelled Optimum 1 and Optimum 2 to emphasise that these areas are not merely characterised by the neutral aspect of their temperature, but that they are also areas which maximise slag fluidity relative to energy inputs. They are optimal engineering solutions for bloomery smelting and may reflect iron-making behaviour within specific socioeconomic contexts. We feel that the term 'optimum' carries a stronger connotation of subjectively better behaviour, relevant to the ancient smelter, than the term 'minimum'. Optimum 1 lies left of the centre of the diagram, at about 50 wt% FeO, 40 wt% $SiO_2$ and 10 wt% $Al_2O_3$; Optimum 2 is situated nearer towards the iron-

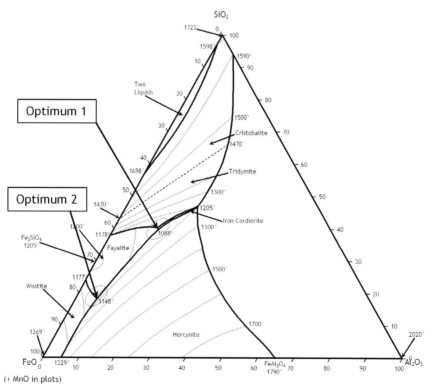

**Figure 1** Ternary diagram presenting liquidus temperatures for the system FeO-$SiO_2$-$Al_2O_3$. The two areas most suitable for bloomery smelting are marked Optimum 1 and Optimum 2. They combine low-temperature melting with low-viscosity compositions.

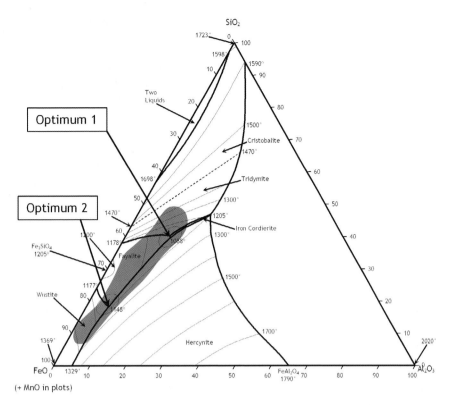

**Figure 2a** Distribution of pre-modern smithing and smelting slag compositions in the system $FeO$-$SiO_2$-$Al_2O_3$. Data reduced by combining suitable oxides (e.g. $FeO$ and $MnO$) and omitting minor compounds (e.g. alkali oxides) to fit the ternary diagram. Smelting and smithing slag compositions overlap near the fayalite composition; smelting slags predominate around Optimum 1, smithing slags scatter around Optimum 2 and towards $FeO$-rich compositions (after Kronz 1998: fig. 6.1).

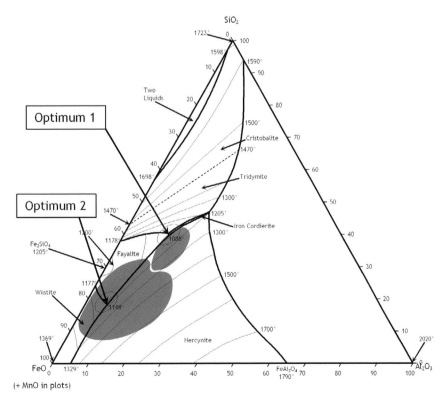

**Figure 2b** Distribution of early medieval smelting slag compositions in the system $FeO$-$SiO_2$-$Al_2O_3$. Data reduced by combining suitable oxides (e.g. $FeO$ and $MnO$) and omitting minor compounds (e.g. alkali oxides) to fit the ternary diagram. Two different slag types cluster around the two optima and overlap near the fayalite composition (after Yalçin and Hauptmann 1995: fig. 11).

213

rich corner of the system, at about 75 wt% FeO, 20 wt% $SiO_2$ and 5 wt% $Al_2O_3$. It is between these two optima that most fayalitic iron slags form, and it appears almost as if the phase diagram dictates the composition that a slag can take at the given temperature frame of 1100–1250 °C. Indeed, it has been argued elsewhere in a different context that under certain conditions the behaviour and composition of a molten charge is closely governed, almost dictated, by the shape of the liquidus surface of the relevant diagram (Rehren 2000a, 2000b). These conditions – namely the availability of one or two solid phases which co-exist with the melt and can contribute more or less to the melt formation as the temperature changes – clearly took place in most prehistoric iron-smelting furnaces. One such solid material is the technical ceramic from which the furnace and any tuyeres are being built. It acts as a reservoir or 'buffer' from which the melt can draw additional material to develop its own composition in line with the particularities of the appropriate phase diagram. Another such buffer is the ore or any free iron oxide left in the slag. Increased ceramic absorption pulls the slag composition towards Optimum 1, excess free iron oxide towards Optimum 2. As a result, the typical slags plot along the line between technical ceramic and ore compositions (Fig. 2a,b). But is this really an unfettered tyranny of the phase diagram, leaving no choice and decision for the human actors? We don't believe so.

First of all, the melting temperature of Optimum 1 is not significantly different from Optimum 2; the system would not favour one over the other when a liquid slag is forming at around 1200 °C. Rather, reaching one or the other of these optima reflects the configurational parameters of the furnace and charge. These in turn are a direct consequence of human decisions regarding furnace design, raw material selection and charge recipe. The typical scenario for slag formation in a bloomery furnace is that an iron-rich charge (the ore) is heated under reducing conditions so that a part of the iron oxide is reduced to iron metal, while another part remains as iron oxide in the system, fluxing the more refractory gangue components silica and alumina. Under less strongly reducing conditions, relatively more iron oxide remains in the system, and the slag approaches Optimum 2, the iron-rich eutectic. There are two alternative ways in which Optimum 1, the iron-poor eutectic, can be reached. One is by operating the furnace under more strongly reducing conditions, removing more iron from the system as metal. The other is to add more alumina and silica to the charge, typically as melting furnace wall material. This can be achieved by using tuyeres protruding into the furnace, designed to melt away as the smelt progresses, to facilitate sufficient slag formation (David *et al.* 1989; Veldhuijzen 2005b; Veldhuijzen and Rehren 2006). The deliberate addition of sand as a flux to the furnace charge enabled the BaPhalaborwa people to reduce high-grade magnetite ores with relative ease (Miller *et al.* 2001). Thus, slag with a composition near Optimum 1 can either indicate a higher yield through more reducing conditions, or a larger contribution to the slag formation from the technical ceramic or intentionally added silica, depending on cultural context.

It is reasonable to assume that the relevant actions were based on human decisions rather than random acts of behaviour. There are different risks and costs associated with these decisions. The former, approaching Optimum 1 through increased removal of iron as metal, would typically include more strongly reducing conditions and hence the production of a carbon-rich steely bloom, carrying the risk of over-carburising and producing cast iron instead. Such a situation is not difficult to achieve as many recent experiments make clear (Crew 2004). Also, it requires much more fuel consumption, thus adding to initial costs. The latter, approaching a low-iron slag through increased sand or ceramic contribution, could reflect the need to add more siliceous flux in order to facilitate smelting of a calcareous (for example Veldhuijzen 2005a, 2005b; Veldhuijzen and Rehren 2006 and in this volume, pp. 189–201) or very rich ore (for example David *et al.* 1989; Miller *et al.* 2001). The costs involved include adding more ballast to the slag volume, increasing both the need for fuel and reducing the amount of iron oxide available to produce metal; due to the reaction of additional iron oxide with the added material, the yield decreases significantly, as demonstrated elsewhere (Joosten 2004; Joosten *et al.* 1998; Veldhuijzen and Rehren 2006). The ores in question, however, although nominally very rich, would otherwise not be smeltable with the available technology, thereby justifying the addition of a flux and accepting a lower yield.

Thus, within the low-temperature slag-forming region of the phase diagram, there are at least three fundamentally different scenarios which result in the clustering of slag compositions in one or the other optimum. These different scenarios are direct reflections of human choices. The degree of reduction is a function of fuel-to-ore ratios and air flow through the furnace; both parameters are directly controlled first by the furnace design, and then the head smelter during the smelt. The contribution of technical ceramic to slag formation is governed by the furnace and tuyere design and the selection of clay used for their production, parameters which have probably developed in response to the choice or availability of a particular ore type and which are enacted by the smelters when building their furnaces. Similarly, the addition of quartz sand as a flux in smelting extremely rich magnetite ore must have evolved in some way. Other cases of adjusting smelting practices to the particular ores include Phoka smelters (Malawi) semi-processing low-grade ores in natural draft furnaces and re-smelting the slag in bellows-driven furnaces. The process produced slags which were virtually wüstite free but contained more silicates and required increased amounts of fuel due to the repeated smelting of the charge (Killick 1990).

A further human-controlled factor influencing the slag composition is the choice of ore or ore blend; this determines – or at least influences – the chemical system within which the slag eventually has to form. Provided that a choice exists, as is often the case (for example Charlton 2007; Crew and Charlton, this volume, pp. 219–25; Gordon and Killick 1993; Ige and Rehren 2003; Miller *et al.* 2001; Rehren 2001), this is another powerful example of human supremacy over the 'tyranny' of the phase diagram.

The main challenge is to isolate and identify these human choices as drivers of evolution within the system-driven constraints (Charlton 2007), and to push the mineralogical slag studies beyond the immediately obvious engineering parameters (Chirikure and Rehren 2006). This not only requires a broad range of analysed materials, including technical ceramics, potential ore samples and even fuel ash material

(Veldhuijzen 2005a; Veldhuijzen and Rehren 2006), but also demands making full use of complementary evidence and information. This can involve, among other things, the assessment of relative costs and availabilities of labour, fuel, ore and metal, historical accounts of special practices, experimental reconstructions and the setting of the furnace sites within the wider sociocultural landscape (Chirikure 2005; Killick 1990). As a starting point, it is still necessary to understand the relevant phase diagrams and the behaviour of the melt as it evolves from ore to slag and metal, not least to identify those driving factors and constraints which have to be taken into account when isolating the variables available for human choice. In many cases, however, it appears that the smelter puts the immanent forces of the system to good use rather than being its powerless subject.

## Current decisions affecting slag analyses

The previous section tried to sketch out a few areas of past human influence on slag formation and composition; here, we aim to look at some current human decisions which influence the part of the total initial slag assemblage we are actually analysing. 'Slag' is a rather general term, much like 'ceramic', and any given smelting site often comprises a multitude of subtypes of slag (Bayley *et al.* 1999; Craddock 1995; Veldhuijzen and Rehren, this volume, pp. 189–201). Most excavations have a clear system dealing with ceramic finds. Non-diagnostic body sherds are often discarded on site with only basic recording; diagnostic rim sherds are retained until fully documented, with some examples kept, while imports or exotic sherds are fully documented and kept in the archive (Orton *et al.* 1993). It is obvious that different types of ceramic finds have different significance, and give different types of information. The same is true for slags (Bachmann 1982); however, often no similarly well-developed system is used for recording and sampling slags for further study. Here lies a major challenge for the future of slag studies: to identify and understand the human factor in selecting slags for analysis. Depositional factors are impossible to control: humans in the past decided where to discard slags, and how. Some may have been recycled in the next smelt for further extraction of metal or to facilitate melt formation (Killick 1990). Some were discarded near the furnace site, others near the smithing hearth. Some formed dedicated slag heaps while others were mixed in with general waste. Similarly, the actions of humans intermediate between the smelters and us may have to be considered. Large slag heaps near suitable means of transport may in later periods have been removed preferentially for re-smelting using new (blast furnace) technology, or in the search for previously irrelevant metals, such as zinc in early lead slags. Slag blocks have been used in other activities such as building house foundations (Humphris 2004; Okafor 1993); some slags were used for ritual purposes etc. (Miller 2002). In contrast, small and scattered slag heaps were often considered unsuitable for reworking, and therefore survived in the archaeological record. Such human choices of the more recent past need to be considered when interpreting the nature and quantity of slag under study.

Our own sampling behaviour also has a potentially strong influence on the results (Fletcher and Locks 1996; Orton 2000). Sampling, of course, and the necessity for a rigorous sampling strategy, is not a new issue in archaeometallurgy (Bachmann 1982; Morton and Wingrove 1969), but we want to draw attention to some more recent observations in this respect. One example for future interesting work is related to the widespread pit furnaces and slag blocks preserved from some of these (Haaland 2004; Okafor 1993). Contrary to the average slag heap which contains jumbled-up waste from an unspecified and often unspecifiable number of smelts, these slag blocks should represent single-event records, or a small number of clearly sequential smelts (Okafor 1993). Thus, not only can the mass of each furnace charge be estimated with much better accuracy, but there is the potential to trace the development of the slag, and therefore the furnace conditions, over the lifetime of a single smelt (Humphris *et al.* 2006). It is reasonable to assume that the bottom of the slag heap represents the earliest slag flow from the charge; this may be richer in fuel ash components than the later material due to an initial pre-heating period which led to an initial accumulation of fuel ash in the furnace. Later, changes in air supply or drastic events such as the breaking open of the furnace to remove the hot bloom are probably reflected in changes of redox conditions in the slag. In essence, it not only matters where we sample slag blocks (top, centre or bottom), but that we are consistent in our sampling. Blocks that represent successive episodes would need different sampling strategy, and offer insight into the variability of ore compositions and consistency of furnace conditions.

We must also be aware that most tap slags (and we would argue that slag blocks are a type of tap slag) are not necessarily representative of the furnace charge overall; they represent by definition the low-melting and low-viscosity fraction of the furnace system which flows most easily. The less liquid, partly crystalline residual material, often rich in hercynitic spinel and/or other high-temperature phases, will remain in the furnace as furnace slag or attached to the bloom, often with a high degree of porosity and more or less intergrown with metallic iron bloom. Depending on the practicalities of bloom retrieval and further processing, this 'crown' or 'gromp' material may end up with the smithing waste rather than the tap slag near the smelting furnace, systematically distorting the chemical picture of the smelting operation based on slag analyses (Chirikure and Rehren 2004; Crew 1991). This is likely to have significant effects on compositional pattern including alumina, chromium, vanadium and other elements that partition strongly into spinels, as well as potash, manganese, lime and those elements that partition preferentially into the low-melting component. In addition, these porous slags are more likely to suffer from corrosion, which preferentially affects the glass phase and fayalite crystals, further concentrating those compounds of the more refractory phases. Furnace slags – here defined as only partly reacted ore with significant free iron oxide and remaining in the furnace – were possibly recycled in the next smelt and may be underrepresented in the archaeological record.

## Costs, risks and efficiency

As briefly mentioned above, decisions carry different risks and costs, which need to be balanced against the potential gains. This is not the place to discuss the extent to which these decisions were made consciously or how much is due to serendipity; such discussions can only be of actual case studies. Factors such as ease of access to fuel and ore, costs of labour and transport, and the value of metal produced will be among the more directly relevant aspects which govern the practical details of iron smelting. It is to be expected that a community with unlimited access to iron ore but in a semi-arid environment will develop an optimum solution for their smelting practice which is different from that of a group with ample fuel supply but far from good quality ore. Similarly, the choice between natural draft and forced draft furnaces, or between different types of ore, is likely to be heavily dependent on both fuel and labour availability, provided the social context and constraints of tradition allow a choice (Gordon and Killick 1993; Killick 1990). As groups and knowledge move across space and time, however, it is to be expected that traditions develop and change, adapting earlier practice to meet current needs and constraints in an evolutionary process. This may be of (slightly) variable transmission, or indeed intentional problem-solving by experimentation, observation of different practice and exchange of ideas.

Evolutionary change implies a feedback loop between changing practice and variable results. The strongest feedback loop develops in an environment of observation of actions and judgement of results. The most likely measure brought to bear in such cumulative processes is probably the perceived efficiency of an individual smelt in relation to the experienced effort put into it. Much technical literature discusses the (lack of) efficiency of the bloomery process, often with some derogatory connotation. The most frequently used measure for efficiency is the remaining iron oxide content in the slag. In brief, much of the technical literature assumes that less iron oxide in the slag equates to a more efficient process. Our decision to select this one parameter, however, is only a very crude approximation of the reality within which past groups of smelters have operated. Their measures of efficiency may have involved all sorts of parameters, from ease of raw material procurement and furnace operation to quality and quantity of metal produced to the perceived effects of the smelt on culturally related events beyond our understanding (Chirikure and Rehren 2006). Residual iron oxide content in the slag will have been the one parameter they were certainly not considering, as it is firmly outside their reality. Our decision to select this one parameter is probably more based on convenience and tradition than comprehension of the reality of past societies; it is time to make the extra effort and improve this, by identifying and discussing parameters relevant to the subjects of our study.

## Conclusions

This paper set out to identify some of the constraints which in the past have governed, or even limited, the study of iron smelting in pre-modern Africa. We noted a distinct lack of coherence and interaction between the two main branches of archaeometallurgy, namely engineering studies and ethnohistorical approaches. A common factor, however, in much of the earlier literature is an underlying assumption of constancy – or shall we say stagnation? – in early iron-smelting technology, despite glaring evidence of extreme diversity which can only be explained by a strong and sustained history of evolution, development and adaptation.

We argue furthermore that the superficially monotonous chemical compositions of many iron slags are technically necessary to obtain a suitably low-melting and fluid medium facilitating bloom formation, but still contain significant and identifiable information about a wide range of human actions and decisions. These human factors as recorded in the slags include both short-term individual decisions affecting only single smelts, and accumulated developments affecting more fundamental parameters such as furnace design and choice of raw materials. The interplay between technological constraints and experimentation, regulating traditions and taboos, and varying political and economic parameters thus results in a multifaceted record. Not every action and decision leaves its trace, and many traces remain undecipherable. However, modern archaeometallurgical research combining established and only seemingly incompatible approaches with awareness for parameters relevant to the subjects of our studies, namely the smelters operating in the past, and critical reflection on our own practices, can unravel details of past practice and contexts previously out of reach of archaeological enquiry.

Of course, none of this is restricted to African slags. European and Asian iron smelters will have operated in very similar networks of technical choices and constraints (Kronz and Keesmann 2005; Paynter 2006) with complex interacting and changing parameters (Pleiner 2000) and possibly with the same level of sophistication, skill and aptitude for change and development as their African counterparts so impressively show. The African record is much richer and therefore better suited to such a study.

## References

Bachmann, H-G. 1982. *The Identification of Slags from Archaeological Sites*. Institute of Archaeology Occasional Paper 6. London: Institute of Archaeology.

Bayley, J., Dungworth, D. and Paynter, S. 1999. *Archaeometallurgy: English Heritage Guidelines for Projects*. English Heritage: London.

Buchwald, V.F. 2005. *Iron and Steel in Ancient Times, Copenhagen*. Det Kongelige: Danske Videnskabernes Selskab.

Charlton, M. 2007. *Ironworking in Northwest Wales: An Evolutionary Analysis*. Unpublished PhD thesis, University College London.

Childs, S.T. 1994. Society, culture and technology in Africa: an introduction. In *Society, Culture and Technology in Africa*, T. Childs (ed.). Philadelphia: MASCA Research Papers in Science and Archaeology 11 (supplement), 6–12.

Childs, S.T. and Schmidt, P. 1985. Experimental iron smelting: the genesis of a hypothesis with implications for African prehistory and history. In *African Iron Working: Ancient and Traditional*, R. Haaland and P. Shinnie (eds). Oslo: Norwegian University Press, 142–64.

Chirikure, S. 2005. *Iron Production in Iron Age Zimbabwe: Stagnation or Innovation?* Unpublished PhD thesis, University College London.

Chirikure, S. and Rehren, Th. 2004. Ores, slags and furnaces: aspects of iron working in the Nyanga complex. *African Archaeological Review* 21: 135–52.

Chirikure, S. and Rehren, Th. 2006. Iron smelting in pre-colonial Zimbabwe: evidence for diachronic change from Swart village and Baranda, northern Zimbabwe. *Journal of African Archaeology* 4(1): 37–54.

Cline, W. 1937. *Mining and Metallurgy in Negro Africa.* Menasha: George Banta.

Craddock, P.T. 1995. *Early Metal Mining and Production.* Edinburgh: Edinburgh University Press.

Crew, P. 1991. The iron and copper slags at Baratti, Populonia, Italy. *Historical Metallurgy* 21(2): 109–15.

Crew, P. 2004. Cast iron from a bloomery furnace. *Historical Metallurgy Society News* 57: 1–2.

David, N., Heiman, R., Killick, D. and Wayman, M. 1989. Between bloomery and blast furnace: Mafa iron smelting technology in northern Cameroon. *African Archaeological Review* 7: 183–207.

Fletcher, M. and Locks, G. 1996. *Digging up Numbers: Elementary Statistics for Archaeologists.* Oxford: Oxford Archaeology Committee.

Gordon, R. and Killick, D. 1993. Adaptation of technology to culture and environment: bloomery iron smelting in America and Africa. *Technology and Culture* 34: 243–70.

Haaland, R. 2004. Iron smelting – a vanishing tradition: ethnographic study of this craft in south western Ethiopia. *Journal of African Archaeology* 2: 65–81.

Herbert, E. 1993. *Iron, Gender and Power: Rituals of Transformations in African Iron Working.* Bloomington, IN: Indiana University Press.

Humphris, J. 2004. *Reconstructing Forgotten Technologies in Great Lakes, East Africa.* Unpublished MA dissertation, Institute of Archaeology, University College London.

Humphris, J., Martinon-Torres, M. and Rehren, Th. 2006. How representative is your slag sample? A pilot study using iron slag from Uganda. Presentation at the BUMA VI Conference, Beijing, September 2006.

Ige, A. and Rehren, Th. 2003. Black sand and iron stone: iron smelting in Modakeke, Ife, south-western Nigeria. *Institute of Archaeo-Metallurgical Studies* 23: 15–20.

Joosten, I. 2004. *Technology of Early Historical Iron Production in the Netherlands.* Amsterdam: Institute for Geo- and Bioarchaeology, Vrije Universiteit.

Joosten, I., Jansen, J.B.H. and Kars, H. 1998. Geochemistry and the past: estimation of the output of a Germanic iron production site in the Netherlands. *Journal of Geochemical Exploration* 62: 129–37.

Kense, J. 1983. *Traditional African Iron Working.* Calgary: Calgary University Press.

Killick, D. 1990. *Technology in its Social Setting: Bloomery Iron-workings at Kasungu, Malawi, 1860–1940.* PhD thesis, Yale University.

Killick, D. 2004a. Social constructionist approaches to the study of technology. *World Archaeology 36: Debates in World Archaeology*: 571–8.

Killick, D. 2004b. Review essay. What do we know about African iron working? *Journal of African Archaeology* 2: 97–113.

Killick, D. and Gordon, R. 1988. The mechanism of iron production in the furnace. In *Proceedings of the 26th International Archaeometry Symposium*, R.M. Farquhar, R.G.V. Hancock and L.A. Pavlish (eds). Toronto: University of Toronto, 120–23.

Kronz, A. 1998. *Phasenbeziehungen und Kristallisationsmechanismen in fayalitischen Schmelzsystemen – Untersuchungen an Eisen- und Buntmetallschlacken.* Friedland: Klaus Bielefeld Verlag.

Kronz, A. and Keesmann, I. 2005. Fayalitische Schmelzsysteme – ein Beitrage zur vorneuzeitlichen Eisen- und Buntmetalltechnologie im Dietzhölztal (Lahn-Dill-Gebiet, Hessen). In *Archäometallurgische Untersuchungen zur Geschichte und Struktur der mittelalterlichen Eisengewinnung im Lahn-Dill-Gebiet (Hessen) (Das Dietzhölztal-Projekt).* Rahden: Marie Leidorf, 403–98.

Miller, D. 2002. Smelter and smith: Iron Age metal fabrication technology in southern Africa. *Journal of Archaeological Science* 29: 1083–131.

Miller, D. and Killick, D. 2004. Slag identification at southern African archaeological sites. *Journal of African Archaeology* 2: 23–49.

Miller, D. and Maggs, T. O'C. 1998. *Pre-colonial Mining and Metallurgy in Africa: A Selected Bibliography.* Rondebosch: Department of Archaeology, University of Cape Town.

Miller, D. and van der Merwe, N.J. 1994. Early metalworking in sub-Saharan Africa: a review of recent research. *Journal of African History* 35: 1–36.

Miller, D., Killick, D. and van der Merwe, N. 2001. Metalworking in the northern Loveld, South Africa, AD 1000–1890. *Journal of Field Archaeology* 28: 401–17.

Morton, G. and Wingrove, J. 1969. Constitution of bloomery slag. Part 1: Roman. *Journal of the Iron and Steel Institute* 207: 1556–64.

Morton, G. and Wingrove, J. 1972. Constitution of bloomery slag. Part 2: Medieval. *Journal of the Iron and Steel Institute* 210: 478–88.

Okafor, E.E. 1993. New evidence on early iron-smelting from south-eastern Nigeria. In *The Archaeology of Africa: Food, Metals and Towns*, T. Shaw, P. Sinclair, B. Andah and A. Okpoko (eds). London: Routledge, 432–48.

Orton, C. 2000. *Sampling in Archaeology.* Cambridge: Cambridge University Press.

Orton, C., Vince, A. and Tyers, P. 1993. *Pottery in Archaeology.* Cambridge: Cambridge University Press.

Paynter, S. 2006. Regional variation in bloomery smelting slag of the Iron Age and Romano-British periods. *Archaeometry* 48: 271–92.

Pleiner, R. 2000. *Iron in Archaeology: The European Bloomery Smelters.* Praha: Archeologický Ústav Avčr.

Quenchon, G. 2002. Les Datations de la metallurgie du fer Termit (Niger): Leur fiabilite, leur signification. In *Aux origines de la metallurgie du fer en Afrique. Un anciente meconneeu. Afrique de l'Quest et afrique Centrale*, H. Bocoum (ed.). Paris: Publications UNESCO, 105–14.

Rehren, Th. 2000a. New aspects of ancient Egyptian glassmaking. *Journal of Glass Studies* 42: 13–24.

Rehren, Th. 2000b. Rationales in Old World glass making. *Journal of Archaeological Science* 27: 1225–34.

Rehren, Th. 2001. Meroe, iron and Africa. *Mitteilungen der Sudanarchäologischen Gesellschaft* 12: 102–9.

Schmidt, P.R. 1978. *Historical Archaeology: A Structural Approach in an African Culture.* Westpoort: Greenwood Press.

Schmidt, P.R. 1997. *Iron Technology in East Africa: Symbolism, Science and Archaeology.* Oxford: James Currey.

Tylecote, R.F. 1962. *The Prehistory of Metallurgy in the British Isles.* London: Edward Arnold.

Tylecote, R.F. 1975. The origin of iron smelting in Africa. *West African Journal of Archaeology* 5: 1–9.

van der Merwe, N.J. and Avery, D.H. 1982. Pathways to steel. *American Scientist* 70: 146–55.

Veldhuijzen, H.A. 2005a. *Early Iron Production in the Levant. Smelting and Smithing at Early 1st Millennium BC Tell Hammeh, Jordan, and Tel Beth-Shemesh, Israel.* Unpublished PhD thesis, University College London.

Veldhuijzen, H.A. 2005b. Technical ceramics in early iron smelting: the role of ceramics in the early first millennium BC iron production at Tell Hammeh (Az-Zarqa), Jordan. In *Understanding People through Their Pottery; Proceedings of the 7th European Meeting on Ancient Ceramics (EMAC '03)*, I. Prudêncio, I. Dias and J.C. Waerenborgh (eds). Lisbon: Instituto Português de Arqueologia (IPA), 295–302.

Veldhuijzen, H.A. and Rehren, Th. 2006. Iron smelting slag formation at Tell Hammeh (Az-Zarqa), Jordan. In *Proceedings of the 34th International Symposium on Archaeometry, 3–7 May 2004, Zaragoza, Spain*, J. Pérez-Arantegui (ed.). Zaragoza: Institución 'Fernando el Católico', 245–50.

Yalçin, Ü. and Hauptmann, A. 1995. Archäometallurgie des Eisens auf der Schwäbischen Alb. *Forschungen und Berichte zur Vor- und Frühgeschichte in Baden-Württemberg* 55 (= Beiträge zur Eisenverhüttung auf der Schwäbischen Alb), 269–309.

## Authors' addresses

- Th. Rehren, Institute of Archaeology, 31–34 Gordon Square, London WC1H 0PY, UK (th.rehren@ucl.ac.uk)
- M. Charlton (michael.f.charlton@gmail.com)
- S. Chirikure, Department of Archaeology, University of Cape Town, Rondebosch 7701, South Africa (shadreck.chirikure@uct.ac.za)
- J. Humphris, Institute of Archaeology, 31–34 Gordon Square, London WC1H 0PY, UK (j.humphris@ucl.ac.uk)
- O.A. Ige, Natural History Museum, Obafemi Awolowo University Ile-Ife, Nigeria (Akin_ige2001@yahoo.com; oige@oauife.edu.ng)
- H.A. Veldhuijzen, Institute of Archaeology, 31–34 Gordon Square, London WC1H 0PY, UK (h.veldhuijzen@ucl.ac.uk)

# The anatomy of a furnace ... and some of its ramifications

*Peter Crew and Michael Charlton*

*ABSTRACT*   Excavation of the medieval bloomery at Llwyn Du has revealed a well-preserved furnace with good evidence for the location of the bellows. Historical records indicate that some 50 kg of raw bloom was made each week, refined to over 30 kg of bar iron. It is argued that this amount of iron and the heavy burning and vitrification of the furnace clay can only be produced by smelting with a large volume of air. A model for the fast smelting of low phosphorus bog ore is proposed, with an air rate of 1200 litres per minute, which would fit the historical, archaeological and analytical data from Llwyn Du.

*Keywords:* iron smelting, furnace, bog ore, experiment, air rate.

## Introduction

During the past 25 years three ironworking sites, of prehistoric and medieval date, have been excavated in northwest Wales. This project, based at the National Park Study Centre, has been carried out by essentially the same team of archaeologists, who have developed some expertise in the difficult business of excavating ironworking sites. In addition, a core team has assisted in carrying out a long series of experiments, over many years, specifically designed to examine the archaeological problems. Analyses of the slags and metal have been made (Salter and Crew 1977; Serneels and Crew 1977) and a major project has recently been completed that studied a large number of the archaeological and experimental slags with the aim of developing an evolutionary model for the smelting technology used at these sites (Charlton 2006).

The interplay between these approaches has been vital. The experiments have used furnaces and bog ores based as far as possible on the archaeological evidence. As we learn more from the experiments, we can excavate the sites with greater understanding, recognising features and deposits which might otherwise have been missed. This has resulted in frequent modification of the experimental furnaces. The experience from the experiments also allows us to be more perceptive in the selection of samples for analysis, in particular for the prehistoric sites, which used a non-tapping technology that makes it difficult to differentiate between smelting and refining slags. The archaeology, experiments and analyses all provide quantitative data which, when combined, can be used to develop more sophisticated production models, thus providing a firmer basis for the interpretation of the archaeology.

As no time limit was placed on this research project, it has been possible to excavate the furnaces and slag deposits with great care and to develop innovative methods both for the quantification of the slags and for magnetic surveys of the sites and ironworking features (Crew 2002). At Bryn y Castell, the excavations lasted for seven seasons, from 1979 to 1985, resulting in complete excavation of the small hillfort and the recovery of some 1.2 tonnes of slag (Crew 1987). At Crawcwellt, 14 seasons of work, from 1986 to 1999, allowed only partial excavation of this large upland settlement. Large-scale magnetic surveys ensured that all the ironworking evidence was examined, however, with the recovery of about 6 tonnes of slag, the largest amount yet known from any prehistoric site in Britain (Crew 1989, 1998a). Llwyn Du is a medieval bloomery of quite different character to the prehistoric sites, with deep deposits and complex stratigraphy. Without the experience gained on those sites it would have been far more difficult to excavate and sample Llwyn Du efficiently. Excavations from 2001 to 2005 have so far examined about half of the site.

The archaeological evidence from furnaces is nearly always incomplete, introducing an element of uncertainty in their interpretation and, of course, in any experimental reconstruction. This paper deals with only a very small part of this project, describing a particularly well-preserved furnace and examining its wider implications.

## Llwyn Du and its furnace

Llwyn Du is a complex and well-preserved site, with some 250 tonnes of smelting and refining debris, furnaces and hearths, and a complex of buildings. There are also massive amounts of charcoal and a remarkably well-preserved collection of wood, including structural material and artefacts, from a charcoal storage area (Crew and Crew 2001).

Historical documentation, from crown rentals, allows comparison with the archaeological evidence (Crew and Crew 1995; Smith 1995). Radiometric dates from charcoal, archaeomagnetic dates from the furnace and dendrochronological dates from the wood and charcoal will enable the chronology of the site to be defined with a degree of precision which is rarely possible. At present, the evidence indicates that there were two main phases of activity, from the late 1350s for about ten years, and from about 1375 for some 30 years.

The core of the site is a magnificent furnace, located by magnetic survey and first uncovered in 1997. It was particularly well preserved by virtue of having been buried in the collapsed slag heap. Excavation of this complex feature was only completed in 2005 (Fig. 1). In its latest phase, the furnace had a stone casing and stakeholes containing the clay structure. Its overall dimensions are some 2 × 1.50 m and the internal diameter is about 40 cm. The walls, therefore, are rather thick, especially on the western side. The tapping arch was originally roofed with flat stone slabs, creating a long tunnel into the furnace. This suggests that repairs were carried out from above, limiting the furnace height to about 1 m above the remains. The furnace was built into a large oval pit which had a slab-covered drainage system. Adjacent to the furnace was a tank lined with sloping stones for which no satisfactory parallels have been found. It is now thought that this may have been used for the temporary storage of hot charcoal when the furnace was emptied. The furnace lay at the southern end of a large wooden building, some 14.5 × 4.5 m, with the refining and smithing area located in the northern part of the building.

Excavation of the eastern side of the furnace showed that the stone casing had been built on a thick deposit of fine charcoal, which had accumulated during the first phase of use, with an inner line of stakeholes from the narrower wall of the first phase of the furnace. It was unclear at this stage from where the furnace was blown, but its much wider western wall had a notable gap in the stone casing. Excavation of this area revealed a triangular wedge of charcoal-rich debris, sitting on a smooth clay base, showing clearly the location of the bellows (Fig. 2). There could have been one large bellows or, our preferred interpretation, two long narrow bellows being used alternately. This arrangement, reminiscent of the blowing arch of later blast furnaces, was clearly intended to get the bellows nozzles as close as possible to the active zone of the furnace to allow maximum penetration of the draft.

Part of the blowing hole survived as a vitrified smooth-sided curve of about 10 cm in diameter (Fig. 3). This is too large in practice – experiment has shown that hot charcoal would be blown back out of a hole this size and it would certainly have been reduced by a circular block tuyere that could then have been repaired or replaced as required.

There was a clear sequence from the vitrified lining, through a 5 cm reduced grey clay zone and with 20 cm of heavily burnt red oxidised clay – a clear indication that this furnace was held at high temperatures for a considerable period of time. The heavily grogged clay lining was vitrified to a depth of 6 cm, but magnetic tests showed that this had been built up by the addition of new clay, trapping metal prills from earlier smelts. Small balls of unbaked clay have been found in the slag dumps, presumably prepared for repairing the furnace lining or the blowing hole. There were also frequent and quite large pieces of metallic debris, several examples of which have been examined:[1] they have proved to be mainly eutectoid steel surrounded by white cast iron, another indication of the high temperature and strongly reducing conditions.

## Site quantification and materials balance

Altogether there is an estimated 250 tonnes of smelting and refining slag at this site. The tap slags are of remarkably consistent character. Most of the slags are fragmentary, but there are enough nearly complete cakes to suggest that the maximum weight of any tap was about 5 kg, though of course there may have been several taps during a smelt. There are clear lenses of charcoal and slag within the heap, indicating cycles of activity – in fact some one-third of this slag heap

**Figure 1** Llwyn Du furnace: final phase with stone casing, the slab-covered drain and, in the foreground, the tank with sloping stones. Scale 1 m.

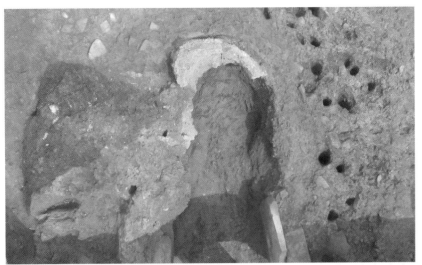

**Figure 2** Llwyn Du furnace: fully excavated, showing location of the bellows through the thick west wall and the northern arc of vitrified lining (internal diameter 40 cm). On the east are the stakeholes from the two phases of the furnace.

**Figure 3** Detail of the 10cm diameter blowing hole through the vitrified lining. The surrounding clay and stones have been removed.

consists of fine waste charcoal, indicating a need for care in volume to weight conversions.

As shown in the ternary diagram (Fig. 4), the slags are high in iron oxide (FeO) and manganese oxide (MnO) and would have had a relatively low melting point. The technical ceramics were all obtained from local clay and this provides a small but clear component of the final slag composition. It is remarkable that, from over 40 tonnes of slag actually excavated, only some 10 kg of discarded lining has been found. This indicates that the furnace was being operated in a very efficient manner, with only minor repairs being required between smelts.

The bog ores found on the site are all rich and rather variable, from hard bright orange ore with 80% or more iron oxide

FeO, to softer dark purple-grey ores, with up to 60% manganese oxide. A notable feature of these ores is that they have a low $P_2O_5$ content, generally less than 0.2%. Analysis of a large number of tap slag pieces shows significant variation in their manganese content, from a few percent to as high as 20%, and it is clear that the manganese-rich ores were systematically used, probably as a deliberate blend with the ores richer in iron oxide. This is a situation where materials balance calculations can be rather unreliable as the underlying mathematics show that the relationships are exponential. Above say 80% metal oxide, a few percent error in the ore grade estimation can make a dramatic difference to the calculated slag production and iron yield. In this case a cautious estimate, based on the

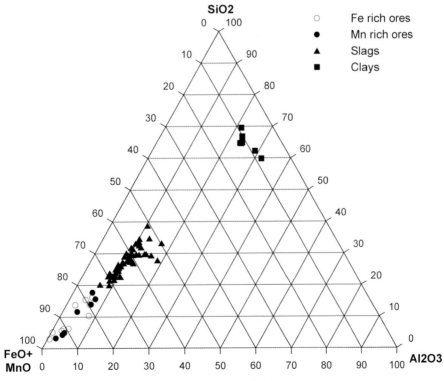

**Figure 4** Ternary diagram for the Llwyn Du ores, slags and clays.

ores and slags from all the phases, suggests a balance with a contribution of about 4% clay and 1% furnace ash, giving an iron yield of about 70%. There is, however, some variation in the balance for the various phases of the slag dump.

It is useful at this point to compare the evidence from Llwyn Du with that from the prehistoric sites at Bryn y Castell and Crawcwellt, where bog ores were being smelted but on a much smaller scale. The prehistoric furnaces are only 25 cm diameter and the clay surrounds are not so heavily burnt. In a small non-tapping furnace, a comfortable smelt of 8 kg ore gives a 2 kg bloom and 5 kg slag. The total slags at the two sites were about 1.2 and 6 tonnes respectively, indicating that some 240 and 1200 smelts would be needed, but this would have been only a relatively small number per year over the life of these settlements.

The story at Llwyn Du is quite different. The historical records indicate that some 50 kg of raw bloom was made each week, which would require six smelts with over 30 kg ore, each producing a bloom of about 8.5 kg and 16 kg of slag. To produce the 250 tonnes of slag, not an enormous amount, still requires an impressive 14,000 smelts, equivalent to 50 years of continuous smelting, six days per week, 48 weeks per year, as given in the crown rentals.

## Models for iron smelting

The quantification for the prehistoric sites is based on early experimental data (Crew 1991). These data have often been quoted, but unfortunately more often misquoted, mainly because there is so little similar data available. To stress yet again, this model *only* applies to the smelting of rich high phosphorus bog ores in low shaft furnaces, without slag tapping. It is very clear, however, that this model cannot explain the scale and intensity of smelting at sites such as Llwyn Du. Furnaces blown at 250 to 300 litres of air per minute do not produce the heavy burning and deep vitrification of the clay observed at Llwyn Du and many other archaeological sites, nor can they even begin to make the large quantities of iron required within an acceptable time.

We have spent years trying to improve our experimental smelting techniques, to make larger blooms more efficiently, but with limited success. Despite many improvements, especially to the air delivery system, we have always come across the fundamental problem that blowing too hard produces more highly carburised metal and, with high phosphorus ores, blooms with so much phosphorus that they are not smithable.

In 2002, two American blacksmiths proposed a new approach to bloomery iron smelting (Sauder and Williams 2002). They wanted to make iron on a large scale from which they could make artworks but they found, as we did, that conditions based on Tylecote's experiments (Tylecote *et al.* 1971) simply cannot produce large blooms of sufficient quality. Then they tried blowing at a very much higher air rate, initially in an attempt to make cast iron, with the result that the furnace conditions changed dramatically, with the hot zone moving to the centre of the furnace and the bloom forming freely in a bath of liquid slag. The whole process is much hotter and much faster when compared with slower smelting regimes,

and there is sufficient heat for reduction to continue in the hearth and to allow slag tapping.

A large number of smelts have been carried out, making well-consolidated blooms with an average weight of 12.5 kg. Their results have been obtained using non-reactive furnace materials, however, with a continuous flow of preheated air through a water-cooled tuyere. These so-called 'minor anomalies' make it difficult to apply this model to archaeological situations much before the 18th century. Despite this, we have to recognise the value of this research, both because it questions received opinion and because it has inspired other experimenters to rethink their approach.

The challenge is, of course, to adapt the Sauder-Williams regime to circumstances based more closely on the archaeological information from Llwyn Du, using a clay furnace and blowing hole that melt, a less sophisticated air delivery system and smelting reducible bog ores. One of our experimental furnaces at Plas Tan y Bwlch is based on that from Llwyn Du and this was modified so that two hand bellows could be used through the same 3 cm diameter blowing hole. Four experiments have been carried out in this furnace with a stroke rate for each bellows of 18–20 per minute, giving an estimated air rate of between 1000 and 1200 litres per minute.

## Experiments 92 to 95

Experiment XP92 was carried out[2] using a blend of FeO and MnO ores with a mean composition when fully roasted of 22% $SiO_2$, 6% $Al_2O_3$, 10% MnO and 60% $Fe_2O_3$, unfortunately not as rich as the archaeological ores. The smelt was quite different from our old regime with much lower blowing rates. The charcoal burned some four times faster, at a mean rate of 13 kg per hour, and 36 kg of ore was smelted in a remarkably quick time of six hours. An important factor was that we had none of the normal problems of blocking of the blowing hole: some clay melted, but the hole stayed clear and hot throughout the smelt.

The result was rather surprising, being a block of cast iron which had formed in the normal position of a bloom and was held in place by the low iron slags (Figs 5 and 6). This has all kinds of interesting implications, not least in providing an explanation for the increasing number of sites all over Europe where cast iron is being found along with bloomery slags (Crew 2004). Clearly the furnace conditions had been very hot and reducing and, as this ore is very reducible anyway, a much lower charcoal-to-ore ratio would have been preferable, perhaps as low as 1:2.

Unfortunately we did not have any bog ore of sufficient quality to test the lower ratio, so XP93 was carried out under the same conditions, but using a high-grade Cumbrian hematite. This is a very hard and poorly reducible ore, so has to be crushed to a small size. Attempts to smelt this ore under slow blowing conditions have only been partially successful at best. The result was a 5 kg flow of very fluid tap slag, which bubbled yellow as it cooled, and a 6 kg bloom which formed vertically in the furnace. The slag tapped too early, however, and the bottom heat was lost, which is why the bloom did not consolidate more or sink further down. This also meant that

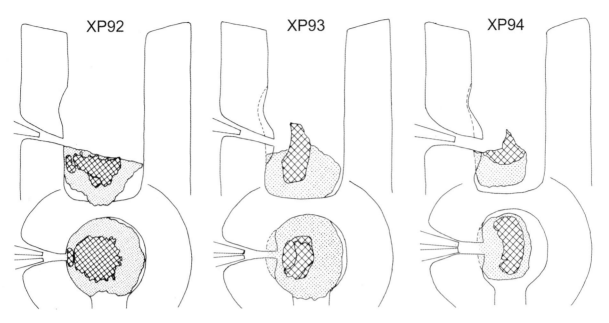

**Figure 5** XP92, 93 and 94: sketch sections and plans of the results (metallic products are shown cross-hatched, slags are stippled). Internal diameter of furnace 35 cm.

**Figure 6** Left: XP92 blowing with two bellows. Centre: XP94 hot bloom removal. Right: XP92 cast iron (scale 25 cm) and XP93 bloom section (scale 10 cm).

the hot zone was deflected onto the furnace wall causing too much clay to melt (Figs 5 and 6).

For XP94, the air system was modified to test the effect of a 5-cm diameter blowing hole and with the nozzles further in; this increased the charcoal rate from 13 to 24 kg per hour, probably due to entrained air through the larger hole. This also caused problems with turbulence and charcoal blowing out of the blowing hole. It seems that about 3 cm diameter is the maximum practical size with these bellows and in this furnace. The ore used was a rich and dense Forest of Dean goethite, crushed to a small size and smelted again at a 1:1 ratio. The 22 kg of ore was smelted

in under three hours, which was certainly too fast as the 'bloom' had a lot of dense magnetic slag and only its upper part was metallic. Again the slag tapped too early and the bloom did not fully sink or separate (Fig. 5). Only a slight shift in operating conditions could have given 7 kg bloom, however, a near optimum yield for this ore. The most important result of this experiment, from a practical point of view, was that the bloom was removed without difficulty (Fig. 6) and the extremely hot furnace repaired successfully, ready for a smelt the following day. This was XP95, an attempt to smelt bog ore dust at a 1:2 ratio, but with disappointing results.

## Discussion

Although there is still much work to do, we feel that we are now close enough to propose a realistic model for fast and repetitive smelting of rich bog ores, with a low phosphorus and high manganese content, which would fit the historical, archaeological and analytical data from Llwyn Du. This would require the smelting of 36 kg of rich bog ore at a 1:2 charcoal/ore ratio, giving a raw bloom of 8.5 kg (60% yield) and 15.5 kg slag. At an air rate of 1200 litres per minute, such a smelt would last for about 5½ hours. A materials balance would require 95% ore, 4% clay and 1% fuel ash, giving a slag with a mean content of 60% FeO plus MnO. The 4% clay is about 1.5 kg, representing only a thin veneer from the hot zone of the furnace, above the blowing hole.

The air delivery system still needs to be improved and perhaps the air rate would need to be higher to achieve a liquid slag bath and good bloom consolidation. The 15.5 kg of slag would be in the form of two taps of about 5 kg each, towards the end of the smelt, with the rest of the slag remaining in the furnace.

In order to carry out successive smelts, the bloom must be removed when hot at the end of a smelt and refined immediately. Even an 8.5 kg bloom is too large to forge and reheat easily and it would need splitting into at least three pieces, which could then be refined separately. This would give the 5.6 kg of bar iron per smelt, equivalent to six 'pieces' of iron of 2 lb each, as described in the records. Then of course, the furnace has to be repaired, ready for the next smelt. The crucial area is the blowing hole, which could be repaired either from above or possibly through the large blowing arch.

Repetitive smelting on this scale requires a continuous supply of ore and charcoal through much of the year. We estimate that a minimum of 10 or even 12 people would be fully employed to maintain the supply of raw materials and to process them. Although the 250 tons of slag at Llwyn Du is not an especially large quantity, it is clear that this would have been a very skilled and well-organised operation.

The implications of fast smelting regimes are becoming more widely recognised as different researchers try to explain their own particular set of circumstances. There are many old excavations where the use of models based on fast smelting might more easily explain their archaeology and technology. One example is the Roman site at Laxton in Northamptonshire, one of a series of sites all over Europe where these very large furnaces were used (Crew 1998b; Jackson and Tylecote 1988). The scale of these furnaces, over 1 m internal diameter, is difficult to appreciate (Fig. 7). They require some 200 kg of charcoal just to fill the lower dome of the furnace to blowing hole level. Our original model for this site, based on a slow smelting regime, required single smelts processing half a ton of ore and charcoal per day, producing some 400 kg of slag and perhaps five blooms of 20 kg each. At Laxton there is an estimated 10,000 tonnes of slag – even smelting on this impressive scale would require 25,000 smelts, or two furnaces operating more or less continuously for 50 years.

However, this model required unrealistic 24-hour smelts which, in any case, would not reproduce the intense subsoil burning, heavy vitrification and free-forming blooms. A fast smelting technique, with 16 kg of charcoal per hour being

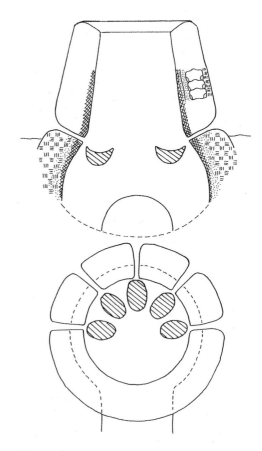

**Figure 7** Laxton furnace: interpretation section and plan with five blowing holes. The furnace is 1 m diameter at the blowing hole level.

burnt at each of the blowing holes, could process this quantity of ore and charcoal in a far more acceptable time and would also reproduce the archaeology more closely. A high degree of skill and organisation is required for this style of smelting, which would have been driven by particular economic pressures. This raises interesting questions about the development of this smelting technology and the mechanism for its transfer across wide areas of Europe.

The fast smelting model, developed by Sauder and Williams and refined here for a specific archaeological circumstance, has a lot to offer in explaining the technology used at some archaeological sites. However, *pace* Sauder and Williams (2002: 130), these techniques should not be applied to 'any form of bloomery'. There are archaeological sites with very small diameter furnaces, presumably producing small steely blooms, and those where high phosphorus ores were smelted, which are still best explained by the use of a slow smelting regime. It seems unlikely that experimental work will be carried out on a sufficient variety of furnaces using a fast smelting model. Such furnaces and smelts require a great deal of preparation and large quantities of materials and manpower, so they are neither easily nor cheaply repeated. Nor is there much latitude if a mistake is made. In the case of the Llwyn Du furnace we still have to achieve the correct conditions for a liquid slag bath and full bloom consolidation, which must await the finding of a suitably rich bog ore.

One important aspect of fast smelting with reducible ores is the lower fuel-to-ore ratio. This significantly reduces the amount of charcoal required and for Llwyn Du has led to the development of a new model for the management and use of the woodland resource.

'Et fornacium magna differentia est.' It is fitting to end with this quotation from Pliny. It is this variety of furnaces, in time and space, smelting different ores, under different technological and economic conditions, which makes the study of ironworking so interesting and rewarding. Of course, the many different furnaces require many different production models to explain how they operated, which we are now slowly beginning to understand.

## Acknowledgements

Thanks are due to many friends and colleagues who have contributed to this project in so many different ways, with the excavations, the experiments and with specialist studies.

## Note

1. By Chris Salter.
2. With help from Thilo Rehren's students from the Institute of Archaeology.

## References

Charlton, M. 2006. *Ironworking in Northwest Wales: An Evolutionary Analysis.* Unpublished PhD thesis, University College London.

Crew, P. 1987. Bryn y Castell Hillfort: a late prehistoric iron working settlement in north-west Wales. In *The Crafts of the Blacksmith. Proceedings of the Symposium of the UISPP Comité pour la Sidérurgie Ancienne, Belfast, 1984*, B.G. Scott and H. Cleere (eds). Belfast: Ulster Museum, 91–100.

Crew, P. 1989. Excavations at Crawcwellt West, Merioneth, 1986–1989: a late prehistoric upland ironworking settlement. *Archaeology in Wales* 29: 11–16.

Crew, P. 1991. The experimental production of prehistoric bar iron. *Historical Metallurgy* 25(1): 21–36.

Crew, P. 1998a. Excavations at Crawcwellt West, Merioneth, 1990–1998: a late prehistoric upland ironworking settlement. *Archaeology in Wales* 38: 22–35.

Crew, P. 1998b. Laxton revisited: a first report on the 1998 excavations. *Historical Metallurgy* 32: 49–53.

Crew, P. 2002. Mapping and dating of prehistoric and medieval ironworking sites in north-west Wales. *Archaeological Prospection* 9(3): 163–82.

Crew, P. 2004. Cast iron from a bloomery furnace. *HMS News* 57: 1–2.

Crew, P. and Crew, S. 1995. Medieval bloomeries in north-west Wales. In *The Importance of Ironmaking: Technical Innovation and Social Change. Proceedings of the Norberg Conference, May 1995*, G. Magnusson (ed.). Jernkontorets Berghistoriska Utskott H58, 43–50.

Crew, P. and Crew, S. (eds) 1997. *Early Ironworking in Europe: Archaeology and Experiment.* Abstracts of the International Conference, September 1997. Occasional Paper 3. Plas Tan y Bwlch, Maentwrog.

Crew, P. and Crew, S. 2001. Excavations at Llwyn Du, Coed y Brenin, Merioneth, 2001: woodland management and charcoal processing at a late 14th century ironworks. *Archaeology in Wales* 41: 83–7.

Jackson, D.J. and Tylecote, R.F. 1988. Two new Romano-British ironworking sites in Northamptonshire: a new type of furnace. *Britannia* 19: 275–98.

Salter, C.J. and Crew, P. 1997. High phosphorus steel from experimentally smelted bog-iron ore. In Crew and Crew 1997, 83–4.

Sauder, L. and Williams, S. 2002. A practical treatise on the smelting and smithing of bloomery iron. *Historical Metallurgy* 36: 122–31.

Serneels, V. and Crew, P. 1997. Ore-slag relationships from experimentally smelted bog-iron ore. In Crew and Crew 1997, 78–82.

Smith, K.E.S. 1995. Iron-working in north-west Wales in the late fourteenth century. *Archaeological Journal* 152: 246–90.

Tylecote, R.F., Austin, J.N. and Wraith, A.E. 1971. The mechanism of the bloomery process in shaft furnaces. *Journal of the Iron and Steel Institute* 209: 342–63.

## Authors' addresses

- Peter Crew, Plas Tan y Bwlch, Snowdonia National Park Study Centre, Maentwrog, Gwynedd LL41 3YU, UK (petercrew@deudraeth.net)
- Michael Charlton (michael.f.charlton@gmail.com)

# Early Chinese ferrous swords from the British Museum collections

## M.L. Wayman and C. Michaelson

*ABSTRACT*  Six iron and steel swords from the collections of the British Museum have been examined metallographically to obtain information on the steelmaking and sword-manufacturing techniques. Four Han dynasty swords were found to have been produced by the piling of carburised steel, yielding banded high carbon steel microstructures. Another Han sword was found to have been produced by casting a cast iron sword and then subjecting it to solid-state decarburisation in a coal- or coke-fired environment. The sixth sword, dating from medieval times, appeared to have been pattern-welded using two different low carbon steels.

*Keywords:* metallurgy, iron, steel, swords, Chinese, microstructure, metallography.

## Introduction

According to current belief, iron smelting began in China in the early 1st millennium BC, probably with the production of bloomery iron. However by at least the middle of the millennium, liquid iron was being produced in Chinese blast furnaces. Some of this iron was being converted to wrought iron by decarburisation, employing a fining or puddling process, but some was used for castings (cast iron). Thus at least for the latter part of the 1st millennium BC in China both cast iron and wrought iron were being produced and used for weapons, tools, agricultural implements, vessels, ornaments, horse trappings and a host of other purposes (for example Bronson 1999; Chen Jianli *et al.* 2003; Han Rubin 1998; Needham 1958; Rawson 1993; Rostoker and Bronson 1990; Wagner 1993, 1999, forthcoming).

Most of the early Chinese cast iron so far identified has been shown to be white cast iron – cast iron where virtually all of the carbon is in the form of hard brittle cementite ($Fe_3C$), making the material strong in compression but too brittle for many direct applications. Some of the early Chinese cast iron objects listed in Han Rubin's (1996) compilation however, are identified as decarburised cast iron and some as malleablised cast iron (see below). While a few grey cast iron (cast iron where some or all of the carbon exists as flakes of graphite) objects have been reported from early times (Bronson 1999; Wagner 1993), it appears likely that grey iron did not come into common use until well after the first appearance of white iron.

Recognising the limitations of cast iron, primarily its brittleness and near inability to be forged, the Chinese applied a number of processes to produce a range of materials that we would now call wrought iron, malleable cast iron and steel, all of which are in some respects improvements over basic cast iron. Some of these processes involved melting the cast iron while others were heat treatments applied to castings at temperatures below their melting points.

If white cast iron is annealed for many hours in a reducing atmosphere at a temperature that is high but still well below its melting temperature, the distribution of the carbon in the cast iron can change. For example, the cementite can be replaced by irregularly shaped clusters of graphite in a matrix of ferrite, pearlite or a mixture of the two depending on the annealing time and temperature as well as the subsequent cooling rate. This process, which is called 'malleablisation', produces a material that exhibits ductility and toughness that are very much improved over those of the original white cast iron. Malleablisation was a Chinese innovation, and one that came into widespread use by the 4th century BC (Wagner 1989; forthcoming: 130).

Alternatively, if the white cast iron is annealed in an oxidising atmosphere the carbon should eventually be completely removed from the iron, so that after cooling the microstructure of the casting would be fully ferritic, creating a material equivalent to a slag-free wrought iron. If this decarburisation process is not carried to completion, steels can be produced. Thus in regions where the carbon content is lowered to 0.8% carbon, a eutectoid (pearlitic) steel would result, while carbon contents below 0.8% would yield hypoeutectoid (ferrite-pearlite) steels. In reality heterogeneous microstructures would be likely, since the solid-state decarburisation relies on carbon diffusion from the centre to the surface of the object, so that a carbon gradient decreasing from the centre of the object towards the surface could normally be expected.

Both malleablisation and solid-state decarburisation are similar processing technologies, representing different stages on a spectrum of heat treatments (i.e. annealing times and temperatures) and in some cases occurring simultaneously. Thus for example it is possible, even likely, that steel formed from white cast iron by solid-state decarburisation can exhibit some graphite precipitated within its ferrite-pearlite microstructure.

As noted above, the Chinese also made use of fining/puddling processes during the 1st millennium BC. Here, by melting cast iron in the presence of excess oxygen (e.g. an air blast) the carbon is progressively oxidised and removed from the liquid iron. As the carbon content falls, the freezing temperature of the liquid rises creating a pasty semi-solid mass, with metallic iron mixed in a slag created by reaction with the furnace walls. The pasty slag-metal mass is then removed from the hearth and hammered to consolidate the mass and express as much of the trapped slag as possible. The resultant product is a solid low carbon iron containing slag inclusions, in other words a wrought iron that is comparable to the product of a bloomery furnace. Although control would be difficult, in principle it would be possible to arrest the operation before all the carbon is removed, thereby creating a steel containing slag inclusions.

Thus there were several routes that early Chinese metallurgists employed to produce steel, including solid-state (incomplete) decarburisation of cast iron and partial fining/puddling. In addition, however, steel was also produced by the carburisation of low carbon iron that had been produced either by solid-state decarburisation or by the fining or puddling of cast iron, or in a bloomery furnace. Furthermore, low carbon iron was also carburised by immersion in a bath of liquid cast iron (a process called 'infusion carburisation', Han Rubin 1998).

Still there remains a great need to improve and consolidate our understanding of the history of Chinese ferrous metallurgy and the interesting technologies employed. With this in mind an integrated programme of metallurgical characterisation has been carried out on a selection of early Chinese ferrous artefacts from the collections of the British Museum, from both the Department of Asia and the Department of Coins and Medals. The parts of this programme that involve cast irons in the form of coins, statuary and other objects are reported elsewhere (Craddock *et al.* 2003; Wayman and Wang 2003; Wayman *et al.* 2004). The present work concerns six swords made of wrought iron or steel; the attributions and lengths of these swords are shown in Table 1. While it is

**Table 1** Sword descriptions.

| Museum reg. no. | Sword | Date | Length |
|---|---|---|---|
| OA+.1137 (1050) | Sword with jade guard | Han dynasty 2nd cent. BC–2nd cent. AD | 92 cm |
| 1940.12-14.279 | Sword with bronze guard | Han dynasty 2nd cent. BC–2nd cent. AD | 95 cm |
| 1940.12-14.280 | Sabre | Han dynasty 2nd cent. BC–2nd cent. AD | 88 cm |
| 1957.10-28.1 | Sword (possibly from Sichuan) | Han dynasty 2nd cent. BC–2nd cent. AD | 102 cm |
| 1915.4-9.73 | Sword with bronze guard | Han dynasty 2nd cent. BC–2nd cent. AD | 91 cm |
| 1993.7-15.1 | Sword with bronze chape and pommel | Medieval 13th–15th cent. AD | 92 cm |

not possible to date these swords with precision, typological study leads to the conclusion that five of the six most likely date from the Han dynasty while the sixth is medieval. The factors considered in this dating include the dimensions of the swords and the extent to which the bronze and jade sword guards are corroded onto the blades.

## Experimental procedures

V-shaped samples of the order of a few cubic millimetres in size were cut from the blades of each of the swords and examined using optical and scanning electron metallography in order to characterise their microstructures. Elemental compositions (notably silicon (Si), manganese (Mn), sulphur (S) and phosphorus (P) contents), both bulk analysis of objects and microanalysis of microstructural constituents such as inclusions, were obtained by energy-dispersive X-ray analysis (EDX) in the scanning electron microscope (SEM). Analysis of standard reference materials confirmed that the lower limits of detection of these elements were approximately 0.1–0.15%.

**Figure 1** The six swords studied in this work. From the top, registration numbers: 1940.12-14.279; 1957.10-28.1; 1915.4-9.73; 1940.12-14.280; 1993.7-15.1; OA+.1137 (1050). The topmost sword is 101.6 cm long (T. Milton/British Museum).

The SEM–EDX results can be considered to have a precision and accuracy of ± 10–20% relative to the values obtained. Carbon contents were estimated from the microstructures. Details of the procedures have been previously reported (Wayman and Wang 2003).

## Results and discussion

Bronze swords are known from the 1st millennium BC, becoming common in the Eastern Zhou period (for example Wagner 1993: 191), while ferrous swords seem to have appeared in the 3rd century BC, and to have become common during the Han dynasty (2nd century BC–2nd century AD). Wagner (1993: table 4.4) lists some 35 pre-Han iron and steel swords and no doubt more have come to light since that information was compiled. These swords typically are of the order of one metre in length, significantly longer than the bronze short swords that preceded them. Double-edged swords with hilts are a common style although ring-pomelled sabres having a single cutting edge also became common in the Han period. Wagner speculates that the earliest ferrous swords were of wrought iron, and that the use of quench-hardened steel was a later development, however a number of the pre-Han and Han swords on which metallography has been performed have been found to exhibit quenched microstructures, at least at their cutting edges.

In the present work six ferrous swords, shown in Figure 1 and described in Table 1, have been studied; five of these are attributed to the Han dynasty and the other to the 13th–15th century AD, as mentioned above. One of the Han swords (1940.12-14.280) is a ring-pomelled sabre with a single cutting edge, while the others are all double-edged. In each case a wedge-shaped sample, extending in depth to about the midline of the blade, was removed from a cutting edge of the sword for metallographic characterisation. Some uncertainty results from the corroded condition of the swords, which means that the existing metal surface is not the original surface of the sword, an effect accentuated at the original cutting edge, which is typically removed by corrosion. The consequence is that it is not always possible to interpret with confidence features related to surface treatments such as carburisation, quenching of the cutting edge and surface working.

All of the six swords studied here were found to be made of steel, but a range of compositions and microstructures are present (Tables 2 and 3). Of the five Han swords, four (OA+.1137(1050), 1940.12-14.279, 1940.12-14.280 and 1957.10-28.1) are similar to each other, with high average carbon contents and with varying amounts of banding in their microstructures. These swords will be discussed first. The fifth Han sword (1915.4-9.73) and the sixth, the medieval sword (1993.7-15.1) are completely different from each other and from the other four, and they will be discussed separately later.

**Table 2** Sword compositions.

| Museum reg. no. | Object | %Si | %P | %S | %C | Silicate inclusions | Sulphide inclusions |
|---|---|---|---|---|---|---|---|
| OA+.1137 (1050) | Sword | nd | nd | nd | 0.8 | yes | no |
| 1940.12-14.279 | Sword | nd | nd | nd | 0.6–0.7 | yes | no |
| 1940.12-14.280 | Sabre | nd | nd | nd | 0.15–0.7 | yes | no |
| 1957.10-28.1 | Sword | nd | nd | nd | 0.6–0.7 | yes | no |
| 1915.4-9.73 | Sword | nd | 0.21 | nd | 0.1–1.0 | very few | yes |
| 1993.7-15.1 | Sword | nd | nd | nd | 0–0.15 | yes | no |

nd = not detected

**Table 3** Sword microstructures.

| Museum reg. no. | %C (av.) | Inclusions (abundance) | Piled? | %P | Microstructure | Grains at cutting edge | Comments |
|---|---|---|---|---|---|---|---|
| OA+.1137 (1050) | ~0.8% banded | silicates (moderate) | likely | nd | pearlite with grain-boundary cementite network | finer equiaxed | possible carburisation at cutting edge |
| 1940.12-14.279 | ~0.6% banded | silicates (moderate) | yes (As-rich bands) | nd | ferrite-pearlite | finer pearlite some bainite | possibly quench at cutting edge |
| 1940.12-14.280 (sabre) | 0.15–0.7% banded | silicates (moderate) | likely | nd | ferrite-pearlite | finer pearlite | cutting edge very hard |
| 1957.10-28.1 | 0.6-0.7% banded | silicates (moderate) | yes (As-rich bands) | nd | ferrite-pearlite | badly corroded | |
| 1915.4-9.73 | centre ~1% edge ~0.1% | sulphides (low) | no | 0.21% | ferrite-pearlite | finer, more irregular | graphite nodules in central region |
| 1993.7-15.1 | 0–0.15% banded | silicates (moderate) | yes | nd, but on area map | ferrite-pearlite | deformed (tip blunted) | folded or pattern welded |

nd = not detected

## The four Han swords with similar compositions and microstructures

All four of these blades have heterogeneous carbon distributions, averaging between approximately 0.6 and 0.8% carbon (i.e. they are all medium to high carbon steels). All have ferrite-pearlite microstructures and all are banded to a greater or lesser degree, the bands lying more or less parallel to the plane of the blade (Fig. 2). Several types of banding were noted, one type being due to variations in carbon content and hence microstructure (different pearlite/ferrite ratios), while other bands revealed by etching are due to compositional effects, specifically localised enrichment in phosphorus or arsenic. The tip areas of the samples, corresponding to however much of the blade edges remain after corrosion, generally have finer grain size with in some cases grain distortion (elongation), presumably a result of mechanical working having been carried out at those locations. In one sword (1940.12-14.279) a minor amount of bainite is present locally below the corrosion layer at the blade edge, suggesting a higher rate of cooling there, specifically a quench to harden the material. None of the other swords shows any definitive evidence of having been quenched, but rather they have the fully pearlite/ferrite microstructures associated with moderate to slow cooling, rather than the more desirable quenching treatment. The pearlite in some of the blades is moderately to heavily spheroidised, sug-

**Figure 2** Transverse section through the blade of sword 1940.12-14.280 showing banding of carbon content. Optical micrograph, nital etch (scale bar: 20 μm).

**Figure 3** Transverse section through the cutting edge of sword 1957.10-28.1 showing that the microstructural banding is parallel to the plane of the blade rather than parallel to the blade surface. Optical micrograph, nital etch (scale bar: 500 μm).

gesting that time was spent around the eutectoid temperature, an effect which is often associated with forging at a temperature in the vicinity of the eutectoid. Away from the cutting edge, all the blades exhibit recrystallised equiaxed grains.

These four Han swords have microstructures consistent with having been hot forged, and doubtless much of the working needed to fabricate the blade was carried out in this manner. In the three double-edged swords, however, the microstructural banding does not follow the surface profile; rather the bands remain parallel to the plane of the blade and run out to intersect the tapered surface as shown in Figure 3. This suggests that the final stages of producing the taper leading down to the cutting edges may have been carried out by removing metal (presumably by filing and/or grinding) rather than by forging. Grinding of a knife edge is mentioned in connection with a late 2nd-century BC knife blade (Wagner 1993: 465). In contrast, in the single-edged sabre the orientation of the banding is parallel or nearly parallel to one and possibly both surfaces, showing that at least some of the tapering was carried out by forging. It must be recalled, however, that some uncertainty exists in the exact determinations of the final treatments of the blade edges because of the loss of surface material to corrosion.

In all four swords, nonmetallic inclusions are abundant, and these too are banded parallel to the plane of the blade. Most of the nonmetallic inclusions are single-phase calcium-aluminium silicate slag particles, generally elongated in the working direction as a result of the hot forging. In addition all exhibit some spherical silica and/or iron silicate particles, possibly coming from a flux used for forge welding. Some inclusions also contain phosphorus. The inclusions appeared to be more abundant in the higher carbon microstructural bands. One of the blades (1940.12-14.279) also contains some calcium/phosphorus-rich inclusions.

Microstructural banding such as that observed here can have several possible origins. Compositional heterogeneities in the raw stock that is subjected to forging tend to be spread out in the forging directions forming bands perpendicular to the most heavily forged direction (e.g. in the plane of a forged plate). Such compositional heterogeneities are expected in solid iron that forms by agglomeration. For example bloomery iron, fined iron and puddled iron can be expected to have banded carbon distributions, and any phosphorus and manganese present will also be heterogeneous. Alternatively banding can arise from the process used to produce a layered material, such as when several different pieces of material are forged together into one (piled). If piling is carried out using different types of material, such as high and low carbon steels, or steels with different phosphorus contents, the banding is normally obvious in the microstructure, being manifest, for example, as bands of different ferrite/pearlite ratios. Additionally, banding can occur if oxidation of the iron, such as occurs during forging and forge welding, is able to concentrate impurity elements at the surfaces which then become the weld interfaces. The latter situation has been observed to give bands enriched in arsenic or nickel at weld interfaces (for example Tylecote 1990; Tylecote and Thomsen 1973). Inclusions can be heterogeneously distributed in materials, leading to banding when they are forged, and they can also be banded when different materials are piled together, for example where the different materials have markedly different inclusion types and/or abundances, typically slag inclusions remaining from the metal production process. Furthermore inclusions can be concentrated at the forge weld interfaces as a consequence of the use of a welding flux to dissolve the iron oxide that forms on heating in air prior to welding. Often sand (silica) is used as a flux leaving iron silicate inclusions or possibly unreacted silica inclusions along the weld interface.

In reality it is often difficult to distinguish between the microstructural effects of piling and the banding that is due to compositional heterogeneities. Banding from piling might be expected to be more regular and to be continuous throughout the object, with welding flux inclusions (silica or iron silicate) marking the weld interfaces. If the object was fabricated by the piling of carburised thin plates then further evidence might be available. For example the carburisation process might leave some inclusion evidence in the form of trapped calcium- and phosphorus-rich inclusions (a residue from the bone ash used as a carburisation accelerator) and slag inclusions from which iron oxide constituents have been reduced by carbon and thus eliminated from the slag microstructures (Rostoker and Dvorak 1988). The hot forging that is a necessary part of the piling operation, however, tends to work towards the homogenisation of composition gradients and hence to remove banding, or at least to make it less regular and distinctive. In some cases it is simply not possible to confidently identify a structure as being piled.

There are several different technologies that could have been used to produce the medium-to-high carbon steels of which these swords are made. Bloomery iron with particularly high carbon could have been used, although this seems less likely than other possibilities. Blast furnace iron could have been fined or puddled, with the process having been arrested before all the carbon had been removed, thereby producing steel. This would then have been suitable for forging into a blade either as a monolithic piece or by forging together with other pieces (piling). Alternatively thin steel plates could have been produced by casting blast furnace iron into thin plates of cast iron that were then decarburised in the solid state with the carburisation being arrested at the appropriate carbon content (there is archaeological evidence (Bronson 1999: 189) for decarburisation of thin cast iron plates). A number of these steel plates could then have been forge-welded together (piled) to produce the blade. Another possibility is that virtually carbon-free iron could have been produced by the bloomery process or from cast iron, either by full solid-state decarburisation or by the fining/puddling process; thin plates of this iron could have been carburised to form steel slabs that were then piled together to form the blade.

The microstructural characteristics of these four swords suggest that they were fabricated by the piling of carburised iron plates. The observed variations in carbon content must be the result of the carburisation process having been insufficient to give a constant high carbon content throughout the thickness of each of the plates, causing a banded carbon distribution when they were forge welded together. The nature and distribution of nonmetallic inclusions gives support to this suggestion and also indicates that the iron plates were made either by the fining/puddling process or by the bloomery process, since slag inclusions would not have been observed

**Figure 4** Transverse section through the cutting edge of sword 1940.12-14.279 showing bright narrow bands parallel to the plane of the blade. SEM–EDX analysis identified these bands as enrichment of arsenic. Optical micrograph, nital etch (scale bar: 500 μm).

in a material formed by solid-state decarburisation of cast iron. The absence of iron oxide constituents in the slag inclusions, and the observation of calcium- and phosphorus-rich inclusions are consistent with the carburisation process as described above. Furthermore the oxidation present at the surfaces of the plates caused in some cases enrichment in some elements, notably arsenic in swords 1940.12-14.279 and 1957.10-28.1 (Fig. 4). The phosphorus banding that was observed in sword 1940.12-14.280 most likely resulted from the piling of iron slabs with different phosphorus contents and hence different responses to carburisation. This type of piled structure is consistent with the so-called 50- and 100-refined steel observed by others (for example Han Rubin 1998; Rostoker *et al.* 1985; Wagner 1993: 282–8; forthcoming: 222–4).

### The fifth Han sword

The fifth Han sword (1915.4-9.73) has a microstructure, and hence a processing history, entirely different from those of the swords described above. This sword exhibits a ferrite-pearlite microstructure with a carbon gradient through the thickness of the blade, the central regions having an average carbon content of about 0.6% with local regions as high as about 1%. The surfaces on the other hand are predominantly ferrite, with minor amounts of grain boundary cementite and spheroidised pearlite giving an estimated carbon content of about 0.1%. The central high carbon regions also contain spheroidal graphite nodules (Fig. 5) and spherical pores, either true porosity or cavities from which graphite has been removed during metallographic sample preparation. The microstructure contains about 0.2% phosphorus and is strikingly clean of nonmetallic inclusions; furthermore those few inclusions that are present are not silicate slag particles. Rather they are predominantly submicron-sized manganese sulphide or iron-manganese sulphide particles (because their sizes are smaller than the SEM analytical volume, the surrounding matrix is contributing iron to the analysis of the inclusion compositions). This microstructure is in sharp contrast to those of the other four Han swords described above.

The observations of the carbon gradient, the absence of banding, the microstructural cleanliness, the tiny sulphide inclusions and the spheroidal graphite at the centre of the

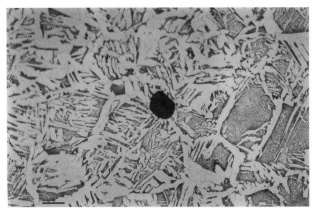

**Figure 5** Transverse section through the centre thickness of the blade of sword 1915.4-9.73 showing a carbon content of approximately 0.5–0.6% and a spheroidal graphite particle. Optical micrograph, nital etch (scale bar: 50 μm).

cross-section show that this is a solid-state decarburised structure. Thus the sword was first cast to shape as a cast iron sword (cast bronze swords are also known in early China). The casting was then subjected to solid-state decarburisation such that some of the carbon was removed. The observed microstructure would have given the sword good properties for purposes other than cutting, but the low carbon content at the surface means that it would not have held a sharp edge unless it had been surface-carburised and given a quenching treatment, and there is no evidence, bearing in mind the uncertainties due to corrosion, that this had been done. Instead the microstructure suggests that the blade edge had been slightly cold-worked to sharpen it, although this would not have been nearly as effective as a proper heat treatment.

The occurrence of the sulphide inclusions may be interpreted as evidence for the use of (sulphur-bearing) coal or coke as fuel for the kiln used for the decarburisation anneal. Coal is known to have been used as fuel in annealing and mould-baking kilns in Han times (Han Rubin 1996). The long annealing time at high temperature would have provided sufficient opportunity for sulphur in the kiln atmosphere to diffuse into the iron and react with the small amount of manganese in solution to produce the sulphide particles.

This is believed to be the first observation of a Chinese sword produced by the solid-state decarburisation or malleabilisation of a casting, and if the typological dating can be believed, it is of great significance.

### The medieval sword

The 13th–15th-century sword (1993.7-15.1) is very different from the Han swords described above. The microstructure of this sword shows it to be a low carbon steel, consisting of equiaxed ferrite grains with spheroidised pearlite on the ferrite grain boundaries, indicating a carbon content of 0.1–0.15%, and an abundance of slag inclusions. There are some regions, however, that have lower carbon and larger ferrite grain size; these regions also contain much lower inclusion density than the other regions. Additionally, although the phosphorus contents of both types of region are relatively low, area mapping of phosphorus by EDX in the SEM revealed that the higher

carbon regions have noticeably higher phosphorus content. These observations taken together suggest that this sword is a composite made from two different materials forged together. Importantly, it was observed here that these two different types of material are not aligned in regular alternating bands parallel to the surfaces of the blade, but rather their orientations are more complex, in places producing a pattern of V-shaped bands with the axis of the V parallel to the blade axis (Fig. 6).

The microstructure is fully recrystallised, and would have been relatively soft; consistent with this, the cutting edge of the blade (i.e. the tip of the sample) appears to have been blunted by cold deformation.

The microstructure of this sword reflects a complex fabrication process involving the folding and/or the forging together of several pieces of different material. A similar type of pattern was observed in a steel sword fragment dating from the 3rd century BC (Li Zhong 1975, translation Wagner 1993: 462) where it was interpreted as the result of folding and forging but here too the overall pattern could not be determined. Techniques that involve the forging together of definite arrangements of strips and twisted rods give rise to the Anglo-Saxon 'pattern-welded' swords that were produced in early Europe. In that case two different materials, frequently carbon steel and phosphoric iron in the form of bars or rods, some of which were twisted, were laid out parallel to each other and forged together to form a blade. Although predominantly associated with the 1st millennium AD, they are also present in 1st millennium BC contexts (Pleiner 1993). The occurrence of a pattern-welded Chinese sword dating to the 13th–15th century AD is not surprising in view of the contacts across Eurasia at that time, and indeed pattern-welded Chinese swords from this period have been previously reported (e.g. Needham 1958, also He Tangkun *et al.* as quoted by Wagner forthcoming: 209 and fig. 610). Furthermore, there have been suggestions that the co-fusion (or infusion carburisation) process could have produced steels with a patterned structure (Wagner forthcoming: 208).

Thus the sword investigated here was made by forging a low carbon phosphoric steel containing an abundance of slag inclusions together with a cleaner, lower carbon, low phosphorus material. The forging pattern is complex and involved folding and/or twisting of the different materials together.

**Figure 6** Transverse section through the thickness of sword 1993.7-15.1, showing the complex pattern of carbon and inclusion distribution. Optical micrograph, nital etch (scale bar: 500 μm).

Although this would not have been a high-quality sword from the perspective of service performance, its complex microstructure would have given it an attractive surface appearance, given proper surface treatment.

## Conclusions

The swords examined here illustrate several of the techniques employed by early Chinese ferrous metallurgists to produce iron and steel, as well as techniques for producing swords. Four of the Han dynasty swords were made of ferrous material produced either by the bloomery process or by the fining of cast iron, and then carburised before fabricating into the final shape by the piling of the carburised iron. The medieval sword had been fabricated by piling and/or folding and forge-welding together two different low carbon steels, possibly in the form of the twisted bars typically used to forge pattern-welded objects. The material in the fifth Han sword had been cast into the final desired shape and then subjected to a solid-state decarburisation process in a coal- or coke-fired kiln. This latter seems to be a uniquely Chinese technology for the production of ferrous swords.

Corrosion made it impossible to ascertain whether or not quenching techniques were used to harden the cutting edges of the swords, although in one sword this did indeed seem to have been done. In some of the swords the final shaping appeared to have been carried out not by forging but by abrasion or filing.

## Acknowledgements

The authors acknowledge their indebtedness to Jessica Rawson, who while Keeper of the Department of Oriental Antiquities (now Department of Asia), greatly expanded the collection of Chinese iron material, very much encouraged our scientific study and has retained her active interest in the project. They are grateful to colleagues at the British Museum, notably Paul Craddock and Janet Lang, for advice and support, and to James Lin for his comments on the dates of the swords. They also wish to express their appreciation to Donald Wagner for helpful comments and for supplying a draft of the manuscript of his forthcoming volume of *Science and Civilization in China*.

## References

Bronson, B. 1999. The transition to iron in ancient China. In *The Archaeometallurgy of the Asian Old World*, V.C. Pigott (ed.). Philadelphia: MASCA Research Papers in Science and Archaeology 16, 177–93.

Chen Jianli, Han Rubin and Wan Xin 2003. The manufacturing technique of ferrous artefacts excavated from Xianbei gravesite at Lamadong village of Liaoning province. In Jett 2003, 70–78.

Craddock, P.T., Wayman, M.L., Wang, H. and Michaelson, C. 2003. Chinese cast iron through twenty-five hundred years. In Jett 2003, 36–46.

Han Rubin 1996. The development of Chinese ancient iron blast furnace. In *The Forum for the Fourth International Conference on the Beginning of the Use of Metals and Alloys (BUMA-IV)*, Shimane, Japan, 151–74.

Han Rubin 1998. Iron and steel making and its features in ancient China. *Bulletin of the Metals Museum* 30: 23–37.

Jett, P. (ed.) 2003. *Scientific Research in the Field of Asian Art. Proceedings of the First Forbes Symposium at the Freer Gallery of Art*. London: Archetype Publications.

Li Zhong 1975. The development of iron and steel technology in ancient China. *Kaogu Xuebao (Acta Archaeologica Sinica)* 2: 1–22.

Needham, J. 1958. *The Development of Iron and Steel Technology in China*. London: Newcomen Society.

Pleiner, R. 1993. *The Celtic Sword*. Oxford: Oxford University Press.

Rawson, J. 1993. Chinese iron technology: the workshop and the factory. Paper presented at the Annual Meeting of the Association for Asian Studies, Los Angeles, March 1993.

Rostoker, W. and Bronson, B. 1990. *Pre-Industrial Iron: Its Technology and Ethnology*. Philadelphia: Archeomaterials Monograph No. 1.

Rostoker, W. and Dvorak, J. 1988. Blister steel = clean steel. *Archeomaterials* 2: 175–86.

Rostoker, W., Notis, M.B., Dvorak, J.R. and Bronson, B. 1985. Some insights on the 'hundred refined' steel of ancient China. *MASCA Journal* 3(4): 99–103.

Tylecote, R.F. 1990. Oxidation enrichment bands in wrought iron. *Historical Metallurgy* 24: 33–8.

Tylecote, R.F. and Thomsen, R. 1973. The segregation and surface enrichment of As and P in early iron artifacts. *Archaeometry* 15: 193–8.

Wagner, D.B. 1989. *Toward the Reconstruction of Ancient Chinese Techniques for the Production of Malleable Cast Iron*. East Asian Institute Occasional Papers no. 4. Copenhagen: University of Copenhagen, East Asian Institute.

Wagner, D.B. 1993. *Iron and Steel in Ancient China*. Leiden: Brill.

Wagner, D.B. 1999. The earliest use of iron in China. In *Metals in Antiquity*, S.M.M. Young, M. Pollard, P. Budd and R. Ixer (eds). BAR International 792. Oxford: Archaeopress, 1–13.

Wagner, D.B. forthcoming. *Science and Civilization in China, 36 (c): Ferrous Metallurgy*, unpublished draft dated 30 March 2005.

Wayman, M.L. and Wang, H. 2003. Cast iron coins of Song Dynasty China: a metallurgical study. *Historical Metallurgy* 37: 6–24.

Wayman, M.L., Lang, J. and Michaelson, C. 2004. The metallurgy of Chinese cast iron statuary. *Historical Metallurgy* 38: 10–23.

## Authors' addresses

• M.L. Wayman, Department of Chemical and Materials Engineering, University of Alberta, Edmonton, Alberta, Canada, T6G 2G6 (mike.wayman@ualberta.ca)
• C. Michaelson, Department of Asia, British Museum, London, WC1B 3DG, UK (cmichaelson@thebritishmuseum.ac.uk)

# Crucible steel in medieval swords

## Alan Williams

*ABSTRACT*   Many Viking-age swords bear an inscription on the blade such as VLFBERHT – thought to be a maker's name. It has become evident that some of these were made of crucible steel. In the past much attention has been focused upon one particular type of crucible steel, *wootz*, and its characteristic surface pattern, indicative of Oriental sword blades of high quality, often called 'Damascus' steel. Crucible steel, however, without any such pattern, was also used for making swords. There is written as well as metallographic evidence of a trade in such steel to western Europe, apparently by the Vikings, and liquid steel was a material familiar to medieval chemists.

*Keywords:* Viking, sword, metallography, Ulfberht, crucible steel, *wootz*, Damascus steel.

## Introduction

During the early Middle Ages in Europe, steel for swords was obtained via the bloomery process. In this process, iron ore was heated with charcoal and reduced to form a solid product, a bloom of iron (parts of which might be steel) together with a slag, which became a liquid at around 1200 °C. The steely parts had then to be somehow separated from the rest of the bloom.

In contrast, an entirely different process, known as crucible steel, developed in Asia. Pieces of bloomery iron were heated for days in sealed crucibles with a carbon-containing material until enough carbon had been absorbed for the steel to be formed as a liquid, at 1300–1400 °C, and so separated entirely from the slag. A special application of crucible steelmaking was the production of 'Damascus' steel.

However it has become clear, especially after archaeological work in Asia, that *wootz* was only a small part, albeit a special part, of the crucible steel industry, described in recent years by various archaeometallurgists such as Bronson (1986), Craddock (1998), Craddock and Lang (2003) and Feuerbach (2003). The earliest production sites so far excavated are Sri Lanka (6th–12th century: Juleff 1997), and in central Asia, Uzbekistan (9th–12th century) and Turkmenistan (8th–9th century). The Mongol invasions of the 13th century may have put an end to this industry. Many Indian sites were documented in the 19th century while they were still in production, especially at Konasamudram, north of Hyderabad, and Mysore. A detailed description of the process at Ghattihosahalli in Karnataka, which remained in use until the late 19th century, and where recent excavations have unearthed crucibles and other metallurgical debris is given by Anantharamu *et al.* (1999). Iron was mixed with wood chips and fired in crucibles of clay tempered with rice charcoal.

The fame of the easily identifiable Damascus steel and the rather inept attempts to copy it in the West by 'pattern-welding' – not to mention attempts to fake it in the East by etching – have tended to overshadow the use of crucible steel in general. It may be that much of the crucible steel made did not undergo the extremely slow cooling which leads to the 'watered-silk' pattern visible on the surface of the most highly prized Damascus blades, and therefore it has not been recognised. If this is so, the quantity of crucible steel employed may well have been considerably underestimated in the past.

## Damascus steel

A good deal has been written about Damascus steel swords (named after their alleged origin in Syria) but it still remains a matter of debate as to where and how the steel for these swords was actually made. These famous blades had a watered silk or damask pattern on their surfaces and were made from a very high carbon crucible steel (*wootz*) which was allowed to cool extremely slowly after liquefaction, so that the cementite (iron carbide, $Fe_3C$) crystals were large enough to form a visible pattern on the surface. These cakes were exported to centres of arms manufacture such as Damascus where they were carefully forged, with some difficulty, into sword blades. Since the melting point of steel falls with increasing carbon content, a lower temperature than usual has to be employed to forge a blade of higher carbon content unless the steel has been softened by prolonged annealing, which would rather defeat the purpose of using it to make a sword. This forging broke up the cementite network left over from the casting, reducing brittleness and producing the characteristic watered-silk pattern on the surface of the blade after polishing and etching. The blade so formed needed no further heat treatment to harden it, although some attempts might be made. No amount of sharpening would have removed the edge.

An article by Panseri (1965) was one of the first to show the microstructure of Damascus steel; this was followed by France-Lanord (1969) and Piaskowski (1978) (see Appendix).

Extensive efforts by Verhoeven and Pendray (Verhoeven *et al.* 1992, 1993, 1998) to recreate this process have led to a series of very detailed papers on its metallurgy. They have suggested (1998) that traces of carbide-forming elements were essential to the banding responsible for the visible pattern. Indeed, Panseri had earlier pointed out that traces of chromium will stabilise cementite.

## Crucible steel

Crucible steel could have been made in one of two ways. First by heating wrought iron with organic matter, as described by al-Tarsusi in the 12th century, or (more easily) by heating wrought iron with cast iron, as noted by al-Biruni (973–1048), who describes the melting together of 'soft iron' and 'hard iron' (presumably cast iron) in covered crucibles for a matter of days. Eventually enough carbon would have been absorbed for the alloy to melt and the broken crucible would yield a cake of cast steel of a convenient size for making a sword blade. He describes how the product was made around Herat, Afghanistan and exported via northern India to Persia and other Muslim lands. The Persians traded in crucible steel; a 7th-century Persian blade from Dailaman, in the British Museum, has recently been analysed (Lang *et al.* 1998), and found to be made of crucible steel. But was any exported to Europe?

Until recently, exports to Europe do not seem to have been recognised. Some years ago, however, a broken sword in the stores of the Württemburg Landesmuseum with an 'Ulfberht' inscription was analysed and found to be made of a hypereutectoid steel of perhaps 1.2% carbon or more (Williams 1977). This bore no resemblance to the microstructure of any other

contemporary western European sword so far observed, but Paul Craddock later suggested that the blade was made from a crucible steel which had undergone sufficient hot-work to obliterate traces of dendrites (although cementite needles are still present mixed with areas of pearlite); and indeed the section suggests that it was folded from a billet in the course of forging. If there was a trade in crucible steel between Asia and Europe a thousand years ago, then we would expect other examples to have been found.

## Ulfberht swords

Around 100 swords with 'Ulfberht', or variants on this name, inlaid into the blade, have been found scattered all over northern Europe (Fig. 1). The largest concentration is in Scandinavia and the Baltic Sea, although it has been suggested that if Ulfbehrt was their maker then on linguistic grounds the source of their manufacture should lie in the Rhineland (Müller-Wille 1970). From the different forms of these swords, Ulfbehrt would have been active for 300 years, so it has been suggested that perhaps this was a family of smiths rather than an individual, or the name was a trade mark of some sort. The analyses of some of these swords are presented below and summarised in Table 1.

### Analyses of Viking Age swords

Quantitative analyses of Viking Age swords were undertaken as early as the 19th century. Lorange (1889) took the view that the quality of a sword depended on its carbon content, so he had three pattern-welded blades (none were Ulfberht swords)

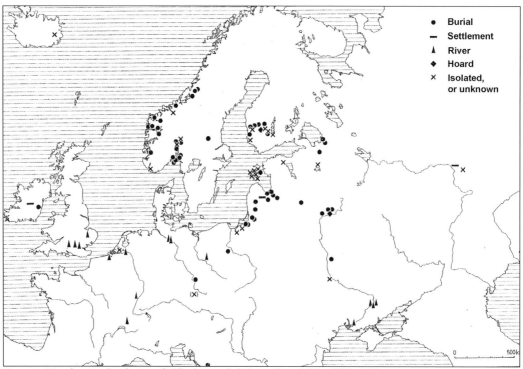

**Figure 1** Map showing distribution of Ulfberht swords (adapted from Müller-Wille 1970).

**Table 1** Summary of results.

| Sword no. | Location | Inventory no. | Carbon content | Position | Microstructure | Hardness |
|---|---|---|---|---|---|---|
| 1 | Oslo | 4690 | > 0.75% | – | – | – |
| 2 | Hamburg | M.1152 | 1.2% | – | – | – |
| 3 | Stuttgart | 1973–70 | ~ 1.2% | edge/core | pearlite + cementite | – |
| 4 | Private collection | – | ~ 0.5% | edge | martensite – core ferrite | 470 |
| 5 | Czech Republic | – | ~ 0.5% | edge | martensite – core ferrite | 460 |
| 6 | Hamburg | 1965/124 | > 1% | edge | pearlite + cementite | 463 |
| | | | | centre | pearlite + cementite | 355 |
| 7 | Solingen | 1973.w.5 | > 0.8% | edge | pearlite + carbides | 260 |
| 8 | Wisbech | 1860.5 | ~ 0.5% | edge | martensite – core ? | 356 |
| 9 | Oxford | 1555.258 | 0% | near edge | ferrite | |

from Norway analysed. Their carbon contents were found to be 0.414%, 0.401% and 0.520% respectively. Some years later, a number of swords were analysed for Petersen (1919) by the engineer K. Refsaas of Trondheim. He set about an extensive series of tests. Due to the condition of the swords he had to be content with samples of 1 g in weight, which were analysed by direct combustion in a stream of oxygen. He was aware of the possible difference in composition between edge and blade, and for fragment 16276, for example, only the inner part with little carbon was preserved. Where possible he attempted to carry out two analyses, but in those specimens marked * in Table 2, only one was achievable.

**Table 2** Results of analyses of nine swords from the Oslo University collection (analyses by K. Refsaas).

| Inventory no. | Fragment | Carbon content | Type |
|---|---|---|---|
| 1977 | small fragment* | 0.48% | D |
| 16276 | small fragment* | 0.20% | D |
| 16519 | larger fragment* | 0.62% | C |
| 13757 | larger fragment | 0.31% | H |
| 14286 | 1/3 sword ? | 0.67% | L |
| 4690 | 10 cm | 0.75% | S |
| 4979 | 15 cm | 0.46% | type 20 |
| 3983 | 9 cm* | 0.43% | AE |
| 11056 | bar | 0.69% | |

Notes: The type given follows the nomenclature developed by Petersen (1919). Where possible Refsaas attempted to carry out two analyses, but in those specimens marked * only one was achievable.

## Catalogue of Ulfberht swords

### Previous metallography of Ulfberht swords

#### 1. Oslo Historical Museum, inv. no. 4690

The highest carbon content of the previous sword analyses by Refsaas was the solitary VLFBERH+T sword from Aker, Hedemarken. I have recently been able to examine this sword, however. Samples for analysis had been taken from the surface of the sword where decarburisation due to forging had taken place. The quoted carbon content was therefore likely to have been considerably underestimated.

#### 2. Museum für Hamburg Geschichte; inv. no. M. 1152 (Type X, 11th century)

Before the First World War, a sword 95.5 cm long with the inscription +VLFBERH+T was dredged from the Alster near Hamburg in northern Germany and given to the Museum für Hamburg Geschichte (Museum for Hamburg History). The sword was published (with an illustration) by Schwietering in 1915, and then in 1951 an analysis (by Dr Schindler) was published in a Festschrift (Jankuhn 1951). It was found to have a carbon content of 1.2%. No other data were published, and the very high carbon content of a hypereutectoid steel was not remarked upon. (Another Ulfberht sword found in the river Elbe in 1957, now in the same museum, also proved to be a hypereutectoid steel; see sword no. 6 below.)

### Recent metallography of Ulfberht swords

#### 3. Württemburg Landesmuseum (WLM) Stuttgart, inv. no.1973-70 (see Fig. 2a–d)

This sword was acquired at auction in 1973, again having been found in a river (at Karlsruhe), with the inscription +VLFBERH+T. The microstructure of a sample taken from a half-section at the broken end consisted of pearlite and cementite, some at grain boundaries and some in needle-like form. It is now evident that this was made from a billet of crucible steel which had been folded twice and forged out into a blade (Williams 1977). Its broken end may indicate that the presence of these cementite needles led to a certain brittleness in the blade.

#### 4. Private British collection (via Jonathan Barrett)

This sword had a worn inscription which might be read as VLFBERN+T. A sample was taken from the broken end of the blade. The microstructure consisted largely of a ferrite core overlaid with a band of martensite corresponding to a cutting edge approximately 6 mm deep. Evidently a thin layer of (perhaps 0.5% carbon) steel was wrapped around a much larger piece of iron, or low carbon steel, and attached by forge-welding. This composite billet was then forged out into a sword blade and hardened by quenching. This gave a hard (423–540, average = 467 VPH) edge on a much less hard core (236 VPH) (Edge and Williams 2003). The resemblance of its microstructure to that of sword no. 5 is remarkably close.

(a)

(b)

(c)

(d)

**Figure 2** (a) Sword from Stuttgart Württemburg Landesmuseum with inscription (inv. no. 1973-70); (b) close-up of inlaid inscription; (c) microstructure: pearlite and cementite (scale bar: 20 μm); (d) microstructure: pearlite areas with grain boundary and needle cementite (scale bar: 5 μm). (Photograph (a) courtesy Stuttgart Württemburg Landesmuseum, Stuttgart.)

**5. Olomouc Regional Museum, Czech Republic – sword from a 10th-century grave at Nemilany with the inscription +VLFBERHT+**

It was examined by the staff of the Conservation Laboratory of the Brno Technical Museum. It had a core of low carbon content (< 235 VPH) and welded-on steel cutting edges, hardened to 320–460 VPH by quenching (Selucká *et al.* 2001).

**6. Museum für Hamburg Geschichte; inv. no.1965/124, Type R (see Fig. 3a–c)**

This sword, with a broken blade, is another river find (recovered from the Elbe in 1957) and is inscribed +VLFBERH+T (Müller-Wille 1970). A sample taken from the damaged edge about 10 cm from the end showed a microstructure of mostly fine pearlite with some cementite at grain boundaries, but no visible slag inclusions. Another sample, taken from near the centre at the broken tip, showed a microstructure of very fine pearlite. This is a steel of perhaps 1% carbon or more and so it is almost certainly a crucible steel.

Microhardness (Vickers, 100 g):
centre   337–388;   average = 355 VPH.
edge    439–476;   average = 463 VPH.

It seems that this blade underwent more hot-working than sword no. 3, and perhaps had a lower carbon content; no cementite was observed in the form of needles within the prior austenite grains, but only at grain boundaries, in a more equiaxed form. By contrast, the next sword (no. 7) seems to have undergone a little too much hot-working.

**7. Solingen, Deutsche Klingenmuseum; inv. no. 1973.w.5 (see Fig. 4a–b).**

This sword has the inscription +VLFBERH+T. A sample taken from the damaged edge showed a microstructure of what seems to have been pearlite which has been mostly divorced into carbide particles, and a network of particles outlining the prior austenitic areas. There were very few slag inclusions.

236

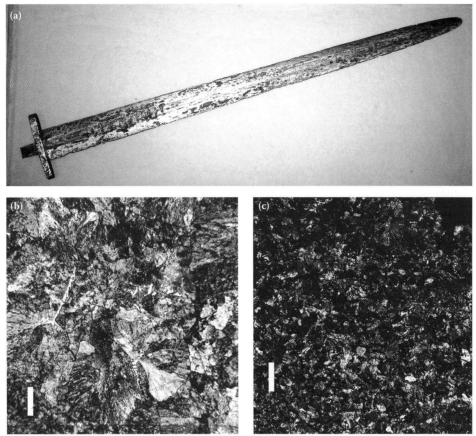

**Figure 3** (a) Sword from Museum für Hamburg Geschichte (1965.124); (b) microstructure of the edge: pearlite areas with grain-boundary cementite (scale bar: 20 µm); (c) microstructure of the centre: very fine pearlite; no ferrite or slag is visible (scale bar: 100 µm). (Photograph (a) courtesy Museum für Hamburg Geschichte, Hamburg.)

Microhardness (Vickers, 100 g):
243–277;     average = 258 VPH.

This seems to be a steel which has been annealed or undergone an excessive amount of hot-working. Since it seems unlikely that any sword would be intentionally softened by annealing, one may speculate that this was another hypereutectoid steel but it was somewhat overheated in working.

### 8. Wisbech and Fenland Museum; inv. no. 1860.5. Type N (see Fig. 5)

This sword was found in the river Nene and has the inscription +VLFBERHTC+ (with L and F inverted) on one side and, surprisingly, another inscription INGEFLRII on the other side. This sword has been described and illustrated recently (Gorman 1999). A sample was detached from a damaged edge, at about 8 cm from the tip. The microstructure consisted of uniform tempered martensite.

Microhardness (Vickers, 100 g load):
329–379;     average = 356 VPH.

This is a sword whose steel edges have been hardened by a successful heat treatment. It is not possible, without sampling the interior of the blade, to know whether the body of the sword was made of iron with steel edges welded on (as sword no. 5 from Nemilany), or whether it was all made of steel.

### 9. Pitt-Rivers Museum, Oxford (inv. no. 1555.2580; on loan to the Ashmolean Museum) (see Fig. 6)

The inscription may be read as +VI¬ IFR I + ⊥. This sword has been described and illustrated recently (Pierce 2004). A sample was detached from a damaged edge, near to the letter R. The microstructure consists of ferrite with a few slag inclusions.

Microhardness (Vickers, 100 g load) = 143 VPH.

It may be hypothesised that this blade did have a hard steel edge at one time, but that repeated sharpening has removed this edge, leaving the iron core. This might perhaps be a reason

**Table 3** Grouping of Ulfberht sword steels.

| Hypereutectoid steels | Eutectoid steels | Quenched steel edges |
|---|---|---|
| 2. Hamburg / Schindler +VLFBERH+T | 1. Oslo / Refsaas VLFBERH+T | 4. Private collection / Williams VLFBERN+T |
| 3. Stuttgart / Williams +VLFBERH+T | | 5. Czech Republic / Selucka +VLFBERHT+ |
| 6. Hamburg / Williams +VLFBERH+T | | 8. Wisbech / Williams* +VLFBERHTC+ |
| 7. Solingen / Williams +VLFBERH+T | | 9. Oxford / Williams** +VI¬ IFR I + ⊥ |

Notes: The data show sword article number, location, analyst and inscription; * the composition of the centre is uncertain; ** the composition of the edge is uncertain

(a)

(b)

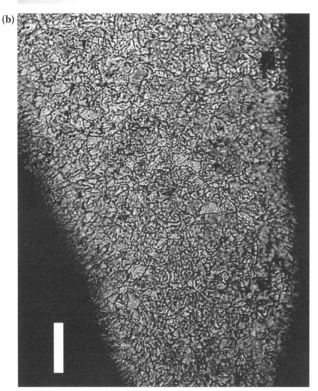

**Figure 4** (a) Sword from Deutsche Klingenmuseum, Solingen (1973. w5); (b) microstructure near the edge: carbide network in a (now somewhat divorced) pearlitic matrix (scale bar: 100 μm). (Photograph (a) courtesy Deutsche Klingenmuseum, Solingen.)

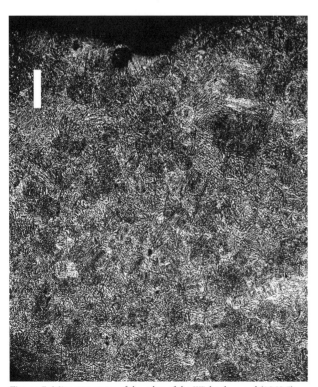

**Figure 6** Microstructure of the Oxford sword (1555.2580): ferrite and slag (scale bar: 50 μm).

**Figure 5** Microstructure of the edge of the Wisbech sword (1860.5): martensite and possibly bainite (scale bar: 100 μm).

why the blade has been bent and seems to have undergone fire damage.

The nine Ulfberht swords so far examined may be said to fall into the three groups summarised in Table 3.

## Discussion of analyses

At least four of the nine swords so far examined would seem to have been made of crucible steel (if the sword from Solingen was indeed an overheated crucible steel) and the inscriptions on these all have similar spellings, +VLFBERH+T.

Crucible steel swords would have been the best available by far, already hard without any need for quenching, possessing an edge that would never fade and of considerable toughness due to the paucity of slag inclusions. Like the more conspicuously patterned, but chemically identical, *wootz* or Damascus blades, they would have fetched a very high price. It would have been well worth importing cakes of crucible steel from Persia to forge these blades. Indeed, during the period of their manufacture, there was a well-established trade route from the Baltic to Persia via the Volga, which was exploited by the Vikings in the 9th and 10th centuries: there are said to be more Samanid (AD 815–1005) silver coins from their Afghan mines found in Sweden than in Persia. After the fall of the Samanids and the rise of the various Russian principalities in the 11th century, the use of this trade route by the Vikings declined (Mitchiner 1987). It is notable that at the same time the manufacture of Ulfberht swords apparently ceases, presumably because the raw material is no longer available.

This would also explain an otherwise puzzling observation by Mack *et al.* (2000) concerning lumps of high carbon steel (between 0.8% and 1.5% carbon), some of which had been melted but not shaped. These were excavated, together with numerous steel tools, from a 9th-century site at Hamwic (precursor to modern Southampton). They tentatively suggested that Saxon steelmaking may have involved decarburising cast iron. I would suggest instead that these were lumps of crucible steel which had been bought from traders, but the local smiths lacked the expertise to forge them and overheated them in so doing.

For the same reason, it would have been very tempting to try to counterfeit these valuable blades. One way, perhaps, was by welding small pieces of bloomery steel onto a billet of iron, and forging that into a blade before quenching it. The sharp edge that could be formed might well deceive the less discerning customer, but with a depth of only a few millimetres it would not have survived many sharpenings. These inferior versions were probably given their (frequently misspelled) copies of +VLFBERH+T inscriptions to help convince the gullible.

## Written evidence concerning crucible steel

There are literary sources to support the contention that swords of crucible steel were appreciated in medieval south-

ern Europe. The Franks valued Saracen swords, and one of Charlemagne's vassals captured *spatha india cum techa de argento parata*. Count Eccard of Mâcon left a *spatha indica* in his will, as well as *tabulas saraciniscas*. It has been suggested by the historian who collected these references (Coupland 1990) that this referred to a Saracen sword, but it may just as well have meant what it said, an Indian sword.

Dinnetz (2001) has collected numerous literary references to these blades in medieval Spain including among others:

- Ibn al-Labbana (11th century) who referred to 'Indian swords'.
- Abu Bakr al-Sayrafi (12th century) who suggested that Indian (*hinduwani*) swords should be used because they were sharper than other swords and better able to pierce the heavy armour worn by the Christian soldiers.
- Al-Zuhri (12th century) who said that Seville produces 'Indian steel'.
- Ibn-Abdun (12th century) who said that makers of scissors, knives, scythes etc. in Seville may only use (materials translated as) steely iron *mudakkar* or cast steel *amal al-tara'ih*.
- Ibn Hudhayl (14th century) who described Frankish swords as *mudakkar* with 'steel edges on an iron body, unlike those of India'.

### Recipes referring to liquid steel

Medieval chemists may have hidden their rational procedures behind the smokescreen of a picturesque vocabulary, far removed from modern 'scientific' accounts, but Indian (or at least an Oriental) steel seems to have been known as a harder steel of lower melting point than European steel. Three references are quoted as representative examples, although this topic clearly requires further study.

Thomas of Cantimpré (1210–1280) mentions a fusible iron in his Encyclopaedia:

*... est et aliud genus ferri in orientis partibus quod vulgare alidea dicitur. Incisionibus aptum est et sit file sicut cuprum (et) argentum. Sed fusile non est sicut ferrum alairum partium...* [and there is another type of iron in Oriental lands which is commonly called *alidea*. It is suitable for cutting and may be (drawn) like copper or silver, but in melting it is not like the iron of other lands] (Ferckel 1927).

William Sedacer (d.1380) was a Catalan compiler of alchemical recipes. His principal work was called *Sedacina*, compiled in 1377 when he was active at the court of Aragon, and a member of the household of the Infante John. In this collection he writes about the properties of different metals:

*Indorum sapientes vero de India inter cetera corpora ferrum elegerunt quia cito facit suum opus et leviter...* [the sages of India choose iron from India from among the other types [of iron] because it is the quickest to work] (Barthélemy 2002).

Since the recipes following this one are all concerned with the liquefaction of iron, usually by adding arsenical compounds, it may be assumed that this means that type of iron with the lowest melting point or crucible steel.

A translation into French from an anonymous 14th-century treatise on metals from southern Italy contains the following statement:

> Iron, of which the Indian one is the best ... Iron can colour silver, cannot be easily separated from gold, and can be 'purified' with acids ... Iron cannot be melted, except with great effort (Colinet 2000: 30).

The editor believes the treatise to have been written in Calabria (a region of Italy then bilingual) *c.* 1377.

## Conclusions

Viking Age swords bearing the inscription +VLFBERH+T were made of a very high carbon steel. The most likely source for this seems to have been the crucible steel industry of Iran, or even India, traded up the Volga by Vikings and forged into sword blades in northern Europe. When this trade route was closed in the 11th century, blades were apparently no longer forged out of this metal. The very desirable qualities of these blades led to many copies being made by European smiths, evidently employing bloomery metal of much lower carbon content, given steel edges and hardened by quenching. These copies bear inscriptions with variant spellings of the maker's name, presumably in an attempt to persuade customers that they were an equally desirable product. Crucible steel also seems to have been known in Muslim Spain and may have been imported from Asia by a Mediterranean route. Perhaps from Spain the knowledge of this unusual 'Indian' steel then spread among Latin writers in 13th-century Europe.

## Appendix: Examples of previous examinations of the microstructures of Damascus steel blades

*(a) Panseri (1965) sectioned two blades*
- Blade 1 (a *kara khorasan* or 'black ground'): the edge showed some martensite; the back showed pearlite mixed with primary cementite.
- Blade 2 (signed by Asadollah, 16th century): showed pearlite, cementite and some graphite.

| | Blade 1 | Blade 2 |
|---|---|---|
| Carbon | 1.62% | 1.42% |
| Silicon | 0.03% | 0.11% |
| Sulphur | 0.007% | 0.04% |
| Phosphorus | 0.09% | 0.03% |

(Microhardnesses were not given.)

*(b) France-Lanord (1969) sectioned a number of blades including two of hypereutectoid steel. Two fragments of blades supposedly dating from the 7th or 6th century BC were analysed.*
- Blade 13: three sections were studied; one transverse section showed a hardness (presumably microhardness) from 300 to 309 at the edges, and 297 VPH in the centre. The microstructure consisted of pearlite areas surrounded by cementite, and, significantly, no inclusions. A longitudinal section showed 397 VPH in the centre and 429 VPH at the point. But, surprisingly, analysis showed a carbon content of only 0.87%.
- Blade 14: the microstructure also consisted of pearlite and cementite, but the cementite crystals were arranged in 'fibres' or 'veins' and dendrites were also present in places; the hardness varied from 339 to 488 VPH. In this specimen also, there were no visible inclusions. Analysis gave a composition of:

| | |
|---|---|
| Carbon | 1.37% |
| Silicon | 0.046% |
| Sulphur | 0.013% |
| Phosphorus | 0.042% |
| Manganese | 0.068% |

Both of these were evidently crucible steels, and no. 14 seems to have had a *wootz* pattern, in addition.

*(c) Piaskowski (1978) summarised the results of previous analyses. He also examined two further broken blades, giving their hardnesses but not their carbon contents:*
- Blade 1, average = 348 VPH (range 227–430)
- Blade 2, average = 366 VPH (range 201–408)

The range of microhardness values showed considerable variation, and the presence of a microconstituent described as *sorbite* near the edge led Piaskowski to believe that an attempt had been made to harden the edge of one blade.

## Acknowledgements

The author would like to thank those enlightened curators for giving permission to analyse swords in their collections: Dr Ralf Wiechmann (Museum for Hamburg History), Dr Barbara Grotkamp (German Blade Museum, Solingen), Professor Volker Himmelein (Württemberg Landesmuseum, Stuttgart), Dr Arthur MacGregor (Ashmolean Museum, Oxford), Dr Michael O'Hanlon (Pitt-Rivers Museum, Oxford), and Dr Robin Hanley and Dr Robert Bell (Fenland Museum, Wisbech). Some of the author's travel expenses were met by the British Academy.

All dates given are Common Era. Microhardness measurements undertaken by the author (Vickers, 100 g) were taken as an average of 10 readings.

## References

Anantharamu, T.R., Craddock, P.T., Nagesh Rao, K., Murthy, S.R.N. and Wayman, M.L. 1999. Crucible steel of southern India. *Historical Metallurgy* 33: 13–25.

Barthélemy, P. 2002. *La Sedacina ou l'oeuvre au crible*. Textes et travaux de Chrysopeia, 8, 2 vols. Milan: Arche.

Bronson, B. 1986. The making and selling of *wootz*: a crucible steel of India. *Archeomaterials* 1: 13–51.

Colinet, A. 2000. *Les Alchimistes Grecs, X: L'anonyme de Zuretti*, Paris: Les Belles Lettres.

Coupland, S. 1990. Carolingian arms and armour in the ninth century. *Viator* 21: 29–50.

Craddock, P.T. 1998. New light on the production of crucible steel in Asia. *Bulletin of the Metals Museum of the Japan Institute of Metals* 29: 41–66.

Craddock P.T. 2003. Cast iron, fined iron, crucible steel. In Craddock and Lang 2003, 231–57.

Craddock, P.T. and Lang, J. (eds) 2003. *Mining and Metal Production through the Ages*. London: British Museum Press.

Dinnetz, M.K. 2001. Literary evidence for crucible steel in medieval Spain. *Journal of Historical Metallurgy* 35: 74–80.

Edge, D. and Williams, A. 2003. Some early medieval swords in the Wallace Collection and elsewhere. *Gladius* 23: 191–210.

Ferckel, C. 1927. Thomas von Chantimpré über die Metalle. In *Festgabe Lippmann – Studien zur Geschichte der Chemie*, J. Ruska (ed.). Berlin: Verlag Springer, 75–80.

Feuerbach, A. 2003. Early Islamic crucible steel production at Merv, Turkmenistan. In Craddock and Lang 2003, 258–66.

France-Lanord, A. 1969. Le fer en Iran au premier millenaire avant J-C. *Revue d'histoire des mines et la métallurgie* 1(1): 75–126.

Gorman, M.R. 1999. Ulfberht: innovation and imitation in early medieval swords. *The XVIth Park Lane Arms Fair Catalogue*, 7–12.

Jankuhn, H. 1951. Ein Ulfberht-Schwert aus der Elbe bei Hamburg. In *Festschrift für Gustav Schwantes*, K. Kersten (ed.). Neumünster: K. Wachholtz, 212–229 (see p. 224 for analysis of the Ulfberht sword).

Juleff, G. 1997. Early Sri Lankan crucible steel. *Historical Metallurgy Society News* 35: 3.

Lang, J., Craddock, P.T. and Simpson, St J. 1998. New evidence for early crucible steel. *Historical Metallurgy* 32: 7–14.

Lorange, A.L. 1889. *Den yngre jernalders svaerd*. Bergen: Grieg.

Mack, I., Mc Donnell, G., Murphy, S., Andrews, P. and Wardley, K. 2000. Anglo-Saxon liquid steel. *Historical Metallurgy* 34: 87–96.

Mitchiner, M. 1987. Evidence for Viking-Islamic trade provided by Samanid silver coinage. *East and West* 37: 139–50.

Müller-Wille, M. 1970. Eine neue Ulfberht – schwert. *Offa* 27: 65–91.

Panseri, C. 1965. Damascus steel in legend and reality. *Gladius* 4: 5–66.

Petersen, J. 1919. *De norske vikingesverd*. Kristiana: Dybwad.

Piaskowski, J. 1978. Metallographic examination of two Damascene steel blades. *Journal for the History of Arabic Science* 2: 3–30.

Pierce, I. 2004. A newly identified Ulfberht inscribed sword in the Ashmolean Museum, Oxford. *The Spring 2004 Park Lane Arms Fair Catalogue*, 14–15.

Schwietering, J. 1915. Ein Ulfberhtschwert des 11 Jahrhunderts. *Zeitschrift für Historische Waffen- und Kostümkunde* 7: 107–8.

Selucká, A., Richtrová, A. and Hložek, M. 2001. Konservace železného meče Ulfberht. *Sborník z konservátorského a restaurátorského semináře*, České Budějovice, 65–8, 103–4.

Verhoeven, J.D., Pendray, A.H. and Peterson, D. 1992. Studies of Damascus steel blades. *Materials Characterisation* 29: 335–41.

Verhoeven, J.D., Pendray, A.H. and Peterson, D. 1993. Studies of Damascus steel blades. *Materials Characterisation* 30: 175 and 187.

Verhoeven, J.D., Pendray, A.H. and Danksch, W.E. 1998. The key role of impurities in ancient Damascus steel blades. *Journal of the Minerals, Metals and Materials Society* 50(9): 58–64.

Williams, A. 1977. Methods of manufacture of swords in medieval Europe. *Gladius* 13: 75–101.

## Author's address

Alan Williams, Conservation Department, The Wallace Collection, London W1U 3BN, UK (alan.williams@wallacecollection.org)

# Index